W9-AGN-698

The Agroecology of Carabid Beetles

The Agroecology of Carabid Beetles

Editor:

JOHN M. HOLLAND

The Game Conservancy Trust, Fordingbridge, Hampshire, UK

Intercept

Andover

British Library Cataloguing in Publication Data
The Agroecology of Carabid Beetles.—

A CIP catalogue record for this book is available from the British Library
ISBN 1–898298–76–9

Published in June 2002 by Intercept Limited,
PO Box 716, Andover, Hampshire SP10 1YG, UK.
Email: intercept@andover.co.uk
Website: www.intercept.co.uk

Cover Photo: *Conservation Headland* by John Holland
Inset Photos (cover): 1 *Bembidion quadrimaculatum*; 2 *Agonum dorsale* by Martin Luff
Frontispiece: Ink pen drawing by Ken Tucker

Typeset in Times by
Ann Buchan (Typesetters), Shepperton, Middlesex.
Printed by St Edmundsbury Press.

Contents

LIST OF CONTRIBUTORS x

PREFACE AND ACKNOWLEDGEMENTS xi

FOREWORD xiii
Henrik Wallin
Swedish Institute for Food and Biotechnology, S1K, Uppsala Science Park, 751 83 Uppsala, Sweden

1 CARABID BEETLES: THEIR ECOLOGY, SURVIVAL AND USE IN AGROECOSYSTEMS 1
John M. Holland
The Game Conservancy Trust, Fordingbridge, Hampshire SP6 1EF, UK
History of carabid research 1
Ecology of carabids in agricultural habitats 5
Population regulation 7
The importance of carabids in agricultural habitats 19
Long-term changes 24
Future research 29
Acknowledgements 30
References 31

2 CARABID ASSEMBLAGE ORGANIZATION AND SPECIES COMPOSITION 41
Martin L. Luff
Department of Agricultural and Environmental Science, University of Newcastle upon Tyne, Newcastle upon Tyne NE1 7RU, UK
Introduction 41
Sampling carabid assemblages 42
Species-independent attributes of carabid assemblages 44
Species composition 57
Body size distributions 65

Effects of environmental factors on assemblage composition 66
Conclusions 70
Acknowledgements 71
References 71

3 CARABID DIETS AND FOOD VALUE 81
Søren Toft and Trine Bilde
Department of Zoology, University of Aarhus, Bldg. 135, DK-8000 Århus C, Denmark
Introduction 81
Methods for recording diets in the field 82
Feeding guilds 82
Larval feeding guilds 84
Life style and foraging 85
Prey detection and attack behaviour 86
Daily and seasonal foraging cycles 86
Diet quality 87
Value of different types of food 89
Food quality in relation to dietary guilds 96
Physiological and biochemical correlates of dietary specialization 98
Food preferences 98
Dietary mixing 100
Food limitation 101
Exploitative competition 101
Food value and pest limitation 102
Conclusion 102
Acknowledgements 103
References 104

4 RELATING DIET AND MORPHOLOGY IN ADULT CARABID BEETLES 111
Joseph Ingerson-Mahar
243 Blake Hall, 93 Lipman Drive, Rutgers University, New Brunswick, NJ 08901, USA
Introduction 111
Gut dissection 125
Conclusion 131
Acknowledgements 131
References 132

5 DIAGNOSTIC TECHNIQUES FOR DETERMINING CARABID DIETS 137
William O.C. Symondson
Cardiff School of Biosciences, Cardiff University, PO Box 915, Cardiff CF10 3TL, UK

Introduction 137
Chemical and radioactive labelling 138
Protein electrophoresis 139
Immunological techniques 140
Molecular detection systems 155
Current recommendations and the future 156
References 158

6 INVERTEBRATE PEST CONTROL BY CARABIDS 165
Keith D. Sunderland
Department of Entomological Sciences, Horticulture Research International, Wellesbourne, Warwickshire CV35 9EF, UK
Introduction 165
Evidence that carabids consume pests (but without proof of predation) 166
Evidence that carabids can kill pests 174
The contribution of carabids to pest control 187
Conclusions 201
Acknowledgements 202
References 202

7 WEED SEED PREDATION BY CARABID BEETLES 215
Josephine Tooley and Gerald E. Brust
Department of Agricultural Botany, School of Plant Sciences, The University of Reading, Building 2, Earley Gate, Reading, Berkshire RG6 6AU, UK and SW Purdue Agricultural Research Center, Purdue University, 4369 North Purdue Road, Vincennes, IN 47591 3043, USA
Introduction 215
Overview of seed predation 216
Most common weed seed predators 216
Adaptations to a seed-feeding habit 217
Methods of detecting predation 219
Weed seed preferences of carabid beetles 220
Factors affecting carabid seed choice 221
Implications for plant populations and weed control 222
Impact of farming systems 224
Acknowledgements 226
References 226

8 IMPACT OF CULTIVATION AND CROP HUSBANDRY PRACTICES 231
Thierry Hance
Unité d'écologie et de biogéographie, Centre de Recherches sur la Biodiversité, Université Catholique de Louvain, Place Croix du Sud, 4–5, B-1348 Louvain-la-Neuve, Belgium

Introduction 231
Crop field organization in space and time: landscape level 232
The soil level 237
Pesticide application 241
Conclusions 242
Acknowledgements 244
References 244

9 CARABIDS AS INDICATORS WITHIN TEMPERATE ARABLE FARMING SYSTEMS: IMPLICATIONS FROM SCARAB AND LINK INTEGRATED FARMING SYSTEMS PROJECTS 251
John M. Holland, Geoff K. Frampton and Paul J. Van den Brink
The Game Conservancy Trust, Fordingbridge, Hampshire SP6 1EF, UK, Biodiversity and Ecology Division, School of Biological Sciences, University of Southampton, Bassett Crescent East, Southampton, Hampshire SO16 7PX, UK and Alterra Green World Research, Department of Water and the Environment, PO Box 47, 6700 AA Wageningen, The Netherlands
Introduction 251
Carabid communities as indicators of farming system effects: evidence from the LINK Integrated Farming Systems Project 252
Results from the LINK Integrated Farming Systems Project 254
Carabid communities as indicators of pesticide effects: long-term evidence from the SCARAB Project 261
Results from the SCARAB Project 263
Discussion 265
Conclusions 273
Acknowledgements 273
References 274

10 NON-CROP HABITAT MANAGEMENT FOR CARABID BEETLES 279
Jana C. Lee and Douglas A. Landis
Department of Entomology, University of Minnesota, 219 Hodson Hall, 1980 Folwell Ave., St. Paul, MN 55108, USA and Department of Entomology, Center for Integrated Plant Systems, Michigan State University, East Lansing, MI 48824, USA
Introduction 279
Functions of non-crop habitat 280
Relevance of non-crop habitats for carabid beetles 282
The influence of non-crop habitats on pest abundance 291
Managing non-crop habitats 293
Future directions 296
Acknowledgements 297
References 298

11 THE SPATIAL DISTRIBUTION OF CARABID BEETLES IN AGRICULTURAL LANDSCAPES 305
C.F. George Thomas, John M. Holland and Nicola J. Brown
Seale-Hayne Faculty of Land, Food and Leisure, University of Plymouth, Newton Abbot, Devon TQ12 6NQ, UK, The Game Conservancy Trust, Fordingbridge, Hampshire SP6 1EF, UK and IACR Long Ashton Research Station, Department of Agricultural Sciences, Long Ashton, Bristol BS41 9AF, UK
Introduction 305
Stable natural habitats: heathland, forests and woodland 308
Semi-stable, semi-natural habitats: hedges, field boundaries and other interstitial habitats 310
Seasonal movement between field boundaries and field centres 314
Spatial distributions within fields 316
Possible causes of spatial variation in population density within fields 320
Spatial distributions at the farm-scale – variation between fields 328
Barriers to movement between fields 332
Metapopulation structure and models 333
Summary, conclusions and future directions 335
Acknowledgements 336
References 337

INDEX 345

Contributors

TRINE BILDE, *Department of Zoology, University of Aarhus, Bldg. 135, DK-8000 Århus C, Denmark*
NICOLA J. BROWN, *IACR Long Ashton Research Station, Department of Agricultural Sciences, Long Ashton, Bristol BS41 9AF, UK*
GERALD E. BRUST, *SW Purdue Agricultural Research Center, Purdue University, 4369 North Purdue Road, Vincennes, IN 47591 3043, USA*
GEOFF K. FRAMPTON, *Biodiversity and Ecology Division, School of Biological Sciences, University of Southampton, Bassett Crescent East, Southampton, Hampshire SO16 7PX, UK*
THIERRY HANCE, *Unité d'écologie et de biogéographie, Centre de Recherches sur la Biodiversité, Université Catholique de Louvain, Place Croix du Sud, 4-5, B-1348 Louvain-la-Neuve, Belgium*
JOHN M. HOLLAND, *The Game Conservancy Trust, Fordingbridge, Hampshire SP6 1EF, UK*
JOSEPH INGERSON-MAHAR, *243 Blake Hall, 93 Lipman Drive, Rutgers University, New Brunswick, NJ 08901, USA*
DOUGLAS A. LANDIS, *Department of Entomology, Center for Integrated Plant Systems, Michigan State University, East Lansing, MI 48824, USA*
JANA C. LEE, *Department of Entomology, University of Minnesota, 219 Hodson Hall, 1980 Folwell Ave., St. Paul, MN 55108, USA*
MARTIN LUFF, *Department of Agricultural and Environmental Science, University of Newcastle upon Tyne, Newcastle upon Tyne NE1 7RU, UK*
KEITH D. SUNDERLAND, *Department of Entomological Sciences, Horticulture Research International, Wellesbourne, Warwickshire CV35 9EF, UK*
WILLIAM O.C. SYMONDSON, *Cardiff School of Biosciences, Cardiff University, PO Box 915, Cardiff CF10 3TL, UK*
C.F. GEORGE THOMAS, *Seale-Hayne Faculty of Land, Food and Leisure, University of Plymouth, Newton Abbot, Devon TQ12 6NQ, UK*
SØREN TOFT, *Department of Zoology, University of Aarhus, Bldg. 135, DK-8000 Århus C, Denmark*
JOSEPHINE A. TOOLEY, *Department of Agricultural Botany, School of Plant Sciences, The University of Reading, Building 2, Earley Gate, Reading, Berkshire RG6 6AU, UK*
PAUL J. VAN DEN BRINK, *Alterra Green World Research, Department of Water and the Environment, PO Box 47, 6700 AA Wageningen, The Netherlands*
HENRIK WALLIN, *Swedish Institute for Food and Biotechnology, SIK, Uppsala Science Park, 751 83 Uppsala, Sweden*

Preface and Acknowledgements

The study of plants and animals in agricultural systems 'agroecology' is now being recognized as a distinct branch of ecology. Although the fauna and flora of agricultural areas have been long studied, it was the recognition that intensive farming methods, and especially pesticides, were causing changes in wildlife that was the catalyst for a more deterministic approach. Carabids play a key part in agroecology because they are one of the most abundant and taxonomically diverse groups of arthropods, with considerable variation in their ecology and environmental requirements. Moreover, they are present both within fields and adjacent non-crop habitats, often with interactions occurring between the two and, consequently, they may be affected by most of the changes in agricultural production. Comprehensive evaluations of carabid biology, habitat requirements, diet, and distribution have now been completed for a number of species. In addition, a variety of methodologies were, and are still being, developed to facilitate these investigations. Considerable research effort has also been devoted to determining the impact on carabids of individual farming practices and the whole system. This was often conducted because of the perceived biocontrol potential of carabids, and this benefit was the driving force behind the development of management practices to increase carabid abundance and diversity on our farms. More recently, how carabids are distributed and the importance of landscape features has been addressed, leading to new insights into their ecology.

Since Hans-Ulrich Thiele's book, there have been an enormous number of publications in scientific journals, while the publications originating from the European Carabidologists' meetings have provided further extensive information on a variety of subjects. In this book, I have attempted to bring together much that is now known of carabids in agroecosystems, drawing on the relevant experts. I believe that the methodologies developed, findings, and their interpretation not only have implications for how we approach the study of agroecology, but for ecological research in other managed habitats.

Presentations on a number of these topics were given in a special session on polyphagous predators at the annual meeting of the Entomological Society of America in Atlanta in 1999, and these were the inspiration for the book, along with my own research.

I am indebted to all the authors for providing such up-to-date and comprehensive reviews. My extra gratitude goes to Martin Luff, Keith Sunderland, Bill Symondson, and George Thomas, who helped with reviewing, along with Peter Chapman, Phil Chiverton, Barbara Ekbom, Bob Froud-Williams, Gabor Lovei, Guy Poppy, Sue Thomas and Linton Winder. My thanks go to all the others who assisted each author with their chapter. Special thanks go to Henrik Wallin for writing the Foreword, assisted by Phil Chiverton. Finally, thank you to Intercept the publishers, for their advice and for being so patient.

John Holland,
The Game Conservancy Trust,
June 2001

Foreword

HENRIK WALLIN

Swedish Institute for Food and Biotechnology, SIK, Uppsala Science Park, 751 83 Uppsala, Sweden

Carabid beetles exist in almost every type of habitat, ranging from tropical rainforests to semi-arid deserts. In temperate regions they are commonly found in woods, meadows and on cultivated land. They are comparatively easy to observe and collect, although recently several sophisticated techniques have been developed to investigate their abundance and behaviour in the field. The biology of carabid beetles has been studied throughout most of the 20th century by entomologists all over the world. However, it is only during the last three decades that their potential beneficial role in agroecosystems has been thoroughly investigated. This book provides an extensive overview of the recent literature from a number of topics relating to the ecology and behaviour of carabid beetles inhabiting agricultural land. In addition, several exciting areas of further research are identified.

One of the distinguishing features of carabids is their great capacity for dispersal on the ground. This feature enables those species living in arable habitats to cope with the many traumatic events associated with modern agricultural operations. For example, the dramatic change in habitat structure caused by tillage, or temporary food shortages caused by applications of agrochemicals. In similarity with many other predatory arthropods, carabids are strongly affected by any fluctuation in prey abundance. Regardless of their prey specialization, all species need food of appropriate nutritional value in order to reproduce and survive.

How do carabids forage in the field? One problem I encountered whilst studying this aspect of carabid behaviour was how to track species that are mostly active at night! By adapting a portable radar system, I finally managed to overcome this obstacle. I was surprised to find that they were capable of moving at high speeds, and frequently altered their behaviour in a three-dimensional way from running on the ground, or climbing on plants or trees, to burrowing into the soil. When active, they were capable of moving across plant rows, and even fields, within hours! Hungry carabids spend a considerable amount of energy on locomotion, and they need to compensate their energy loss by feeding more often when they are active.

From the individual farmer's perspective, it is important to know if predatory carabids are capable of suppressing pest insect populations below economic injury levels. The authors in this book use several approaches to answer this crucial question in a justified manner by referring to many new and important field observations. It is evident that the most important role of carabid beetles is being part of an assemblage – including many other predatory arthropods – that, under favourable conditions, may prevent or suppress pest outbreaks. It is therefore necessary that they need to be present in agricultural habitats even when pest insects are scarce or absent from the crop. Under such conditions they not only act as beneficial predators, they also serve as important dietary items for many birds inhabiting farmland. However, the increased fragmentation of the agricultural landscape in conjunction with a growing urban, as well as suburban, environment has a negative effect on the natural flora and fauna. Predatory carabid beetles occurring on cultivated land are, of course, faced with these large-scale effects in which alternative prey items are becoming scarce, and where the area of suitable aggregation sites during winter is declining.

It is not easy for the individual farmer to make a proper balance between short-term management decisions and long-term sustainability. Government programmes that compensate farmers for their loss of crop income when establishing or restoring non-crop habitats are welcomed. After all, agriculture is an important first link in the human food chain. Current food safety-related issues involve the quality of the food we eat, which is ultimately affected by the way our food is being produced. Intensified food customer demands, and specific requirements set by the food industry, may further promote the implementation of a more sustainable agriculture in the near future. This, in turn, should benefit carabid beetles inhabiting farmland. Proper information delivery is important to facilitate the transition towards integrated farming practices. In this context, predatory carabid beetles should serve as an excellent model for the implementation of extensive ecological knowledge in promoting the conservation of biodiversity on arable land.

Henrik Wallin

1
Carabid Beetles: Their Ecology, Survival and Use in Agroecosystems

JOHN M. HOLLAND

The Game Conservancy Trust, Fordingbridge, Hampshire SP6 1EF, UK

History of carabid research

Carabid beetles are one of the most studied insect families. They are also one of the most diverse families; the latest worldwide count revealed 32,561 species in 1859 genera (Lorenz, 1998). Initially, it was their aesthetic appeal, the ease with which they could be collected, and their relative ease of identification which fostered carabidology and lifelong research by several dedicated individuals. Their ecology and distribution was first extensively studied and summarized by Lindroth (1949) and translated from German into English (Lindroth, 1992). Lindroth's publications covered the taxonomy and ecology of beetles from most of northern Europe and a substantial part of North America. The next carabidologist to display the same lifetime commitment was Thiele, who published a monograph on carabids in 1977 (Thiele, 1977), which covered the period from the late 1950s until the mid-1970s. He focused on the ecology of beetles in natural and managed habitats, but also included sections on cultivated areas. Since Thiele and others first started their investigations of carabids in agricultural areas, there have been extensive publications and reviews regarding most aspects of carabid ecology and their role in agroecosystems. In 1969, the first European Meeting of Carabidologists was held, which have since continued triennially. These meetings have attracted a worldwide participation on a diverse range of subjects, often with a high proportion of studies reported on carabids in cultivated areas. The publications from these meetings (Den Boer, 1971; Den Boer *et al.*, 1979; Erwin *et al.*, 1979; Brandmayr *et al.*, 1983; Den Boer *et al.*, 1986, 1987; Stork, 1990; Desender *et al.*, 1994) are some of the main sources of information since Thiele's book. A number of reviews have also been published (Allen, 1979; Luff, 1987; Larochelle, 1990), with some focusing just on carabids of agricultural areas and the impact of farming practices (Kromp, 1999; Holland and Luff, 2000). No doubt the recognition that crop pests were consumed by some carabids (Symondson, this volume) and consequently that they had potential for pest control, instigated much of this research (reviewed in Sunderland, this volume). More recently, there has been an

Abbreviations: SADIE, Spatial Analysis by Distance Indicies

The Agroecology of Carabid Beetles

emphasis on their role as *bioindicators* in response to chemical pollutants (e.g. pesticides and heavy metals) or management regimes (e.g. crop, moorland, dune, and grassland management), and even their response to restricted human access in military training areas (Mossakowski *et al.*, 1990).

Carabids are also proving to be useful organisms in the study of landscape ecology. Their specific environmental requirements, combined with their preference for dispersal by walking rather than flight, means that they disperse sufficiently to populate new areas but that the scale of their movements is not too large to monitor. The reclamation of the 'Zuider Zee' in the Netherlands provided the ideal experimental design in which to test many of these theories, and also address some fundamental ecological questions. Carabids were intensively studied there by den Boer and others over many years (Den Boer, 1970, 1977, 1985, 1987, 1990). Characterizing carabids by their habitat preference has also been tried (e.g. Luff *et al.*, 1992 for grassland habitats). Carabids were classified according to site management, soil moisture and bulk density, and altitude.

There have also been attempts to characterize habitats according to species assemblages because the type of habitat chosen by individual species can, in some cases, be very specific. Much of this work has been conducted in natural or grassland habitats (Luff and Rushton, 1989; Maelfait and Desender, 1990; Luff *et al.*, 1992). For example, two types of lowland grassland were identified from their characteristic carabid assemblage (Luff and Rushton, 1989). This zoogeographical approach was also tested by Dufrêne *et al.* (1990) across a range of cultivated and natural habitats, but not arable fields. Carabids were found to have potential as bioindicators for the five habitat groups tested (peatbogs, coastal, grasslands, forests and semi-aquatic habitats). The species that can be expected in cultivated areas and adjacent non-crop habitats are now relatively well known and reviewed by Luff (this volume). They are now also the subjects of population genetic studies to investigate the impacts of habitat fragmentation and population isolation on dispersal, gene flow and distribution. The spatial distribution of carabids is reviewed in Thomas *et al.* (this volume).

Since the 1980s there has been increasing interest in the impact of pesticides, especially insecticides, on carabids. Moreover, given the ubiquitous distribution of some carabid species in agricultural areas, and particularly across northern Europe, they have been included as a test organism for pesticide risk assessment. Most of the initial methodologies were developed by the IOBC (International Organization for Biological and Integrated Control of Noxious Animals and Plants), and more latterly in Europe by the BART group (Beneficial Arthropod Regulatory Testing group) (Dohmen, 1998). A combination of laboratory, semi-field and field tests is now used to assess the impact of new pesticides on carabids and is a requirement for pesticide registration. Although much of this work is not widely available, the realization that carabids were susceptible to pesticide use and the relative ease with which they could be monitored spawned further research outside of the agrochemical industry (reviewed by Holland and Luff, 2000). Unfortunately, many field studies rely solely on the results of pitfall trapping, which can give misleading results because carabid movement is strongly affected by hunger levels (Mols, 1987). Pesticides may initially create a deluge of available prey, followed by a dearth. There may also be a range of sub-lethal effects (Jepson, 1989) followed by reinvasion, the speed and extent of which is largely dependent on the size of the treated area (Duffield and Aebischer,

1994; Holland *et al.*, 2000). Consequently, quantitative results from field-scale studies must be treated with some caution, although there is no doubt that a broad range of insecticides are acutely toxic to carabids. Further research on the sub-lethal effects and the consequences for long-term population abundance and distribution is needed if we are to truly understand the impact of insecticides.

The susceptibility of some carabid species to insecticides, herbicide use through modification of plant cover and microclimate, and soil cultivation has ensured that they are also frequently monitored in farming system studies (Holland *et al.*, 1994). However, these studies have frequently found that differences between farming systems are relatively small compared to that between years, fields and farms (Holland and Luff, 2000). This is because carabids exhibit considerable natural temporal and spatial variation, as discussed in Thomas *et al.* (this volume). Moreover, pitfall trapping is usually the sole means of monitoring, with comparisons often being made between different crops between which sampling efficacy can vary (Sunderland *et al.*, 1995). There are exceptions where ground photoeclectors or emergence traps (Büchs *et al.*, 1997; Idinger and Kromp, 1997) have been used in farming system studies but not fenced pitfall traps, the technique recommended by Sunderland *et al.* (1995).

SAMPLING TECHNIQUES

Considerable research effort has been spent investigating the vagaries of pitfall trapping with periodic summaries (Luff, 1975; Adis, 1979; Sunderland *et al.*, 1995). Observations of beetle behaviour at pitfall traps highlighted many of their inadequacies (Halsall and Wratten, 1988). As a consequence, the limitations of pitfall traps are well documented, but they continue to remain the most widely used sampling technique. Sometimes they are appropriate, but there are many examples of when they are not. For example, when making comparisons between different vegetation types, as demonstrated experimentally using ants in manipulated grassland (Melbourne, 1999), or for comparing the toxicity of pesticides when the sub-lethal effects on activity are unknown (Luff, 1987). Many different forms of trap are used, containing a variety of trapping solutions, some of which are strong attractants (Holopainen and Varis, 1986). More species are captured using a killing agent, although dry traps may capture a different range of species, confounding any comparisons using the two techniques (Weeks and McIntyre, 1997). In addition, within dry traps there is the risk of inter- and intra-predation. *Pterostichus melanarius* were not found to predate on each other when confined in pitfall traps for up to 24 hours but predated heavily on *Bembidion lampros*, even when cover for the smaller beetle was provided (*Figure 1.1*). Ethylene glycol is the most widely used preservative solution, although it is attractive to many species and is also toxic to mammals. A safer alternative is propylene glycol, which captures a similar range of species to ethylene glycol (Weeks and McIntyre, 1997). Having a standardized trap design, as advocated by Adis (1979), would facilitate comparisons between studies, but is unlikely ever to be achieved. Moreover, if the aim is to estimate species composition, then a range of different trap types of varying diameter and construction material should be used because each type can be species-specific. Pitfall traps also continue to be used in isolation, although other techniques which provide an estimate of density would, in many cases, greatly aid interpretation of the findings.

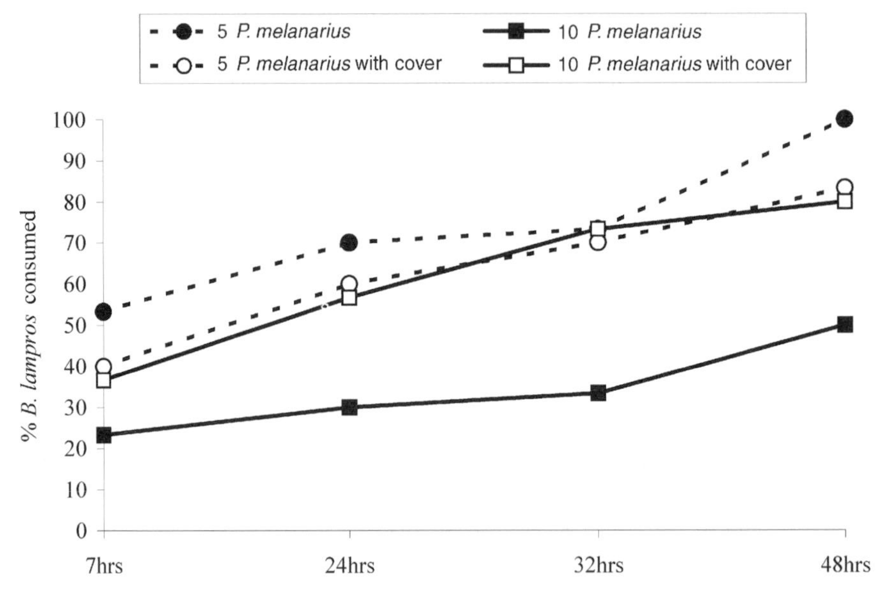

Figure 1.1. Predation of *Bembidion lampros* by *Pterostichus melanarius* in dry pitfall traps with and without cover.

Fenced pitfall traps are the simplest way of estimating density and can be correlated with capture from pitfall traps (Holland and Smith, 1999). However, because vegetation coverage and beetle activity may affect the numbers captured using unfenced pitfall traps, being activity dependent, the relationship to the fenced traps may change through a season. Fenced pitfall traps are relatively quick and easy to set in the field, and consequently are less time-consuming compared to other density estimation techniques such as mark–release–recapture, soil flooding (Desender and Segers, 1985), microhabitat removal (Edwards and Fletcher, 1971), or the combination of suction sampling, ground searching, isolating and removal trapping, as used by Sunderland *et al.* (1987). Fewer species may be collected in fenced pitfall traps compared to photoeclectors (Funke, 1971), which are comprised of an enclosure fitted with a pitfall trap and a light trap, but they are cheaper to construct and consequently suitable for more extensive studies. More of the smaller carabid species were captured in fenced, compared to unfenced, pitfall traps (Mommertz *et al.*, 1996). The smaller species were most frequently caught by Holland and Smith (1999), but in both trap types. Suction sampling is the only widely adopted method used to provide an estimate of invertebrate density. The efficiency of this technique has been evaluated for a range of invertebrate taxa (Duffey, 1980; Hand, 1986; Mommertz *et al.*, 1996) and has been found to have a number of limitations. The suction power limits the technique to smaller species and is also affected by the vegetation cover (Hand, 1986) and the species' vertical stratification (Sunderland and Topping, 1995). Some of these limitations can be overcome by using a combination of suction sampling and surface searching, which has been shown to provide accurate estimates of spider densities (Sunderland and Topping, 1995). Alternatively, repeated sampling within a frame using the stronger suction of a narrower nozzle can be used to collect larger species but sample sorting times are high at up to one hour per sample (Dinter, 1995). Suction sampling, however, unless repeated many times, still only provides data at a snapshot in time. Furthermore, because suction sampling cannot be used when foliage is wet, usage is often limited to daylight, thus it may underestimate nocturnal species, which include most of the Carabidae found in agricultural habitats (Thiele, 1977). Thus, fenced pitfall traps, although more laborious to set up than suction sampling, overcome many of the limitations outlined above and provide cleaner samples that are quicker to sort and identify. Carabid sampling is also discussed by Luff (this volume).

Ecology of carabids in agricultural habitats

The biology and ecology of carabids in agricultural areas has been reviewed since Thiele by Luff (1987) and later by Lövei and Sunderland (1996). Carabidae found in agricultural areas were considered to be remnants of the original forest fauna, with some additional immigrants from the steppes (Thiele, 1977, p 26). However, a gradient appears to exist with hygrophilic species originating from riparian habitats and tolerant of low temperatures predominating in maritime climates, whilst more thermophilic species, more typical of the steppe habitats, predominate in continental climate areas. As a consequence, the species assemblage in continental climate areas is richer. However, in both regions a typical assemblage of agricultural areas has been identified and is discussed more fully in Luff (this volume).

In Europe and other temperate zones, most Carabidae are univoltine (Thiele, 1971), producing one generation per year, although adults may survive more than one season. They may hibernate either as adults or larvae, although in rare cases also as eggs. Initially, they were classified by Larsson (1939) as either: (1) 'autumn breeders' that reproduce in the autumn and hibernate as larvae (2) 'spring breeders with autumn activity' which hibernate as adults, reproduce in spring and the new generation is active (but does not breed) in the autumn before overwintering (3) 'spring breeders without autumn activity' that hibernate as adults and reproduce in spring, but the new generation adults are inactive until the following year. However, because many species reproduce in summer, the term autumn breeder was deemed inaccurate and Lindroth (1949) proposed the terms 'larval hibernators' or 'adult hibernators'. These classifications have been amended, especially as it is now known that some species may reproduce more than once per year and survive more than one year. Thus, even the five annual rhythms described by Thiele (1977, p 248) do not account for all species. Instead, Den Boer and Den Daanje (1990) proposed a simpler system of summer and winter larvae, although this still remains to be tested for a greater range of species and climates. Even this may be too rigid a classification because there is evidence that some species possess sufficient flexibility in their life cycles to alter the time of year at which they breed according to habitat disturbance (Fadl and Purvis, 1998), annual differences in weather patterns, habitat type and geographic location (Makarov, 1994). The life history traits for some of the most common species found in arable land are given in *Table 1.1*.

Table 1.1. Environmental preferences and breeding behaviour of some carabids typically found in agricultural habitats (A = autumn; S = spring/summer).

Species	Breeding period	Temperature			Humidity			Light			% day active
		Cold	Eury	Warm	Hygro	Eury	Xero	Dark	Eury	Light	
Pterostichus madidus	A		X		X			X			15–30
P. melanarius	A		X		X			X			variable
Poecilus cupreus	S			X			X		X		> 45
Loricera pilicornis	S	X			X					X	0–15
Nebria brevicollis	A		X			X		X			0–15
Asaphidion curtum	S		X			X			X		95
Nebria salina	A	X				X		X			0–15
Stomis pumicatus	S		X		X			X			0–15
Calathus fuscipes	A		X				X	X			23
Harpalus rufipes	A			X			X	X			0–15
*Broscus cephalotes**	A			X			X	X			> 45
Carabus auratus	S			X		X			X		> 45
Agonum dorsale	S			X			X			X	0–15
Bembidion lampros	S		X		X					X	97
Trechus quadristriatus	A	X				X		X			0

* Only found in agricultural areas in eastern Europe
(Sources: Thiele, 1977; Luff, 1978; Desender *et al.*, 1984; Kegel, 1990; Luff, 1998; Turin, 2000)

Population regulation

SURVIVAL OF DIFFERENT LIFESTAGES

Whether a species can survive in agricultural areas depends on many interacting factors, some of which we can identify through monitoring or experimental manipulation; others may yet await discovery. Most research has focused on the requirements of adults and the abiotic and biotic factors influencing their survival. Less is known about larvae, partly because of taxonomic difficulties but more so because of the practicalities of research on soil-living organisms. In addition, their more specific environmental requirements create difficulties when trying to rear them under laboratory conditions. Lövei and Sunderland (1996) considered the larval stage to be the most vulnerable because it has poor mobility, weak chitinization and is therefore vulnerable to extremes of temperature and moisture, while also requiring an adequate food supply. Furthermore, this stage has the longest development time. Similarly, pupae are only weakly sclerotized and can last for long periods, although Lövei and Sunderland considered it to be a less vulnerable stage than the larvae or eggs. Survival of the latter can be maximized by ensuring they are placed in optimum environmental conditions and sometimes they may be enclosed by a small cocoon.

The population dynamics of carabids in disturbed agricultural habitats is poorly understood because their low population density and relatively high mobility makes them difficult to study (Luff, 1987). Although Thiele (1977) did not consider egg mortality to be important, more recent investigations suggest that it could be quite high, with reports of 67% mortality (Heessen, 1981). Larval mortality can also be high. Research on *Pterostichus oblongopunctatus* combined with computer simulations predicted up to 96% mortality for pupae and larvae (Heessen and Brunsting, 1981), with cannibalism being an important regulatory component at higher densities (Brunsting and Heessen, 1984). Survival of adults is thought to be relatively high. The mortality rate of *Harpalus rufipes* was approximately 10% per month for each of three months (Luff, 1980), as was also found for other carabids by Grüm (1975), but was highest during the breeding season. Life table analysis, as far as I am aware, has not been systematically investigated for any species typical of agricultural areas. Such studies would, in any case, have limited applicability outside the study area because of the variability in agricultural practices between farms, localities and countries. Considerable variation between species may be expected, depending on their phenology and environmental requirements. For example, where beetles hibernate, and whether they do so as adults or larvae, will determine their vulnerability to cultivation practices and insecticide treatments carried out during the autumn. Species hibernating in the field margins as adults and/or larvae are less likely to be affected, provided sufficient suitable habitat is available. Those that hibernate within the field, usually as larvae (although a few species, e.g. *Trechus quadristriatus*, remain active as adults through the winter), will be most affected by autumn operations. Some carabid species may escape the effects of autumn farming practices by burying themselves deeply when hibernating. *Harpalus* spp. and *Pterostichus* spp. have been found at 45 cm or more below the soil surface (Briggs, 1965).

Despite the disruptions of agricultural operations, the populations of carabids in arable crops have been found to be relatively constant. Pitfall trapping conducted from 1973–1981 in an arable field showed that the peak catch of *H. rufipes*, *H. aeneus*, *P. madidus*, *P. melanarius* and *N. brevicollis* remained relatively constant, as did the standard deviation of the yearly maximum $\log_e$ (Luff, 1982). Similar values were also calculated by Luff using data obtained by Jones (1979). The standard deviations of $\log_e$ population size were at the lower end of the range given by Williamson (1972) for insect populations, indicating that carabids exhibit quite stable populations. The low rate of population change was attributed to one or more density feedback mechanisms (Luff, 1982).

DISPERSAL ABILITY

To a large extent, the survival of individual species is dependent on their dispersal ability. Poorly dispersing species have generally decreased in abundance, especially if they have specific habitat requirements such as those shown by *Abax parallelepipedus* and *P. oblongopunctatus*. Den Boer (1977) estimated that local populations of such species survive for 40–50 years in disturbed areas, but if the locations of suitable habitat changed faster than this, then they may become extinct. These he classified as L-species (Den Boer, 1985, 1986), having low turnover of 1–2%. Such species are often associated with non-crop habitats within agricultural areas and may only utilize the edges of cultivated fields. The majority of species inhabiting agricultural fields have greater dispersal ability, often by flight, are generally eurytopic, and are thus better adapted to living in unstable or temporary habitats. As a consequence, they have a high turnover (10%), and have been classified as T-species by Den Boer. Species typical of arable fields are included in this group (e.g. from the genera *Amara*, *Pterostichus*, *Agonum* and includes species such as *Loricera pilicornis*, *Nebria brevicollis*, *Harpalus rufipes* and *Asaphidion flavipes*) (Den Boer, 1987).

Besides the suitability of environmental conditions, carabid survival in agricultural habitats is also affected by parasites and predators, inter- and intra-specific competition, dispersal ability and nutrition (reviewed by Toft, this volume). These topics are discussed below, drawing largely from extensive studies by Thiele (1977) and more recent publications, of which there have been relatively few.

ENVIRONMENTAL REQUIREMENTS

Following mark–release–recapture experiments, Thiele (1964) concluded that carabids were sharply divided between field and woodland or hedgerow species. Later studies contradicted this with evidence that a number of species overwinter within field margins (Pollard, 1968; Desender *et al.*, 1981; Desender, 1982; Sotherton, 1985), moving into the crop during spring and summer (Wallin, 1985; Coombes and Sotherton, 1986; Jensen *et al.*, 1989; Thomas *et al.*, 1991; Dennis and Fry, 1992). Other species may be found in both habitats during their peak activity periods (Lyngby and Nielsen, 1980; Wallin, 1987), or may prefer the field margins for breeding if the crop is unsuitable (Desender and Alderweireldt, 1988). Some species prefer the habitat provided by the edge of the field margin (Lyngby and Nielsen, 1980). Carabidae were classified into the following four groups by Fournier and Loreau (1999): (1) species restricted to the hedge; (2) species preferring the hedge; (3) species preferring the crop; (4) species unaffected by the hedge. Similarly, French and

Elliot (1999) classified ground beetles as habitat generalists, wheat specialists, grassland specialists, or boundary specialists. However, it is highly unlikely that any one species would be confined to any of these habitats because carabids respond to environmental conditions rather than vegetation type. But, to some extent, a range of habitat preference would be expected given that over 400 species of Carabidae are associated with arable farmland (Allen, 1979). The crop edge and the field margin also provide a greater diversity of habitats compared to the mid-field, and this may explain why numbers remain higher at the crop edges compared to mid-field throughout the year (Coombes and Sotherton, 1986; Holland *et al.*, 1999). Further discussion on the use of non-crop habitats is given in Lee and Landis and in Thomas *et al.* (both this volume).

Where beetles locate themselves is largely determined by their environmental requirements, although this may differ between lifestages, with larvae having the most specific requirements. However, the majority of studies have focused on the adult requirements, largely because the larvae are not easy to study. The importance of abiotic and biotic factors was investigated, amongst others, by Thiele (1977), who identified temperature, humidity, light and substrate characteristics to be the most influential. Of the latter, chemical factors, pH, the substrate structure and moisture-holding characteristics were the most important. However, as stated by Thiele, we do not know whether the distribution of a species is governed by one, several, or the combined action of these factors, some of which are interrelated. Moreover, an experimental approach in which one factor is manipulated, keeping others constant, can only be used comparatively for closely related species. Early researchers such as Lindroth and Thiele used this approach. More recently, relationships between habitat factors have been sought by sampling naturally occurring distributions and using a multivariate analysis approach (Stork, 1990).

It has been long recognized that adult carabids can exhibit quite distinct preferences for a certain range of temperatures and humidities. However, the majority of carabids frequenting cultivated areas were found to be eurytopic, showing no great preference, although they exhibit varying degrees of plasticity and within-day variation (*Table 1.1*). Other species, such as *H. rufipes*, although warm preferrent, survive extremes of these by also being dark preferrent and are able to survive temperature extremes by seeking out shade. In general, those species with a metallic, shining cuticle, are most active on warm, dry days, in contrast to small, dark coloured beetles (e.g. *Bembidion* species), which are active on cooler, darker days (Kegel, 1990). To a large extent, these differences in preferred microclimate explain the differences found in species composition between cereal and root crops, as discussed in Hance (this volume). Differences are greater between species than between the sexes, although females may also alter their temperature requirements as they approach the time for oviposition and select conditions that are most suitable for the development through egg, pupal and larval stages. This has been demonstrated for *Angoleus nitidus*, a hygrophilic species of coastal areas (Atienza *et al.*, 1996). Females selected areas of taller vegetation, which offered more limited extremes of temperature and a greater degree of physical protection, whereas the males selected areas of higher moisture more typical of the species. The opposite occurred for *P. oblongopunctatus* because during development the larvae require a higher temperature than the adults and oviposition sites are chosen accordingly. Indeed, several species have been shown to vary their activity rhythm seasonally and daily in response to environmental conditions as they seek those sites that are most favourable. Temperature also has a strong influence on daily activity: generally, activity increases with temperature for diurnal species, but is negatively correlated for nocturnal species (Kegel, 1990).

Carabids differ widely in their tolerance to light intensity. The daily rhythmicity of carabids is discussed in some detail in Thiele and Weber (1968), who reported that some species exhibited circadian rhythmicity, but that this did not always occur in all individuals of the same species. They also found, in a study of 67 carabid species, that 60% were night active and 22% predominantly day active, whilst the remainder varied their preference. Similarly, Luff (1978) found that most of the common species found in temperate agricultural areas were nocturnal. Light–dark cycles may also influence activity. Activity was lower on cloudy days when temperatures were too low to instigate activity (Luff, 1978; Alderweireldt and Desender, 1990). *Notiophilus biguttatus* showed a bimodal activity pattern, decreasing activity in the summer in response to high temperatures, but returning to a unimodal pattern of activity in cooler weather by increasing activity at midday (Luff, 1978). Colour and size have also been found to be important for regulating body temperature (Wilmer and Unwin, 1981) and explains why this is a good indicator of diurnal activity. Diurnal species, on the whole,

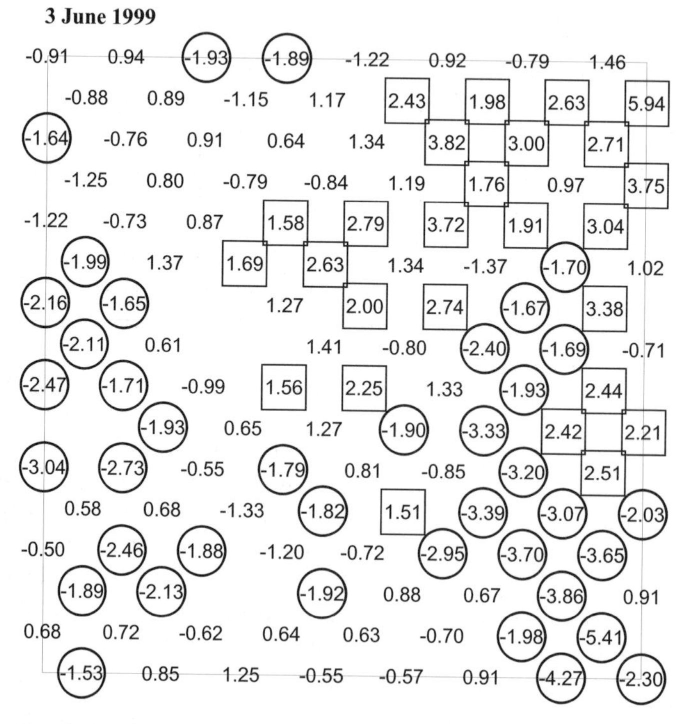

Figure 1.2. Distribution of carabid larvae within a cereal field on three occasions. Data presented for each sampling point are cluster indices produced using SADIE (Spatial Analysis by Distance IndiciEs) method described in Perry *et al.* (1999). Above-average clustering at each sample unit into patches of greater than average neighbouring counts is measured by the clustering index, v_i. Below-average clustering creating gaps of less than average neighbouring counts is measured by the clustering index, v_j. Strong clustering into patches is indicated by units surrounded by circles with $v_i < -1.5$. Strong evidence of gaps is indicated by units surrounded by squares with $v_j > 1.5$.

continued

Figure 1.2 cont.

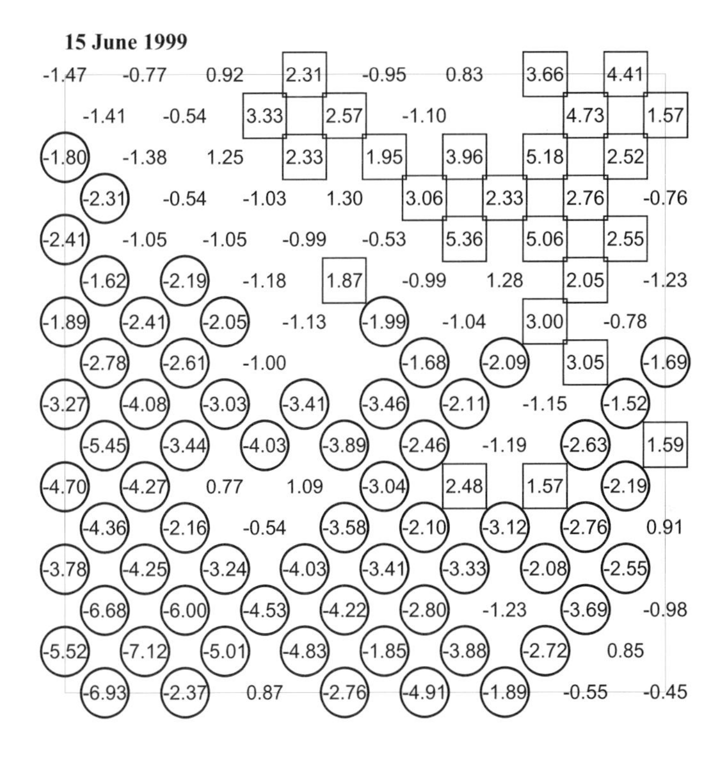

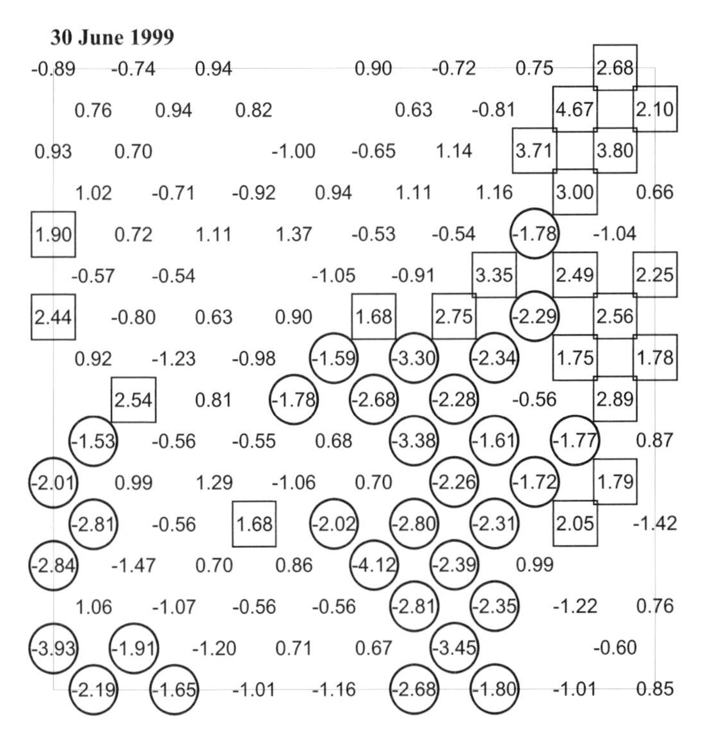

have highly reflective, metallic or black cuticles, allowing them to tolerate higher radiation levels. In contrast, nocturnal species were brown or yellow (Kegel, 1990). Many day-active species are also smaller, as this renders them less prone to overheating. Interestingly, the larval overwintering species were mostly active at night, whereas the adult overwintering species had two peaks of activity: one early in the night, the other during the afternoon.

Of all their environmental requirements, soil moisture is likely to be the most influential, as has been shown in studies of carabid assemblages in other habitats (Luff, 1996). The temporal variation in soil moisture also varies among different soil types and is therefore likely to cause variation in the suitability of different areas for underground lifestages, although this has not been specifically demonstrated. Larvae have been shown to be quite sensitive to soil temperature, with survival decreasing at temperatures above 10°C, but not increasing at lower temperatures (Luff, 1994). Moisture has also been shown to be an important factor, with higher survival rates where moisture is greater. The selection of areas with higher moisture may be especially important for those species with summer larvae which are therefore exposed to higher temperatures during development. In contrast, those species with winter larvae may prefer dryer areas to avoid drowning (Murdoch, 1967). Carabids may also use other environmental features in order to select the most appropriate habitat. One example is their positive or indifferent response towards directed light and silhouettes. Species favouring more open habitats show a positive response towards directed light, whereas those seeking wetter habitat (forest inhabitants) are indifferent, but show a positive orientation towards silhouettes, beneath which there may exist more shaded habitat (Andersen, 1995).

Adult stages of carabids typical of cultivated land exhibit quite distinct spatial distributions within fields (Hengeveld, 1979; Holopainen, 1995; Thomas *et al.*, 1998; Holland *et al.*, 1999) and soil moisture was thought (Hengeveld, 1979) to be the governing factor, although there may be other, possibly interacting, factors beyond our perception. When the distribution of carabid larva was measured at 128 points within a cereal field using a suction sampler, they also only occurred within particular areas of the field (*Figure 1.2*) and this, again, may be a reflection of preferences for soil moisture. Mark–release–recapture studies have revealed that adult *P. melanarius* restrict their movement to within a 20 m area during their peak activity period (Thomas *et al.*, 1998), suggesting that areas of suitable habitat are quite localized. Distributions may, however, vary through time (Thomas *et al.*, 1997) as the habitat and environment undergo changes. Early in the season, roadside verges harboured a greater number and diversity of Carabidae than the adjacent crop but, as the crop canopy converged, the beetles favoured the crop (Varchola and Dunn, 1999).

Other soil characteristics may also have some effect on carabids. Soil characteristics were ranked by Holopainen *et al.* (1995) in the following decreasing order of importance to Carabidae: soil clay content>soil type>soil water content>soil organic content>soil pH. Similarly, Thiele (1977) captured a greater number and diversity of carabids in clay than sandy soil, although he suggested that it was not just the environmental conditions which were more favourable, but that the food abundance was greater. Beetles captured from heavy soils were also heavier on average than those from light sandy soils. Soil pH may be important for species selecting specialist habitats, e.g. highly acidic peat bogs. Only five of fifteen aggregating species were found to have any correlation with soil pH (Gruttke and Weigmann, 1990). The species tested were found in a ruderal ecosystem, although some of those tested also frequent agricultural habitats and two of those showing a correlation with soil pH, *Agonum dorsale* and *Nebria brevicollis*, are common in farmland. However, in cultivated soils pH is maintained within a narrow range to maximize crop productivity and is unlikely to show much variation. In arable crops, soil particle size distribution may also have some influence (Meissner, 1984). Substrate may be important for psammophilic (digging) carabids such as *Harpalus* and *Dyschirrus* species. Lindroth demonstrated that they had a preference for fine sand but, for other species which showed a similar preference, it was the microclimate associated with the soil type that was important.

Soil moisture and microclimate, and thereby carabid numbers and diversity at soil level, may also be determined by other factors such as crop and weed cover. A number of studies have revealed greater carabid numbers in weedy crops (Speight and Lawton, 1976; Purvis and Curry, 1984; Powell *et al.*, 1985; Kromp, 1989; Pavuk *et al.*, 1997). Dense crops create a more humid climate beneath the canopy that is favoured by hygrophilic species, e.g. *Harpalus pensylvanicus* and *Trechus quadristriatus*, although the latter species was not found to favour undersown spring barley over non-undersown (Vickerman, 1978). More *P. melanarius*, *Harpalus rufipes* and *Agonum dorsale* were found in the undersown spring barley. Carabids showed no preference for undersown crops in May and June but did do so in July, when the undersown grass crop was of sufficient density to alter the microclimate (*Figure 1.3*) (McIlwraith, 1999). Similarly, species richness was higher later in the season when weeds were allowed to develop (Pavuk *et al.*, 1997). On the whole, fewer species preferred dense stands and more carabids favoured a crop environment which provided open areas, where they may make use of higher temperatures, and adjacent more heavily vegetated areas of higher humidity (Honek, 1988; Armstrong and McKinlay, 1997). Carabid larvae were, however, reduced by the application of herbicide (Powell *et al.*, 1985) and this may have been because weedy areas have a

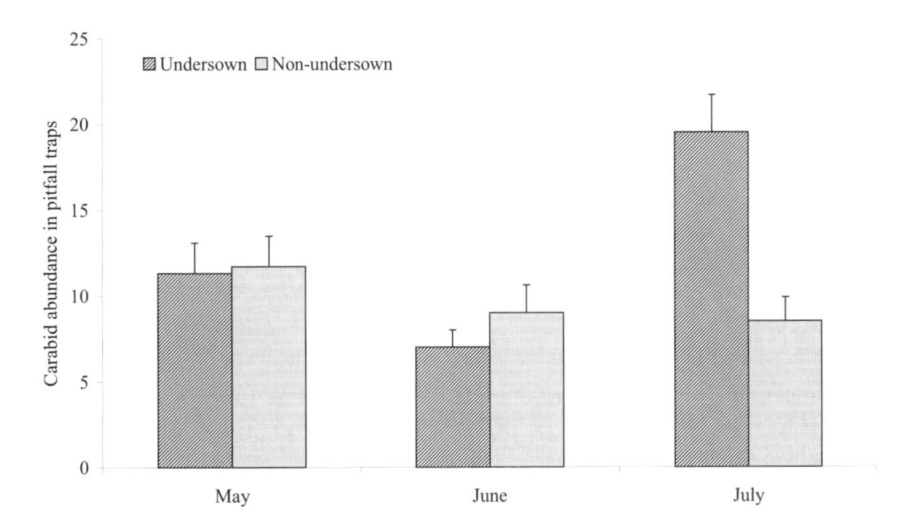

Figure 1.3. Carabid activity-density in undersown and non-undersown spring barley.

higher humidity that is more suitable for oviposition and larval development. Weedy crops may also supply seed resources and thus be favoured by spermophagous groups such as *Amara* and *Harpalus* species (Tooley and Brust, this volume). Other arthropod prey may also be higher where weeds are present (Speight and Lawton, 1976; Shelton and Edwards, 1983).

The conditions created by vegetation also determine the suitability of a site for overwintering, as first demonstrated with *Bembidion* spp. in field margins (Andersen, 1985b) and for *Demetrias atricapillus* in beetle banks (Thomas *et al.*, 1992). This explains why overwintering beetles exhibit strong aggregation in overwintering habitats (Thomas *et al.*, 1991). The tussock-forming grasses (e.g. *Dactylus glomerata* and *Holcus lanatus*) buffered extremes of temperature, unlike the mat-forming species (e.g. *Lolium perenne* and *Agrostis stolonifera*) (Luff, 1965). Luff (1966) revealed that this buffering can reduce winter mortality.

PATHOGENS, PARASITISM AND PREDATION

This subject is one of the least researched areas of carabidology because it is not generally assumed to be significant in population regulation (Luff, 1987). Carabids inhabiting agricultural areas are attacked by a limited number of fungi, protozoa, nematodes, mites and parasitoids (Hymenoptera and Diptera). Bacterial and viral diseases are seldom mentioned, although this may be because they have not been intensively researched. Details of these are given in Thiele (1977) and a bibliography was produced by Larochelle (1978), with a brief biological account in Luff (1987). However, organisms attacking larvae or adults are mentioned more often, although egg and pupal stages may be more vulnerable. Egg mortality was higher in fresh litter (87%) compared to sterilized soil (7%) (Heessen, 1981), suggesting that attack by pathogens might be important for population regulation. The pupae and larval stages may also be vulnerable because they have weak chitinization. Adult carabids are attacked by non-specific fungi if maintained on the same substrate for too long, but such infections are rare on carabids in the wild. They may play a small role in limiting survival of overwintering adults and larvae (Kmitowa and Kabacik-Wasylik, 1971; Riedel and Steenberg, 1998) but only very low infection levels of entomopathic fungi have been found (Steenberg *et al.*, 1995; Riedel and Steenberg, 1998), suggesting strong resistance to infection. Reducing the risk of infection may be a criterion in the selection of overwintering habitat. Artificially applied pathogens used for pest control have no impact on adult carabids (Kabacik-Wasylik and Kmitowa, 1973; Young and Hamm, 1985; Riedel and Steenberg, 1998). In Norway, riparian *Bembidion* spp. were attacked by ectoparasitic fungi and endoparasitic nematodes, which in a moist site reached infection levels of 41.1 and 11.4% respectively (Andersen and Skorping, 1991). However, in an open site away from a river, infection levels were only 0.8 and 0.4%, indicating the importance of humidity for infection. Infection of *P. madidus* by the nematode *Mermis nigrescens* rose from 1% to 61% between 1966–1972 in a garden environment (Luff, 1973), but parasitism was negatively associated with peak log catching rate, suggesting that parasitism was inversely density dependent.

A few parasites have evolved specifically on Coleoptera, and even Carabidae, notably Podopolipidae (Acari) and Proctotrupoidea (Hymenoptera). The former are thought to have no impact on population regulation (Thiele, 1977). The most frequently found parasite is *Phaenoserphus viator* (Hymenoptera), which mainly attacks winter larvae of some carabids common in cultivated crops (*Nebria brevicollis*, *Pterostichus melanarius*, *P. madidus*, *Agonum dorsale* and *Calathus fuscipes*). Summer active larvae were predominantly parasitized by *P. pallipes* (Critchley, 1973). Larval parasitism of *H. rufipes* by *Proctotrupes gladiator* reached 20% (Luff, 1976b) but, in comparison, less than 1% of *N. brevicollis* larvae were parasitized by *Phaenoserphus* species (Luff, 1976a). Levels of parasitism appear to differ widely, and occur sporadically in different locations and years, but are not thought to have much impact on population regulation. Adults may also be parasitized. Parasitism by the brachonid *Microtonus caudatus* on *H. rufipes* varied between 17 and 27%, but there was no density-dependent response by the parasitoid (Luff, 1976b).

Carabids are also preyed upon by a wide range of vertebrates. Larochelle published bibliographies of mammalian species (Larochelle, 1975b), amphibians and reptiles (Larochelle, 1975a) and birds of North America (Larochelle, 1975c) and of Europe and Asia (1980). Most vertebrates, with perhaps the exception of rodents, are present at too low a density in agricultural areas to have much impact on carabid populations, although occasional bouts of concentrated feeding may have some effect on a small scale. Exclusion experiments in the USA revealed that rodents could influence population regulation (Parmenter and MacMahon, 1988). In northern temperate areas, shrews and mice are likely to be the most abundant vertebrates. Shrews may have the greatest potential to regulate carabid populations because they show a preference for larger invertebrates; 60–82% of the prey of the common shrew (*Sorex araneus*) was found to be in the size range 6–10 mm (Churchfield, 1982), which encompasses many carabid species found in agricultural crops. Moreover, a positive correlation has been found between the availability of beetles and their incidence in the diet of common shrews (Churchfield, 1982). Wood mice are common in cereal fields where they consume arthropods mainly in the spring, whilst in sugar beet fields they eat arthropods, predominantly larvae, mainly in the autumn (Green, 1979). The house mouse is also common, and sometimes as abundant as the wood mouse, in agricultural land (Delany, 1961). It is again an opportunistic feeder, responding to the localized abundance of prey, which includes carabids (Berry, 1968). The only evidence that small mammals can regulate carabid numbers comes from exclusion experiments in grassland (Churchfield *et al.*, 1991). A greater range of vertebrate predators will be present in hedgerows and woodland, but the extent to which these regulate carabids overwintering in margins, or inhabiting margins during the summer, is unknown. The value of carabids as dietary items is discussed in more detail later in this chapter.

DEVELOPMENT AND REPRODUCTION

Carabids may occupy several different ecological niches as they develop through their different lifestages, and the conditions within these niches will determine their survival rate. If, for example, temperatures are higher, then less energy is needed to maintain essential bodily functions. Evidence of this was found by Van Dijk (1983), who reported a longer reproductive period and greater egg production at higher temperatures. Similarly, with *Carabus auronitens* egg production was retarded at low temperatures, even when food was readily available (Althoff *et al.*, 1994). For *Notiophilus biguttatus*, egg dry weight was greater at 10°C compared to 20°C (Ernsting and Isaaks, 1994). Such experiments, where fecundity is only measured at a limited number of temperatures, must be treated with some caution. Whether experiments detect a positive or negative effect of temperature will depend on the temperature regimes tested and the response of the species. For example, if the species has an optimum temperature, with fecundity decreasing at lower and higher temperatures, a positive response to temperature may be detected if the highest temperature tested is the optimum. If, however, the optimum is the lowest temperature tested, then a negative response to increasing temperature will be found.

For many animals, however, it is food quality and availability that regulates development, overall size, and fecundity. Larval feeding has been shown to determine adult size of a number of carabid species (Nelemans, 1988; Ernsting *et al.*, 1992; Van Dijk, 1994). In many insects, adult size determines reproductive capacity, but this is not always true for carabids. The Harpalini, for example, produce larger eggs than a larger Pterostichine (Luff, 1981). For *Poecilus cupreus* (Bommarco, 1998a) and *P. versicolor* (Van Dijk, 1983) it was adult feeding which had the greatest influence on reproductive capacity. Similarly, egg production was positively correlated with the availability of food for *Agonum dorsale* (Basedow, 1994) and for *B. lampros*, *P. melanarius* and *P. cupreus,* although the quality of food was also influential (Wallin *et al.*, 1992). However, a negative relationship between number of eggs laid and egg size would appear to negate the effect of greater food availability because larval survival was positively correlated with egg size; larger eggs producing larger first instars have a better survival rate. But with *Calathus melanocephalus* in heathland, food was not found to be a limiting factor, although if food was in super-abundance, egg production increased with temperature. Oviposition rate may also respond to satiation levels (Van Dijk, 1982). The activity of *P. versicolor* and *C. melanocephalus* is influenced by hunger levels; but with *P. versicolor*, in contrast to *C. melanocephalus*, no eggs are laid during periods of high locomotory activity relating to feeding, rather it is when they are satiated and inactive that oviposition occurs. Thus, a high nutritional status appears to be a stimulus for oviposition.

FOOD SUPPLY, INTER- AND INTRASPECIFIC EFFECTS

Interspecific competition may occur through two broad mechanisms: through competition for resources (Alley, 1982) such as food, and through interference (Keddy, 1989). Interspecific competition in carabid beetle assemblages was reviewed in detail by Niemelä (1993), although relatively few of the studies discussed were conducted in agroecosystems. Of the 32 papers reviewed by Niemalä, evidence for interspecific competition in carabid assemblages or resource partitioning was only found in half, and competitive exclusion was not demonstrated. Indeed, he found that only two studies convincingly demonstrated interspecific competition. Criticism was also expressed of the way in which interspecific competition was evaluated, some researchers selecting systems to study in which they expected interspecific competition to be found *apriori*, and the methodology used was often flawed because interacting factors were ignored or eliminated. Moreover, the literature may be biased because cases where competition was detected are more likely to be reported than those showing no effect.

Interspecific competition has long been regarded as being insignificant in carabid population regulation (Lindroth, 1949; Thiele, 1977; Den Boer, 1985; Hengeveld, 1985), although, for forest inhabiting carabids at least, Lenski (1982) and Loreau (1986) argue that species richness, abundance and spatial distribution are largely governed by interspecific competition. However, in later experiments Loreau (1990) found no evidence for interspecific competition. Müller (1987) argues that differences in the timing of teneral emergence are greater between species of similar size than between those of different sizes because there is competition between species. Also, the niches occupied by species with similar body sizes are different, thus allowing them to use different resources. This then overcomes interspecific competition. The spatial separation of species according to body size was attributed to interspecific competition by other researchers (Davis, 1987; Erikstad *et al.*, 1989). However, Bengtson (1980) and Van Zant *et al.* (1978), although agreeing that relationships between body size and distribution occurred, could find no evidence to support the theory of competition. Despite the claims of different researchers, Wolda (1989) concluded that the role of intercompetition on carabid assemblages was unresolved.

Intraspecific competition may also occur through cannibalism and if resources (notably food and shelter) become limited. Larval cannibalism is known and where densities have been manipulated experimentally, density-dependent regulation has been shown to occur (Brunsting and Heessen, 1984). Adults also exhibit cannibalism when confined in the laboratory, but its importance in the field is unknown. Moreover, in the laboratory, intraspecific predation is usually confined to the robust individuals preying on weaker or injured beetles. Inter- and intraspecific predation of eggs may also occur given the polyphagous nature of many species, but this has not been reported.

It is well known that the availability and quality of food influences the survival and reproduction of individuals, and thereby population growth (e.g. Lenski, 1982). This has been demonstrated for Carabidae (Van Dijk, 1979; Mols, 1988). Toft and Bilde (this volume) discuss the quality of food with respect to growth and survival of individuals. Competition for food can create a shortage, and this is thought to be a mechanism by which predator populations are regulated. This may occur through interspecific and intraspecific competition and has been observed for carabids in field experiments (Lenski, 1982). Extensive studies by Den Boer (1980), however, suggest otherwise, as no evidence was found of competitive exclusion in a range of habitats, including agricultural ones. He suggests that cases of competitive exclusion are rare and only occur where resources or food are extremely limited or the species very specialized.

The polyphagous nature of many of the larger carabids inhabiting agricultural areas is well documented. Dissections frequently reveal the remains of other species in the guts of these beetles, confirming interspecific predation. However, manipulative experiments have shown no evidence of interspecific competition (Loreau, 1990). This was tested by confining two species in an enclosure. Species were chosen that had similar diets, annual activity cycles and environmental requirements, although one species (*Abax parallelepipedus*) was better suited to the habitat compared to the other (*Pterostichus madidus*). *Abax parallelepipedus* exhibited some population regulation; but although *P. madidus* did less well, it was not affected by the other species. Loreau suggested that dominant species will approach equilibrium conditions and competitive regulation because they are well adapted to the environment and can exploit the resources to their maximum potential. Other less abundant species are in non-equilibrium being in sub-optimum conditions, so are rarely affected by competition, following the theory of core and satellite species (Hanski, 1982). Competition may also be avoided if individual species actively seek out their most preferred habitat niche.

Intraspecific competition could occur if food was limited, although it is assumed that, because carabids occur at relatively low densities, their food should remain super-abundant. However, gut dissections have revealed that a relatively high proportion (up to 50%) of different species have empty guts (Sunderland, 1975; Holland and Thomas, 1997). Moreover, a number of studies have found that food is limited in agricultural areas and, as a consequence, carabids may not reach their full reproductive potential. Food supply has been shown to affect the overwintering survival of the spring breeder, *Demetrius atricapillus*. Adults collected from the field during the winter contained food in their guts, and mortality levels in the laboratory were lower (19.6%) when food was provided, compared to those which were starved (49.4%) (Thomas *et al.*, 1992). *D. atricapillus* does not enter obligate diapause, unlike many other carabid species, but alternates between activity and inactivity; thus, food availability during winter can be important for the survival of some species. Similarly, Bommarco (1999) found that food was limiting when he measured the development of *P. cupreus* in spring barley, which had a lower abundance of arthropods compared with a perennial ley. In spring barley, their egg load, Energy Reserves Index (ERI), which is calculated by comparing body weight and size, stored fat, and live body weight, were lower, indicating a low feeding rate. Manipulating the density of *P. cupreus* had no effect on the abundance of their prey community composition. Furthermore, *P. cupreus* collected from the field had a lower ERI compared to those reared in the laboratory (Bommarco, 1998b). Similarly, Baars and Van Dijk (1984) reported a negative relationship between density and fecundity, and implied that population regulation may occur through food shortage.

Agricultural practices such as the application of insecticides which remove prey, or habitat destruction, may have sufficient impact to create unfavourable conditions for Carabidae, but these may not be long lasting if reinvasion or dispersal occurs relatively quickly. There is evidence that many agricultural practices are detrimental to carabid populations and these are reviewed by Holland and Luff (2000) and Hance (this volume). In reality though, it will be a combination of many factors that will determine population size and distribution.

When the feeding rate and fecundity of *P. cupreus* were examined in Swedish landscapes of varying complexity, considerable variation was found (Bommarco, 1999). The localities differed in field size, amount of field edge, percentage of annual crops and farming practice. Fecundity was higher in those localities that were farmed organically and had smaller fields with more perennial cropping, although more *P. cupreus* were found in The Netherlands from areas with larger fields compared to a landscape of small fields (Den Nijs *et al.*, 1998). At one locality, poor food quality was responsible for the lower fecundity, as has been demonstrated in laboratory experiments (Wallin *et al.*, 1992; Toft, 1995). In addition, beetle size was negatively correlated with field size. This was thought to be a result of poor larval feeding conditions, as found for *Nebria brevicollis* (Nelemans, 1988), *N. biguttatus* (Ernsting *et al.*, 1992) and *Calathus melanocephalus* and *Pterostichus versicolor* (Van Dijk, 1994). However, prey densities may not always be higher in areas farmed organically. Collembola favoured conventionally-grown wheat because the dense crop created more humid conditions at the soil surface, encouraging plurivoltine epigeal Collembola, despite four insecticide applications (Basedow, 1994). Moreover, because predator densities were lower in the conventionally grown wheat, more prey was available per predator. This, however, only increased the fecundity of the spring breeding *A. dorsale*, whereas that of *P. melanarius*, an autumn breeder, was unaffected.

Interspecific competition may also be inferred from spatial seperation. Four species of *Poecilus* were found to exist within the same field but differentiated themselves spatially, thereby avoiding competition (Kegel, 1994). Other studies of spatial pattern further support these findings; *Bembidion lampros* and *P. madidus* predominating in different areas of the same field (Holland *et al.*, 1999). However, differences in abiotic or biotic requirements which are themselves spatially differentiated may also cause these patterns to occur. The spatial distribution of carabids is reviewed in more detail in Thomas *et al.* (this volume).

These findings indicate that the population dynamics of carabids within agricultural areas is extremely complicated and almost impossible to predict. Moreover, because each species has such specific abiotic and biotic requirements, these must first be considered before trying to identify intra- and interspecific interactions. Many of the factors that might influence the population dynamics of carabids in agroecosystems are summarized in *Figure 1.4*, although this is by no means a comprehensive list and some of the possible interactions are not yet fully understood, if at all. If the controversy over interspecific competition is to be resolved, then detailed investigations as outlined by Niemelä (1993) are needed and, even then, may not be sufficient because of the complex interactions which influence beetle population development.

The importance of carabids in agricultural habitats

PEST CONTROL

The potential of carabids for pest control has long been recognized by observant farmers. Early agricultural scientists in the nineteenth century and later often mention the beneficial habits of carabids in their textbooks. Carabidae are typically polyphagous, having extremely varied diets, a proportion of which may be crop pests. The diet of Carabidae in Europe and North America has been periodically reviewed (Thiele, 1977; Allen, 1979; Hengeveld, 1980; Luff, 1987; Larochelle, 1990; Lövei and Sunderland, 1996; Kromp, 1999; Toft and Bilde, this volume) and the pests they eat include: a range of aphid species; dipteran eggs, larvae and pupae; coleopteran eggs and larvae; lepidopteran pests; and slugs. Most of the literature concentrates on adult diet. Allen (1979) concluded that they were opportunistic feeders and consequently that their diet was likely to vary considerably. Many larvae are thought to be more carnivorous, with a more restricted diet range than adults (Lövei and Sunderland 1996). Despite this longstanding knowledge and the wide range of recent studies demonstrating the value of carabids for pest control (reviewed by Sunderland, this

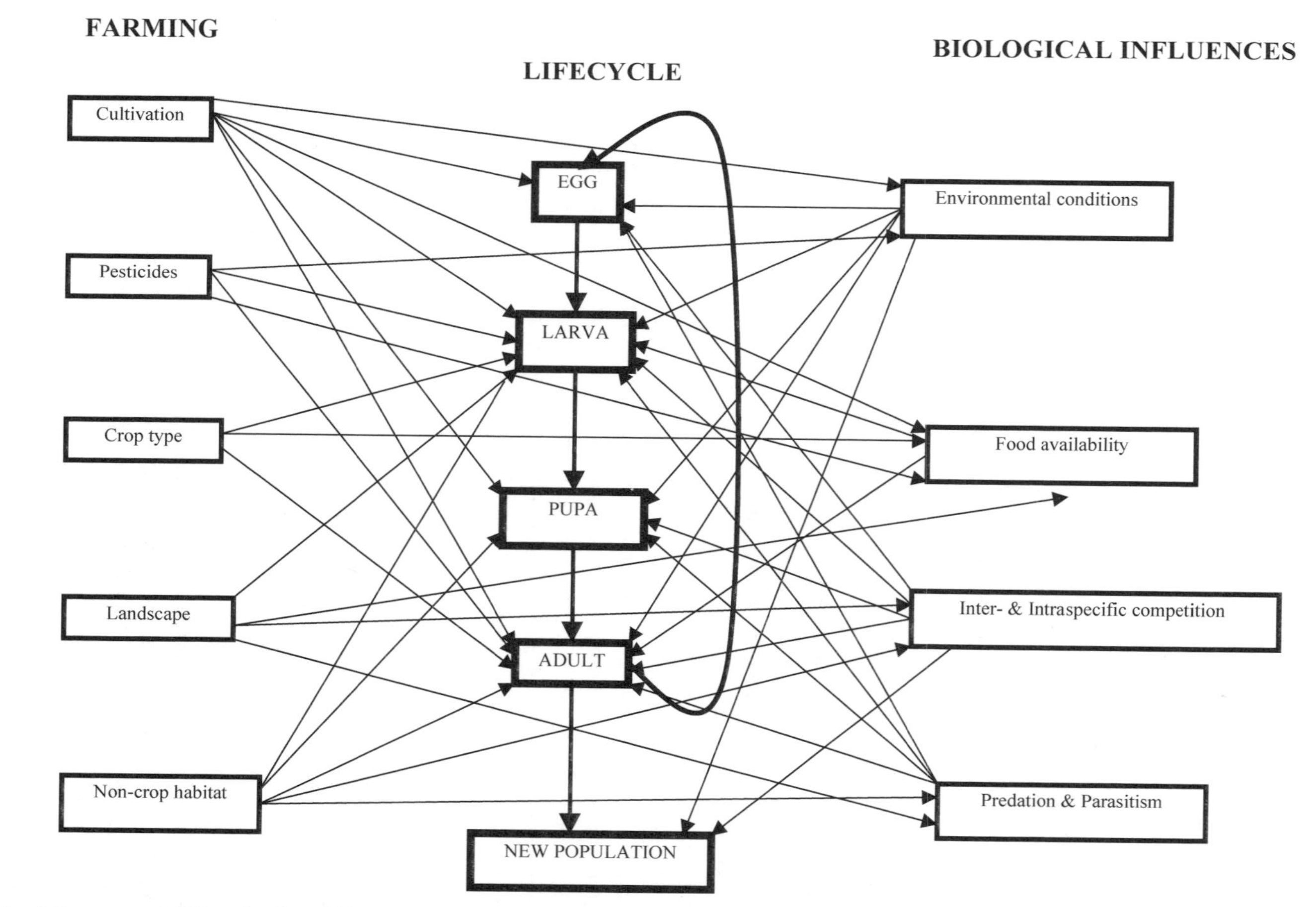

Figure 1.4. Influences on the lifecycle of carabids in farmed landscapes.

volume), it is only in the last decade that arable farmers have again become aware of natural control processes and have considered them in their crop management decisions.

In recent years, the falling value of arable crops, combined with pressure from environmental organizations and consumer groups, has driven farmers to look more closely at integrated crop management and integrated farming techniques. Of these, lower insecticide usage and choice of selective insecticides (considered in Holland and Luff), non-inversion tillage (see Hance, this volume) and augmenting non-crop habitat (Lee and Landis, this volume) are likely to have the greatest impact on Carabidae (Holland and Luff, 2000). Adoption of integrated farming practices does, in some cases, encourage Carabidae (e.g. El Titi and Ipach, 1989; Winstone *et al.*, 1996), but these may need to be more extensive in order to have a noticeable impact on pest management. Moreover, as Sunderland (this volume) concludes, carabids alone may not prevent pest outbreaks but are an important component of the benficial fauna.

CARABIDS AS FOOD

Emphasizing the value of carabids for pest control is likely to gain most interest from farmers. However, carabids have other valuable roles in the agricultural landscape. Carabids are important dietary items for many birds inhabiting farmland, especially for nestlings which require a protein-rich diet. They are also consumed by some mammals, amphibians and reptiles (Larochelle, 1975a,b, 1980), although the latter two groups are mainly confined to non-crop habitats in agricultural areas. Because most carabids are active at night or in low light, they are mainly consumed by nocturnally active vertebrates, as demonstrated in Spain (Hernandez *et al.*, 1991). Interestingly, Thiele (1977) regarded all these groups as carabid predators and was concerned with the impact they may have on carabid populations, in contrast to the present day, when they are regarded as an exploitable food source for other species.

Birds

A review by Larochelle (1980) revealed that 203 Eurasian bird species and subspecies preyed upon 183 species of Carabidae. For most birds, Carabidae formed only a minor part of their diet, but for a few corvid and raptor species they are an important dietary component. In North America, Larochelle (1975c) reported that carabids constituted at least 7% of the diet of 11 species (eastern bluebird, western bluebird, common grackle, eastern meadowlark, western meadowlark, California shrike, loggerhead shrike, vesper sparrow, starling and red-headed woodpecker). Carabids are also important dietary items for the wild turkey (Hamrick and Davis, 1971), red-shouldered hawk (Stewart, 1949), American kestrel (Philips, 1977), and many species of shrike (Lefranc, 1997).

Recently, public and political attention has been drawn to the widespread declines in farmland birds in western Europe (Tucker and Heath, 1994; Fuller *et al.*, 1995). One of the most important factors causing these declines is a shortage of invertebrate food, especially for chicks which need a protei- rich diet. Indeed, grey partridge chick survival was shown to be strongly correlated with the abundance of insects, with small

Table 1.2. European farmland bird species, their conservation status and the importance of carabids in their diet (1 = present in diet but not quantified as being important; 2 = present in diet quantified as being important).

Bird species	Importance of carabids in diet	Conservation status		
		Countries occupied	% Countries declining	% Countries increasing
Herbivorous feeding				
Grey partridge	2	11	100	0
Red-legged partridge	1	3	67	0
Pheasant	2	11	36	18
Quail	1	11	45	9
Skylark	2	11	100	0
House sparrow	1	11	45	0
Tree sparrow	1	11	45	18
Chaffinch	1	11	0	9
Brambling	1	No data		
Linnet	1	11	55	9
Yellow hammer	1	11	45	0
Cirl bunting	1	5	20	0
Corn bunting	1	11	82	0
Invertebrate feeding				
Stone curlew	2	4	100	0
Lapwing	2	11	64	18
Golden plover	2	No data		
House martin	1	11	45	9
Sand martin	1	11	64	9
Swallow	1	11	100	0
Meadow pipit	1	11	27	18
Yellow wagtail	2	11	55	9
Pied wagtail	2	11	0	18
Blackbird	1	11	9	18
Song thrush	1	11	27	0
Mistle thrush	1	11	9	27
Robin	1	11	0	0
Dunnock	1	11	27	9
Wren	1	11	0	0
Spotted flycatcher	1	11	36	0
Red-backed shrike	2	10	80	0
Starling	2	11	36	9
Hobby	1	10	10	40
Little owl	2	11	100	0
Barn owl	1	11	64	0
Kestrel	2	No data		

(Sources: Wilson *et al.*, 1996; Campbell *et al.*, 1997)

Carabidae being the most important (Potts and Aebischer, 1991). Initially, extensive research on the ecology of the grey partridge revealed the impact that intensive farming methods, and especially pesticides, were having on invertebrates (Potts, 1986). This has been shown to hold true for a number of other farmland bird species (Wilson *et al.*, 1996; Campbell *et al.*, 1997). Carabidae were shown to be present in the diet of 34 farmland bird species (*Table 1.2*) common to farmland in north-western Europe and for 11 they are an important dietary component; these were not in the list previously compiled by Larochelle (1980). For these species, all but one are declining in more countries than they are increasing in (Wilson *et al.*, 1996). The list excludes corvids and shrikes, which also feed on carabid adults and larvae (Larochelle, 1980). Corvid numbers have increased considerably in the last few decades. In contrast, shrikes have declined substantially over the last 30 years in the majority of European countries, and now all species are of Conservation Concern (Tucker and Heath, 1994). Agricultural intensification and loss of habitat are thought to be the cause and, because they depend heavily on beetles, pesticides have been especially implicated (Lefranc, 1997). Carabids may also feature in the diet of woodland species (e.g. nuthatch), which may forage in fields and their margins.

The identification of which carabid species are most preyed upon by birds has received only limited attention. In general, it is the larger species from the genera *Agonum*, *Amara*, *Carabus*, *Harpalus* and *Pterostichus* that are most frequently recorded as being eaten by birds (Larochelle, 1980). Nocturnal birds, such as little owls, tawny owls, barn owls and stone curlew, feed upon carabids because such a large proportion of them are also night active. Of these, little owls (88%) contained the highest proportion of carabids, with most of these being from the genera *Pterostichus*, *Carabus*, *Harpalus* and *Abax* (Thiele, 1977). However, many carabids of agricultural land are diurnal and also fall prey to ground-feeding birds. Small diurnal carabid beetles (e.g. *Trechus quadristriatus*, *Bembidion* spp.) are an important component in the diet of grey partridge, red-legged partridge, and pheasant chicks (Green, 1984; Hill, 1985).

Mammals

Mammalian species inhabiting or foraging in fields and their margins will consume carabids. Some of these species are of high conservation status, for example bats are protected in all European countries (Stebbings, 1991). Some bat species (e.g. greater horseshoe *Rhinolophus ferrumequinum*; Serotine *Eptesicus serotinus*; mouse-eared bat *Myotis myotis*) specialize in eating large beetles. Flying carabids may be taken by bats, but the mouse-eared bat will scoop carabids from the soil surface while flying (Arlettaz, 1996). Carabids are also known to be important in the diet of mammals inhabiting non-crop areas adjacent to surrounding fields and long-term set-aside. For some of these species (hedgehogs, shrews and mice), carabids can be important dietary items. Carabids were found in 60% of hedgehog stomachs from East Anglia, UK and were the second most abundant prey item, although they accounted for only 8% by weight (Yalden, 1976). Carabids were consumed most during June, coinciding with the availability of *H. rufipes* and *Pterostichus* spp., which were the two most frequently found species. Other large and medium-sized carabids were also identified, e.g. *Carabus* spp., *Cychrus caraboides*, *Nebria* spp., *L. pilicornis*, *Leistus* spp. *Harpalus affinis* and *Amara communis*. *Abax parallelepipedus* was not found, although this species was found to be acceptable prey in feeding trials (Dimelow, 1963). The water shrew was found to consume large numbers of carabids from the genera *Agonum* and *Pterostichus* (Murdoch, 1966). The yellow-necked field mouse was found to have fed upon carabid adults, but more so on the larvae (Obrtel, 1973). Beetles made up 20% of the diet of common, pygmy and water shrews (Churchfield, 1982, 1984, 1985). Small mammals may even take beetles from dry pitfall traps, and this must be taken into account when estimating population size or measures taken to inhibit their activities (Stapp, 1997). Small mammals are also frequently captured in wet pitfalls if a protective guard or funnel is not used (Baars, 1979). Moles consume large numbers of insect larvae, especially during the summer (Larkin, 1948), which may include carabid larvae.

Amphibians and reptiles

Carabids are consumed by a number of different frog and toad species because they concentrate on more active invertebrates. In North America, 62 species of carabid were found in the stomachs of six frog species (Larochelle, 1974a), while one toad (*Bufo americans*) was found to have fed upon 98 carabid species (Larochelle, 1974b). Predation by amphibians is, however, likely to be largely confined to the non-crop habitats where annual crops predominate, but predation may be more extensive in less intensively grazed pasture. In upland pastures, the amphibian, *Salamandra lanzai*, feed mostly on carabids (Andreone *et al.*, 1999). A literature survey by Larochelle (1975a) revealed that 76 species and subspecies of amphibians and reptiles were recorded as predators of carabids in North America.

Long-term changes

There is evidence that the abundance and diversity of Carabidae are declining in the long term since farming started to become more intensive in the 1950s. A 50–81% reduction in activity-density for carabids and formicids combined was found over a 30-year period in Germany (Heydemann and Meyer, 1983). The same authors found a 48–85% decrease in species numbers. Even over relatively short time periods, changes can be seen. Basedow (1987) found an 81% decrease in trapping rates and a 90% decrease in biomass for sampling conducted from 1971–1974, compared to 1978–1983. Two further studies comparing the relative abundance of carabids captured during the 1950s with that of the late 1970s until the early 1980s also found evidence of a shift in species dominance, with some becoming more abundant whilst others have become rarer (Croy, 1987; Körner, 1990). Species with the poorest dispersal power have declined the most (Desender and Turin, 1989), although these were not always the smallest. The larger carabids such as *Carabus* spp. suffered the most, disappearing from some localities (Lebrun *et al.*, 1987; Körner, 1990), even though members of this genus are capable of moving further than some smaller species (Grüm, 1983). These declines coincide with the intensification of farming and the widespread use of pesticides. Monitoring conducted within the LINK Integrated Farming System Project in the UK from 1992–1997 revealed that carabid numbers and diversity were very low at most of the sites, suggesting that intensive arable farming was having adverse effects (Holland *et al.*, 1998). In contrast, sampling carried out in Belgium in 1963 and 1986 using the same methodology revealed that carabid species diversity and community structure had not changed, and has even increased in relative value (Lebrun *et al.*, 1987).

The most extensive, long-term survey of arthropods is that carried out by The Game Conservancy Trust. Every year since 1970 suction samples have been taken in mid-June from 100 cereal fields covering 62 km^2 of arable farmland in West Sussex. The 'Sussex Study', as it is now known, was initially started as part of a research programme looking into the reason behind changes in the productivity and distribution of the grey partridge since the First World War. Changing farming practices were thought to be responsible, especially the effects of pesticides, because an adequate supply of protein-rich arthropods is essential for the survival of grey partridge chicks (Potts and Vickerman, 1974). Although suction samplers only collect small carabids available on the soil surface at the time of sampling, the technique has the advantage

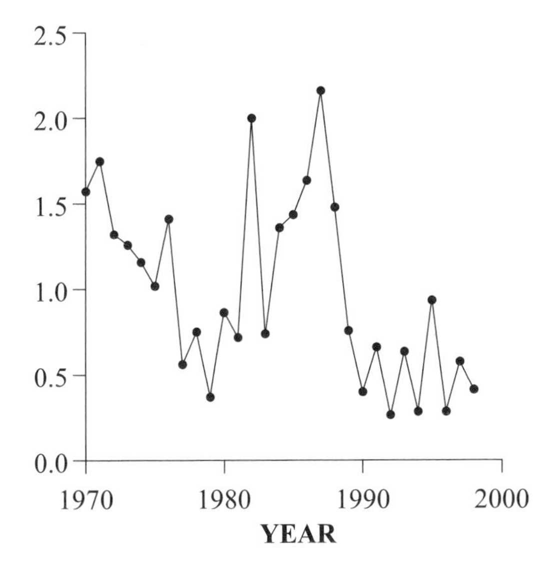

Figure. 1.5. Decline in Carabidae over the last 30 years in Sussex, UK based upon suction samples taken from approximately 100 cereal fields in June each year.

of providing an estimate of density. Over this 30-year period the number of carabids, with some exceptions, declined until the start of the 1990s (*Figure 1.5*). Since then, their densities have remained relatively constant, but at only one third of their peak in 1970. The carabids collected were largely comprised of species from the genera *Bembidion*, *Trechus* and *Demetrias*. The declines in invertebrates has been attributed to a combination of greater inputs of agrochemicals, the loss of grass leys from the rotation, and the enlargement of fields (Potts, 1997; Ewald and Aebischer, 2000). *Figures 1.6–1.8* illustrate the spatial distribution of Carabidae and some taxa over this period in cereal fields. From these it is apparent that in the 1970s and 1980s carabids were present in most cereal fields, with a quarter of the fields having more than 0.25 carabids per sample. However, by the 1990s numbers per field were considerably lower, and in many none were collected. The individual taxa show a similar trend, although with some variation. *Demetrias* species (*Figure 1.7*) declined consistently through the decades, while *Bembidion lampros* (*Figure 1.8*) showed the most marked decline in their distribution in the 1990s. As both these groups overwinter within field margins, they may be particularly vulnerable to the loss of field margins and the consequent increase in field size. Five of the six farms in the study area have adopted intensive arable production techniques, whereas one has remained more traditional, combining both arable and livestock production. The impact of these changes on hedgerow length and field size was appraised by Aebischer (1991). On the more intensive farms, between 1970 and 1989, the length of hedgerows per km^2 decreased from 6.9 to 5.2 km while field size increased from 12.5 to 15.3 ha.

At a landscape scale, some species have expanded their ranges, as found in Sweden (Lindroth, 1972) and Norway (Andersen, 1987), although others have contracted their ranges or become extinct. In Sweden, climate change was thought to be responsible (Lindroth, 1972), whereas in Norway the expansion of cereal cropping was responsible for the increases of five species (Andersen, 1985a), whilst others

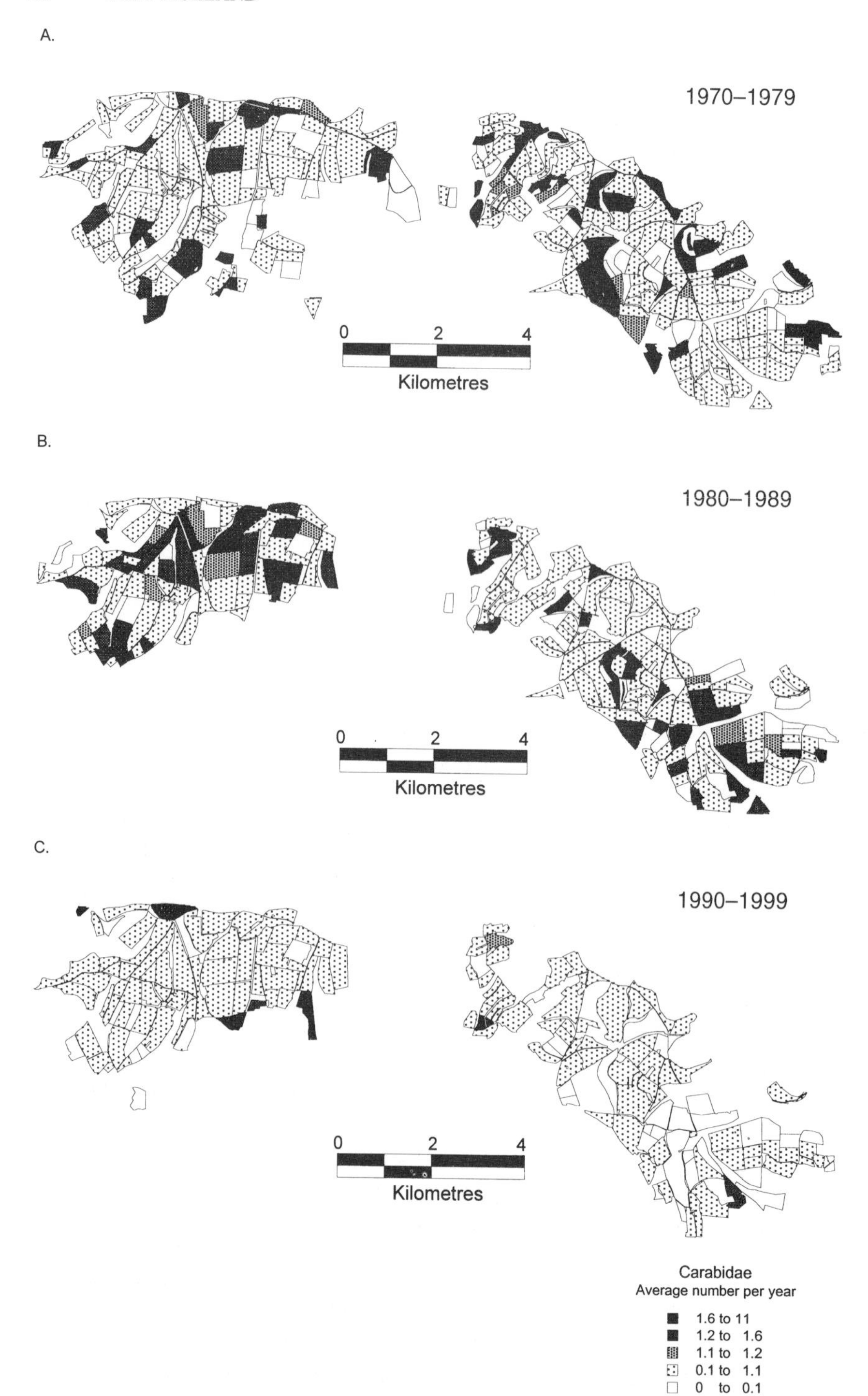

Figure 1.6. Distribution of Carabidae across the Sussex Study area over three decades based upon suction samples taken from approximately 100 cereal fields in June each year.

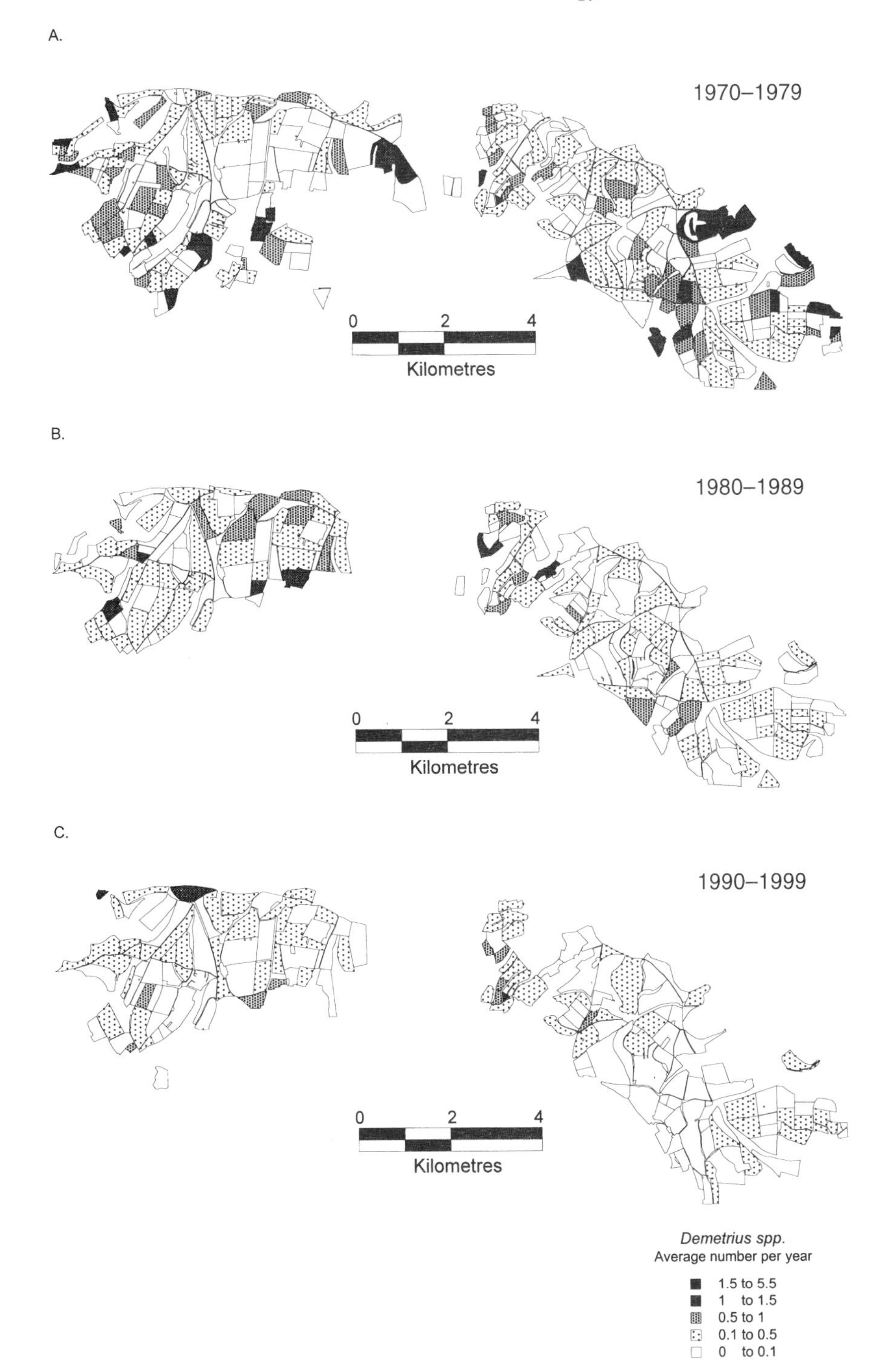

Figure 1.7. Distribution of *Demetrias* species across the Sussex Study area over three decades based upon suction samples taken from approximately 100 cereal fields in June each year.

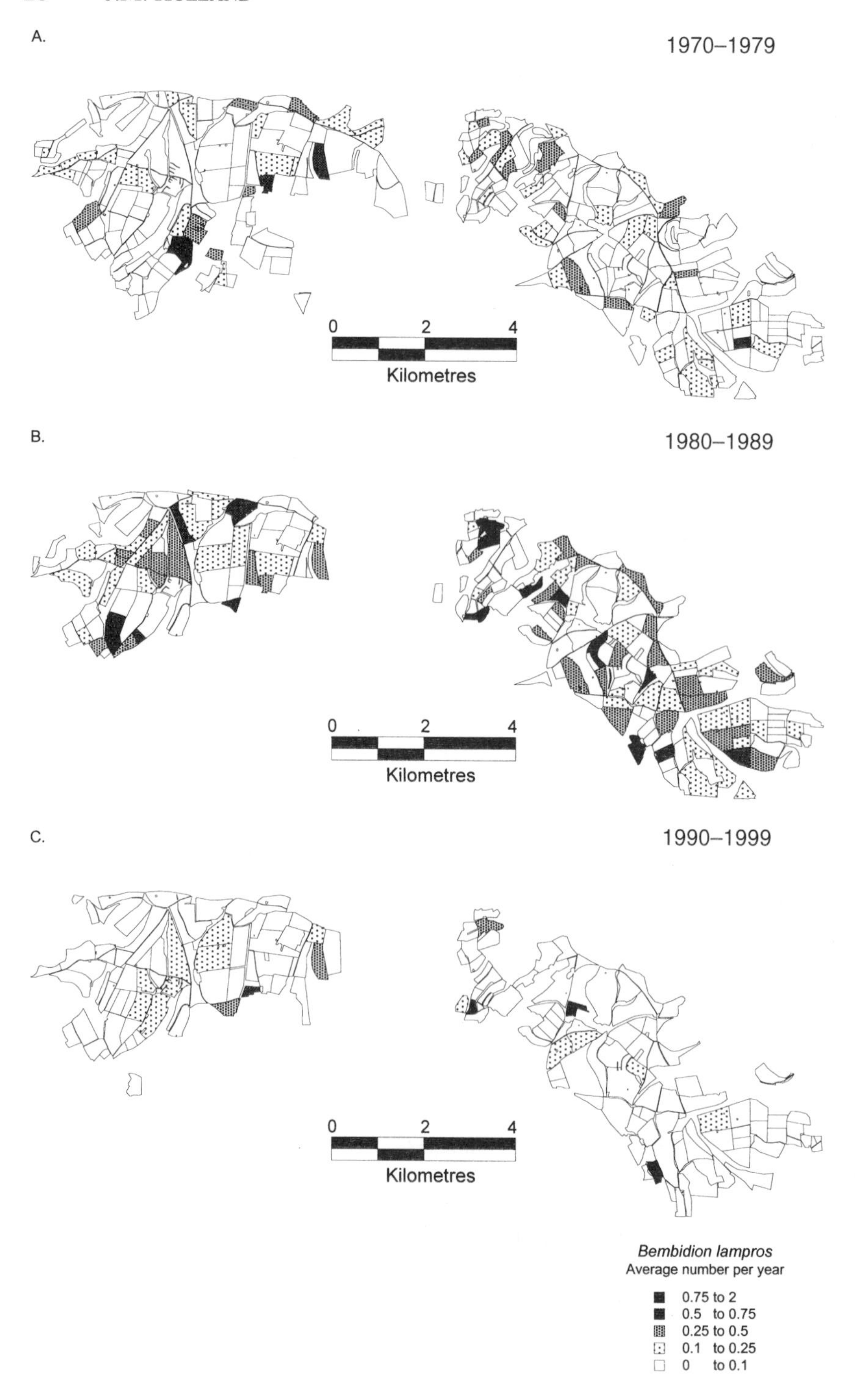

Figure 1.8. Distribution of *Bembidion lampros* across the Sussex Study area over three decades based upon suction samples taken from approximately 100 cereal fields in June each year.

were encouraged by the building of roads and railways which facilitated dispersal (Lindroth, 1972). The extensive transport of soil in, for example, plant containers, has also facilitated the spread of carabids between countries and continents. *P. melanarius* is thought to have been introduced to North America in this way in ship's ballast, and is now relatively widespread (Niemelä and Spence, 1999). By 1994 there were 47 European species of Carabidae which had become established in Canada (Spence, 1994). Their success was attributed largely to their flexibility in habitat use and competitiveness.

Farming systems are always changing, mainly in response to economic pressure but sometimes also from political or public pressure. The widespread switch from plough-based cultivation systems to non-inversion soil tillage in North America in response to widespread soil erosion is one example. In Europe, the low value of most arable crops has made many farmers look at ways of reducing production costs and switching to non-inversion soil tillage and using fewer agrochemicals. Both these practices are advocated within an integrated crop management programme which, even though widely advocated, has not seen much uptake, although most farmers incorporate some integrated practices. There is evidence that the adoption of appropriate farming practices (Hance, this volume) and systems (El Titi and Ipach, 1989) can encourage greater species diversity and abundance, but often this just selects a different assemblage which is better adapted to the conditions created (Holland and Luff, 2000).

Future research

Despite such an extensive and diverse array of research studies, a number of fundamental questions still remain with respect to carabids in agroecosystems. In addition, although research on one or a few species may indicate what is controlling their survival, the results may not be widely applicable because each species appears to have its own specific, and sometimes complex, environmental requirements. However, some broad principles apply and these may have applicability to other epigeal arthropods, such as the Staphylinidae, but their ecology, in contrast to Carabidae, is only poorly understood.

Some of the important questions that remain to be answered on carabid ecology and population regulation are outlined below. The list is by no means comprehensive, but may provide some food for thought.

1. Research in the UK on six farms with a wide geographic spread revealed that diversity and abundance is extremely varied, but also very low on some farms (Holland *et al.*, 1998). More widespread monitoring is needed to determine whether, and to what extent, carabids are declining on farmland. Those farms with greater abundance were, incidentally, mixed farms, suggesting that the presence of pasture is important.
2. In some cases, sampling in localized patches may fail accurately to measure population densities because carabids are heterogeneously distributed across arable areas. Further research is needed on sampling strategies to identify the most accurate methods of estimating population density within the landscape.
3. Most studies of spatial pattern have looked at distributions found within fields,

but larger-scale evaluations are needed to gauge how far these extend across the landscape. Such patterns may be the result of past population regulation or agricultural activities, although the beetles may also be actively responding to the most favourable environmental conditions or food availability. If the most important regulatory factors governing beetle distributions can be identified, appropriate management practices can then be designed.

4. Large-scale spatial studies are also needed to identify to what extent carabids are dispersing across the landscape. Do carabids utilize their flying ability to avoid adverse conditions created by farming practices or to reinvade following such practices? In Norway, the majority of new species found in the last 20 years was predominantly those with the ability to fly (Andersen, 1987).
5. There is some evidence that carabid diversity and abundance is declining in agricultural areas, but this needs more widespread evaluation. What is their capacity to suffer repeated disturbances and is this influenced by the landscape structure? The presence of roads/railways/hedgerows may facilitate dispersal for some species but inhibit the movement of others.
6. We know that the survival rate of the various lifestages differs but not which mortality-causing factors are most important. To what extent do parasitism and pathogens regulate numbers of dormant and active lifestages? Do carabids actively try to reduce their impact by selecting environmental conditions which discourage infection? This may be especially important in the selection of oviposition and overwintering sites.
7. Intra- and interspecific predation occurs, but how important this is for population regulation in agroecosystems remains to be verified. Does, for example, the presence of a voracious predator such as *P. melanarius* discourage or regulate smaller species? How important is predation or cannabilism of eggs, larva and pupa, and how does the availability of other food sources affect this? An application of an insecticide may reduce alternative food, encouraging cannabilism.
8. What are the most effective ways to encourage carabids in agricultural crops? To achieve this we need further information on which farming practices are having the greatest impact on carabids. Does mechanical cultivation cause mortality? What are the long-term effects of insecticide applications on, for example, fecundity and development? How important is weed cover for providing preferred environmental conditions, food and cover? Should we concentrate on the provision of adequate non-crop habitats which are only important for some species, or should we also be encouraging more sympathetic crop management practices? Beetle banks are one of the best examples where additional non-crop habitat has been provided and, although they support high overwintering numbers (Thomas *et al.*, 1991), there is scant evidence that mid-field densities are encouraged in the long term. Integrated farming should encourage mid-field densities, but results are highly variable (Holland and Luff, 2000).

Acknowledgements

I wish to thank Drs Martin Luff (University of Newcastle) and George Thomas (Plymouth University) for their valuable comments on this manuscript. I am also grateful to Dr Julie Ewald (The Game Conservancy Trust) for compiling the figures from the Sussex Study. The studies on the spatial distribution of carabids were funded by the BBSRC and were a collaborative project with Dr Linton Winder (Plymouth University) and Professor Joe Perry (IACR Rothamsted).

References

ADIS, J. (1979). Problems of interpreting arthropod sampling with pitfall traps. *Zoologischer Anzeiger Jena* **202**, 177–184.

AEBISCHER, N.J. (1991). Twenty years of monitoring invertebrates and weeds in cereal fields in Sussex. In: *The ecology of temperate cereal fields*. Eds. L.G. Firbank, N. Carter, J.F. Darbyshire and G.R. Potts, pp 305–331. Oxford: Blackwell.

ALDERWEIRELDT, M. AND DESENDER, K. (1990). Microhabitat preferences of spiders (Araneae) and carabid beetles (Coleoptera, Carabidae) in maize fields. *Mededelingen van de Faculteit Landbouwwetenschappen, Rijksuniversiteit Gent* **55**, 501–510.

ALLEN, R.T. (1979). The occurrence and importance of ground beetles in agricultural and surrounding habitats. In: *Carabid beetles: their evolution, natural history and classification*. Eds. T.L. Erwin, G.E. Bell and D.R. Whitehead, pp 485–506. The Hague: Junk.

ALLEY, T.R. (1982). Competition theory, evolution and and the concept of ecological niche. *Acta Biotheoretica* **31**, 165–179.

ALTHOFF, G.H., HOCKMAN, P., KLENNER, M., NIEHUS, F.J. AND WEBER, F. (1994). Dependence of running activity and net reproduction in *Carabid auronitens* on temperature. In: *Carabid beetles: ecology and evolution*. Eds. K. Desender, M. Duferne, M. Loreau, M. Luff and J.P. Maelfait, pp 95–100. Dordrecht: Kluwer Academic Publishers.

ANDERSEN, A. (1985a). *Agonum dorsale* (Pontoppidan) (Col., Carabidae), an expanding species in Norway. *Fauna Norvegica* Ser. B **32**, 52–57.

ANDERSEN, A. (1985b). Carabidae and Staphylinidae (Col.) in swede and carrot fields in south-eastern Norway. *Fauna Norvegica* Ser. B **32**, 12–27.

ANDERSEN, J. (1987). Qualitative changes in the Norwegian carabid beetle fauna during the present century. *Acta Phytopathologica et Entomologica Hungarica* **22**, 35–44.

ANDERSEN, J. (1995). A comparison of pitfall trapping and quadrat sampling of Carabidae (Coleoptera) on river banks. *Entomologica Fennica* **6**, 65–77.

ANDERSEN, J. AND SKORPING, A. (1991). Parasites of carabid beetles: prevalence depends on habitat selection of the host. *Canadian Journal of Zoology* **69**, 1216–1220.

ANDREONE, F., MICHELIS, S. AND CLIMA, V. (1999). A montane amphibian and its feeding habits: *Salamandra lanzai* (Caudata, Salamandridae) in the Alps of northwestern Italy. *Italian Journal of Zoology* **66**, 45–49.

ARLETTAZ, R. (1996). Feeding behaviour and foraging strategy of free-living mouse-eared bats, *Myotis myotis* and *Myotis blythii*. *Animal Behaviour* **51**, 1–11.

ARMSTRONG, G. AND MCKINLAY, R.G. (1997). Vegetation management in organic cabbages and pitfall catches of carabid beetles. *Agriculture, Ecosystems and Environment* **64**, 267–276.

ATIENZA, J.C., FARINOS, G.P. AND ZABALLOS, J.P. (1996). Role of temperature in habitat selection and activity patterns in the ground beetle *Angoleus nitidus*. *Pedobiologia* **40**, 240–250.

BAARS, M.A. (1979). Catches in pitfall traps in relation to mean densities of carabid beetles. *Oecologia* **41**, 25–46.

BAARS, M.A. AND VAN DIJK, T.S. (1984). Population dynamics of two carabid beetles at a Dutch heathland. II. Egg production and survival in relation to density. *Journal of Animal Ecology* **53**, 389–400.

BASEDOW, T. (1987). Der Einfluß gesteigerter bewirtschaftungsintensität im getreidebau auf die laufkäfer (Coleoptera: Carabidae). *Mitteilungen aus der Bioloischen Bundesanstalt für Land-u Forstwirtschaft Berlin-Dahlem*, **235**, 123pp.

BASEDOW, T. (1994). Phenology and egg production in *Agonum dorsale* and *Pterostichus melanarius* (Col., Carabidae) in winter wheat fields of different growing intensity in

Northern Germany. In: *Carabid beetles: ecology and evolution*. Eds. K. Desender, M. Duferne, M. Loreau, M. Luff and J.P. Maelfait, pp 101–108. Dordrecht: Kluwer Academic Publishers.

BENGTSON, S.-A. (1980). Species assemblage and coexistence of Faroe Island ground beetles (Coleoptera: Carabidae). *Entomolgia Generalis* **6**, 251–266.

BERRY, R.J. (1968). The ecology of an island population of the house mouse. *Evolution* **18**, 468–483.

BOMMARCO, R. (1998a). Reproduction and energy reserves of a predatory carabid beetle relative to agroeosystem complexity. *Ecological Applications* **8**, 846–853.

BOMMARCO, R. (1998b). Stage sensitivity to food limitation for a generalist arthropod predator, *Pterostichus cupreus* (Coleoptera: Carabidae). *Environmental Entomology* **27**, 863–869.

BOMMARCO, R. (1999). Feeding, reproduction and community impact of a predatory carabid in two agricultural habitats. *Oikos* **87**, 89–96.

BRANDMAYR, P. (1983). *Ecology of carabids: the synthesis of field study and laboratory experiment*. Wageningen: Pudoc.

BRANDMAYR, P., DEN BOER, P.J. AND WEBER, F. (Eds.) (1983). *Ecology of carabids: the synthesis of field study and laboratory experiment*. Wageningen: Pudoc.

BRIGGS, J.B. (1965). Biology of some ground beetles (Coleoptera, Carabidae) injurious to strawberries. *Bulletin of Entomological Research* **56**, 79–93.

BRUNSTING, A.M.H. AND HEESSEN, H.J.L. (1984). Density regulation in the carabid beetle *Pterostichus oblongipunctatus* F. *Journal of Animal Ecology* **53**, 751–760.

BÜCHS, W., HARENBERG, A. AND ZIMMERMANN, J. (1997). The invertebrate ecology of farmland as a mirror of the intensity of the impact of man? – an approach to interpreting results of field experiments carried out in different crop management intensities of a sugar beet and an oil seed rape rotation including set-aside. *Biological Agriculture and Horticulture* **15**, 83–108.

CAMPBELL, L.H., AVERY, M.I., DONALD, P., EVANS, A.D., GREEN, R.E. AND WILSON, J.D. (1997). A review of the indirect effects of pesticides on birds. JNCC Report No. 227.

CHURCHFIELD, S. (1982). Food availability and the diet of common shrew, *Sorex araneus*, in Britain. *Journal of Animal Ecology* **51**, 15–28.

CHURCHFIELD, S. (1984). Dietary separation of three shrew species inhabiting water-cress beds. *Journal of Zoology* **204**, 211–228.

CHURCHFIELD, S. (1985). Feeding ecology of the European water shrew. *Mammal Review* **15**, 13–21.

CHURCHFIELD, S., HOLLIER, J. AND BROWN, V.K. (1991). The effects of small mammal predators on grassland invertebrates, investigated by field exclosure experiment. *Oikos* **60**, 283–290.

COOMBES, D.S. AND SOTHERTON, N.W. (1986). The dispersal and distribution of predatory Coleoptera in cereals. *Annals of Applied Biology* **108**, 461–474.

CRITCHLEY, B.R. (1973). Parasitism of the larvae of some Carabidae (Coleoptera). *Journal of Entomology* **48**, 37–42.

CROY, P. (1987). Faunisitsch-ökologische Untersuchungen der Carabidaen im Umfeld eines industriellen Ballungsgebeites. *Entomologische Nachrichten und Berichte* **31**, 1–9.

DAVIS, L. (1987). Long adult life, low reproduction and competition in two sub-Antarctic carabid beetles. *Ecological Entomology* **12**, 149–162.

DELANY, M.J. (1961). The ecological distribution of small mammals in north-west Scotland. *Proceedings of the Zoological Society of London* **26**, 283–295.

DEN BOER, P.J. (1970). On the significance of dispersal power for populations of carabid beetles. *Oecologia* **4**, 1–28.

DEN BOER, P.J. (1971). Stabilization of animal numbers and the heterogeneity of the environment: The problem of persistence of sparse populations. In: *Dynamics of populations*. Eds. P.J. Den Boer and G.R. Gradwell, pp 77–97. Wageningen: Pudoc.

DEN BOER, P.J. (1977). Dispersal power and survival: carabids in a cultivated countryside. *Miscellaneous papers LH Wageningen* **14**, 1–19.

DEN BOER, P.J. (1980). Exclusion or coexistence and the taxonomic or ecological relationship between species. *Netherlands Journal of Zoology* **30**, 278–306.

DEN BOER, P.J. (1985). Fluctuations of density and survival of carabid populations. *Oecologia* **67**, 322–330.

DEN BOER, P.J. (1986). What can carabid beetles tell us about dynamics of populations? In: *Carabid beetles. Their adaptions and dynamics*. Eds. P.J. Den Boer, M.L. Luff, D. Mossakowsi and F. Weber, pp 315–330. Stuttgart: Fischer.

DEN BOER, P.J. (1987). On the turnover of carabid populations in changing environments. *Acta Phytopathologica Entomologica Hungarica* **22**, 71–83.

DEN BOER, P.J. (1990). Density limits and survival of local populations in 64 carabid species with different powers of dispersal. *Journal of Evolutionary Biology* **3**, 19–48.

DEN BOER, P.J. AND DEN DAANJE, W. (1990). On life–history tactics in carabid beetles: are there only spring and autumn breeders? In: *The role of ground beetles in ecological and environmental studies*. Ed. N.E. Stork, pp 247–258. Andover: Intercept.

DEN BOER, P.J., THIELE, H.-U. AND WEBER, F. (Eds.) (1979). On the evolution of behaviour in carabid beetles. *Miscellaneous Papers LH Wageningen* **18**, 1–222.

DEN BOER, P.J., LUFF, M.L., MOSSAKOWSKI, D. AND WEBER, F. (Eds.) (1986). *Carabid beetles. Their adaptions and dynamics*. 551 pp. Stuttgart/New York: Fischer Verlag.

DEN BOER, P.J., LÖVEI, G.L., STORK, N.E. AND SUNDERLAND, K.D. (Eds.) (1987). Proceedings of the 6th European carabidologists meeting. *Acta Phytopathologica Entomologica Hungarica* **22**, 1–458.

DEN NIJS, L.J.M.F., BOESWINKEL, E., ZIJP, J.P.H., SMEDING, F.W. AND BOOIJ, C.J.H. (1998). Methods to study the effect of ecological infrastructure on aphids and their enemies on winter wheat. *Proceedings of the Experimental and Applied Entomology Section of the Netherlands Entomological Society* **9**, 93–98.

DENNIS, P. AND FRY, G.L.A. (1992). Field margins: can they enhance natural enemy population densities and general arthropod diversity on farmland. *Agriculture, Ecosystems and Environment* **40**, 95–115.

DESENDER, K. (1982). Ecological and faunal studies on Coleoptera in agricultural land. II. Hibernation of Carabidae in agro–ecosystems. *Pedobiologia* **23**, 295–303.

DESENDER, K. AND ALDERWEIRELDT, M. (1988). Population dynamics of adult and larval carabid beetles in a maize field and its boundary. *Journal of Applied Entomology* **106**, 13–19.

DESENDER, K. AND SEGERS, R. (1985). A simple device and technique for quantitative sampling of riparian beetle populations with some carabid and staphylinid abundance estimate from different riparian habitat (Coleoptera). *Revue d'Ecologie et de Biologie du Sol* **22**, 497–506.

DESENDER, K. AND TURIN, H. (1989). Loss of habitats and changes in the composition of ground and tiger beetle fauna in four West European countries since 1950 (Coleoptera: Carabidae, Cicinelidae). *Biological Conservation* **48**, 277–294.

DESENDER, K., MAELFAIT, J.P., D'HULSTER, M. AND VANHERCKE, L. (1981). Ecological and faunal studies on Coleoptera in agricultural land. I. – Seasonal occurrence of Carabidae in the grassy edge of pasture. *Pedobiologia* **22**, 379–384.

DESENDER, K., MERTENS, J., D'HULSTER, M. AND BERBIERS, P. (1984). Diel activity patterns of Carabidae (Coleoptera), Staphylinidae (Coleoptera) and Collembola in a heavily grazed pasture. *Revue d'Ecologie et de Biologie du Sol* **24**, 347–361.

DESENDER, K., DUFRENE, M., LOREAU, M., LUFF, M.L. AND MAELFAIT, J.P. (Eds.) (1994). *Carabid beetles: ecology and evolution*. Dordrecht: Kluwer Academic Publishers.

DIMELOW, E.J. (1963). Observations on the feeding of the hedgehog (*Erinaceus europeaus*). *Proceedings of the Zoological Society of London* **141**, 291–309.

DINTER, A. (1995). Estimation of epigeic spider population densities using an intensive D-vac sampling technique and comparison with pitfall trap catches in winter wheat. In: *Arthropod natural enemies in arable land. I. Density, spatial heterogeneity and dispersal*. Eds. S. Toft and W. Riedel, pp 23–32. Aarhus: Aahrus Unversity Press.

DOHMEN, G.P. (1998). Testing side-effects of pesticides on carabid beetles: a standardized method for testing ground-dwelling predators in the laboratory for registration purposes. In: *Ecotoxicology: pesticides and beneficial organisms*. Eds. P.T. Haskell and P. McEwen, pp 98–106. London: Kluwer Academic Publishers.

DUFFEY, E. (1980). The efficiency of the Dietrick Vacuum Sampler (DVAC) for invertebrate population studies in different types of grassland. *Bulletin of Ecology* **11**, 421–431.

DUFFIELD, S.J. AND AEBISCHER, N.J. (1994). The effect of spatial scale of treatment with dimethoate on invertebrate population recovery in winter wheat. *Journal of Applied Ecology* **31**, 263–281.

DUFRÊNE, M., BAGUETTE, M., DESENDER, K. AND MAELFAIT, J.P. (1990). Evaluation of carabids as bioindicators: a case study in Belgium. In: *The role of ground beetles in ecological and environmental studies*. Ed. N.E. Stork, pp 377–382. Andover: Intercept.

EDWARDS, C.A. AND FLETCHER, K.E. (1971). A comparison of extraction methods for terrestrial arthropods. In: *Methods of quantitative soil ecology: population, production and energy flow*. Ed. J. Phillipson, pp 50–185. IBP Handbooks, 18. Oxford: Blackwell.

EL TITI, A. AND IPACH, A. (1989). Soil fauna in sustainable agriculture – results of an integrated farming system at Lautenbach, FRG. *Agriculture, Ecosystems and Environment* **27**, 561–572.

ERIKSTAD, K.E., BYRKJEDAL, I. AND KALS, J.A. (1989). Resource partitioning among seven carabid species on Hardangervidda, southern Norway. *Annales Zoologici Fennici* **26**, 113–120.

ERNSTING, G. AND ISAAKS, J.A. (1994). Egg size variation in *Notiophilus biguttatus* (Col., Carabidae). In: *Carabid beetles: ecology and evolution*. Eds. K. Desender, M. Dufrêne, M. Loreau, M. Luff and J.P. Maelfait, pp 133–138. Dordrecht: Kluwer Academic Publishers.

ERNSTING, G., ISAAKS, A.A. AND BERG, M.P. (1992). Life cycle and food availability indicies in *Notiophilus biguttatus* (Coleoptera: Carabidae). *Ecological Entomology* **17**, 33–42.

ERWIN, T.L. (1979). *Carabid beetles: their evolution, natural history and classification*. The Hague: Junk.

ERWIN, T.L., BALL, G.E., WHITEHEAD, D.L. AND HALPERN, A.J. (Eds.) (1979). *Carabid beetles: their evolution, natural history and classification*. The Hague: Junk.

EWALD, J.A. AND AEBISCHER, N.J. (2000). Trends in pesticide use and efficacy during 26 years of changing agriculture in southern England. *Environmental Monitoring and Assessment* **64**, 493–529.

FADL, A. AND PURVIS, G. (1998). Field observations on the lifecycles and seasonal activity patterns of temperate carabid beetles (Coleoptera: Carabidae) inhabiting arable land. *Pedobiologia* **42**, 171–183.

FOURNIER, E. AND LOREAU, M. (1999). Effects of newly planted hedges on ground-beetle diversity (Coleoptera, Carabidae) in an agricultural landscape. *Ecography* **22**, 87–97.

FRENCH, B.W. AND ELLIOTT, N.C. (1999). Temporal and spatial distribution of ground beetle (Coleoptera: Carabidae) assemblages in grasslands and adjacent wheat fields. *Pedobiologia* **43**, 73–84.

FULLER, R.J., GREGORY, R.D., GIBBONS, D.W., MARCHANT, J.H., WILSON, J.D., BAILLIE, S.R. AND CARTER, N. (1995). Population declines and range contractions among farmland birds in Britain. *Conservation Biology* **9**, 1425–1442.

FUNKE, W. (1971). Food and energy turnover of leaf-eating insects and their influence on primary production. *Ecological Studies* **2**, 81–93.

GREEN, R. (1979). The ecology of wood mice *Apodemus sylvaticus* on arable farmland. *Journal of Zoology, London* **188**, 357–377.

GREEN, R.E. (1984). The feeding ecology and survival of partridge chicks (*Alectoris rufa* and *Perdix perdix*) on arable farmland in East Anglia. *Journal of Applied Ecology* **21**, 817–830.

GRÜM, L. (1975). Mortality patterns in carabid populations. *Ekologia Polska* **23**, 649–665.

GRÜM, L. (1983). Home range estimates as applied to study carabid dispersal. In: *Ecology of carabids: the synthesis of field study and laboratory experiment*. Eds. P. Brandmayr, P.J. Den Boer and F. Weber, pp 55–58. Wageningen: Pudoc.

GRUTTKE, H. AND WEIGMANN, G. (1990). Ecological studies on the carabid fauna (Coleoptera) of a ruderal ecosystem in Berlin. In: *The role of ground beetles in ecological and environmental studies*. Ed. N.E. Stork, pp 181–189. Andover: Intercept.

HALSALL, N.B. AND WRATTEN, S.D. (1988). Video recording of aphid predation by Carabidae in a wheat crop. *1988 Brighton Crop Protection Conference – Pests and Diseases* **3**, pp 1047–1052.

HAMRICK, W.J. AND DAVIS, J.R. (1971). Summer food items of juvenile wild turkeys. *Proceedings of the 25th Annual Conference of the Southeastern Association of Game and Fish Commissioners*, pp 85–89.

HAND, S.C. (1986). The capture efficiency of the Dietrick vacuum insect net for aphids on grasses and cereals. *Annals of Applied Biology* **108**, 233–241.

HANSKI, I. (1982). Dynamics of regional distribution: the core and satellite species hypothesis. *Oikos* **38**, 210–221.

HEESSEN, H.J.L. (1981). Egg mortality in *Pterostichus oblongopunctatus* (Fabricius) (Col. Carabidae). *Oecologia* **50**, 233–235.

HEESSEN, H.J.L. AND BRUNSTING, A.M.H. (1981). Mortality of larvae of *Pterostichus oblongopunctatus* (Fabricius) (Col. Carabidae) and *Philonthus decorus* (Coleoptera: Carabidae). *Netherlands Journal of Zoology* **31**, 729–745.

HENGEVELD, R. (1979). The analysis of spatial patterns of some ground beetles (Col. Carabidae). In: *Spatial and temporal analysis in ecology*. Eds. R.M. Cormack and J.K. Ord, pp 333–346. Fairland: International Co-operative Publishing House.

HENGEVELD, R. (1980). Qualitative and quantitative aspects of the food of ground beetles: a review. *Netherlands Journal of Zoology* **30**, 555–563.

HENGEVELD, R. (1985). Methodology of explaining differences in dietary composition of carabid beetles by competition. *Oikos* **45**, 37–49.

HERNANDEZ, A., ALEGRE, J., SALGADO, J.M. AND GUTIERREZ, A. (1991). The role of Coleoptera in the nutrition of some vertebrates in the northeast of Spain. *Elytron* **5**, 231–237.

HEYDEMANN, B. AND MEYER, H. (1983). Auswirkungen der intensivkultur auf die fauna in agrarbiotopen. *Schriftenreihe Deutscher Rat für Landespflege und Wirtschaft* **42**, 174–191.

HILL, D.A. (1985). The feeding ecology and survival of pheasant chicks on arable farmland. *Journal of Applied Ecology* **22**, 645–654.

HOLLAND, J.M. AND LUFF, M.L. (2000). The effects of agricultural practices on Carabidae in temperate agroecosystems. *Integrated Pest Management Reviews* **5**, 109–129.

HOLLAND, J.M. AND SMITH, S. (1999). Sampling epigeal arthropods: an evaluation of fenced pitfall traps using mark–release–recapture and comparisons to unfenced pitfall traps in arable crops. *Entomologia Experimentalis et Applicata* **91**, 347–357.

HOLLAND, J.M. AND THOMAS, S.R. (1997). Assessing the role of beneficial invertebrates in conventional and integrated farming systems during an outbreak of *Sitobion avenae*. *Biological Agriculture and Horticulture* **15,** 73–82.

HOLLAND, J.M., FRAMPTON, G.K., CILGI, T. AND WRATTEN, S.D. (1994). Arable acronyms analysed – a review of integrated farming systems research in Western Europe. *Annals of Applied Biology* **125**, 399–438.

HOLLAND, J.M., COOK, S.K., DRYSDALE, A., HEWITT, M.V., SPINK, J. AND TURLEY, D. (1998). The impact on non-target arthropods of integrated compared to conventional farming: results from the LINK Integrated Farming Systems Project. *1998 Brighton Crop Protection Conference – Pests and Diseases* **2**, pp 625–630.

HOLLAND, J.M., PERRY, J.N. AND WINDER, L. (1999). The within-field spatial and temporal distribution of arthropods in winter wheat. *Bulletin of Entomological Research* **89**, 499–513.

HOLLAND, J.M., WINDER, L. AND PERRY, J.N. (2000). The impact of dimethoate on the spatial distribution of beneficial arthropods and their reinvasion in winter wheat. *Annals of Applied Biology*, **136**, 93–105.

HOLOPAINEN, J.K. (1995). Spatial distribution of polyphagous predators in nursery fields. In: *Arthropod natural enemies in arable land. I. Density, spatial heterogeneity and dispersal.* Eds. S. Toft and W. Riedel, pp 213–220. Aarhus: Aahrus Unversity Press.

HOLOPAINEN, J.K. AND VARIS, A.L. (1986). Effects of a mechanical barrier and formalin preservative on pitfall catches of carabid beetles (Coleoptera, Carabidae) in arable fields. *Journal Applied Entomology* **102**, 440–445.

HOLOPAINEN, J.K., BERGMAN, T., HAUTALA, E.L. AND OKSANEN, J. (1995). The ground beetle fauna (Coleoptera: Carabidae) in relation to soil properties and foliar fluoride content in spring cereals. *Pedobiologia* **39**, 193–206.

HONEK, A. (1988). The effect of crop density and microclimate on pitfall trap catches of Carabidae, Staphylinidae (Coleoptera), and Lycosidae (Araneae) in cereal fields. *Pedobiologia* **32**, 233–242.

IDINGER, J. AND KROMP, B. (1997). Ground photoeclector evaluation of different arthropod groups found in unfertilised, inorganic and compost-fertilised cereal fields in Eastern Austria. *Biological Agriculture and Horticulture* **15**, 171–176.

JENSEN, T.S., DYRING, L., KRISTENSEN, B., NIELSEN, B.O. AND RASMUSSEN, E.R. (1989). Spring dispersal and summer habitat distribution of *Agonum dorsale* (Col. Carabidae). *Pedobiologia* **33**, 115–165.

JEPSON, P.C. (1989). The temporal and spatial dynamics of pesticide side-effects on non-target invertebrates. In: *Pesticides and non-target invertebrates*. Ed. P.C. Jepson, pp 95–128. Wimborne: Intercept Ltd.

JONES, M.G. (1979). The abundance and reproductive activity of common Carabidae in a winter wheat crop. *Ecological Entomology* **4**, 31–43.

KABACIK-WASYLIK, D. AND KMITOWA, K. (1973). The effect of single and mixed infections of entomopathogenic fungi on the mortality of the Carabidae (Coleoptera). *Ekologia Polska* **21**, 645–655.

KEDDY, P.A. (1989). *Competition*. London: Chapman and Hall.

KEGEL, B. (1990). Diurnal activity of carabid beetles living on arable land. In: *The role of ground beetles in ecological and environmental studies*. Ed. N.E. Stork, pp 65–76. Andover: Intercept.

KEGEL, B. (1994). The biology of four sympatric *Poecilus* species. In: *Carabid beetles: ecology and evolution*. Eds. K. Desender, M. Dufrene, M. Loreau, M.L. Luff and J.P. Maelfait, pp 157–163. Dordrecht: Kluwer Academic Publishers.

KMITOWA, K. AND KABACIK-WASYLIK, D. (1971). An attempt at determining the pathogenicity of two species of entomopathic fungi in relation to Carabidae. *Ekologia Polska* **19**, 727–733.

KÖRNER, H. (1990). Der Einfluß der pflanzenschutzmittel auf die faunnenvielfalt der agrarlandschaft (unter besonderer berücksichtigung der gliederfüßler der oberfläche der felder). *Bayerisches Landwirtschaftliches Jahrbuch* **67**, 375–496.

KROMP, B. (1989). Carabid beetle communities (Carabidae, Coleoptera) in biologically and conventionally farmed agroecosytems. *Agriculture, Ecosystems and Environment* **27**, 241–251.

KROMP, B. (1999). Carabid beetles in sustainable agriculture: a review on pest control efficacy, cultivation impacts and enhancement. *Agriculture, Ecosystems and Environment* **74**, 187–228.

LARKIN, P.A. (1948). Ecology of mole (*Talpa europaea*). PhD Thesis, University of Oxford.

LAROCHELLE, A. (1974a). Carabid beetles (Coleoptera: Carabidae) as prey of North American frogs. *Great Lakes Entomologist* **7**, 147–148.

LAROCHELLE, A. (1974b). The American toad as champion carabid beetle collector. *Pan-Pacific Entomology* **50**, 203–204.

LAROCHELLE, A. (1975a). A list of amphibians and reptiles as predators of Carabidae. *Carabologia* **3**, 99–103.

LAROCHELLE, A. (1975b). A list of mammals as predators of Carabidae. *Carabologia* **3**, 95–98.

LAROCHELLE, A. (1975c). A list of North American birds as predators of carabid beetles. *Carabologia* **1**, 153–163.

LAROCHELLE, A. (1978). Catalogue des parasites et phoretiques animaux des Coleopteres Carabidae (les Cicindelini compris) du monde. *Cordulia* **4**, 1–7, 69–75.

LAROCHELLE, A. (1980). A list of birds of Europe and Asia as predators of carabid beetles including Cicindelini (Coleoptera: Carabidae). *Cordulia* **6**, 1–19.

LAROCHELLE, A. (1990). The food of carabid beetles. *Québec: Fabreries Supplement* 5, Association des Entomologistes Amateurs du Québec.

LARSSON, S.G. (1939). Entwicklungstypen und entwicklungszeiten der danischen Carabiden. *Entomologiske Meddelelser* **20**, 277–560.

LEBRUN, PH., BAGUETTE, M. AND DUFRÊNE, M. (1987). Species diversity in a carabid community: comparison of values estimated at 23 year intervals. *Acta Phytopathologica Entomologica Hungarica* **22**, 165–173.

LEFRANC, N. (1997). Shrikes and the farmed landscape in France. In: *Farming and birds in Europe*. Eds. D.J. Pain and M.W. Pienkowski, pp 236–268. London: Academic Press.

LENSKI, R.E. (1982). Effects of forest cutting on two *Carabus* species: evidence for competition for food. *Ecology* **63**, 1211–1217.

LINDROTH, C.H. (1949). *Die Fennoskandischen Carabidae. Eine tiergeographische Studie. I. Spezieller Teil.* Goteborg: Elanders Boktryckeri Aktiebolag.

LINDROTH, C.H. (1972). Changes in the Fennoscandian ground-beetle fauna (Coleoptera, Carabidae) during the twentieth century. *Annales Zoologici Fennici* **9**, 49–64.

LINDROTH, C.H. (1992). *Ground beetles (Carabidae) of Fennoscandia: a Zoogeographic study: Part 1. Specific knowledge regarding the species.* 630 pp. Andover: Intercept.

LOREAU, M. (1986). Niche differentiation and community organisation in forest carabid beetles. In: *Carabid beetles: their adaptions and dynamics.* Eds. P.J. Den Boer, M.L. Luff, D. Mossakowsi and F. Weber, pp 465–487. Stuttgart: Fischer.

LOREAU, M. (1990). Competition in a carabid beetle community: a field experiment. *Oikos* **58**, 25–38.

LORENZ, W. (1998). *Systematic list of extant ground beetles of the World (Insecta Coleoptera 'Geadephaga': Trachypachidae and Carabidae incl. Paussinae, Cicindelinae, Rhysodinae).* Tutzing, Germany, (privately published by the author, after meeting the Zoological Codes's criteria), 503 pp.

LÖVEI, G.L. AND SUNDERLAND, K.D. (1996). Ecology and behaviour of ground beetles (Coleoptera: Carabidae). *Annual Review of Entomology* **41**, 231–256.

LUFF, M.L. (1965). The morphology and microclimate of *Dactylis glomerata* tussocks. *Journal of Ecology* **53**, 771–787.

LUFF, M.L. (1966). Cold hardiness of some beetles living in grass tussocks. *Entomologia Experimentalis et Applicata* **9**, 191–199.

LUFF, M.L. (1973). The annual lifecycle and activity pattern of *Pterostichus madidus* (F.) (Coleoptera: Carabidae). *Entomologica Scandinavica* **4**, 259–273.

LUFF, M.L. (1975). Some features influencing the efficiency of pitfall traps. *Oecologia* **19**, 345–357.

LUFF, M.L. (1976a). Notes on the biology and development stages of *Nebria brevicollis* (F.) (Col., Carabidae) and on their parasities, *Phaenoserphus* spp. (Hym., Proctotrupidae). *Entomologist's Monthly Magazine* **111**, 249–255.

LUFF, M.L. (1976b). The biology of *Microctonus caudatus* (Thomson), a braconid parasite of the ground beetle Harpalus rufipes (Degeer). *Ecological Entomology* **1**, 111–116.

LUFF, M.L. (1978). Diel activity patterns of some field Carabidae. *Ecological Entomology* **3**, 53–62.

LUFF, M.L. (1980). The biology of the ground beetle *Harpalus rufipes* in a strawberry field in Northumberland. *Annals of Applied Biology* **94**, 153–164.

LUFF, M.L. (1981). Diagnostic characteristics of the eggs of some Carabidae. *Entomologia Scandinavica Supplementu* **15**, 317–327.

LUFF, M.L. (1982). Population dynamics of Carabidae. *Annals of Applied Biology* **101**, 164–170.

LUFF, M.L. (1987). Biology of polyphagous ground beetles in agriculture. *Agricultural Zoology Reviews* **2**, 237–278.

LUFF, M.L. (1994). Starvation capacities of some carabid larvae. In: *Carabid beetles: ecology and evolution.* Eds. K. Desender, M. Dufrêne, M. Loreau, M. Luff and J.P. Maelfait, pp 171–175. Dordrecht: Kluwer Academic Publishers.

LUFF, M.L. (1996). Use of carabids as environmental indicators in grasslands and cereals. *Annales Zoologici Fennici* **33**, 185–195.

LUFF, M.L. (1998). *Provisional atlas of the ground beetles (Coleoptera, Carabidae) of Britain.* Huntingdon: Institute of Terrestrial Ecology.

LUFF, M.L. AND RUSHTON, S.P. (1989). The ground beetle and spider fauna of managed and unimproved pasture. *Agriculture, Ecosystems and Environment* **25**, 195–205.

LUFF, M.L., EYRE, M.D. AND RUSHTON, S.P. (1992). Classification and prediction of grassland habitats using ground beetles (Coleoptera, Carabidae). *Journal of Environmental Management* **35**, 301–305.

LYNGBY, J.E. AND NIELSEN, H.B. (1980). The spatial distribution of carabids (Coleoptera: Carabidae) in relation to a shelterbelt. *Entomologiske Meddelelser* **48**, 133–140.

MAELFAIT, J.P. AND DESENDER, K. (1990). Possibilities of short-term carabid sampling for site assesment studies. In: *The role of ground beetles in ecological and environmental studies.* Ed. N.E. Stork, pp 217–225. Andover: Intercept.

MAKAROV, K.V. (1994). Annual reproductive rhythms of ground beetles: a new approach to the old problem. In: *Carabid beetles: ecology and evolution*. Eds. K. Desender, M. Dufrêne, M. Loreau, M. Luff and J.P. Maelfait, pp 177–182. Dordrecht: Kluwer Academic Publishers.

MCILWRAITH, A. (1999). The value of undersowing to beneficial arthropods in spring barley fields. Unpublished MSc Thesis, University of Reading.

MEISSNER, R.L. (1984). Zur biologie und Ökologie der ripicolen Carabiden *Bembidion femoratum* Sturm und *B. punctatum* Drap. II. Die Substratbindung. *Zeitschrift für Morphologie Ökologie der Tiere Zeitschrift für Morphologie Ökologie der Tiere* **1**, 369–383.

MELBOURNE, B.A. (1999). Bias in the effect of habitat structure on pitfall traps: An experimental evaluation. *Australian Journal of Ecology* **24**, 228–239.

MOLS, P.J.M. (1987). Hunger in relation to searching behaviour, predation and egg production of the carabid beetles *Pterostichus coerulescens* L.: results of simulation. *Acta Phytopathologica Entomologica Hungarica* **22**, 187–205.

MOLS, P.J.M. (1988). Simulation of hunger, feeding and egg production in the carabid beetle *Pterostichus coerulescens*. *Agricultural University Wageningen Papers* **88**, 1–96.

MOMMERTZ, S., SCHAUER, C., KOSTERS, N., LANG, A. AND FILSER, J. (1996). A comparison of D-Vac suction, fenced and unfenced pitfall trap sampling of epigeal arthropods in agroecosystems. *Annales Zoologici Fennici* **33**, 117–124.

MOSSAKOWSI, D., FRÄMBS, H. AND BARO, A. (1990). Carabid beetles as indicators of habitat destruction caused by military tanks. In: *The role of ground beetles in ecological and environmental studies*. Ed. N.E. Stork, pp 237–246. Andover: Intercept.

MÜLLER, J.K. (1987). Period of adult emergence in carabid beetles: an adaption for reducing competition? *Acta Phytopathologica Entomologica Hungarica* **22**, 409–415.

MURDOCH, W.W. (1966). Aspects of the population dynamics of some marsh Carabidae. *Journal of Animal Ecology* **35**, 127–156.

MURDOCH, W.W. (1967). Life history patterns of some British Carabidae (Coleoptera) and their ecological significance. *Oikos* **18**, 25–32.

NELEMANS, M.N.E. (1988). Surface activity and growth of larvae of *Nebria brevicollis* (F.) (Coleoptera: Carabidae). *Netherlands Journal of Zoology* **38**, 74–95.

NIEMELÄ, J. (1993). Interspecific competition in ground-beetle assemblages (Carabidae): what have we learned? *Oikos* **66**, 353–355.

NIEMELÄ, J. AND SPENCE, J.R. (1999). Dynamics of local expansion by an introduced species: *Pterostichus melanarius* III. (Coleoptera, Carabidae) in Alberta, Canada. *Diversity and Distributions* **5**, 121–127.

OBRTEL, R. (1973). Animal food of *Apodemus flavicollis* in a lowland forest. *Zoologicke Listy* **22**, 15–30.

PARMENTER, R.R. AND MACMAHON, J.A. (1988). Factors influencing species composition and population sizes in a ground beetle community (Carabidae): predation by rodents. *Oikos* **52**, 350–356.

PAVUK, D.M., PURRINGTON, F.F., WILLIAMS, C.E. AND STINNER, B.R. (1997). Ground beetle (Coleoptera: Carabidae) activity density and community composition in vegetationally diverse corn agroecosystems. *American Midland Naturalist* **138**, 14–28.

PERRY, J.N., WINDER, L., HOLLAND, J.M. AND ALSTON, R.D. (1999). Red-blue plots for detecting clusters in count data. *Ecology Letters* **2**, 106–113.

PHILIPS, J.R. (1977). Carabid beetle remains in an American kestrel nest. *Bird-Banding* **48**, 371.

POLLARD, E. (1968). Hedges III. The effect of removal of the bottom flora of a hawthorn hedge on the Carabidae of the hedge bottom. *Journal of Applied Ecology* **5**, 125–139.

POTTS, G.R. (1986). *The partridge: pesticides, predation and conservation*. London: Collins.

POTTS, G.R. (1997) Cereal farming, pesticides and grey partridges. In: *Farming and birds in Europe*. Eds.D.J. Pain and M.W. Pienkowski, pp 151–177. London: Academic Press.

POTTS, G.R. AND AEBISCHER, N.J. (1991). Modelling the population dynamics of the Grey Partridge: conservation and management. In: *Bird population studies: their relevance to conservation management*. Eds. C.M. Perrins, J.D. Lebreton and G.J.M. Hirons, pp 373–390. Oxford: Oxford University Press.

POTTS, G.R. AND VICKERMAN, G.P. (1974). Studies on the cereal ecosystem. *Advances in Ecological Research* **8**, 107–197.

POWELL, W., DEAN, D.A. AND DEWAR, A. (1985). The influence of weeds on polyphagous arthropod predators in winter wheat. *Crop Protection* **4**, 298–312.

PURVIS, G. AND CURRY, J.P. (1984). The influence of weeds and farmyard manure on the activity of Carabidae and other ground-dwelling arthropods in a sugar beet crop. *Journal of Applied Ecology* **21**, 271–283.

RIEDEL, W. AND STEENBERG, T. (1998). Adult polyphagous coleopterans overwintering in cereal boundaries: winter mortality and susceptibility to the entomopathogenic fungus *Beauveria bassiana*. *Biocontrol* **43**, 175–188.

SHELTON, M.D. AND EDWARDS, C.R. (1983). Effects of weeds on the diversity and abundance of insects in soybeans. *Environmental Entomology* **12**, 296–298.

SOTHERTON, N.W. (1985). The distribution and abundance of predatory Coleoptera overwintering in field boundaries. *Annals of Applied Biology* **106**, 17–21.

SPEIGHT, M.R. AND LAWTON, J.H. (1976). The influence of weed cover on the mortality imposed on artificial prey by predatory ground beetles in cereal fields. *Oecologia* **23**, 211–223.

SPENCE, J.R. (1994). Success of European carabid species in Western Canada: preadaption for synanthropy? In: *The role of ground beetles in ecological and environmental studies*. Ed. N.E. Stork, pp 129–141. Andover: Intercept.

STAPP, P. (1997). Small mammal predation on darkling beetles (Coleoptera: Tenebrionidae) in pitfall traps. *The Southwestern Naturalist* **43**, 352–355.

STEBBINGS, R.E. (1991). Bats: Order Chiroptera. In: *The handbook of British mammals*. Eds. G.B. Corbet and S. Harris, pp 86. Oxford: Blackwell Scientific Publications.

STEENBERG, T., LANGER, V. AND ESBJERG, P. (1995) Entomopathogenic fungi in predatory beetles (Col: Carabidae and Staphylinidae) from agricultural fields. *Entomophaga* **40**, 77–85.

STEWART, R.E. (1949). Ecology of a nesting red-shouldered hawk population. *Wilson Bulletin* **61**, 26–35.

STORK, N.E. (Ed.) (1990). *The role of ground beetles in ecological and environmental studies*. Andover: Intercept.

SUNDERLAND, K.D. (1975). The diet of some predatory arthropods in cereal crops. *Journal of Applied Ecology* **12**, 507–515.

SUNDERLAND, K.D. AND TOPPING, C.J. (1995). Estimating population densities of spiders in cereals. In: *Arthropod natural enemies in arable land. I. Density, spatial heterogeneity and dispersal*. Eds. S. Toft and W. Riedel, pp 13–22. Aarhus: Aahrus University Press.

SUNDERLAND, K.D., CROOK, N.E., STACEY, D.L. AND FULLER, B.T. (1987). A study of feeding by polyphagous predators on cereal aphids using ELISA and gut dissection. *Journal of Applied Ecology* **24**, 933.

SUNDERLAND, K.D., DE SNOO, G.R., DINTER, A., HANCE, T., HELENIUS, J., JEPSON, P., KROMP, B., SAMU, F., SOTHERTON, N.W., ULBER, B. AND VANGSGAARD, C. (1995). Density estimation for beneficial predators in agroecosystems. In: *Arthropod natural enemies in arable land. I. Density, spatial heterogeneity and dispersal*. Eds. S. Toft and W. Riedel, pp 133–164. Aarhus: Aarhus University Press.

THIELE, H.-U. (1964). Experimentelle unterschungen über die ursachen der biotopbindung bei carabiden. *Zeitschrift für Morphologie Ökologie der Tiere* **53**, 387–452.

THIELE, H.-U. (1971). Die steuerung der Jahresrhythmik von carabiden in feldheken? *Zoologische Jahrb Systematics* **78**, 341–371.

THIELE, H.-U. (1977). *Carabid beetles in their environments*. Berlin: Springer-Verlag.

THIELE, H.-U. AND WEBER, F. (1968). Tagesrhythmen der Aktivität bei Caribiden. *Oecologia (Berl.)* **2**, 7–18.

THOMAS, C.G., GREEN, F. AND MARSHALL, E.P. (1997). Distribution, dispersal and population size of the ground beetles, *Pterostichus melanarius* (Illiger) and *Harpalus rufipes* (Degeer) (Coleoptera, carabidae), in field margin habitats. *Biological Agriculture and Horticulture* **15**, 337–352.

THOMAS, C.F.G., PARKINSON, L. AND MARSHALL, E.J.P. (1998). Isolating the components of activity-density for the carabid beetle *Pterostichus melanarius* in farmland. *Oecologia* **116**, 103–112.

THOMAS, M.B., WRATTEN, S.D. AND SOTHERTON, N.W. (1991). Creation of 'island' habitats in farmland to manipulate populations of beneficial arthropods: densities and emigration. *Journal of Applied Ecology* **28**, 906–917.

THOMAS, M.B., MITCHELL, H.J. AND WRATTEN, S.D. (1992). Abiotic and biotic factors influencing the winter distribution of predatory insects. *Oecologia* **89**, 78–84.

TOFT, S. (1995). Value of the aphid *Rhopalosiphum padi* as food for cereal spiders. *Journal of Applied Ecology* **32**, 552–560.

TUCKER, G.M. AND HEATH, M.F. (1994). *Birds in Europe: their conservation status.* BirdLife conservation series no 3. Cambridge: BirdLife International.

TURIN, H. (2000). *De Nederlandse Loopkevers. Verspreiding en Oecologie [The Netherlands ground beetles. Distribution and ecology]*, Utrecht: KNNV Uitgeverij.

VAN DIJK, T.S. (1979). On the relationship between reproduction, age and survival in two carabid beetles: *Calathus melanocephalus* L. and *Pterostichus coerulescens* L. (Coleoptera: Carabidae). *Oecologia* **40**, 63–80.

VAN DIJK, T.S. (1982). Individual variability and its significance for the survival of animal populations. In: *Environmental adaption and evolution*. Eds. D. Mossakowsi and G. Roth, pp 233–51. Stuttgart: Fischer.

VAN DIJK, T.S. (1983). The influence of food and temperature on the amount of reproduction in carabid beetles. In: *Ecology of carabids: the synthesis of field study and laboratory experiment*. Eds. P. Brandmayr, P.J. Den Boer and F. Weber, pp 105–123. Wageningen: Pudoc.

VAN DIJK, T.S. (1994). On the relationship between food, reproduction and survival of two carabid beetles: *Calathus melanocephalus* and *Pteristichus versicolor*. *Ecological Entomology* **19**, 263–270.

VAN ZANT, T., POULSON, T.L. AND KANE, T.C. (1978). Body size differences in carabid cave beetles. *American Naturalist* **112**, 229–234.

VARCHOLA, J.M. AND DUNN, J.P. (1999). Changes in ground beetle (Coleoptera: Carabidae) assemblages in farming systems bordered by complex or simple roadside vegetation. *Agriculture, Ecosystems and Environment* **73**, 41–49.

VICKERMAN, G.P. (1978). The arthropod fauna of undersown grass and cereal fields. In: *Scientific proceedings of the Royal Dublin Society*, pp 155–165.

WALLIN, H. (1985). Spatial and temporal distribution of some abundant carabid beetles (Coleoptera: Carabidae) in cereal fields and adjacent habitats. *Pedobiologia* **28**, 19–34.

WALLIN, H. (1987) Distribution, movements and reproduction of carabid beetles (Coleoptera: Carabidae) inhabiting cereal fields. Plant Protection Reports, Dissertations 15. Uppsala.

WALLIN, H., CHIVERTON, P.A., EKBOM, B.S. AND BORG, A. (1992). Diet, fecundity and egg size in some polyphagous predatory carabid beetles. *Entomologia Experimentalis et Applicata* **65**, 129–140

WEEKS, R.D. AND MCINTYRE, N.E. (1997). A comparison of live versus kill pitfall trapping techniques using various killing agents. *Entomologia Experimentalis et Applicata* **82**, 267–273.

WILLIAMSON, M. (1972). *The analysis of insect populations*. London: Arnold.

WILMER, P.G. AND UNWIN, D.M. (1981). Field analysis of insect heat budgets: reflectance, size and heating rates. *Oecologia* **50**, 250–255.

WILSON, J.D., ARROYO, B.E. AND CLARK, S.C. (1996). The diet of bird species of lowland farmland: a literature review. BBSRC-NERC Ecology and Behaviour Group. University of Oxford, Unpublished report to the Department of Environment and English Nature.

WINSTONE, L., ILES, D.R. AND KENDALL, D.J. (1996). Effects of rotation and cultivation on polyphagous predators in conventional and integrated farming systems. *Aspects of Applied Biology* **47**, 111–117.

WOLDA, H. (1989). Comment on article 'On testing temporal niche differentiation in carabid beetles' by M. Loreau and the 'Comment on the article of M. Loreau' by P.J. Den Boer. *Oecologia* **81**, 99.

YALDEN, D.W. (1976). The food of the hedgehog in England. *Acta Theriologica* **21/30**, 401–424.

YOUNG, O.P. AND HAMM, J.J. (1985). Compatibility of two fall armyworm pathogens with the predaceous beetle *Calosoma sayi* (Coleoptera: Carabidae). *Entomological Science* **20**, 212–218.

2

Carabid Assemblage Organization and Species Composition

MARTIN L. LUFF

Department of Agricultural and Environmental Science, University of Newcastle upon Tyne, Newcastle upon Tyne NE1 7RU, UK

Introduction

In any agricultural environment, the carabid fauna consists of a number of species, each characterized by features such as body size, phenology and breeding cycle, diel activity pattern, larval and adult dietary needs, dispersal ability and preferred microclimatic conditions. The assemblage of carabid species is therefore, to a large extent, a result of these species-dependent attributes. Depending on the extent to which these features overlap between species, there may also be some interspecific interactions, such as competition or predation, by one carabid species on another. The present chapter is not defined by these potential interactions, but simply includes all those species of Carabidae that can be found in the agricultural habitat, whether or not they interact directly with one another.

The biology and ecology of Carabidae were reviewed by Thiele (1977), concentrating on the way in which environmental factors led to habitat selection by affecting the beetles' physiology and behaviour. Thiele (l.c.) included a table of the most frequently occurring carabid species in European agricultural regions, based on 32 published lists from 29 arable habitats extending from the UK to Belo-Russia. He concluded that the basic fauna was surprisingly uniform; 26 species occurred in at least one third of the region, and eight species were represented in more than two thirds. This list was updated using 31 later surveys by Luff (1987), and the fauna of eastern Europe considered by Lövei and Sárospataki (1990), since when Lövei and Sunderland (1996) have further reviewed carabid biology. Kromp (1999) reviewed the role of carabids in sustainable agriculture and Holland and Luff (2000) considered the impacts of agricultural practices on temperate Carabidae.

Abbreviations: CA, Correspondence Analysis; CCA, Canonical Correspondence Analysis; DCA, Detrended Correspondence Analysis; PCA, Partial Correspondence Analysis; PCCA, Partial Canonical Correspondence Analysis; DCCA, Detrended Canonical Correspondence Analysis; MAFF, Ministry of Agriculture, Fisheries and Food; LINK IFS, LINK Integrated Farming Systems; SCARAB, Seeking Confirmation About Results At Boxworth; TALISMAN, Towards A Low Input System Minimizing Agrochemicals And Nitrogen

---

The Agroecology of Carabid Beetles

In North America, the equivalent role of ground beetles was recognized in the early classic book by Balduf (1935). Rivard (1964, 1966) assessed carabid assemblages in Canada, and Allen (1979) has later reviewed the impact of carabids in agricultural situations based largely on North American studies. Further restricted surveys are those of Alderweireldt and Desender (1994) covering work in Belgium, and Yano *et al.* (1995) who reviewed Carabidae from paddy fields of the world. There have been few studies of agricultural carabids other than from the more temperate regions of the world. Horne and Edward (1997) found 28 species of Carabidae in agricultural crops in Victoria, Australia, although Bishop and Blood (1980) found only eight species among the 23 ground-living arthropods trapped in cotton in S.E. Queensland.

Information on the biology, phenology, distributions and dispersal of individual north European species of Carabidae was given at length in the work of Lindroth (1945), later translated and republished (Lindroth, 1992). This has been further updated for many species of that region by Turin (2000), who also characterized the carabid assemblages of 33 habitat types within the Netherlands, based on multivariate analyses of pitfall catches.

The aim of this chapter is to update and summarize what is now known of the characteristics of carabid assemblages, with particular reference to those of arable land in northern Europe and North America. Grassland carabids are generally not considered. The majority of studies that have listed the carabids of agricultural land have used pitfall trapping as their collecting method. This has implications for the relative importance of different species of beetle, and the method will first be reviewed and discussed. Any beetle assemblage can be characterized firstly by attributes, such as species richness and diversity, which are species independent. Other attributes, such as mean body size and impact on potential prey, clearly depend, at least partly, on the species composition of the carabids present. Both the species independent and dependent attributes will therefore be considered. In addition, data on some possible effects of geographical, cultural and environmental factors on carabid assemblages will also be reviewed.

Sampling carabid assemblages

Because temperate ground beetles move largely by running on the soil surface, they have the potential to be trapped by any container into which they may fall and from which they are then unable to escape. This is the principle of the ubiquitous pitfall trap, whose uses were reviewed by Adis (1979) and limitations discussed by Sunderland *et al.* (1995). The catch is a species-specific, largely undetermined (but see Kuscha *et al.*, 1987) function of density, activity and trappability, all of which may vary independently in response to environmental factors. This means that the catch of any one species is only an approximate analogue of its population density (Baars, 1979; Ulber and Wolf-Schwerin, 1995) and also that the relative numbers of each species in the trapped assemblage do not necessarily reflect their relative true abundance. Desender and Maelfait (1986) advocated trapping within enclosures to estimate the relative abundances of coexisting species, but the range of species trapped within an enclosure may differ from those outside (Holopainen and Varis, 1986; Ulber and Wolf-Schwerin, 1995; Mommertz *et al.*, 1996; Holland, 1998; Holland and Smith, 1999).

Differences in the catch/abundance relationship between species result in part from differential mobility of each species; in general, larger carabids are more mobile and therefore more likely to meet the trap. The intrinsic hunger level of each species will also influence its catch (Wallin and Ekbom, 1994), as much surface movement is foraging for food, so that more hungry beetles will be caught than satiated ones. A further complication is the differing phenologies of each species, so that they will be active for differing periods of time in different parts of the season. The trappability of most carabids is also low (Halsall and Wratten, 1988; Hawthorne, 1995), although if this remains constant between species it will not affect the resulting assemblage composition, except by reducing the probability of capturing any of the less abundant species. It must be borne in mind, however, that pitfalling may be flawed when comparing crops with differing penetrability of ground cover, because this can affect the relative capture rates of each species, and hence also assemblage composition in the traps.

However, if all that is required is a species list and some idea of relative numbers of species across a number of sites, or in different seasons, these problems in relating pitfall catch to actual densities and assemblage composition are offset by the practicality and cost efficiency of the technique. For this reason, Spence and Niemelä (1994) concluded that pitfall trapping remained the only practical method for large-scale surveys of carabid assemblages. Similarly, Hawthorne (1995) concluded that pitfall traps could be used to assess the apparent preferences of ground beetle species for differing habitats such as cereal field headlands, although the traps gave little information about actual population densities. It is in this context that pitfall trapping can be considered as a practicable way of surveying or assessing carabid assemblages in agricultural crops, and continues to be widely used.

Many features of the traps used have varied between studies, and although Adis (1979) recommended standardizing the method, different workers have used traps of varying numbers, size, shape, material, spacing, and with or without a range of preservatives. The effects of some of these factors were considered by Luff (1975). Glass traps have the lowest probability of escape by trapped beetles, and small traps (< 5 cm diameter) catch a higher proportion of the smaller species in comparison to larger traps (> 20 cm diameter). Trap spacing can also affect catch, as close spacing can cause local depletion (Digweed *et al.*, 1995). In order to assess comprehensively the ground beetle assemblage of any crop, sampling should cover as much as possible of the beetles' seasonal activity periods (usually late spring to mid-autumn) and should be continuous, rather than opening traps intermittently during this period. This maximizes the probability of trapping species that only have a short activity period, especially if this 'window' of activity is very dependent on short-term weather conditions. Seasonal cultivation, or other agronomic practices, may limit the trapping period in agricultural crops. In order to minimize any small-scale local heterogeneity in trap efficiency and beetle activity (see Luff, 1986), it is better to pool the catches of a number of small traps than to use fewer larger ones.

A suggested pitfall trapping protocol (Luff, 1996) for carabid survey work is to use 5–10 small (7.5 cm diameter) plastic pots spaced at not less than 2 m apart. Although less efficient than glass traps, plastic ones are cheaper, more easily transported and less breakable. The traps can be set into matching cylinders or outer pots that remain permanently in the soil, so as to minimize soil disturbance, and avoid the 'digging in' effect on the catch (Digweed *et al.*, 1995). Each set of 5–10 traps constitutes one sampling unit, their catch being pooled in the field at each visit. Frequency of emptying can range from weekly to monthly, depending on the travel and other costs involved. The trap efficiency generally declines as the traps become contaminated with dirt, rainwater, and partially preserved catch, but time and labour costs militate against too frequent servicing of the traps. For assessing assemblages, a preservative should be used in the traps so as to allow as long a time as possible between servicings, and to enable easier identification of the beetles caught. This may lead to further variations in trappability between species (Scheller, 1984; Waage, 1985; Holopainen, 1990; Lemieux and Lindgren, 1999). Luff (1996) and Lemieux and Lindgren (1999) both found ethylene glycol (in the form of vehicle antifreeze) to be an efficient and cheap preservative. It is, however, attractive to grazing farm stock (Marshall and Doty, 1990) and is toxic to mammals, including humans (Hall, 1991). Alternatives suggested by Lemieux and Lindgren (1999) are propylene glycol (less toxic and also available in commercial antifreezes, but often more expensive) or salt solution (less effective). Picric acid and formaldehyde have also been used, but both are toxic and unpleasant to work with.

Although many pitfall trapping techniques have been used previously to assess carabid assemblages in the field, most, if not all, have at least indicated successfully which species of ground beetle predominate in terms of their activity/abundance in particular regions, crops and agricultural habitats. Acceptance of this fact enables the results of such work to be used to consider ground beetle assemblage structure and composition in the following sections.

Species-independent attributes of carabid assemblages

Characteristics that are independent of the species present (typological parameters, Penev, 1996) provide less information about the assemblage than those which depend on the assemblage composition (individualistic parameters, Penev, 1996). But the species of carabids present in any particular agricultural ecosystem will be limited by the geographical limits to the species' distributions, and can only be drawn from the local species pool. Therefore, in order to review the organization of ground beetle assemblages across differing continents or biogeographical regions, only species-independent features can be used.

SPECIES RICHNESS

The first parameter to characterize any assemblage is 'how many species are there?' Because of the limitations of sampling, and in particular some of the features of pitfall trapping as discussed above, there is no easy answer to this question. The number of species actually trapped will depend on sampling effort, in particular how many traps are used and for how long. Given, for reasons already mentioned, that traps should be operated for the whole activity season of the beetles, the number of traps used remains the main variable that affects numbers of species caught. Both Obrtel (1971) and Niemelä *et al.* (1986) show that the 5–10 traps over a whole season, as recommended above, appear to catch most species.

Table 2.1 shows the species richness from 108 half-field treatment plots, operated

Table 2.1. Mean species richness (with s.e.) of carabids from the MAFF SCARAB experiment, from individual treatments, years, fields, farms and overall.

Data source	N	mean	s.e.
treatments	108	15.06	0.95
years	54	17.39	0.89
fields	7	25.00	2.59
farms	3	28.67	7.47
overall	1	32.00	–

for six or seven years between 1990 and 1996 on two or three fields of three farms in the UK's MAFF SCARAB (Seeking Confirmation About Results At Boxworth) experiment. In this experiment, eight whole fields were divided into two treatments, subjected respectively to two different pesticide regimes representing current farm practice and a reduced input approach. Individual treatments within each year caught a mean of 15 species (range 4–25), about 50% of the total 32 recorded. Combining the two treatments in each year only increased this by a little over two more species, but combining all years (six or seven according to which farm) to give totals per field raised this to 25 species, resulting from variation in species from season to season. Further pooling of the data by whole farms (two or three fields per farm) and then combining these three farms for the whole experiment both increased the species richness by similar amounts, indicating both local and large-scale spatial variations in the species occurrences.

Where, as in this case, replicated sets of pitfalls were used, various species accumulation models, often based on the relative abundances of each species in each sample (reviewed in Colwell and Coddington, 1994), can be used to estimate true species richness, and to show how species richness increases with sample size. Several such models were applied to the 42 samples from just one farm (High Mowthorpe) in the SCARAB dataset. The models used were: ACE – abundance-based coverage estimator; ICE – incidence-based coverage estimator (both from Chazdon *et al.*, 1998); Chao's richness estimator (Chao, 1984); First-order Jackknife richness estimator (Heltshe and Forrrester, 1983) and Coleman richness expectation (Coleman *et al.*, 1982). The latter produces a species accretion curve that is virtually identical to the more familiar rarefaction technique, but is computationally much more efficient (Colwell and Coddington, 1994). It also estimates sample species richness from the pooled species total, rather than total species richness from samples (Colwell, 1999).

The resulting estimated total species richness curves (*Figure 2.1*) all suggest the total species richness to be about 32–33 species, and that between 20 and 30 samples would be sufficient to estimate this total. The Coleman curve, in contrast, simply shows the smoothed actual progression to the observed total of 28 species on that farm. However, *Table 2.1* has already indicated that the actual species total for the whole experiment (not just one farm) was 32. The total species richness models correctly estimated this using the samples from just one farm of the three.

Within this dataset, the actual catch (i.e. numbers of individuals) of carabid beetles within each treatment/field/year combination varied from 25 to 5487. This is an

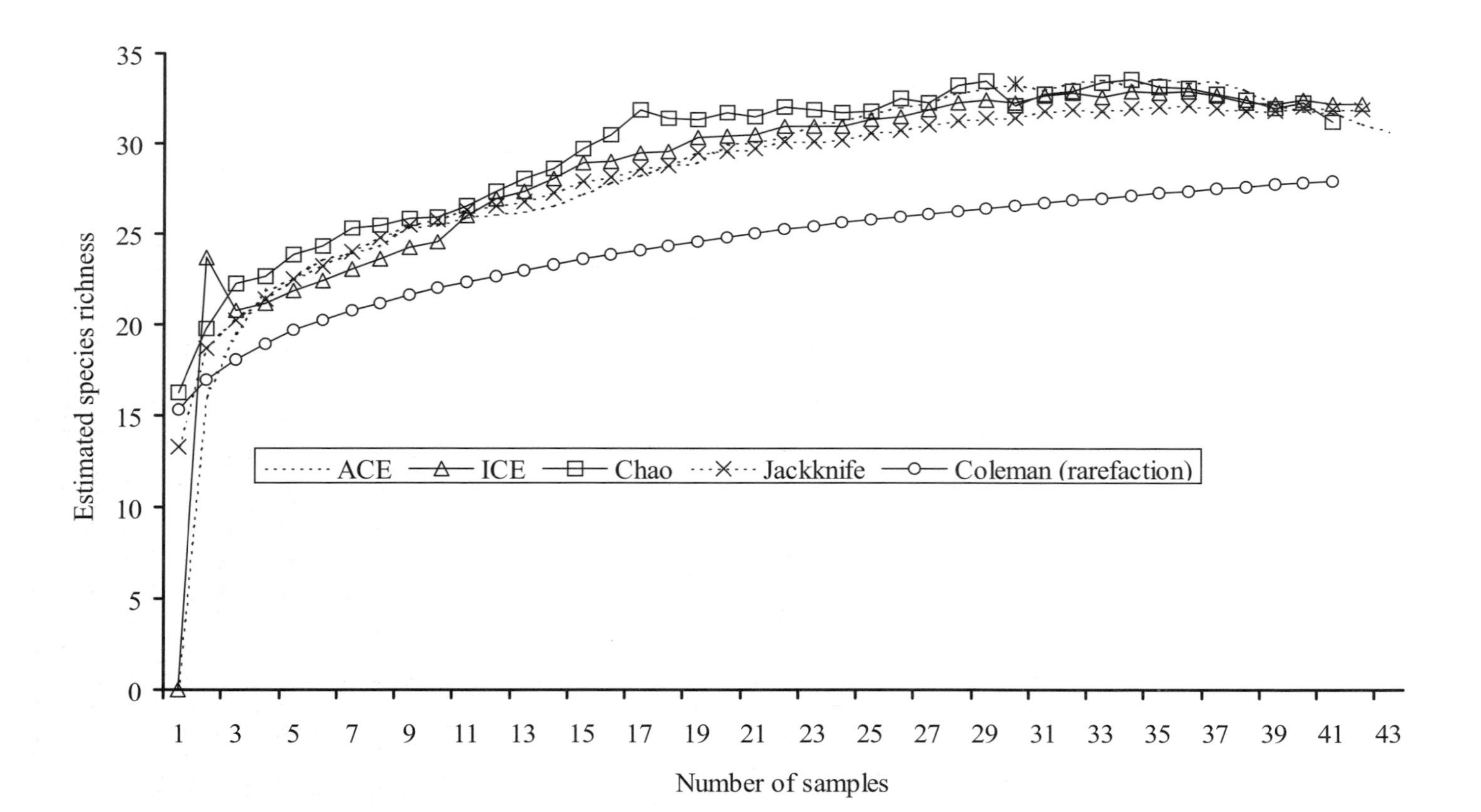

Figure 2.1. Species accretion curves for carabid assemblages from 42 season samples at High Mowthorpe, UK in the MAFF SCARAB experiment. For further details of the models used, see text, and Colwell (1999).

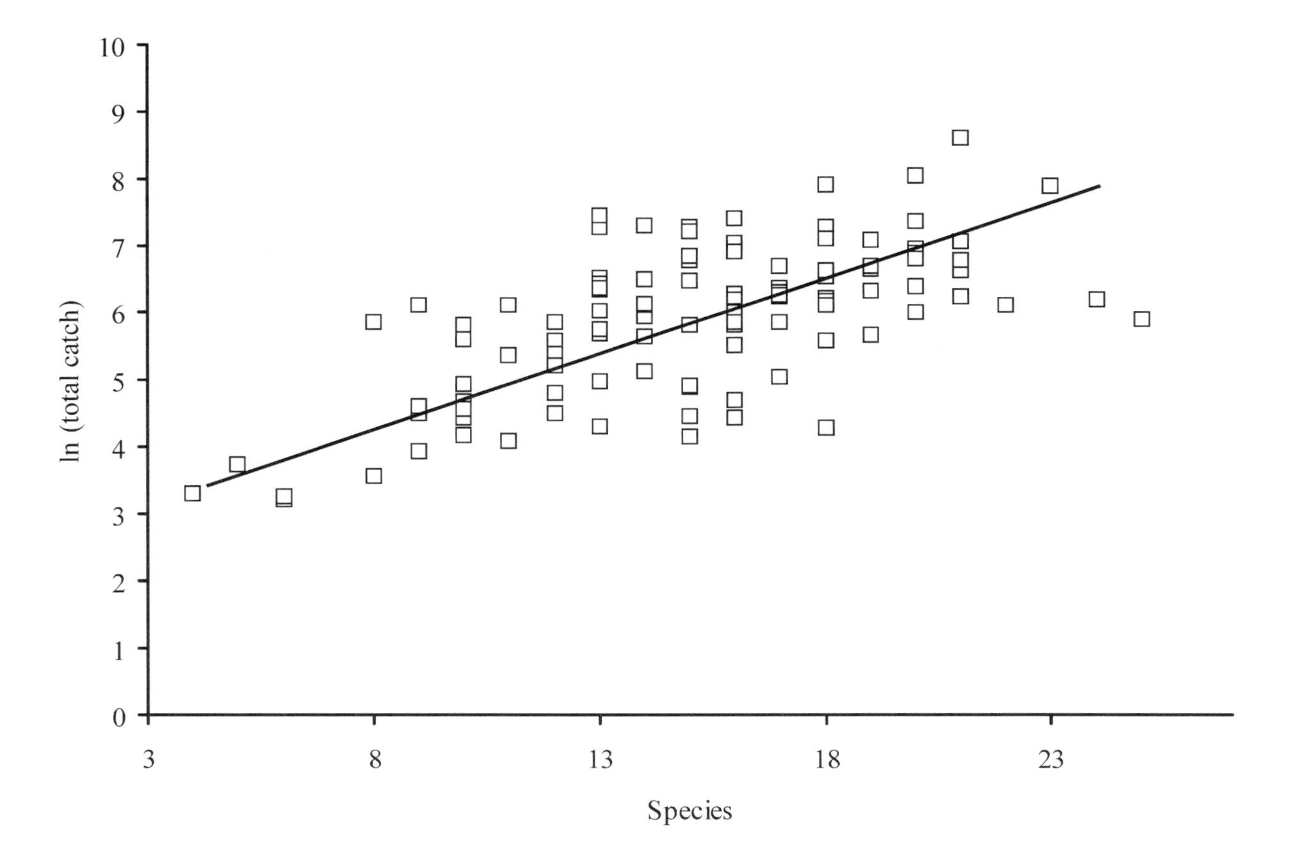

Figure 2.2. Total catch (logarithmic scale) plotted against observed species richness for all 108 treatment/field/year/farm species lists in the MAFF SCARAB experiment. Trend line is eye-fitted.

Table 2.2. Species richness from 119 sets of carabid pitfall data from individual field-seasons throughout the world.

Region	N	Individuals	Species			
			mean	s.e	min	max
N/W Europe	56	1826.8	24.85	1.45	7	52
Cent/E Europe	38	4207.9	29.42	1.38	13	50
N America	22	3758.5	29.95	2.22	19	49
Japan	3	1542.3	26.33	1.76	23	29
Overall	119	2937.1	27.29	0.93	7	52

Published sources: Andersen (1982, 1985); Armstrong (1995); Barney and Pass (1986); Boiteau (1983); Boivin and Hance (1994); Borg (1973); Davies (1963); Ferguson and McPherson (1985); Gautsch *et al.* (1980); Geiler (1967); Hokkannen and Holopainen (1986); Holopainen *et al.* (1995); House and All (1981); Ishitani (1996); Ishitani *et al.* (1994); Jarošík and Hůrka (1986); Kegel (1991); Kelly and Curry (1985); Kennedy (1994); Kromp (1989); Kromp and Steinberger (1992); Levesque and Levesque (1994); Lövei (1984); Müller (1968); Novak (1964, 1968); Petruska (1967, 1971, 1974); Purvis and Bannon (1992); Sekulić (1976); Szwejda (1984); Varvara *et al.* (1989); Wallin *et al.* (1981); Wiedenmann *et al.* (1992).

alternative measure of sampling effort, and is clearly related positively to the numbers of species caught (*Figure 2.2*). Points above the overall trend have many individuals per species and have low diversity, and *vice versa*. Species richness can thus be increased, either by having more replicated sample sites, or by increasing the catch at any one (or more) sites, e.g. by using a larger number of traps.

On a larger scale, *Table 2.2* summarizes data from 119 published lists of carabids from individual field/season combinations worldwide, subdivided into four geographical regions. There is remarkable consistency, all regions averaging just below 30 species, despite the considerable range of species richness from one list to another within regions. Although the means and minima from northern/western Europe are lower than those for central/eastern Europe, there is not quite any overall significant difference between regions ($F_{3,115} = 2.21$, $0.1 > P > 0.05$). The relationship between log-catch and species richness from this dataset (*Figure 2.3*) is less consistent than for the SCARAB UK data. There is a suggestion of a divergence above about 30 species; many lists of 30–50 species have fewer individuals than the general trend. This dichotomy separates a few fields with exceptionally large numbers (usually of a single dominant species – see below), and hence low diversity, from the majority.

Species diversity was calculated from the log-series alpha (Taylor *et al.*, 1976) as this does not depend on the evenness of abundance among the species; other indices such as Shannon's and Simpson's are affected by high levels of dominance of the commonest species (Booij, 1994), although they have been proposed as useful indicators of long-term changes in carabid communities (Jarošík, 1991). Despite the dichotomy shown in *Figure 2.3*, alpha was more or less normally distributed (*Figure 2.4*), but with a slight 'tail' of higher values. These represent fields with larger numbers of species per unit of catch, and correspond to the lower line in *Figure 2.3*. It seems, however, that calculating alpha as a measure of species diversity from these agricultural sites adds little information to that obtained from species richness, as the two measures are highly correlated ($R_{117} = 0.841$, $P < 0.01$) (*Figure 2.5*).

In conclusion, despite wide variations in the numbers of carabid species recorded in individual fields in any particular season, the seasonal mean catch averages about 30 carabid species throughout temperate cropping regions where carabids are a dominant part of the soil surface fauna. It is remarkable that this is usually such a small

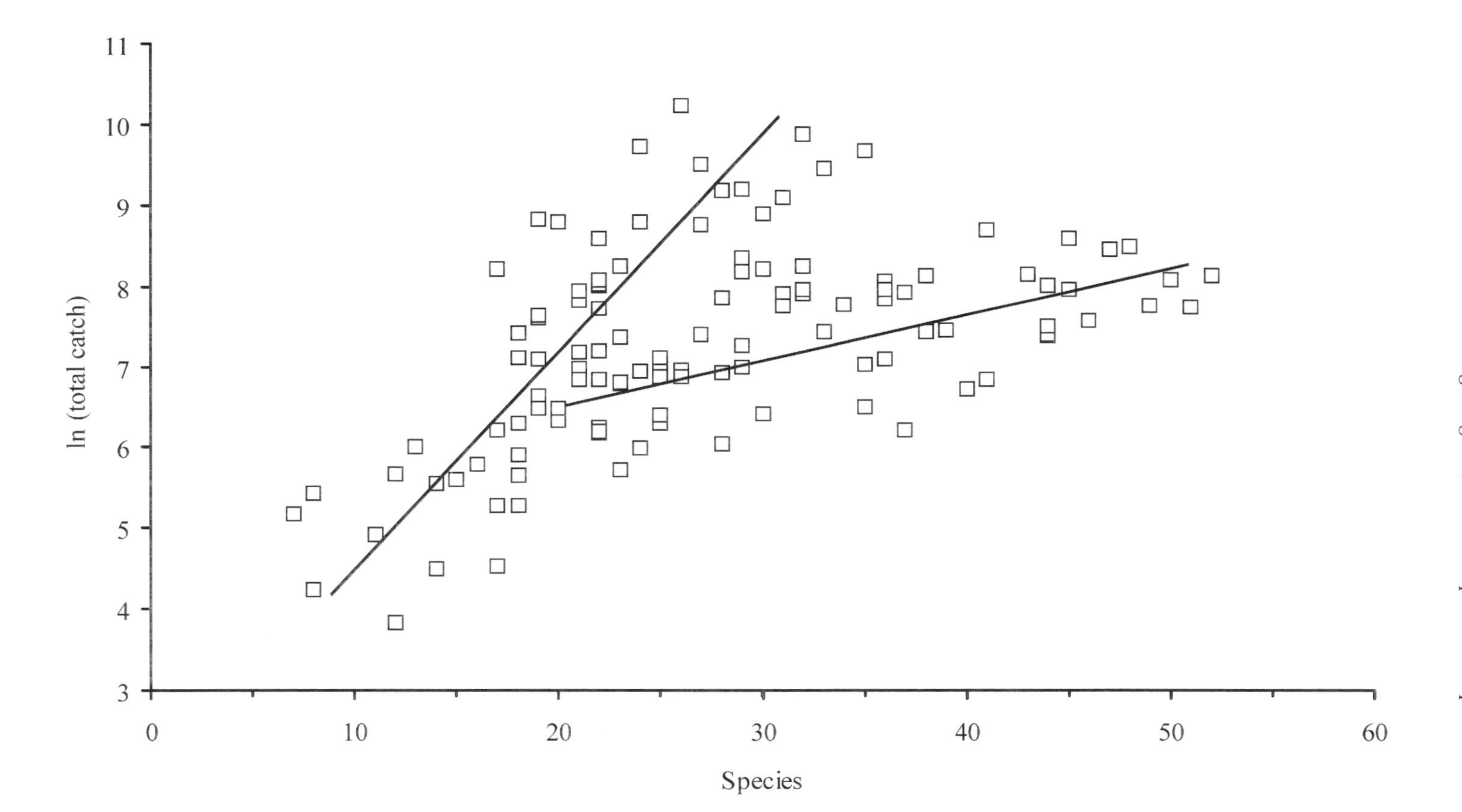

Figure 2.3. Total catch (logarithmic scale) plotted against observed species richness for 119 published species lists of Carabidae throughout the world, with eye-fitted trend lines.

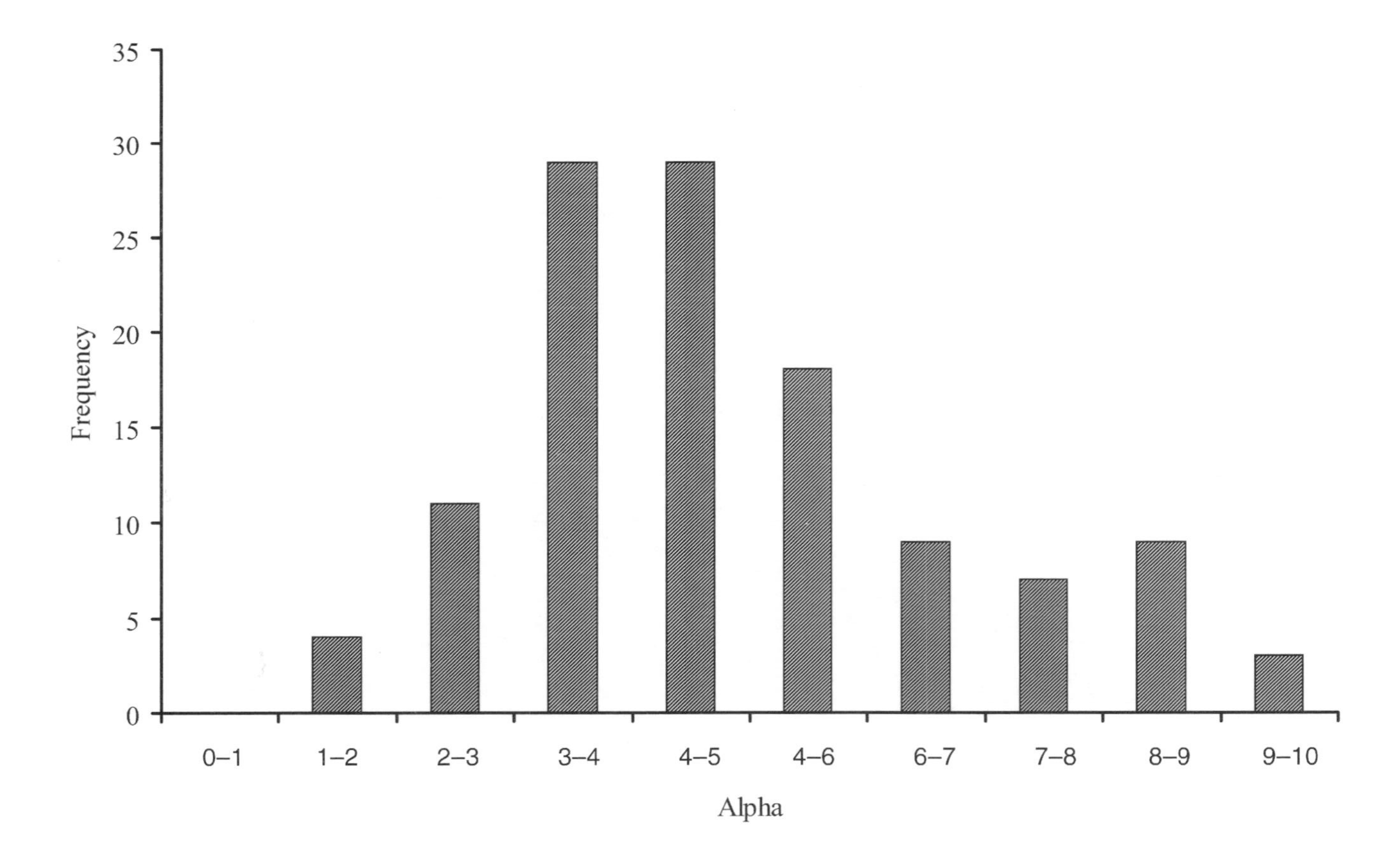

Figure 2.4. Frequency distribution of Fisher's log-series Alpha index of diversity, from 119 published species lists of Carabidae throughout the world.

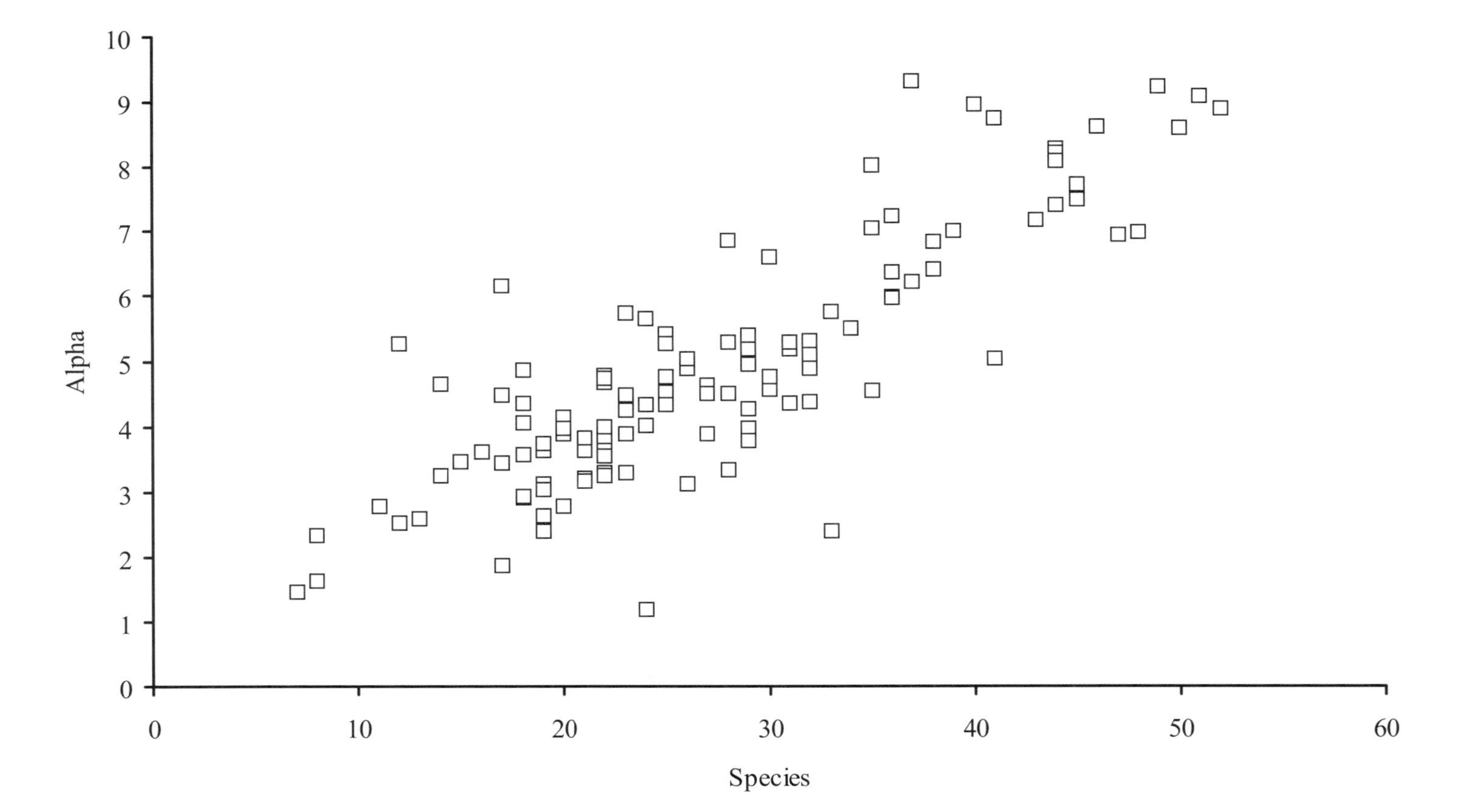

Figure 2.5. Fisher's log-series Alpha index of diversity plotted against observed species richness, from 119 published species lists of Carabidae throughout the world.

proportion of the total carabid fauna of any particular region. The British and Scandinavian faunas contain almost 350 and over 400 species respectively (Luff, 1998; Lindroth, 1992) and many European countries, as well as North America (about 1700 species), have substantially more (Allen, 1979; Turin, 1981, 2000). This suggests that the agricultural field conditions must suit only a small subset of the potential carabid fauna. Turin *et al.* (1991) and Turin (2000) analysed 30 years of pitfall trapping from all habitats in the Netherlands, and ranked agricultural habitats as just one of 33 habitat sub-types, based on the carabid assemblages of each. They concluded that most ground beetles of agricultural land were widely distributed, eurytopic species, many with apparently high tolerance of disturbance and chemical pollution. Only 53 species in the Dutch fauna of 380 species fell into this category.

DOMINANCE AND EVENNESS

As mentioned in the previous section, not all species in any one catch will be equally represented. As large carabids can move further and often faster than smaller ones, they often tend to dominate much of the catch, especially in open agricultural habitats (Tonhasca, 1993), and many 'less dominant' species may be present in only small numbers. For purposes of defining a characteristic assemblage of any site, it seems reasonable to include all those species making up 95% of the total catch of individuals in the traps. This will include these numerically 'dominant' species at the expense of the rarer ones. A further measure of the 'dominance' or lack of evenness among the catch is the proportion accounted for by the five most abundant species in the total catch. *Table 2.3* presents these parameters for the 119 world species lists used in the preceding section. On average, the five dominant species comprise nearly 85% of the catch, and 10 species comprise 95% or all the 'significant' part of the carabid fauna. *Figure 2.6* plots the proportion comprising 95% of the catch against total species richness. At low species richness, about 0.8 (8 of 10 species) comprised 95%, at the middle about 0.3 (10 of 30), and in the most diverse assemblages, 0.2 (10 of 50). The remaining part of the catch (from 2–40 species, average about 20) comprises species that occur in small numbers that can be considered as a 'non-significant' component of the fauna of any site. *Figure 2.7* shows how the overall cumulative catch increases with each species added to the list.

There are, of course, considerable variations about the average curve in *Figure 2.7*. The minimum contribution of the five most abundant species (54%) and maximum species to comprise 95% (18) in *Table 2.3* both relate to the most 'even' set of carabid catches in the dataset. These are from Kromp and Steinberger (1992), trapped in a wheat field near Vienna, Austria. Rather than one dominant species, the catches of the most abundant five species were all within 3 orders of magnitude (301, 323, 127, 110, 97). In contrast, the maximum contribution of the five dominant species (97%) and fewest species comprising 95% of the catch again both come from a single species list (Novak, 1964). This study of carabids in sugar beet in the former Czechoslovakia found 15,467 of the dominant species (*Pterostichus melanarius*), followed by 2400, 805, 245 and 161 of the next four commonest species respectively. The range within these five species is nearly 100-fold. Interestingly, the same field sampled in the previous year had a much less extreme abundance profile, with the most abundant five species comprising a more 'normal' 82% of the catch; the year following was,

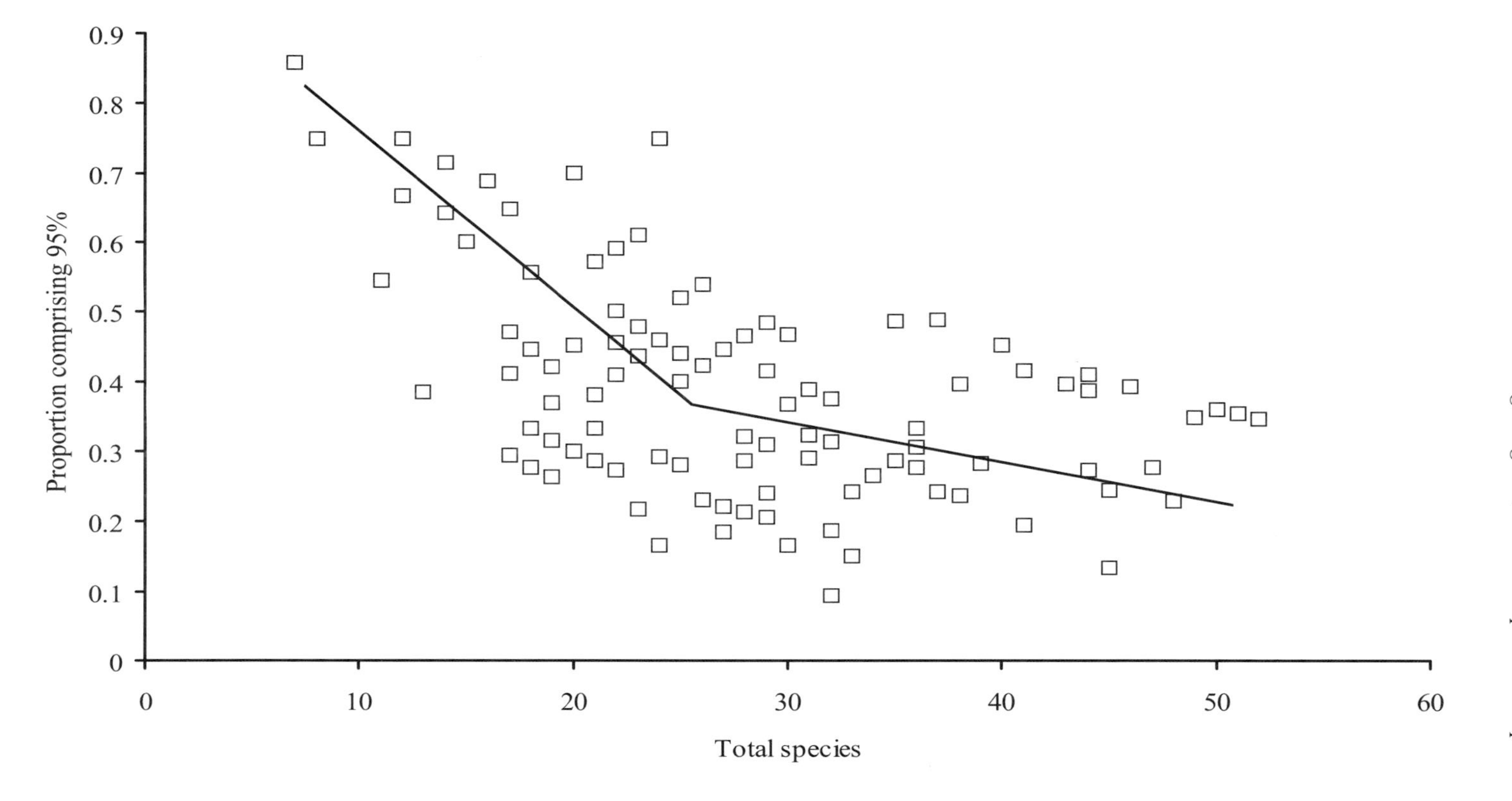

Figure 2.6. The proportion of carabid species that comprise 95% of the total numbers caught, plotted against observed species richness, from 119 published species lists of Carabidae throughout the world.

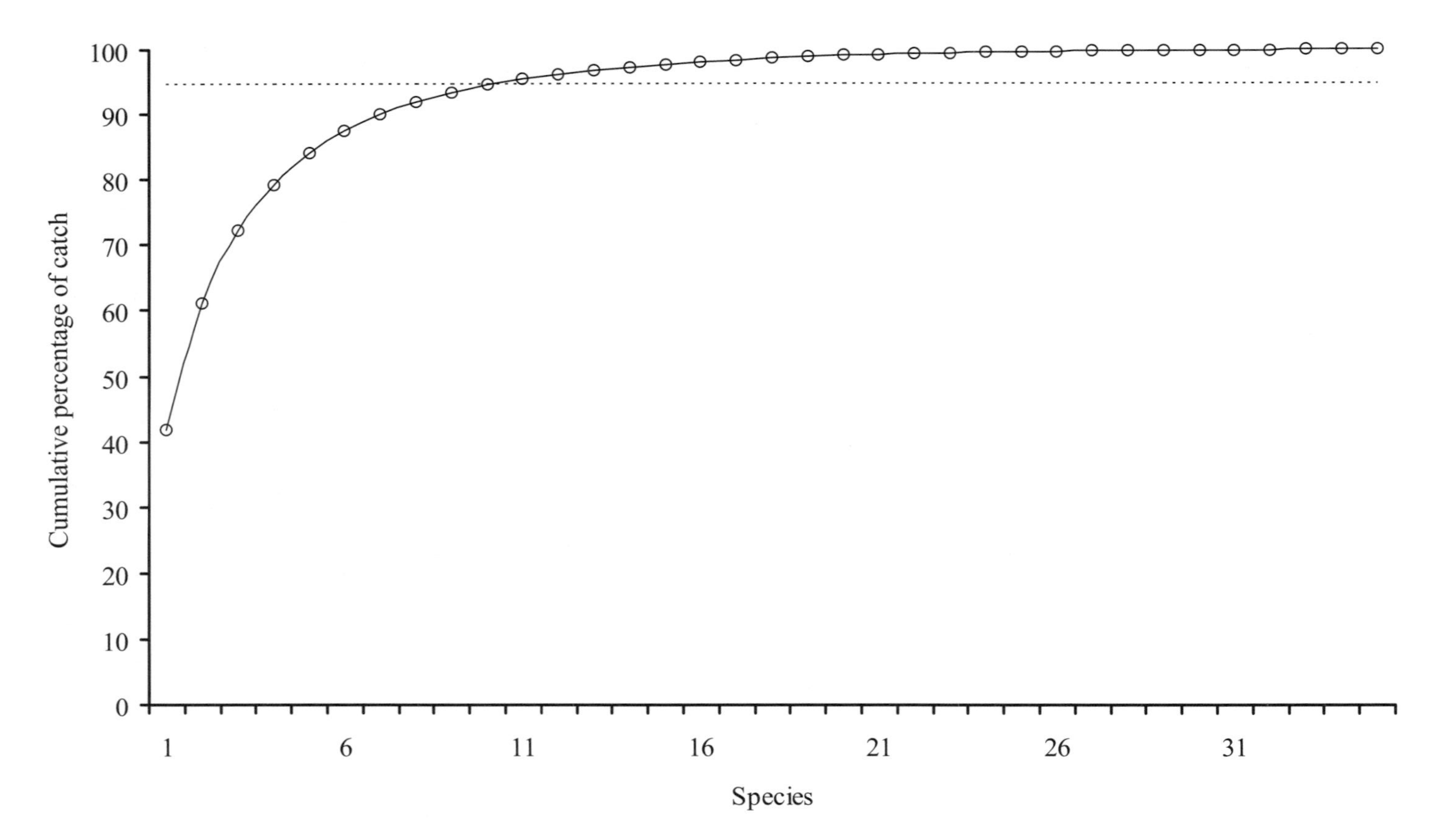

Figure 2.7. The cumulative percentages of total catch accounted for by increasing species numbers, based on 119 published species lists of Carabidae throughout the world.

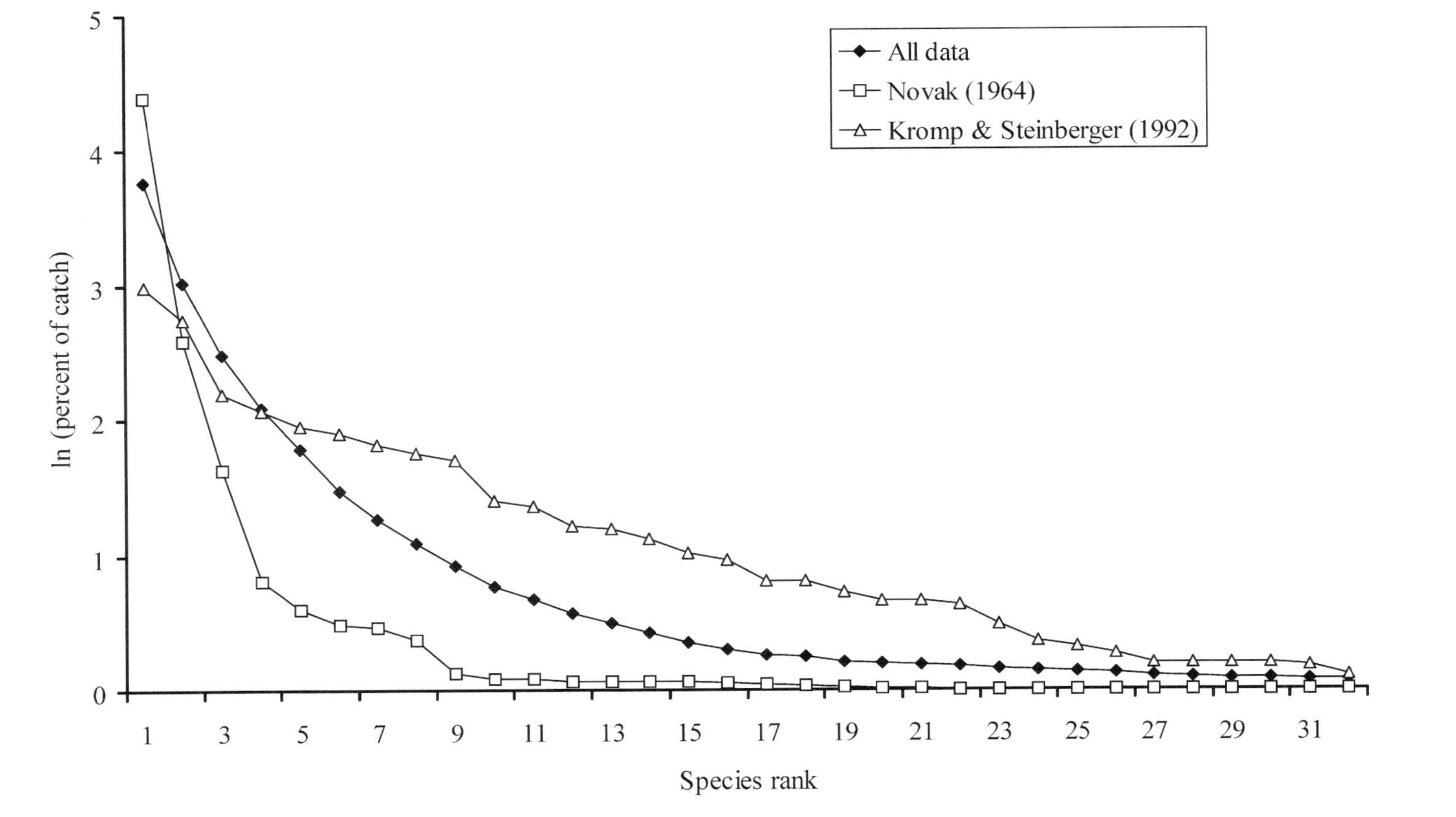

Figure 2.8. Percentage of total catch (logarithmic scale) plotted against species rank, for 119 published species lists of Carabidae throughout the world. Also shown are the most even list in this data (Kromp and Steinberger, 1992) and the least even (Novak, 1964).

Table 2.3. Dominance parameters calculated from the 119 carabid pitfall datasets used in *Table 2.2.*

Parameter	mean	s.e	min	max
% of catch comprising the five most abundant species	84.30	0.881	54.19	97.54
Number of species comprising 95% of catch	9.92	0.351	3	18
95% species / total species	0.39	0.0137	0.09	0.86

however, almost as extreme, but with greatly reduced catches. *Figure 2.8* shows the overall relationship between catch (logarithmic scale) and species rank, together with these two extreme cases: most carabid datasets from agricultural situations should lie somewhere between these two extremes.

EFFECTS OF MANAGEMENT

Holland and Luff (2000) have reviewed the impact of agricultural practices on Carabidae in temperate ecosystems, including factors that affect species diversity and richness. These findings are merely summarized here; effects on species composition are considered at the end of the following section.

The crop type itself is less important in affecting carabid diversity than whether the crop is winter sown or spring sown. Most studies (e.g. Hance *et al.*, 1990; Booij and Noorlander, 1992; Booij, 1994; Holland *et al.*, 1994) suggest that spring-sown root crops have lower diversity than winter-sown crops such as cereals and oil seed rape. But spring-sown cereal crops may have either a more or less diverse carabid fauna than winter-sown cereals (Gardner *et al.*, 1999). These differences are probably linked primarily to the timing of cultivation (Holland and Luff, 2000). In Hungary, Lövei (1984) showed that the carabids of a maize monoculture were more diverse than in maize crops as part of a rotation (although overall the rotation had slightly more species). This was attributed to the unfavourable bare period between harvesting the previous wheat crop in June and sowing the maize in the following May.

The method of cultivation may also affect total carabid numbers and diversity. Several studies (e.g. Brust *et al.*, 1985; House and Parmelee, 1985; Stinner and House 1990; Brust 1994; Digweed *et al.*, 1995; Heimbach and Garbe 1995) have shown increased ground beetle numbers following a change from ploughing to non-inversion tillage. However, this effect does not always occur, or may even be reversed (Barney and Pass, 1986). It is possible in the latter case that the disturbance caused by ploughing could stimulate increased movement and dispersal by the beetles, leading to greater pitfall catches.

The impact of pesticides on ground beetle numbers is very complex (see Holland and Luff, 2000 for more extensive discussion). Carabids are potentially at high risk from the short-term lethal effects of many agrochemicals (Jepson, 1989). However, in many cases their populations recover rapidly by re-invasion from adjacent unsprayed habitats (Thacker and Dixon, 1996; Holland *et al.*, 1999). Long-term monitoring of agricultural Carabidae on a landscape scale in southern England did not show consistent changes in diversity resulting from intensive farming (Aebischer, 1991), although it should be pointed out that this study was based on suction sampling rather than pitfall trapping. It therefore was based on the smaller-bodied species more readily collected by that technique. In Germany, Basedow (1990) considered two long-term studies and showed that intensive 'insurance spraying' of insecticides reduced carabid species richness, and that larger species, particularly *Carabus auratus*, did not re-occur when pesticide levels were reduced.

The most dramatic effect on numbers of carabids in any crop might be expected to result from combining changes in both cultivation practices and agrochemical use through the adoption of alternative farming systems such as integrated, organic or bio-dynamic systems. Reduction in overall agrochemical inputs increases ground beetle diversity and species richness (Cárcamo *et al.*, 1995, Ellsbury *et al.*, 1998), and Booij (1994) found that species richness was higher in integrated and organic systems than in conventional ones. However, the overall effects of any farming system can be variable, and depend as much on the timing of cultivation of each particular crop. Andersen and Eltun (2000) found that year-to-year variations in species richness and diversity completely masked any differences between conventional and biological systems in Norway. One problem seems to be that the farming system acts at the level of the individual species (Holland and Luff, 2000); beneficial effects on some species can be negated by adverse effects on others. As a result, there may be little or no overall effect on total carabid catch, or on species richness, despite changes in species composition. Holland and Luff (2000) summarized published results of the relative effect (compared to conventional farming) of three types of alternative farming systems on 24 common carabid species. Overall, there were 66 cases of species increase in abundance, 18 cases of decrease, and 5 showing no obvious effect. The impact of alternative farming systems thus seems likely to increase overall carabid numbers, even if it does not always increase diversity or species richness of these beetles.

Species composition

The previous sections have demonstrated that, within any season's catch of carabid beetles, a relatively few species will predominate, and therefore be potentially of greatest importance in that ecosystem. Thiele (1977) commented that the 'dominant' agricultural carabid fauna was rather uniform across Europe. This is considered in more depth and on a wider geographical basis in *Table 2.4*. This is based on the five most abundant Carabidae in each of 94 rankings from 82 published sources. Also included are three farms from the UK's MAFF SCARAB experiment referred to in *Table 2.1*, and a further two farms from the MAFF TALISMAN (Towards a Lower Input System Minimizing Agrochemicals and Nitrogen) (Gardner *et al.*, 1999). This experiment applied one to three reduced input treatments in a randomized block layout at three farm sites. *Table 2.4* is subdivided into the same regions as in *Table 2.2*.

Note that, in the previous section, *Tables 2.2* and *2.3* were based on separate species lists from each field/year combination (when authors had sampled in more than one field or year). Using each of these lists separately in the present section would have biased the species list in favour of these multiple sampling studies, as the same species tended to occur repeatedly over most years or fields within each study. *Table 2.4*, in contrast, is based on the pooled data from each study. The only exceptions are where (as in the SCARAB and TALISMAN experiments referred to earlier) more than one clearly separated geographical site was sampled; in these cases, an overall list was compiled for each such site. Although many authors have sometimes listed only their

Table 2.4. Frequency of occurrences of the five most abundant Carabidae found in 92 published and unpublished species lists, listed by geographical region. The species are arranged taxonomically. Each genus is followed (in parentheses) by the number of occurrences of that genus in the dominant five species of any list.

TRIBE and Genus	Species	Species occurrences			
		N/W Europe	Cent/E Europe	N America	Japan
CICINDELINI					
Megacephala (1)	*virginica*			1	
CARABINI					
Carabus (6)	*auratus*	2			
	granulatus		1		
	nemoralis			1	
	scheidleri		1		
	ullrichi		1		
Calosoma (5)	*alternans*			1	
	auropunctatum	1	1		
	calidum			2	
NEBRIINI					
Nebria (8)	*brevicollis*	8			
NOTIOPHILINI					
Notiophilus (1)	*biguttatus*	1			
ELAPHRINI					
Elaphrus (1)	*riparius*	1			
LORICERINI					
Loricera (12)	*pilicornis*	10	2		
SCARITINI					
Clivina (14)	*collaris*	1			
	fossor	11		2	
Dyschirius (2)	*globosus*		1		
	globulosus			1	
Scarites (1)	*subterraneus*			1	
BROSCINI					
Broscus (1)	*cephalotes*	1			
PATROBINI					
Patrobus (1)	*atrorufus*	1			
TRECHINI					
Trechus (36)	*discus*	1			
	micros	1			
	quadristriatus	26	2		
	secalis	6			
BEMBIDIINI					
Bembidion (81)	*aeneum*	2			
	bipunctatum	1			
	bruxellense	1			
	castor			1	
	femoratum	2			
	guttula	1			
	lampros	29	10	1	
	lunulatum	1			
	nigripes			1	
	obscurellum			1	
	obtusum	6			
	properans	2	1		
	quadrimaculatum	7		8	
	rapidum			1	
	tetracolum	5			
Asaphidion (3)	*flavipes*	3			

continued

Table 2.4. cont.

TRIBE and Genus	Species	Species occurrences			
		N/W Europe	Cent/E Europe	N America	Japan
PTEROSTICHINI					
Poecilus (40)	*chalcites*			4	
	cupreus	4	12		
	lepidus	1			
	lucublandus			8	
	punctulatus	1	1		
	sericeus		4		
	versicolor	3	2		
Pterostichus (72)	*adstrictus*			3	
	corvus			1	
	madidus	2			
	melanarius	35	13	12	
	niger	2			
	pensylvanicus			1	
	samurai				1
	scitulus			1	
	strenuus	1			
AMARINI					
Amara (26)	*apricaria*	1			
	bifrons	1			
	carinata			1	
	cf. lunicollis			1	
	chalcites				2
	consularis		1		
	cupreolata			4	
	familiaris	1			
	farcta			1	
	ingenua		1		
	interstitialis	1			
	littoralis			1	
	obesa			4	
	ovata	1			
	quenselii			1	
	similata	1			
	spreta	1			
	torrida			1	
	tricuspidata	1			
PLATYNINI					
Agonum (50)	*assimile*	1			
	chalcomus				1
	cupreum			3	
	dorsale	22	8	1	
	magnus				1
	muelleri	2	2	2	
	octopunctatum			1	
	placidum			5	
	punctiforme			1	
Calathus (24)	*ambiguus*	1	1		
	fuscipes	4	4	1	
	melanocephalus	10			
	mollis	1			
	opaculus			2	
Dolichus (3)	*halensis*		1		2
Synuchus (2)	*nivalis*	1	1		

continued

Table 2.4. cont.

TRIBE and Genus	Species	Species occurrences			
		N/W Europe	Cent/E Europe	N America	Japan
HARPALINI					
Anisodactylus (7)	*dulcicollis*			1	
	santaecrucis			3	
	signatus		3		
Evarthrus (3)	*alternans*			1	
	seximpressus			1	
	sodalis			1	
Harpalus (75)	*aeneus*	3	1	3	
	ampertatus			2	
	caliginosus			1	
	chalcentus				1
	compar			2	
	distinguendus		3		
	eous				1
	erraticus			2	
	fallax			1	
	griseus				1
	jureceki				2
	micans				1
	pensylvanicus			12	
	rufipes	17	15	3	
	sinicus				1
	tridens				3
Stenolophus (4)	*comma*			4	
Trichocellus (1)	*cognatus*			1	
CHLAENIINI					
Chlaenius (1)	*micans*				1
LEBIINI					
Apristus (1)	*subsulcatus*			1	
Microlestes (1)	*linearis*			1	
BRACHININI					
Brachinus (1)	*explodens*		1		
Pserosophus (1)	*jessoensis*				2

Published sources additional to those used in *Tables 2.2* and *2.3*: Andersen and Eltun (2000); Basedow (1990); Belakova (1962); Boivin and Hance (1994); Cárcamo and Spence (1994); Cárcamo (1995); Cárcamo *et al.* (1995); Coaker and Williams (1963); Desender and Alderweireldt (1990); Dritschilo and Erwin (1982); Duffield and Baker (1990); Dunn (1982); Ekbom and Wiktelius (1985); Fadl and Purvis (1998); Fan *et al.* (1993); Finlayson and Campbell (1976); Frank (1971); French and Elliot (1999); Górny (1971); Grégoire-Wibo (1983); Hance and Grégoire-Wibo (1987); Hicks (1970); Holopainen and Varis (1986); Hsin *et al.* (1979); Huusela-Veistola (1996); Jones (1976); Kegel (1990); Kirk (1971); Kiss *et al.* (1994); Knauer and Stachow (1987); Kromp (1990); Lester and Morrill (1989); Lövei and Sárospataki (1990); Müller (1972); Pauer (1975); Pfiffner and Niggli (1996); Pietraszko and DeClercq (1981); Powell *et al.* (1985); Purvis and Curry (1984); Rivard (1964, 1966); Scheller (1984); Sekulić *et al.* (1987); Stassart and Grégoire-Wibo (1983); Szel *et al.* (1997); Togashi *et al.* (1992); Van Dinther and Mensinck (1971); Wallin (1985); Weiss *et al.* (1990); Yano *et al.* (1995).

most abundant species, sometimes without actual abundances or numbers caught, such data still provides the information need to compile the table of species composition of the dominant agricultural carabids.

No fewer than 119 species in 32 genera occur in this survey of only the five most abundant species in each list. The most frequently occurring genera are *Bembidion* (81 occurrences, 15 species), *Harpalus* (75, 16) and *Pterostichus* (72, 9), followed by *Agonum* (50, 9), *Poecilus* (40, 7), *Trechus* (36, 4) and *Amara* (26, 19). All of these are well known as common genera in European agricultural fields (Thiele, 1977; Luff, 1987; Lövei and Sárospataki, 1990). However, *Table 2.4* shows that species of

Pterostichus, *Harpalus*, *Agonum* and *Amara* are also among the dominant carabids of some fields in both North America and Japan. Some of the commonest European agricultural carabids now occur across the northern hemisphere, so that *Pterostichus melanarius, Bembidion quadrimaculatum, B. lampros, Agonum dorsale, Calathus fuscipes, Harpalus aeneus,* and *H. rufipes* all feature on both sides of the Atlantic. *Bembidion* species are also found in North American agricultural lists, but not in Japan; *Trechus* species do not figure in agricultural lists outside Europe, although the Trechini are well represented worldwide faunistically. The only European species also found in the few Japanese lists seen is *Dolichus halensis*. However, most of the genera common in Japanese agricultural habitats also occur in the Northern Hemisphere, with *Harpalus* species being particularly predominant in the Japanese agricultural fauna. The 28 Australian Carabidae found by Horne and Edward (1997) in crops in Victoria are not included in *Table 2.4*. The only genus common to this and European/American studies was *Clivina*, but species analogous to *Harpalus* and *Pterostichus* were found. In other Australian studies, Horne (1992) showed the life histories and ecology of the genus *Notonomus* in Victoria to be remarkably similar to that of *Pterostichus* in Europe. It should be remembered, however, that all these studies in *Table 2.4* used pitfall trapping, so that genera such as *Demetrias* and *Dromius* (Lebiini) which are largely found up on the vegetation, and which can climb readily out of pitfall traps (until killed by the preservative), are under-represented.

A further feature of interest is the occurrence in *Table 2.4* of species that are not normally associated with agricultural situations. Thus, Holopainen *et al.* (1995) found *Patrobus atrorufus* and *Pterostichus niger* among the three commonest species in spring cereal fields in Finland. Both these are more usually typical of damp, often shaded habitats in UK (Luff, 1998) and the Netherlands (Turin *et al.*, 1991; Turin, 2000). In Alberta, Canada, Frank (1971) recorded *Pterostichus adstrictus* and *Trichocellus cognatus* among the commonest five species in a cereal field, yet in Europe these are boreal species associated mainly with *Calluna* heath and upland moors. But *P. adstrictus* was also found (although not among the dominant five species) in vegetable fields at Tromsø in Norway (Andersen, 1985), so its preferred habitat would seem to be determined more by latitude and climate than by vegetation type.

The overall impression is that the dominant carabid fauna of agricultural crops is restricted to relatively few genera. Most of these have some species that occur in more than one geographical region, or there may be ecologically comparable species in the fauna of each region. The remainder of this section summarizes existing ecological information about many of the taxa listed above as most numerous in agricultural fields. Further information, especially on diets and dispersal, occurs in later chapters of this book.

The Carabini, represented by *Carabus* and *Calosoma* are large (> 15 mm body length) beetles. *Calosoma* (caterpillar hunters) are specialist predators as both larvae and adults on lepidopterous larvae. *Carabus* are generalist predators as both larvae and adults on soft-bodied prey including lepidopterous larvae, earthworms and slugs. Many *Carabus* are flightless: some *Calosoma* species can disperse long distances by flight, while others are confined to movements on the ground and in the crop (Wallin, 1991). Although both spring- and autumn-breeding species of *Carabus* exist, those species found commonly in arable land are spring breeders. Their large size means that in cooler climates they may require more than one season to complete development, with a correspondingly long larval period in the soil. For this reason, they only dominate the arable carabid fauna in regions such as central and eastern Europe, and continental North America where development can be completed in one season. In the more westerly regions of Europe (including Britain) and Japan, Carabini are present, but seldom abundant, in agricultural crops.

A single species of Nebriini, *Nebria brevicollis*, is found commonly in some field situations in north-western Europe, where the soil is heavy and climate is damp. This is a medium-sized (12 mm) autumn-breeding species, more usually associated with grasslands and woodland. It feeds on invertebrates, particularly dipteran larvae. It is fully-winged, but seldom disperses by flight (Nelemans, 1987). Similarly, the sole Notiophiline in *Table 2.4*, *Notiophilus biguttatus*, is a small (5 mm) beetle that is widespread in north and western Europe, but more usually associated with woodlands. It is a spring-breeding predator on small invertebrates such as Collembola. The Loricerine, *Loricera pilicornis* (7 mm) also occurs occasionally among the dominant carabids of UK, Ireland and Scandinavia. It is a typical species of grasslands, preferring crops that are cool or damp. It is a specialist predator on Collembola in both adult and larval stages, breeds in the spring, and flies readily.

The Scaritini are fossorial, predatory carabids, represented by two genera of small beetles, and one of large or very large species. *Clivina* are small (5 mm) cylindrical beetles, of which one species, *C. fossor*, occurs occasionally as a dominant species in both Europe and North America. It is usually subterranean in light soils, probably predatory in all stages, although the related American *C. impressifrons* will feed on germinating seeds when invertebrate prey is scarce (Pausch and Pausch, 1980). *C. fossor* is wing-dimorphic, and probably seldom flies. It appears to breed throughout the season. Several even smaller (2–4 mm) species of the large genus *Dyschirius* occur occasionally among the dominant carabids in eastern Europe and North America. Most are specialist predators on *Bledius* species (Staphylinidae). The third genus, *Scarites*, comprises large (15–30 mm) beetles, with only *S. subterraneus* in North America occurring as a dominant agricultural species. The genus occurs worldwide, however, and has been suggested as a biological control agent in New Zealand grasslands (Cameron and Butcher, 1979). The single Broscine, *Broscus cephalotes,* is a similar large (20 mm) burrowing, predatory beetle that is among the commonest carabids of some agricultural sites in the Netherlands. It is one of a number of species, including *Amara spreta* (Van Dinther and Mensink, 1971) and *Calathus mollis* that are coastal in the UK, but not in mainland Europe (see Luff and Eyre, 2000).

The Trechini, with the large genus *Trechus*, are small-sized (3–5 mm) species, of which the commonest agricultural species is *T. quadristriatus.* The genus is mainly found as a common part of field carabid faunas in northern and western Europe. In Scandinavia, *T. secalis* also occurs regularly in field lists. Trechini are predatory in all stages, autumn–winter breeding and can sometimes fly, although *T. quadristriatus* is wing-dimorphic, and *T. secalis* brachypterous.

The Bembidiini comprise mainly the very large genus *Bembidion*, small (3–5 mm) species, the most widespread of which in fields is *B. lampros. B. quadrimaculatum* is often dominant both in north-west Europe and in North America, where it occurs as the sub-species *oppositum.* The very small *B. obtusum* also occurs regularly in northern Europe. *Bembidion* are predatory in all stages, and breed in the spring, or even earlier in the winter (*B. obtusum*). Most (but not all) can fly; *B. obtusum* is wing-dimorphic. *Bembidion* species tend to be most active in crops with little shade and some bare ground.

The most frequently found dominant agricultural ground beetles are the Pterostichini, with 112 occurrences among the five most abundantly caught species. *Poecilus*, formerly a sub-genus of *Pterostichus*, is represented primarily by *P. cupreus* in much of Europe (especially in the east), but with several species, especially *P. sericeus*, also occurring among the commonest species regularly in central/eastern Europe. In North America, the genus is represented by the closely related *P. chalcites* and *P. lucublandus*. They are moderately large (11–13 mm), usually bright metallic-coloured and spring-breeding species, predatory in all stages. They are spring/summer breeders, capable of flight, and found in warm, often sunny conditions, hence their predominance in more continental climates.

Pterostichus itself is a large genus, occurring among the dominant carabid faunas of arable land throughout Europe, North America and, in one case, also Japan. They are mainly large (12–16 mm) beetles; although smaller species occur, they are not normally among the dominant species in arable fields. *P. melanarius* (= *P. vulgaris*) is one of the commonest agricultural carabids in both Europe and N. America, where it has been introduced. It is in competition there with the local Carabidae, and this may have considerable impact on the native fauna (Spence and Spence, 1988), although it has not affected the forest carabid species (Niemelä and Spence, 1999). *P. melanarius* is predatory in all stages, summer–autumn breeding (and partially biennial at the northern limits of its range). It is seldom capable of flight, most individuals being brachypterous, and occurs on all types of soils, and in grasslands as well as arable habitats. In Europe, *P. melanarius* can excessively dominate the carabid assemblage (see the example in the previous section). When it is very numerous, the diversity of other (smaller) species appears to decline, possibly because they are also preyed on by *P. melanarius*, or due to losing out in competition for prey. *Pterostichus* species tend to be omnivores, feeding mainly as predators but also eating already dead prey, as well as even including some plant material in the diet.

The Amarini include the most species-rich genus in *Table 2.4*, namely *Amara*, with 19 species represented. Yet none of these was found in more than one geographical region, and all the 10 dominant species in European lists only occur once each. The only species that seem to be regular rather than casual dominants are the American *A. cupreolata* and *A. obesa*. *Amara* are small–medium sized (6–11 mm) carabids. Some larger species, such as *A. aulica*, now in the genus *Curtonotus,* do not figure in the list of 'dominant' agricultural species. Their larvae are primarily spermophagous, but adults will eat both plant and invertebrate food. Most species are typically found in open, often sunny situations, with a high proportion of bare, often weedy soil. Long distance dispersal is by flight. Both spring- and autumn-breeding species occur. Also in the same tribe is the genus *Zabrus*. Although not in the commonest five species in any of the lists used in *Table 2.4*, *Z. tenebrioides* is a spring-breeding, phytophagous carabid whose larvae and adults feed on young seedlings of grasses and cereals. It can be a serious pest in eastern Europe, through to Russia and central Asia.

The Platynini were previously considered part of the Pterostichini by European workers. The tribe includes the genus *Agonum*, medium sized (6–10 mm) species, well represented among the dominant field carabids in all regions in *Table 2.4*. The genus is often divided into different genera, such as *Anchomenus* and *Platynus*, in continental literature. They are predatory as both larvae and adults, and all are capable of flight, as far as is known. They are usually spring breeders. The commonest European species, which has also been introduced to N. America, is *A. dorsale*, typical of light soils in warmer situations. It is well known as an aphid feeder in north and west Europe. *A. muelleri* is found in cooler sites with more shade, or on heavier soils, including in America, but the commonest N. American species is the native *A. placidum*. Two species are also listed among the dominant Japanese arable carabids.

Calathus are medium–large sized (6–12 mm) species usually common in grasslands, but occurring in arable crops, especially when more shaded, or on lighter soils. The commoner European species, *C. fuscipes* and *C. melanocephalus*, are autumn breeders, but the American *C. opaculus* breeds earlier in the season. They are predatory in all stages. *C. fuscipes* has reduced wings and is therefore flightless, but other species of the genus are often polymorphic for wing size, and flight can occur. Also in the Platynini is *Dolichus halensis*, a diurnal medium-sized species that breeds in the summer. It occurs in agricultural fields from southern and eastern Europe across central Asia to China and Japan.

The Harpalini are well represented in all four regions in *Table 2.4*. The genera *Anisodactylus* and *Harpalus* are moderately large (8–12 mm), mostly nocturnal beetles, but some metallic-coloured species in both genera can be diurnal. These are omnivorous species that will feed as adults on both seeds and invertebrates. Their larvae are primarily spermophagous. Both *A. signatus* in eastern Europe and *A. santaecrucis* in N. America are often among the dominant Carabidae in agricultural crops. Species of *Harpalus* occur commonly in crops in all regions studied. In both western and eastern Europe, the predominant species is *H. rufipes* (often treated in the genus *Pseudophonus*, sometimes under the pseudonym *pubescens*), which has also been introduced into N. America. However, the closely related *H. pensylvanicus* is more often dominant there, as well as *H. erraticus*. *H. aeneus* (= *affinis*) also occurs not uncommonly both in Europe and N. America. Most species of *Harpalus* can fly. They are commonest on light soils with open, warm conditions. *H. rufipes* and *H. pensylvanicus* breed primarily in late summer, but *H. aeneus* breeds from early summer onwards. Both *Anisodactylus* and *Harpalus* seem somewhat flexible in their phenology; thus *H. rufipes* may switch from annual life cycles in the south to being partly or wholly biennial further north (Luff, 1980), and *A. signatus* is effectively biennial in Hungary (Fazekas *et al.*, 1997). As many as seven species of *Harpalus* occur among the dominant Japanese agricultural ground beetles, making this the commonest genus in their dominant fauna.

Two further harpaline genera occur among the N. American common carabids in particular. *Evarthrus*, with three species each occurring once in the list, are medium-sized species occurring in the southern and central parts of N. America. They are midsummer breeders. A further common American species is *Stenolophus comma* (formerly in the genus *Agonoderus*), a smaller (5–7 mm), spring-breeding species that flies readily.

A single species of Chlaeniine, *Chlaenius micans*, has been recorded as a dominant Japanese species; these are medium to moderately large (10–16 mm) predatory carabids, mostly associated with water, and hence pre-adapted to the conditions in paddy fields. In the Lebiini, the American dominant carabids include one instance each of *Apristus* and *Microlestes*, both small (< 5 mm) predatory genera about whose biology little is known. In western Europe, the Lebiine *Demetrias atricapillus* may occur commonly in cereal crops although, as mentioned earlier, it is less frequently caught in pitfall traps. It is a small (5 mm) species that climbs plants readily to feed on soft prey such as aphids on the plants. It is spring breeding, overwintering in grass tussocks and litter adjacent to the crop.

Finally, two species of bombardier beetles (Brachinini) have occurred as dominants in field crops, *Brachinus explodens* in eastern Europe and *Pserosophus jessoensis* in Japan. These are predatory as adults, but are believed to have ectoparasitic larvae living in association with immobile prey such as beetle pupae.

Body size distributions

The range of body sizes in the species discussed above extends from above 20 mm (*Megacephala*, *Calosoma*, some *Carabus* species) to less than 3 mm (some *Bembidion* species, *Microlestes*). As the size of any predator is closely related to the size of its preferred prey (Wheater, 1988), the distribution of body sizes among the carabid assemblage will have a direct impact on the role of that assemblage in pest management. As discussed above, *Carabus* species are intolerant of disturbance in the larval stages, so that they are only abundant in arable fields in warmer climates where their development can be completed within one year. This is part of a general tendency for the mean body size of arable carabids to be smaller in the cooler, more northern regions, where *Bembidion* and *Trechus* predominate. Examples are fields sampled by Andersen (1982) in Norway, where 77% of 609 carabids caught were less than 5 mm in length; Szwejda (1984) in Poland, 70.3% of 2209; Kennedy (1994) in Ireland, 66% of 2795 (increasing to 95% of 1120 in just winter samples of the previous year); Wallin (1985) in Sweden, at least 57% of 18,056. In contrast, more southern (in the Northern Hemisphere) or continental assemblages may have high proportions of larger species. Examples are Varvara *et al.* (1989) in Romania, with 96% of 3182 caught being larger than 10 mm; Ishitani *et al.* (1994) in Japan, 76% of 1427; House and All (1981) in south-eastern USA, 45% of 2003, 25% of which were even larger than 20 mm. These are, however, selected cases, and a more detailed comparative analysis, outside the scope of this chapter, would be needed to prove any statistically significant geographical trends in body size.

Šustek (1987) and Blake *et al.* (1994) have proposed that carabid body size is an indicator of habitat disturbance. The more disturbed habitats have a carabid fauna of smaller average body size. In arable crops, some measure of disturbance is always present, although intensity of management will affect this. The mean body size measure as used by Blake *et al.* (1994) will also be heavily biased by a predominance of a single large species, such as occurs with *Pterostichus melanarius* (see above). Body length of individual species has also been used as an indicator of crop management intensity and this declined over time as the quality of the habitat deteriorated (Köhler and Stumpf, 1992), again probably because of increased disturbance. Büchs *et al.* (1997) also used this technique but only found an increase in body size where no pesticides and little fertilizer were used, compared to a range of higher input levels.

Effects of environmental factors on assemblage composition

The aim of this section is to review some studies that have considered the effect of environmental factors of the overall make-up or composition of the carabid assemblages in agricultural systems. Work that has considered individual species is covered at length by Holland and Luff (2000), as well as later in this book. The approach used is usually to carry out multivariate analyses of the site (or field)–species–environment matrix, often resulting in an ordering of sites/fields and species in two or more dimensions that can be represented by two dimensional ordination plots. Early analyses have used principal components analysis (PCA), e.g. Booij and Noorlander, (1988), which assumes linear relationships between species' responses and the ordination axis scores (Ter Braak, 1988). The more widely used ordination method is correspondence analysis (CA), often detrended (DCA) to remove any 'arch effect' where the 2nd axis scores are a non-linear function of the 1st axis scores (Hill, 1979). CA (and DCA) are based on a unimodal response model between each species and the axis scores. They also enable plotting of both sample and species scores on the same ordination axes.

Effects of environmental factors can be tested by relating them to the ordination axis scores. This is 'indirect gradient analysis', in which the environmental data is not itself incorporated into the ordination analyses. The alternative is 'direct gradient analysis', which incorporates the environmental data into the ordinations as constraining variables. This results in canonical correspondence analysis (CCA or DCCA, with or without detrending, respectively) (Ter Braak, 1986). The ordination axes are constrained to be linear combinations of the environmental variables that maximize the separation of the species scores. Particular environmental variables can be eliminated from the analysis by incorporating them as covariables; this results in partial CCA (here referred to as pCCA) (Ter Braak, 1988).

There have been many such analyses of the carabid fauna of grasslands (e.g. Siepel *et al.*, 1989; Asteraki *et al.*, 1992; Blake *et al.*, 1996; review in Luff, 1996). These techniques have been less extensively used in agricultural situations, but *Table 2.5* summarizes 10 studies in relation to the factors being studied. This serves to show the diversity of the few analyses that have been done on agricultural carabids. Only Armstrong (1995) considered samples from more than one locality, and the locality effects overrode any effect of cropping system (organic versus conventional). Within one locality, but sampling over more than one season, year effects were found to be less than the differences in beetle assemblage from crop to crop by Booij and Noorlander (1988) and Hance and Grégoire-Wibo (1987), but more important than crop by French and Elliott (1999) and Sanderson (1994). It must be remembered that, in several cases, the various environmental factors themselves are correlated. This is indicated in the biplots generated by CCA, but their effects cannot be fully separated other than by successive partialling out of each variable by pCCA. Only two of the studies in *Table 2.5* have done this.

Luff and Sanderson (1992) attempted to analyse a very large dataset of Carabidae from 149 experimental cereal trials provided by eight different institutions in England. The full data comprised 77 species of ground beetles. The mean starting and ending dates of sampling were 10th April and 27th July respectively, so only pitfall catches between these dates were included; this reduced the species list by two.

Table 2.5. Summary of 10 studies using ordination analyses of Carabidae in agricultural environments. Factors are listed in decreasing order of their perceived effect on the assemblage. Separate analyses are separated by '/'.

Reference	Country	Analysis	Crop(s)	Factors analysed, in decreasing order of effect
Armstrong (1995)	Scotland	DCA	Potatoes	Locality, cropping system
Asteraki (1994)	England	DCA	Various margins	Unsown botanical diversity, margin age
Boivin and Hance (1994)	Canada	CA	Carrots	Crop vs. uncultivated, soil moisture
Booij and Noorlander (1988)	Netherlands	PCA	Various	Crop, cropping system, year
Desender and Alderweireldt (1990)	Belgium	DCA	Grass, maize	Crop, rotation vs. monoculture
French and Elliot (1999)	USA	CCA / pCCA	Grass, wheat	Season, year / crop, edge vs. centre
Hance and Grégoire-Wibo (1987)	Belgium	CA	Various	Crop, year / organic matter, insecticide level / crop, previous crop
Holopainen *et al.* (1995)	Finland	CCA	Cereals	Soil type, soil moisture, foliar fluoride level
Kiss *et al.* (1994)	Hungary	CA	Wheat	Edge vs. centre
Sanderson (1994)	England	pCCA	Various	Year, cropping system, field

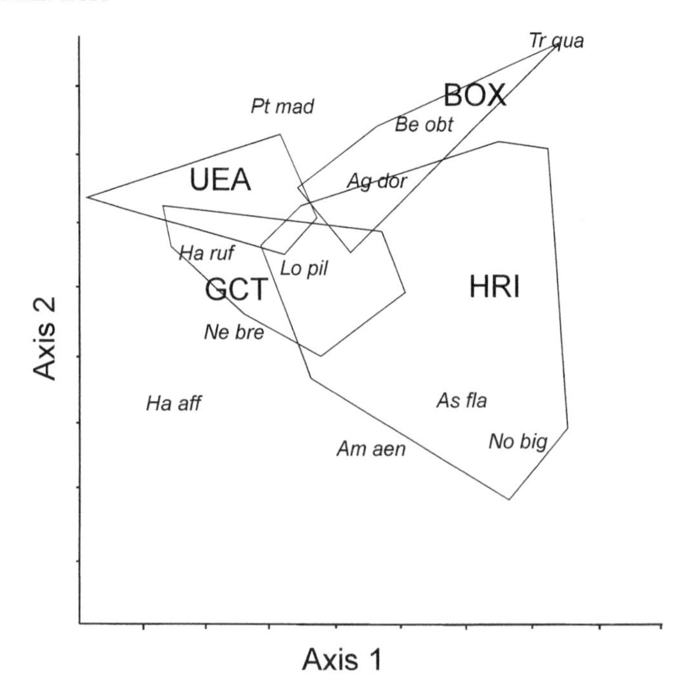

Figure 2.9. DCA ordination of experimental cereal trials provided by eight different institutions in England. Individual trials are not shown; polygons enclose those of the four major institutions that comprised 126 of the total 149 trials. Selected common species are shown; see text for details. Institution codes: BOX = ADAS Boxworth; GCT = Game Conservancy Trust; HRI = Horticultural Research Institute; UEA = University of East Anglia (simplified from Luff and Sanderson, 1992).

Because of the limited environmental data that were in common for the whole dataset, only sampling year and institute (an approximate proxy for sampling locality) were included in CCA. A simplified DCA ordination of samples (*Figure 2.9*) shows that the sample points largely separate out according to the institutes that provided the data. The common species that also contributed substantially to the ordination are shown. Species predominant in any one institute's samples overly that area of the ordination space. Thus, *Trechus quadristriatus* and *Bembidion obtusum* occurred predominantly in the Boxworth data, *Asaphidion flavipes* and *Notiophilus biguttatus* in samples from the Horticultural Research Institute. CCA of the same dataset suggested that year effects were even more important than the institute, although year and institute were themselves strongly correlated, as different institutes carried out their work over different, only partially overlapping, periods of time.

More detailed large-scale ordination analyses were carried out by Gardner *et al.* (1999) using data from the SCARAB and TALISMAN experiments referred to earlier, together with the LINK IFS (LINK Integrated Farming Systems) experiment which sampled Carabidae from six farms using a split field design of paired or quartered fields. Conventional and integrated farming systems were applied to each field. Analyses of variance were used to test the environmental factors against ordination axis scores (indirect gradient analyses – see above), whilst DCCA/CCA (as appropriate for each dataset), together with partialling out of variables as co-variates, was used in direct gradient analyses. A full account is in preparation for

Table 2.6. Summary of significant environmental variables in analyses of variance of axis 1 and axis 2 CA scores from MAFF SCARAB, TALISMAN and LINK IFS large-scale experiments (from data in Gardner *et al.*, 1999).

Experiment	Variables analysed	Axis 1	Axis2
SCARAB	C S Y	C*, Y*, SY*, CSY**	C**
TALISMANBoxworth only	I T Y	T**, Y*, IY*	Y**, I*, IY**
TALISMANWheat, 2 farms only	I S T Y	Y**	I*
LINK IFS	C S T Y	C**	C**, Y**, CY**, CSY**

Variables: C = crop; I = crop timing (autumn or spring sown); S = site; T = treatment; Y = year. Interactions are shown by two or more variables together (e.g. CY = crop × year)
Significance levels: * = P < 0.05; ** = P < 0.01

Table 2.7. Percentages of the total variance explained by each environmental variable, and shared with co-variables in CCA analyses from MAFF SCARAB, TALISMAN and LINK IFS large-scale experiments (from data in Gardner *et al.*, 1999).

Experiment	Variable	Variance explained	Co-variable	Variance shared
SCARAB	Site	24.3	Crop	6.3
	Crop	19.1	Timing	7.0
	Year	15.7		
	Timing	12.9		
	Treatment	1.0		
TALISMAN	Year	31.4	Crop	5.2
	Site	24.5	Treatment	2.6
	Crop	19.5		
	Treatment	9.4		
	Timing	3.8		
LINK IFS	Crop	19.9	Timing	7.6
	Site	12.3	Year	2.0
	Timing	9.1	Site	3.0
	Year	8.9		
	Treatment	0.4		

publication, but *Tables 2.6–2.7* summarize the significant effects and their ranking in each experiment and analysis.

Indirect analyses using ANOVA (*Table 2.6*) show the significant effects to be year of sampling (five cases), crop (four cases), timing (spring vs. autumn sown – two cases) and treatment (one case). Only the last is a within-field effect, all the others differed between fields. The sampling site (i.e. farm) was only significant as an interaction term together with crop and year. These results suggest that the carabid assemblages were affected primarily by the particular crop grown in each field in each year, and less by what farm they occurred on, or by the pesticide or other treatment variation between plots in each field.

A summary of the DCCA/CCA analyses is given in *Table 2.7*. The SCARAB analyses indicated that of the main environmental variables, site explained the largest percentage of the variance in the carabid dataset. Part of this variance was also shared with crop. Year and time of sowing explained around 16% and 13% of the variance respectively when fitted as single factors. However, 7% of the timing effect was also shared with crop. Several of the crops, potatoes, sugar beet, beans, were only sown in the spring, thus in this analysis the factor timing was confounded with differences between crops, as well as different sowing dates.

In the TALISMAN analyses, the five main environmental variables, site, crop, year, treatment and timing, each accounted for a significant percentage of the variance associated with the carabid sample ordination. Year had the largest individual effect (over 30% of the variance), although approximately 5% of this was shared with crop. Similarly, around 3% of the variance accounted for by treatment was shared with site. This suggests that the treatment effect varied from site to site; *Table 2.6* shows that ANOVA of the data from one of the three farms (Boxworth) showed a significant effect of treatment on axis 1 CA scores.

Results from the IFS analyses indicated that, individually, crop accounted for the largest proportion of the overall variance (20%), although 7% of this was shared with crop timing. Crop and timing were confounded in this experiment and could not be separated as virtually no crops were sown in both autumn and spring. Site was not significant in explaining the overall variation in the dataset, although it was significant with respect to the first ordination axis. When fitted as a co-variable with year, the explanatory power of year was increased, suggesting a strong interaction between year and site. The treatment effect was extremely small.

Within these three datasets, site, crop, year and time of sowing were all important in determining the relative abundance of carabid species present. For SCARAB and TALISMAN, between-site differences, including to some extent variation in cropping regime, were the principal factors associated with the first ordination axis. Although no specific measurements were taken, it is likely that the different soil types present at each site contributed largely to these differences. Among the SCARAB sites, for example, one (Drayton) occurs on heavy clay soils compared to the other two. For the LINK IFS dataset, crop accounted for a significant proportion of the variation associated with ordination Axis 1. Crop, or its interactions, also accounted for a significant proportion of the variation associated with the second ordination axis in each of the three datasets. This result accords with those from other studies where crop type has been seen to have a greater influence on the density and diversity of Carabidae than the farming system applied to that crop (Booij, 1994; Holland *et al.*, 1998). Interpretation of such data is often complicated by variation in crop rotations applied at each site in different years (Jones, 1979; Sanderson, 1994; Holland *et al.*, 1998). This can result in confounding between crop, site and year effects, which in some cases, most notably TALISMAN in this study, make it impossible to relate the observed variation in the carabid assemblage to specific factors or their interactions.

Conclusions

Carabid ground beetles are a significant part of the fauna of the agricultural environment throughout temperate ecosystems. Most of our knowledge of their species richness and assemblage composition has been derived from pitfall trapping; this should be supplemented wherever possible by further methods of sampling, as outlined in Sunderland *et al.* (1995). Despite the limitations of such trapping, and considerable variations in species richness on a crop, field, farm and geographical scale, a mean species richness of about 30 carabid species applies across most areas studied. Species richness and 'diversity' are usually highly correlated. Many carabid assemblages from farming systems are highly dominated by a few excessively numerous species, so that the five most dominant species usually account for about 85% of the total catch, and about 10 species comprise 95%. The effect of intensive management will generally reduce carabid diversity and species richness, although contrasting effects on individual species may mask some overall changes.

The temperate agricultural carabid fauna is dominated by relatively few genera: *Bembidion*, *Pterostichus* (including *Poecilus*), *Harpalus*, *Agonum*, *Trechus* and *Amara* are the dominant genera, despite differences in species from region to region. Most of these are largely or wholly predatory, but *Pterostichus* species may also be scavengers, and both *Harpalus* and *Amara* are partly seed feeders, either as adults or larvae. The species composition at any particular farm site appears to be determined by geographical position, local soil and crop conditions, and seasonal variability; small-scale management effects are usually subsidiary to these main factors, which make prediction of any particular carabid assemblage difficult or impossible. Only actual sampling in the field can answer the question as to what ground beetles are present in the crop, and hence to further assessment of their potential benefit in that system.

Acknowledgements

I am grateful to Dr John Holland and Dr Henrik Wallin for helpful comments and suggestions on drafts of this chapter. Dr Sarah Gardner kindly allowed results from analyses of the MAFF experiments to be included. Species accretion models were calculated using the computer program 'EstimateS', provided by Dr Robert Colwell, University of Connecticut, at http://viceroy.eeb.uconn.edu/estimates.

References

ADIS, J. (1979). Problems of interpreting arthropod sampling with pitfall traps. *Zoologischer Anzeiger Jena* **202**, 177–184.

AEBISCHER, N.J. (1991). Twenty years of monitoring invertebrates and weeds in cereal fields in Sussex. In: *The ecology of temperate cereal fields*. Eds. L.G. Firbank, N. Carter, J.F. Darbyshire and G.R. Potts, pp 305–331. Oxford: Blackwell.

ALDERWEIRELDT, M. AND DESENDER, K. (1994). Belgian carabidological research on high-input agricultural fields and pastures: a review. In: *Carabid beetles: ecology and evolution*. Eds. K. Desender, M. Dufrene, M. Loreau, M.L. Luff and J.P. Maelfait, pp 409–415. Dordrecht: Kluwer.

ALLEN, R.T. (1979), The occurrence and importance of ground beetles in agricultural and surrounding habitats. In: *Carabid beetles: their evolution, natural history and classification*. Eds. T.L. Erwin, G.E. Ball and D.R. Whitehead, pp 485–507. The Hague: Junk.

ANDERSEN, A. (1982). Carabidae and Staphylinidae (Col.) in swede and cauliflower fields in south-eastern Norway. *Fauna norvegica, Series B* **29**, 49–61.

ANDERSEN, A. (1985). Carabidae and Staphylinidae (Col.) in swede and carrot fields in northern and south-western Norway. *Fauna norvegica, Series B* **32**, 12–27.

ANDERSEN, A. AND ELTUN, R. (2000). Long-term developments in the carabid and staphylinid (Col., Carabidae and Staphylinidae) fauna during conversion from conventional to biological farming. *Journal of Applied Entomology* **124**, 51–56.

ARMSTRONG, G. (1995). Carabid beetle (Coleoptera: Carabidae) diversity and abundance in organic potatoes and conventionally grown seed potatoes in the north of Scotland. *Pedobiologia* **39**, 231–237.

ASTERAKI, E. (1994). The carabid fauna of sown conservation margins around arable fields. In: *Carabid beetles: ecology and evolution*. Eds. K. Desender, M. Dufrene, M. Loreau, M.L. Luff and J.P. Maelfait, pp 229–233. Dordrecht: Kluwer.

ASTERAKI, E.J., HANKS, C.B. AND CLEMENTS, R.O. (1992). The impact of chemical removal of the hedge-base flora on the community structure of carabid beetles (Col., Carabidae) and spiders (Araneae) of the field and hedge bottom. *Journal of Applied Entomology* **113**, 398–406.

BAARS, M.A. (1979). Patterns of movement of radioactive carabid beetles. *Oecologia* **44,** 125–140.

BALDUF, W.W. (1935). *The bionomics of entomophagous Coleoptera*. Hampton: Classey (1969 reprint edition).

BARNEY, R.J. AND PASS, B.C. (1986). Ground beetle (Coleoptera: Carabidae) populations in Kentucky alfalfa and influence of tillage. *Journal of Economic Entomology* **79**, 511–517.

BASEDOW, T. (1990). Effects of insecticides on Carabidae and the significance of these effects for agriculture and species number. In: *The role of ground beetles in ecological and environmental studies*. Ed. N.E. Stork, pp 115–125. Andover: Intercept.

BELAKOVA, A. (1962). Carabidenfauna der drei feldwirtschaftlichen Kulturen (Rüben, Winterweizen und Luzern). *Acta Facultatis Rerum Naturalium Universitatis Comenianae, Zoologia* **7**, 94–118. [In Romanian, German summary.]

BISHOP, A.L. AND BLOOD, P.R.B. (1980). Arthropod ground strata composition of the cotton ecosystem in south-eastern Queensland, and the effect of some control strategies. *Australian Journal of Zoology* **28**, 693–697.

BLAKE, S., FOSTER, G.N., EYRE, M.D. AND LUFF, M.L. (1994). Effects of habitat type and grassland management practices on the body size distribution of carabid beetles. *Pedobiologia* **38**, 502–512.

BLAKE, S., FOSTER, G.N., FISHER, G.E.J. AND LIGERTWOOD, G.L. (1996). Effects of management practices on the carabid faunas of newly established wildflower meadows in southern Scotland. *Annales Zoologici Fennici* **33**, 139–147.

BOITEAU, G. (1983). Activity and distribution of Carabidae, Arachnida and Staphylinidae in New Brunswick potato fields. *Canadian Entomologist* **115**, 1023–1030.

BOIVIN, G. AND HANCE, T. (1994). Phenology and distribution of carabid beetles (Coleoptera: Carabidae) in muck-grown carrots in southwestern Quebec. In: *Carabid beetles: ecology and evolution*. Eds. K. Desender, M. Dufrene, M. Loreau, M.L. Luff and J.P. Maelfait, pp 417–424. Dordrecht: Kluwer.

BOOIJ, C.H.J. AND NOORLANDER, J. (1988). Effects of pesticide use and farm management on carabids in arable crops. *British Crop Protection Council Monographs* **40**, 119–126.

BOOIJ, C.H.J. AND NOORLANDER, J. (1992). Farming systems and insect predators. *Agriculture, Ecosystems and Environment* **40**, 125–135.

BOOIJ, K. (1994). Diversity patterns in carabid assemblages in relation to crops and farming systems. In: *Carabid beetles: ecology and evolution*. Eds. K. Desender, M. Dufrene, M. Loreau, M.L. Luff and J.P. Maelfait, pp 425–431. Dordrecht: Kluwer.

BORG, A.K.E. (1973). The occurrence of carabids in a strawberry field (Col. Carabidae). *Entomologiske Tidskrift* **94**, 56–58. [In Swedish, English summary.]

BRUST, G.E. (1994). Natural enemies in straw-mulch reduce Colorado potato beetle populations and damage in potatoes. *Biological Control* **4**, 163–169.

BRUST, G.E., STINNER, B.R. AND MCCARTNEY, D.A. (1985). Tillage and soil insecticides effects on predator-black cutworm (Lepidoptera: Noctuidae) interactions in corn agroecosystems. *Journal of Economic Entomology* **78**, 1389–1392.

BÜCHS, W., HARENBERG, A. AND ZIMMERMANN, J. (1997). The invertebrate ecology of farmland as a mirror of the intensity of the impact of man? – an approach to interpreting results of field experiments carried out in different crop management intensities of a sugar beet and an oil seed rape rotation including set-aside. *Biological Agriculture and Horticulture* **15**, 83–108.

CAMERON, P.J. AND BUTCHER, C.F. (1979). Pitfall trapping and soil sampling for Scarabaeidae (Coleoptera) and possible predators in some Northland and Auckland pastures. *Proceedings of the 2nd Australasian Conference on Grassland Invertebrate Ecology*, pp 113–117.

CÁRCAMO, H.A. (1995). Effect of tillage on ground beetles (Coleoptera: Carabidae): a farm-scale study in central Alberta. *Canadian Entomologist* **127**, 631–639.

CÁRCAMO, H.A. AND SPENCE, J.R. (1994). Crop type effects on the activity and distribution of ground beetles. *Environmental Entomology* **23**, 684–692.

CÁRCAMO, H.A., NIEMELÄ, J.K. AND SPENCE, J.R. (1995). Farming and ground beetles: effects of agronomic practice on populations and community structure. *Canadian Entomologist* **127**, 123–140.

CHAO, A. (1984). Non-parametric estimation of the number of classes in a population. *Scandinavian Journal of Statistics* **11**, 265–270.

CHAZDON, R.L., COLWELL, R.K., DENSLOW, J.S. AND GUARIGUATA, M.R. (1998). Statistical methods for estimating species richness of woody regeneration in primary and secondary rain forests of NE Costa Rica. In: *Forest biodiversity research, monitoring and modeling: conceptual background and Old World case studies*. Eds. F. Dallmeier and J.A. Comiskey, pp 285–309. Paris: Parthenon Publishing.

COAKER, T.H. AND WILLIAMS, D.A. (1963). The importance of some Carabidae and Staphylinidae as predators of the cabbage root fly, *Erioischia brassicae* (Bouché). *Entomologia Experimentalis et Applicata* **6**, 156–164.

COLEMAN, B.D., MARES, M.A., WILLIG, M.R. AND HSIEH, Y.-H. (1982). Randomness, area, and species richness. *Ecology* **63**, 1121–1133.

COLWELL, R.K. (1999). *User's guide to EstimateS 5. Statistical estimation of species richness and shared species from samples*. Storrs, University of Connecticut.

COLWELL, R.K. AND CODDINGTON, J.A. (1994). Estimating terrestrial diversity through extrapolation. *Philosophical Transactions of the Royal Society (Series B)* **345**, 101–118.

DAVIES, T.G. (1963). Observations on the ground beetle fauna of Brassica crops. *Plant Pathology* **12**, 7–11.

DESENDER, K. AND ALDERWEIRELDT, M. (1990). The carabid fauna of maize fields under different rotation regimes. *Mededelingen van de Fakulteit der Landbouwwetenschappen der Rijksuniversiteit te Gent* **55**, 493–500.

DESENDER, K. AND MAELFAIT, J.P. (1986). Pitfall trapping within enclosures: a method for estimating the relationship between the abundances of co-existing carabid species (Coleoptera: Carabidae). *Holarctic Ecology* **9**, 245–250.

DIGWEED, S.C., CURRIE, C.R., CÁRCAMO, H.A. AND SPENCE, J.R. (1995). Digging out the 'digging-in effect' of pitfall traps: Influences of depletion and disturbance on catches of ground beetles (Coleoptera: Carabidae). *Pedobiologia* **39**, 561–576.

DRITSCHILO, W. AND ERWIN, T.L. (1982). Responses in abundance and diversity of cornfield carabid communities to differences in farm practices. *Ecology* **63**, 900–904.

DUFFIELD, S.J. AND BAKER, S.E. (1990). Spatial and temporal effects of dimethoate use on populations of Carabidae and their prey in winter wheat. In: *The role of ground beetles in ecological and environmental studies*. Ed. N.E. Stork, pp 95–104. Andover: Intercept.

DUNN, G.A. (1982). Ground beetles (Coleoptera: Carabidae) collected by pitfall trapping in Michigan small grain fields. *Great Lakes Entomologist* **15**, 37–38.

EKBOM, B.S. AND WIKTELIUS, S. (1985). Polyphagous arthropod predators in cereal crops in central Sweden, 1979–1982. *Zeitschritft für angewandte Entomologie* **99**, 433–442.

ELLSBURY, M.M., POWELL, J.E., FORCELLA, F., WOODSON, W.D., CLAY, S.A. AND RIEDELL, W.E. (1998). Diversity and dominant species of ground beetle assemblages (Coleoptera: Carabidae) in crop rotation and chemical input systems for the Northern Great Plains. *Annals of the Entomological Society of America* **91**, 619–625.

FADL, A. AND PURVIS, G. (1998). Field observations on the lifecycles and seasonal activity patterns of temperate carabid beetles (Coleoptera: Carabidae) inhabiting arable land. *Pedobiologia* **42**, 171–183.

FAN, Y., LIEBMAN, M., GRODEN, E. AND ALFORD, A.R. (1993). Abundance of carabid beetles and other ground-dwelling arthropods in conventional versus low-input bean cropping systems. *Agriculture, Ecosystems and Environment* **43**, 127–139.

FAZEKAS, J., KÁDÁR, F., SÁROSPATAKI, M. AND LÖVEI, G.L. (1997). Seasonal activity, age structure and egg production of the ground beetle *Anisodactylus signatus* (Coleoptera: Carabidae) in Hungary. *European Journal of Entomology* **94**, 473–484.

FERGUSON, H.J. AND MCPHERSON, R.M. (1985). Abundance and diversity of adult Carabidae in four soybean cropping systems in Virginia. *Journal of Entomological Science* **20**, 163–171.

FINLAYSON, D.G. AND CAMPBELL, C.J. (1976). Carabid and staphylinid beetles from agricultural land in the Lower Fraser Valley, British Columbia. *Journal of the Entomological Society of British Columbia* **73**, 10–20.

FRANK, J.H. (1971). Carabidae (Coleoptera) of an arable field in central Alberta. *Quaestiones Entomologicae* **7**, 237–252.

FRENCH, B.W. AND ELLIOTT, N.C. (1999). Temporal and spatial distribution of ground beetle (Coleoptera: Carabidae) assemblages in grasslands and adjacent wheat fields. *Pedobiologia* **43**, 73–84.

GARDNER, S.M., LUFF, M.L., RIDING, A. AND HOLLAND, J.M. (1999). *Evaluation of carabid beetle populations as indicators of normal field ecosystems*. MAFF Project Report. MAFF, London.

GAUTSCH, VON O., MUNGENAST, F. AND THALER, K. (1980). Carabidae (Insecta, Coleoptera) im Kulturland des Innsbrucker Mittelgebirges (900m NN, Nordtirol, Österreich). *Anzeiger für Schädlingskunde, Pflanzen- und Umweltschutz* **53**, 149–155.

GEILER, H. (1967). Die Coleopteren des Luzerne-Epigaions von Nordwestsachsen. *Faunistiche Abhandlunden der Staatliches Museum für Tierkunde in Dresden* **2**, 19–36.

GÓRNY, M. (1971). Untersuchungen über die Laukäfer (Col., Carabidae) der Feldschutzhecke und angrenzenden Feldkulturen. *Polskie Pismo Entomologiczne* **41**, 387–415. [In Polish, German summary.]

GRÉGOIRE-WIBO, C. (1983). Incidences écologiques de traitements phytosanitaires en culture de betterave sucrière. II. Acariens, Polydesmes, Cryptophagides et Carabides. *Pedobiologia* **25**, 93–108.

HALL, D.W. (1991). The environmental hazard of ethylene glycol in insect pit-fall traps. *Coleopterists' Bulletin* **45**, 193–194.

HALSALL, N.B. AND WRATTEN, S.D. (1988). The efficiency of pitfall trapping for polyphagous predatory Carabidae. *Ecological Entomology* **13**, 293–299.

HANCE, T. AND GRÉGOIRE-WIBO, C. (1987). Effect of agricultural practices on carabid populations. *Acta Phytopathologia et Entomologica Hungarica* **22**, 147–160.

HANCE, T., GRÉGOIRE-WIBO, C. AND LEBRUN, P. (1990). Agriculture and ground-beetles populations. *Pedobiologia* **34**, 337–346.

HAWTHORNE, A. (1995). Validation of the use of pitfall traps to study carabid populations in cereal field headlands. In: *Arthropod natural enemies in arable land. I. Density, spatial heterogeneity and dispersal.* Eds. S. Toft and W. Riedel, pp 61–75. Aarhus: Aarhus Unversity Press.

HEIMBACH, U. AND GARBE, V. (1995). Effects of reduced tillage systems in sugar beet on predatory and pest arthropods. *Acta Jutlandica* **71**, 195–208.

HELTSHE, J. AND FORRESTER, N.E. (1983). Estimating species richness using the jackknife procedure. *Biometrics* **39**, 1–11.

HICKS, S.D. (1970). 3.008 causal [*sic*] and pitfall trap collections of ground beetles from three locations in the Ottawa district, Ontario. *Coleopterists' Bulletin* **24**, 51–52.

HILL, M.O. (1979). *DECORANA: A FORTRAN program for detrended correspondence analysis and reciprocal averaging.* Ithaca, New York: Section of Ecology and Systematics, Cornell University.

HOKKANEN, H. AND HOLOPAINEN, J.K. (1986). Carabid species and activity densities in biologically and conventionally managed cabbage fields. *Zeitschrift für angewandte Entomologie* **102**, 353–363.

HOLLAND, J.M. (1998). The effectiveness of exclusion barriers for polyphagous predatory arthropods in wheat. *Bulletin of Entomological Research* **88**, 305–310.

HOLLAND, J.M. AND LUFF, M.L. (2000). The effects of agricultural practices on Carabidae in temperate agroecosystems. *Integrated Pest Management Reviews* **5**, 109–129.

HOLLAND, J.M. AND SMITH, S. (1999). Sampling epigeal arthropods: an evaluation of fenced pitfall traps using mark–release–recapture and comparisons to unfenced pitfall traps in arable crops. *Entomologia Experimentalis et Applicata* **91**, 347–357.

HOLLAND, J.M., HEWITT, M.V. AND DRYSDALE, A. (1994). Predator populations and the influence of crop type and preliminary impact of integrated farming systems. *Aspects of Applied Biology* **40**, 217–224.

HOLLAND, J.M., COOK, S.K., DRYSDALE, A., HEWITT, M.V., SPINK J. AND TURLEY, D. (1998). The impact on non-target arthropods of integrated compared to conventional farming: results from the LINK Integrated Farming Systems Project. *1998 Brighton Crop Protection Conference – Pests and Diseases* **2**, 625–630.

HOLLAND, J.M., WINDER, L. AND PERRY, J.N. (1999). The impact of dimethoate on the spatial distribution of beneficial arthropods. *Annals of Applied Biology* **136**, 93–105.

HOLOPAINEN, J.K. (1990). Influence of ethylene glycol on the numbers of carabids and other soil arthropods caught in pitfall traps. In: *The role of ground beetles in ecological and environmental studies*. Ed. N.E. Stork, pp 339–341. Andover: Intercept.

HOLOPAINEN, J.K. AND VARIS, A.L. (1986). Effects of a mechanical barrier and formalin preservative on pitfall catches of carabid beetles (Coleoptera, Carabidae) in arable fields. *Journal of Applied Entomology* **102,** 440–445.

HOLOPAINEN, J.K., BERGMAN, T., HAUTALA, E.L. AND OKSANEN, J. (1995). The ground beetle fauna (Coleoptera: Carabidae) in relation to soil properties and foliar fluoride content in spring cereals. *Pedobiologia* **39**, 193–206.

HORNE, P.A. (1992). Comparative life histories of two species of *Notonomus* (Coleoptera: Carabidae) in Victoria. *Australian Journal of Zoology* **40**, 163–171.

HORNE, P.A. AND EDWARD, C.L. (1997). Preliminary observations on awareness, management and impact of biodiversity in agricultural ecosystems. *Memoirs of the Museum of Victoria* **56**, 281–285.

HOUSE, G.J. AND ALL, J.N. (1981). Carabid beetles in Soybean agroecosystems. *Environmental Entomology* **10**, 194–196.

HOUSE, G.J. AND PARMELEE, R.W. (1985). Comparison of soil arthropods and earthworms from conventional and no-tillage agroecosystems. *Soil Tillage Research* **5**, 351–360.

HSIN, C.-Y., SELLARS, L.G. AND DAHM, P.A. (1979). Seasonal activity of carabids and their toxicity of carbofuran and terbufos to *Pterostichus chalcites*. *Environmental Entomology* **8**, 154–159.

HUUSELA-VEISTOLA, E. (1996). Effects of pesticide use and cultivation techniques on ground beetles (Col., Carabidae) in cereal fields. *Annales Zoologici Fennici* **33**, 197–205.

ISHITANI, M. (1996). Ecological studies on ground beetles (Coleoptera: Carabidae, Brachinidae) as environmental indicators. *Miscellaneous Reports of the Hiwa Museum for Natural History* **34**, 1–110. [In Japanese, English summary]

ISHITANI, M, WATANABE, J. AND YANO, K. (1994). Species composition and spatial distribution of ground beetles (Coleoptera) in a forage crop field. *Japanese Journal of Entomology* **62**, 275–283.

JAROŠÍK, V. (1991). Are diversity indices of carabid beetle (Col., Carabidae) communities useful, redundant or misleading? *Acta Entomologica Bohemoslovica* **88**, 273–279.

JAROŠÍK, V. AND HŮRKA, K. (1986). Die Coleopterenfauna des Rapsfelds. *Vestník Československé Společnosti Zoologické* **50**, 192–212.

JEPSON, P.C. (1989). The temporal and spatial dynamics of pesticide side-effects on non-target invertebrates. In: *Pesticides and non-target invertebrates*. Ed. P.C. Jepson, pp 95–128. Wimborne: Intercept.

JONES, M.G. (1976). The carabid and staphylinid fauna of winter wheat and fallows on a clay with flints soil. *Journal of Applied Ecology* **13**, 775–791.

JONES, M.G. (1979). The abundance and reproductive activity of common Carabidae in a winter wheat crop. *Ecological Entomology* **4,** 31–43.

KEGEL, B. (1990). Diurnal activity of carabid beetles living on arable land. In: *The role of ground beetles in ecological and environmental studies*. Ed. N.E. Stork, pp 65–76. Andover: Intercept.

KEGEL, B. (1991). *Freiland- und Laboruntersuchungen zur Wirkung von Herbiziden auf epigäiische Arthropoden, inbesondere der Laufkäfer (Col.: Carabidae).* Berlin: Institut für Biologie der technischen Universität.

KELLY, M.T. AND CURRY, J.P. (1985). Studies on the arthropod fauna of a winter wheat crop and its response to the pesticide methiocarb. *Pedobiologia* **28**, 413–421.

KENNEDY, T.F. (1994). The ecology of *Bembidion obtusum* (Ser.) (Coleoptera, Carabidae) in winter wheat fields in Ireland. *Biology and Environment: Proceedings of the Royal Irish Academy* **94B**, 33–40.

KIRK, V.M. (1971). Ground beetles in cropland in South Dakota. *Annals of the Entomological Society of America* **64**, 238–241.

KISS, J., KADAR, F., TOTH, I., KOZMA, E. AND TOTH, F. (1994). Occurrence of predatory arthropods in winter wheat and in the field edge. *Ecologie* **25**, 127–132.

KNAUER, N. AND STACHOW, U. (1987). Aktivitäten von Laufkäfern (Carabidae Col.) in einem intensiv wirtschaftenden Ackerbaubetrieb – Ein Beitrag zur Agarökosystemanalyse. *Journal of Agronomy and Crop Science* **159**, 131–145.

KÖHLER, F. AND STUMPF, H. (1992). Die Käfer der Wahner Heide in der Nederrheinischen Bucht bei Köln (Insecta: Coleoptera). *Decheniana-Beihefte (Bonn)* **31**, 499–593.

KROMP, B. (1989). Carabid beetle communities (Carabidae, Coleoptera) in biologically and conventionally farmed agroecosytems. *Agriculture, Ecosystems and Environment* **27**, 241–251.

KROMP, B. (1990). Carabid beetles (Coleoptera, Carabidae) as bioindicators in biological and conventional farming in Austrian potato fields. *Biology and Fertility of Soils* **9**, 182–187.

KROMP, B. (1999). Carabid beetles in sustainable agriculture: a review on pest control efficacy, cultivation impacts and enhancement. *Agriculture, Ecosystems and Environment* **74**, 187–228.

KROMP, B. AND STEINBERGER, K-H. (1992). Grassy field margins and arthropod diversity: a case study on ground beetles and spiders in eastern Austria (Coleoptera: Carabidae; Arachnida: Aranei, Opiliones). *Agriculture, Ecosystems and Environment* **40**, 71–93.

KUSCHA, V., LEHMANN, G. AND MEYER, U. (1987). Zur Arbeit mit Bodenfallen. *Beitrage für Entomologie* **37**, 3–27.

LEMIEUX, J.P. AND LINDGREN, B.S. (1999). A pitfall trap for large-scale trapping of Carabidae: comparison against conventional design, using two different preservatives. *Pedobiologia* **43**, 245–253.

LESTER, D.G. AND MORRILL, W.L. (1989). Activity density of ground beetles (Coleoptera: Carabidae) in alfalfa and sainfoin. *Journal of Agricultural Entomology* **6**, 71–76.

LEVESQUE, C. AND LEVESQUE, G.-Y. (1994). Abundance and seasonal activity of ground beetles (Coleoptera: Carabidae) in a raspberry plantation and adjacent sites in southern Quebec. *Journal of the Kansas Entomological Society* **67**, 73–101.

LINDROTH, C.H. (1945). *Die Fennoskandischen Carabidae. Eine tiergeographische Studie. I. Spezieller Teil.* Goteborg: Elanders.

LINDROTH, C.H. (1992). *Ground beetles (Carabidae) of Fennoscandia. A zoogeographic study. Part 1. Specific knowledge regarding the species.* Intercept: Andover.

LÖVEI, G.L. (1984). Ground beetles (Coleoptera: Carabidae) in two types of maize fields in Hungary. *Pedobiologia* **26**, 57–64.

LÖVEI, G.L. AND SÁROSPATAKI, M. (1990). Carabid beetles in agricultural fields in Eastern Europe. In: *The role of ground beetles in ecological and environmental studies*. Ed. N.E. Stork, pp 87–93. Andover: Intercept.

LÖVEI, G.L. AND SUNDERLAND, K.D. (1996). Ecology and behaviour of ground beetles (Coleoptera: Carabidae). *Annual Review of Entomology* **41**, 231–266.

LUFF, M.L. (1975). Some features influencing the efficiency of pitfall traps. *Oecologia* **19**, 345–357.

LUFF, M.L. (1980). The biology of the ground beetle *Harpalus rufipes* in a strawberry field in Northumberland. *Annals of Applied Biology* **94**, 153–164.

LUFF, M.L. (1986). Aggregation of some Carabidae in pitfall traps. In: *Carabid beetles. Their adaptations and dynamics*. Eds. P.J. den Boer, M.L. Luff, D. Mossakowski and F. Weber, pp 385–397. Stuttgart: Fischer.

LUFF, M.L. (1987). Biology of polyphagous ground beetles in agriculture. *Agricultural Zoology Reviews* **2**, 237–278.

LUFF, M.L. (1996). Use of Carabids as environmental indicators in grasslands and cereals. *Annales Zoologici Fennici* **33**, 185–195.

LUFF, M.L. (1998). *Provisional atlas of the ground beetles (Coleoptera, Carabidae) of Britain*. Huntingdon: Institute of Terrestrial Ecology.

LUFF, M.L. AND EYRE, M.D. (2000). Factors affecting the ground beetles (Coleoptera: Carabidae) of some British coastal habitats. In: *British saltmarshes*. Ed. B. Sherwood, pp 235–245. Cardigan: Forrest Text.

LUFF, M.L. AND SANDERSON, R.A. (1992). *Analysis of data on cereal invertebrates*. MAFF Project Report. MAFF: London.

MARSHALL, D.A. AND DOTY, R.L. (1990). Taste responses of dogs to ethylene glycol, propylene glycol and ethylene glycol-based antifreeze. *Journal of the American Veterinary Medical Association* **12**, 1599–1602.

MOMMERTZ, S., SCHAUER, C., KÖSTERS, N., LANG, A. AND FILSER, J. (1996). A comparison of D-Vac suction, fenced and unfenced pitfall trap sampling of epigeal arthropods in agroecosystems. *Annales Zoologici Fennici* **33**, 117–124.

MÜLLER, G. (1968). Faunistische-ökologische Untersuchungen der Coleopterenfauna der küstennahen Kulturlandschaft bei Greifswald. Teil I. Die Carabidenfauna benachbarter Acker- und Weideflächen mit dazwischenliegendem Feldrain. *Pedobiologia* **8**, 313–339.

MÜLLER, G. (1972). Faunistische-ökologische Untersuchungen der Coleopterenfauna der küstennahen Kulturlandschaft bei Greifswald. Die Wirkung der Herbizide UVON-Kombi (II) und ELBANIL (III) auf die epigäische Fauna von Kulturflächen. *Pedobiologia* **12**, 169–211.

NELEMANS, N.M.E. (1987). Possibilities for flight in the carabid beetle *Nebria brevicollis* (F.). *Oecologia* **72**, 502–509.

NIEMELÄ, J. AND SPENCE, J.R. (1999). Dynamics of local expansion by an introduced species: *Pterostichus melanarius* Ill. (Coleoptera, Carabidae) in Alberta, Canada. *Diversity and Distributions* **5**, 121–127.

NIEMELÄ, J., HALME, E., PAJUNEN, T. AND HAILA, Y. (1986). Sampling spiders and carabid beetles with pitfall traps: the effect of increased sampling effort. *Annales Entomologici Fennici* **52**, 109–111.

NOVAK, B. (1964). Saisonmässiges Vorkommen und Synökologie der Carabiden auf Zuckerrübenfeldern von Haná (Col. Carabidae). *Acta Universitatis Palackianae Olomucensis. Facultas Rerum Naturalium* **13**, 101–251. [In Czech, German summary.]

NOVAK, B. (1968). Bindungsgrad der Imagines einiger Feldcarabiden-Arten an die Lebensbedingungen in einem Winterweizenbestand (Col. Carabidae). *Acta Universitatis Palackianae Olomucensis. Facultas Rerum Naturalium* **28**, 99–131.

OBRTEL, R. (1971). Number of pitfall traps in relation to the structure of the catch of soil surface Coleoptera. *Acta Entomologica Bohemoslovaca* **68**, 300–309.

PAUER, R. (1975). Zur Ausbreitung der Carabiden in der Agrarlandschaft, unter besonderer Berücksichtigung der Grenzbereiche verschiedener Feldkulturen. *Zeitschrift für angewandte Zoologie* **62**, 457–489.

PAUSCH, R.D. AND PAUSCH, L.M. (1980). Observations on the biology of the slender seedcorn beetle, *Clivina impressifrons* (Coleoptera: Carabidae). *Great Lakes Entomologist* **13**, 189–194.

PENEV, L. (1996). Large-scale variation in carabid assemblages, with special reference to the local fauna concept. *Annales Zoologici Fennici* **33**, 49–63.

PETRUSKA, F. (1967). Carabiden als Bestandteil der Entomofauna der Rübenfelder in der Uničov-Ebene. *Acta Universitatis Palackianae Olomucensis. Facultas Rerum Naturalium* **25**, 121–243. [In Czech, German summary.]

PETRUSKA, F. (1971). The influence of the agricultural plants on the development of the populations of Carabidae living in the fields. *Acta Universitatis Palackianae Olomucensis. Facultas Rerum Naturalium* **34**, 151–191. [In Czech, English summary.]

PETRUSKA, F. (1974). On the dispersion of some species of the group of Carabidae in a field growing sugarbeet. *Acta Universitatis Palackianae Olomucensis. Facultas Rerum Naturalium* **47**, 145–178. [In Czech, English summary.]

PFIFFNER, L. AND NIGGLI, U. (1996). Effects of bio-dynamic, organic and conventional farming on ground beetles (Col. Carabidae) and other epigaeic arthropods in winter wheat. *Biological Agriculture and Horticulture* **12**, 353–364.

PIETRASZKO, R. AND DECLERCQ, R. (1981). Carabidae of agricultural land in Belgium. *Parasitica* **37**, 45–58.

POWELL, W., DEAN, D.A. AND DEWAR, A. (1985). The influence of weeds on polyphagous arthropod predators in winter wheat. *Crop Protection* **4**, 298–312.

PURVIS, G. AND BANNON, J.W. (1992). Non-target effects of repeated methiocarb slug pellet application on carabid beetle (Coleoptera: Carabidae) activity in winter-sown cereals. *Annals of Applied Biology* **121**, 401–422.

PURVIS, G. AND CURRY, J.P. (1984). The influence of weeds and farmyard manure on the activity of Carabidae and other ground-dwelling arthropods in a sugar beet crop. *Journal of Applied Ecology* **21**, 271–283.

RIVARD, I. (1964). Observations on the breeding periods of some ground beetles (Coleoptera: Carabidae) in Eastern Ontario. *Canadian Journal of Zoology* **42**, 1081–1084.

RIVARD, I. (1966). Ground beetles (Coleoptera: Carabidae) in relation to agricultural crops. *Canadian Entomologist* **98**, 189–195.

SANDERSON, R.A. (1994). Carabidae and cereals: a multivariate approach. In: *Carabid beetles: ecology and evolution*. Eds. K. Desender, M. Dufrêne, M. Loreau, M.L. Luff and J.P. Maelfait, pp 457–463. Dordrecht: Kluwer.

SCHELLER, H.V. (1984). The role of ground beetles (Carabidae) as predators on early populations of cereal aphids in spring barley. *Zeitschrift für angewandte Entomologie* **97**, 451–463.

SEKULIĆ, R. (1976). Contribution to the Carabidae family of maize culture on chernozem in central Bačka. *Acta entomologica Jugoslavica* **12**, 35–48. [In Yugoslavian, English summary.]

SEKULIĆ, R., ČAMPRAG, D., KEREŠI, T. AND TALOŠI, B. (1987). Fluctuations of carabid population density in winter wheat fields in the region of Bačka, northern Yugoslavia (1961–1985). *Acta Phytopathologica et Entomologica Hungarica* **22**, 265–271.

SIEPEL, H., MEIJER, J., MABELIS, A.A. AND DEN BOER, M.H. (1989). A tool to assess the influence of management practices on grassland surface macrofaunas. *Journal of Applied Entomology* **108**, 271–290.

SPENCE, J.R. AND NIEMELÄ, J.K. (1994). Sampling carabid assemblages with pitfall traps – the madness and the method. *Canadian Entomologist* **126**, 881–894.

SPENCE, J.R. AND SPENCE, D.H. (1988). Of ground beetles and men: introduced species and the synanthropic fauna of Western Canada. *Memoirs of the Entomological Society of Canada* **144**, 151–168.

STASSART, P. AND GRÉGOIRE-WIBO, C. (1983). Influence du travail du sol sur les populations de carabides en grande culture, resultats preliminaires. *Mededelingen van de Fakulteit der Landbouwwetenschappen der Rijksuniversiteit te Gent* **48**, 465–474.

STINNER, B.R. AND HOUSE, G.J. (1990). Arthropods and other invertebrates in conservation-tillage agriculture. *Annual Review of Entomology* **35**, 299–318.

SUNDERLAND, K.D., DE SNOO, G.R., DINTER, A., HANCE, T., HELENIUS, J., JEPSON, P., KROMP, B., LYS, J.-A., SAMU, F., SOTHERTON, N.W., TOFT, S. AND ULBER, B. (1995). Density estimation for invertebrate predators in agroecosystems In: *Arthropod natural enemies in arable land. I. Density, spatial heterogeneity and dispersal.* Eds. S. Toft and W. Riedel, pp 133–162. Aarhus: Aarhus University Press.

ŠUSTEK, Z. (1987). Changes in body size structure of carabid communities (Coleoptera, Carabidae) along an urbanisation gradient. *Biológia (Bratislava)* **42**, 145–156.

SZEL, G., KADAR, F. AND FARAGO, S. (1997). Abundance and habitat preference of some adult-overwintering ground beetle species in crops in western Hungary (Coleoptera: Carabidae). *Acta Phytopathologica et Entomologica Hungarica* **32**, 369–376.

SZWEJDA, J. (1984). Carabidae as a component of entomofauna occurring on onion field. *Polskie Pismo Entomologiczne* **54**, 391–402. [In Polish, English summary.]

TAYLOR, L.R., KEMPTON, R.A. AND WOIWOD, I.P. (1976). Diversity statistics and the log-series model. *Journal of Animal Ecology* **45**, 255–272.

TER BRAAK, C.J.F. (1986). Canonical correspondence analysis: a new eigenvector technique for multivariate direct gradient analysis. *Ecology* **67**, 1167–1179.

TER BRAAK, C.J.F. (1988). *CANOCO – a FORTRAN program for canonical community ordination by partial detrended canonical correspondence analysis, principal compo-*

nents analysis and redundancy analysis. Wageningen: Groep Landbouwwiskunde.

THACKER, J.R.M. AND DIXON, J. (1996). Modelling the within-field recovery of carabid beetles following their suppression by exposure to an insecticide. *Annales Zoologici Fennici* **33**, 225–231.

THIELE, H.-U. (1977). *Carabid beetles in their environments.* Berlin: Springer-Verlag.

TOGASHI, I., TAKA, J. AND NAKATA, K. (1992). Arthropod fauna occurring in a fire burnt field in Ishikawa prefecture (Part 3). Ground beetle fauna occurring in the fire burnt Japanese radish and Kôbo Sawa millet fields. *New Entomology* **41**, 59–62. [In Japanese, English summary.]

TONHASCA, A. (1993). Carabid beetle assemblage under diversified agricultural systems. *Entomologia Experimentalis et Applicata* **68**, 279–285.

TURIN, H. (1981). Provisional checklist of the European ground-beetles. *Monographieën van de Nederlandse Entomologische Vereniging* **9**, 1–249.

TURIN, H. (2000). *De Nederlandse Loopkevers. Verspreiding en Oecologie [The Netherlands ground beetles. Distribution and ecology].* Utrecht: KNNV Uitgeverij.

TURIN, H., ALDERS, K., DEN BOER, P.J., VAN ESSEN, S., HEIJERMAN, T., LAANE, W. AND PENTERMAN, E. (1991). Ecological characterization of carabid species (Coleoptera, Carabidae) in the Netherlands from thirty years of pitfall sampling. *Tijdschrift voor Entomologie* **134**, 279–304.

ULBER, B. AND WOLF-SCHWERIN, G. (1995). A comparison of pitfall trap catches and absolute density estimates of carabid beetles in oilseed rape fields. In: *Arthropod natural enemies in arable land. I. Density, spatial heterogeneity and dispersal.* Eds. S. Toft and W. Riedel, pp 77–86. Aarhus: Aarhus University Press.

VAN DINTHER, J.B.M. AND MENSINK, F.T. (1971). Use of radioactive phosphorus in studying egg predation by carabids in cauliflower fields. *Mededelingen van de Fakulteit der Landbouwwetenschappen der Rijksuniversiteit te Gent* **36**, 283–293.

VARVARA, M., DONESCU, D. AND VARVARA, V. (1989). Contributions to the knowledge of carabid beetles in potato crops in the Bîrsei Country. *Symposium Entomofagii şi rolul lor în păstrarea echilibrului natural*, pp 95–101.

WAAGE, B.E. (1985). Trapping efficiency of carabid beetles in glass and plastic pitfall traps containing different solutions. *Fauna Norvegica Series B* **32**, 33–36.

WALLIN, H. (1985). Spatial and temporal distribution of some abundant carabid beetles (Coleoptera: Carabidae) in cereal fields and adjacent habitats. *Pedobiologia* **28**, 19–34.

WALLIN, H. (1991). Movement patterns and foraging tactics of a caterpillar hunter inhabiting alfalfa fields. *Functional Ecology* **5**, 740–749.

WALLIN, H. AND EKBOM, B. (1994). Influence of hunger level and prey densities on movement patterns in three species of *Pterostichus* beetles (Coleoptera: Carabidae). *Environmental Entomology* **23**, 1171–1181.

WALLIN, H., WIKTELIUS, S. AND EKBOM, B.S. (1981). Occurrence and distribution of beetles in a spring barley field. *Entomologiske Tidskrift* **102**, 51–56. [In Swedish, English summary.]

WEISS, M.J., BALSBAUGH, E.U., FRENCH, E.W. AND HOAG, B.K. (1990). Influence of tillage management and cropping systems on ground beetle (Coleoptera: Carabidae) fauna in the northern Great Plains. *Environmental Entomology* **19**, 1388–1391.

WHEATER, C.P. (1988). Predator-prey size relationships in some Pterostichini (Coleoptera: Carabidae). *Coleopterists' Bulletin* **42**, 237–240.

WIEDENMANN, R.N., LARRAIN, P.L. AND O'NEIL, R.J. (1992). Pitfall sampling of ground beetles (Coleoptera: Carabidae) in Indiana soybean fields. *Journal of the Kansas Entomological Society* **65**, 279–291.

YANO, K., ISHITANI, M. AND YAHIRO, K. (1995). Ground beetles (Coleoptera) recorded from paddy fields of the world: a review. *Japanese Journal of Systematic Entomology* **1**, 105–112.

Illustrations

Photographs supplied by Martin Luff

Photograph 1. Larvae of *Pterostichus melanarius*, a large nocturnally active species found in northern Europe and North America. Sometimes the dominant species in fields of shade-providing crops such as oilseed rape.

Photograph 2. Pupa of *Pterostichus melanarius*, ventral view, in a cavity in the soil.

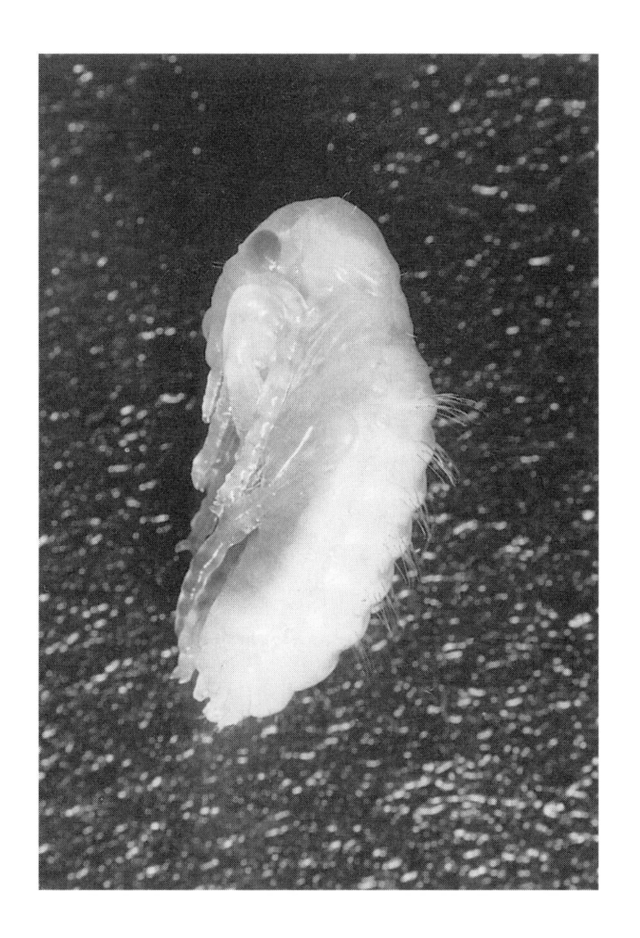

Photograph 3. Pupa of *Pterostichus melanarius*, lateral view, in a cavity in the soil.

Photograph 4. Adult of *Carabus violaceus*, a large species found in relatively undisturbed fields in Europe.

Photograph 5. Adult of *Carabus nemoralis*, a large species found in spring and autumn in gardens and field margins in both Europe and North America.

Photograph 6. Adult of *Notiophilus biguttatus*, a small European species of diurnal carabid that feeds on Collembola.

Photograph 7. Adult of *Nebria brevicollis*, a moderate-sized species that is predominant in northern and western Europe.

Photograph 8. Adult of *Loricera pilicornis*, a carabid found on damp soils in northern Europe.

Photograph 9. Adult of *Bembidion quadrimaculatum*, a very small carabid sometimes abundant in open fields in both north-western Europe and North America.

3
Carabid Diets and Food Value

SØREN TOFT AND TRINE BILDE

Department of Zoology, University of Aarhus, Bldg. 135, DK-8000 Århus C, Denmark

Introduction

Carabid beetles are the dominant ground-living invertebrate predators in many ecosystems. Because of their often considerable size and their potential role in food-web dynamics and pest control, the food and feeding habits of carabid beetles may have been investigated in more detail than of any other group of predatory arthropod. This is also evidenced by numerous reviews (Forbes, 1883; Lindroth, 1949/1992; Thiele, 1977; Hengeveld, 1980b,c; Dajoz, 1987; Luff, 1987; Lövei and Sunderland, 1996) and extensive compilatory works (e.g. Larochelle, 1990). Feeding type, overall morphology and habitat choice show great variation (Forsythe, 1982a, 1983, 1987, 1991; Evans and Forsythe, 1985) and have led to the creation of extensive life form classifications (Sharova, 1981). A diversity of feeding specializations has been recognized. The most important deviation from the basic generalist predatory life style is the predominantly granivorous feeding of many species, but specializations to various animal prey have also been extensively described. In spite of the amount of work already done, there are numerous gaps in our knowledge and understanding, even as concerns common, relatively well-studied and economically important species. This review will build on the framework of feeding types established by earlier reviewers based on studies of stomach contents and natural history observations, but also focus on recent results concerning the quality of various food types to carabids with different feeding habits and specializations. Aspects related to food quantity (availability) will be largely neglected. The approach will expose large fields of research which have not previously been investigated. The review will concentrate on temperate species mostly from agricultural or forest habitats, whereas 'exotic' specializations in all senses of the word will also be neglected.

Few, if any, carabid species have diets restricted to a single food type or prey taxon: they are all polyphagous to various degrees. Preferences and adaptations to certain types of food have anyway evolved. These feeding specializations should be understood in a relative sense. Species have adapted to obtain the greater part of their food from a restricted range of food types that probably allow them to maintain a high fitness ('essential food'). But they also accept many other food types they encounter

The Agroecology of Carabid Beetles

during their search that they can handle; these are eaten in lower amounts and may contribute more or less to beetle fitness ('supplementary food'). There is no absolute distinction between essential and supplementary food types, rather there exists a graded series of food qualities.

Methods for recording diets in the field

Interest in the feeding habits of carabid beetles has arisen from two main sources: 1) out of an interest in the carabid beetles themselves (what do they eat? how do they obtain their food? what is their role in the biological community?); and 2) out of an interest in the population dynamics of particular prey species (usually a pest). The methods used for studying carabid diets reflect this double purpose. They are described in detail elsewhere in this book (Symondson, this volume; Ingerson-Mahar, this volume); therefore only details pertaining to the recording of diet composition will be briefly mentioned here. Some methods are intended to record all sorts of food in the proportion eaten, though this is rarely achieved in full (e.g. microscopical analysis; iso-enzyme electrophoresis; serological or DNA analysis of gut contents). Others are aimed at recording all the different species that consume a certain food type (e.g. serological methods). Microscopical analysis of gut contents is only amenable for species that swallow the sclerotized parts of the prey (fragment feeders), while serological or molecular methods are needed for species with extra-oral digestion.

Feeding guilds

Thiele (1977) distinguished three feeding types based on the method of handling prey. These are equivalent to the groupings adopted by Forsythe (1982a, 1983) and Evans and Forsythe (1985): 1) fluid feeders with extra-oral digestion, 2) fragment feeders with no extra-oral digestion, and 3) mixed fluid and fragment feeders. Most species belong to the latter group. Thiele (1977) also classified carabids according to main food types into: polyphagous predators; oligophagous predators (mollusc, Collembola, and caterpillar specialists); and phytophagous carabids. The range of food types found in beetle stomachs includes both essential and supplementary foods. Since few studies of stomach contents make any attempt to quantify consumption rates, but rank prey by the proportion of beetles in which they are found, these studies tend to exaggerate the degree of polyphagy and underestimate the dependence on essential foods. Thus, based on stomach contents alone, Hengeveld (1980b) was able to distinguish only generalist and specialist feeders, the latter divided into Collembola and mollusc (snail and slug) feeders.

The following list of feeding guilds should not be viewed as a rigid classification system, because single species may comply to various degrees with more than one of the groups. Grouping stresses the differences between major adaptive forms but neglects variability within the groups. The feeding guilds have been recognized by combining several types of evidence: stomach contents, field and laboratory observations of feeding, and morphological analyses (e.g. structure of mouthparts and digestive tract). Physiological data on food utilization and food value are scant, as will be clear below, but are needed in order to precisely delimit the degree of feeding specialization of a species. The morphological variation has been documented extensively by Forsythe (1982a, 1983), Evans and Forsythe (1985) and others, see Ingerson-Mahar (this volume). Data from many different sources indicate that feeding specializations are at least partly characteristic of tribes or even higher taxonomic levels (Hengeveld, 1981), suggesting that they have long evolutionary histories.

GENERALIST CARNIVORES

The vast majority of carabid beetles are generalist zoophages with different restrictions. They accept an extremely wide range of invertebrate prey, including earthworms, gastropods and arthropods, and they include plant material to various degrees (Lindroth, 1949/1992; Cornic, 1973; Luff, 1974; Sunderland, 1975; Sota, 1985). Scavenging is also common in this group. It has been recognized repeatedly that diet width of generalist carabids is correlated with the size of the animal (Loreau, 1983b; Pollet and Desender, 1987). For example, Tod (1973) noted that only large species included slugs in their diet. The extreme dietary width of *Pterostichus melanarius* was exemplified by Pollet and Desender (1985, 1986) who recorded 49 prey taxa at family level, including slugs, earthworms and a wide selection of insects. True broad-spectrum generalist carnivores are found among species of *Carabus*, *Abax*, large *Pterostichus*, and others (Thiele, 1977; Hengeveld, 1980b; Loreau, 1983a,b, 1994).

GENERALIST INSECTIVORES

Many smaller carabid species include slugs and worms to a very limited extent, but still accept a wide variety of insect prey, e.g. *Agonum dorsale*, *Trechus*, *Bembidion*, *Calathus*, amongst others (Davies, 1953; Mitchell, 1963; Zhavoronkova, 1969). They also include seeds and vegetation to only a minor degree, but scavenging on dead insects may be widespread. Even in *Cicindela* the proportion of dead insects may be 25% (Dreisig, 1981).

MOLLUSC SPECIALISTS

It has long been recognized that species of the tribe Cychrini prey predominantly on snails and slugs (Tod, 1973; Greene, 1975; Digweed, 1993) and show morphological adaptations to snail predation (Ingerson-Mahar, this volume). Several *Carabus* spp. have also been reported as predominantly mollusc feeders (Scherney, 1959, 1961), but most species of this genus also prey to a large extent on insects and thus may better be grouped as generalist zoophages. Predation by carabids on gastropods has been reviewed by Symondson (in press).

MICROARTHROPOD SPECIALISTS

Notiophilus spp., *Loricera pilicornis*, *Nebria brevicollis*, *Leistus ferrugineus* and others are known from stomach analyses to prey predominantly on Collembola and mites, whilst other soil-living arthropods serve as supplementary food (Davies, 1953; Anderson, 1972; Sunderland, 1975). Detailed studies of their morphology and prey-catching behaviour have revealed intricate adaptations to the capture of Collembola with high escape abilities (Bauer, 1981; Hintzpeter and Bauer, 1986). Species of this group do not accept seeds or other plant material (Goldschmidt and Toft, 1997).

CATERPILLAR SPECIALISTS

Calosoma spp. prey predominantly on caterpillars of Lepidoptera and tenthredinid wasps and the literature contains little information on supplementary food types (cf. Larochelle, 1990). The relative monophagy of these species is confirmed by their populations fluctuating with caterpillar abundance (Weseloh, 1985).

GRANIVORES

Species of the genera *Harpalus* (s.lat.), *Amara*, *Zabrus*, *Synuchus*, *Ditomus*, and *Diachromus* are predominantly seed eaters. Their preference for seeds and supplementary feeding on insects has been documented both by stomach analyses (Skuhravy, 1959), laboratory experiments (Hagley *et al.*, 1982; Brandmayr, 1990; Jørgensen and Toft, 1997a,b) and observations (Schremmer, 1960; Trautner *et al.*, 1988).

OTHER SPECIALISTS

Family Carabidae worldwide contains several myrmecophiles and termitophiles, and other specialized habits are likely to be discovered. Some myrmecophiles seem to live predominantly with ants.

SUPPLEMENTARY FEEDING ACTIVITY

Scavenging on dead insects is no doubt a quantitatively important feeding mode for many carabids (Lindroth, 1949/1992; Dawson, 1965). Many carabids can be raised on meat but vertebrate scavenging is not likely to be of importance in nature. Carabids and staphylinids are often found in numbers at dead mammals or birds, but they may as well have been attracted because of the Diptera and other insects that gather at a corpse.

Herbivory other than granivory (consumption of plant leaves, fruits, pollen, fungi, etc.) is a common supplementary feeding mode as unspecified plant material is often found in gut dissections (e.g. Dawson, 1965). Apart from the true granivore beetles, inclusion of various plant materials in the diet of carabids seems to be supplementary feeding only (see below).

Larval feeding guilds

Carabid larvae seem to fit into the same feeding guilds as the adults. Though diet composition of carabid larvae rightfully is considered much neglected, all available evidence shows a close correspondence between larval and adult feeding habits. Broadly speaking, if the adult is a generalist, so is the larva; if the adult is a specialist, the larva has specialized on the same food type. The feeding techniques may differ, however. Carabid larvae are generally fluid feeders, including those that are fragment feeders as adults, e.g. *Nebria* (Spence and Sutcliffe, 1982).

In *Pterostichus madidus* both larvae and imagines are broad generalist insectivores, but whereas the imagines include plant material in their diet, the larvae do not (Luff, 1974). Schelvis and Siepel (1988) found that larvae of two *Pterostichus* spp. consumed a spectrum of small, soft-bodied invertebrates. Thus, soil-living insect larvae (Diptera, other Coleoptera), mites and earthworms formed the larger part of the diet, and cannibalism and mutual interspecific predation was substantial. These larvae must be characterized like the adults as generalist carnivores/insectivores; however, the exact menu is modified according to the smaller size, lower power and more edaphic life style of the larvae. According to Scherney (1961), three *Carabus* spp. larvae preferred earthworms and insect larvae, as eaten by the adults. Whereas the adults of *Abax parallelepipedus* eat a broad range of prey, the larvae could be mass reared without cannibalism on earthworms alone (Symondson, 1994).

Notiophilus and *Loricera* both have larvae specialized on Collembola, and they show morphological and behavioural adaptations to capture of Collembola which differ widely from those of the adults. Adult *Notiophilus* are diurnal hunters relying on good eyesight and high agility (Bauer, 1981); the larvae use mainly tactile cues to detect the same prey (Bauer, 1982). In *Loricera*, the adult beetles use their antennal setae to entrap Collembola (Hintzpeter and Bauer, 1986); the larvae use adhesive maxillary galea (Bauer and Kredler, 1988).

Larval Cychrini are snail specialists like the adults, though they do not have the characteristic body shape that is supposed to facilitate attack on snails by the adults (Greene, 1975).

Harpalus rufipes and *Amara similata* larvae were able to complete the full development on seeds alone (Luff, 1980; Jørgensen and Toft, 1997a,b), and are thus granivorous as the adults. *H. rufipes* was also able to develop on a pure insect diet, whereas *A. similata* was not. Thiele (1977) suggested that carabid larvae might be more specialized than the imago, though he recognized that extreme specialization had not evolved. Except for the effect of smaller size limiting the range of potential prey, there is little support for this suggestion (see also below).

Life style and foraging

Evans and Forsythe (1984) pointed out the different solutions to the trade-off between agility and powerfulness observed among the ground-living Carabidae. At one end of the spectrum are agile surface runners (e.g. *Cicindela*), at the other, soil-digging species (e.g. *Clivina*); in between are the larger proportion of species which push themselves through litter or vegetation. Arboreal climbers have adapted in other directions. All through this diversity of life style and habitat occupancy, being an opportunistic carnivore/insectivore is the basic feeding mode, though beetles will obviously encounter different types of prey and the exact diets will differ. Thus, the soil-digging *Clivina fossor* eats ‘a variety of animal food’ (Pollet and Desender, 1987): Collembola and soil-inhabiting invertebrates, such as insects, nematodes and enchytraeids. *Cicindela* takes ‘mostly small insects’: ants, beetles, spiders, Heteroptera nymphs, Collembola and Lepidoptera larvae (Dreisig, 1981). Arboreal *Dromius* eats ‘softbodied insects and insect larvae’ (Trautner, 1984).

If feeding modes are associated with a restricted habitat range, it is via the habitat distribution of the essential food. Thus, granivores are mainly searching for seeds on the soil surface (Harpalini) or both on the surface and on low plants (Amarini). Slug feeders are also mostly restricted to near-surface layers. The caterpillar specialist *Calosoma sycophanta* climbs trees where caterpillars may occur in outbreak abundance.

Prey detection and attack behaviour

Wheater (1989) found good correspondence between feeding behaviour, type of prey and the senses used for prey detection in species of various feeding guilds. All species tested responded with attack when touching prey (tactile cues). Diurnal generalist insectivores (*Cicindela*, *Calosoma*, *Scarites*) used mainly vision, however, for detecting movable prey, while nocturnal generalist insectivores (*Pterostichus* spp.) used olfactory cues from prey, detected by the antennae. *Abax parallelepipedus* may be both day and night active and use both types of cues. *Cychrus* and two *Carabus* spp. (mollusc specialists) responded positively to contact with slug mucus, i.e. they probably used gustatory cues detected by the palps. Snail/slug specialists are thought to be able to follow mucus tracks.

In spite of their relatively narrow diet, the Collembola hunters use a variety of sensory modes for prey detection. *Notiophilus* have prominent eyes and are predominantly visual hunters (Bauer, 1981), but they also respond to gustatory cues from prey (Ernsting *et al.*, 1985). *Leistus* is a tactile hunter (Bauer, 1985), and Collembola specialist larvae seem to be the same (Bauer, 1982; Bauer and Kredler, 1988). One can hypothesize that the divergent strategies evolved to cope with Collembola are directed towards different types of Collembola, or to hunting in different microhabitats. Loreau (1983b) found *Leistus* preying predominantly on entomobryid Collembola, whereas *Notiophilus* preyed mostly on isotomids. Evolutionary constraints may also be involved: thus it may be difficult for larvae of holometabolous insects without composite eyes to evolve into efficient visual hunters.

Carabid responses to olfactory cues seem to be quite unspecific and do not necessarily indicate high food preference. In the studies of Kielty *et al.* (1996), the Collembola-specialist *Nebria brevicollis* was attracted to the odour of live Collembola. *Pterostichus melanarius* responded positively to live Collembola, live aphids and aphid alarm pheromone; however, in feeding tests, both Collembola and aphids were accepted in rather low amounts (Bilde and Toft, 1997b). Also, *Harpalus rufipes* responded to aphid alarm pheromone and wheat extract. This species is unable to utilize aphids (Jørgensen and Toft, 1997b) and is not phytophagous to any extent (Goldschmidt and Toft, 1997).

Daily and seasonal foraging cycles

Pollet and Desender (1986) demonstrated a coincidence of diel activity patterns of two species (*Loricera pilicornis*, *Bembidion properans*) and their most important prey groups as revealed by stomach analyses. However, this result tells little about possible adaptations of the beetles to secure an optimal diet. Diurnal beetles may take advantage of the possibility for basking in the sun, thereby achieving a high body temperature and agility that allow them to overpower otherwise difficult prey. Mitchell (1963) noted that *Bembidion lampros* stomachs were always empty outside the warm season, whereas *Trechus quadristriatus* fed in all seasons. *B. lampros* is diurnal, with a high threshold temperature for feeding, 9°C (Mitchell, 1963; Sørensen, 1996). *Cicindela hybrida* has a lower feeding threshold at 20°C and a maximal capture rate at 35°C (Dreisig, 1981). Probably, diurnally and seasonally restricted activity are correlated. The feeding rate of *B. lampros* increased steeply at higher temperatures, compared to the nocturnal staphylinid *Tachyporus hypnorum* (Sørensen, 1995). Day-active species may thus be able to compensate for a short feeding time by a very high feeding rate at the optimal temperature.

A result of nocturnal activity may be that the carabid is predatorily active at a time when day-active prey (e.g. flies) are least active and least ready to escape an attack due to low temperature. P.D. Kruse (unpublished) found in the laboratory that *Calathus fuscipes* were able to catch more fruit flies at 5°C than at higher temperatures. This effect was apparent only in darkness; under light conditions, the nocturnal beetles were inactive. Flies may often be high-quality prey (see below), so adaptations that secure a higher proportion of flies in the diet may have great fitness value. A whole category of prey (flying insects) may be more available to less agile predators at the time of the day (or during the seasons) when they are least active.

Seasonal variation in diets has been demonstrated in most extensive studies of stomach contents, primarily reflecting fluctuations in availability (Dawson, 1965; Koehler, 1976; Loreau, 1983a; Lukasiewicz, 1996). Skuhravý (1959) found *Poecilus cupreus* to take mainly plant material early in the season and insect food during summer. Cornic (1973) found the same pattern in *Harpalus affinis* and *H. rufipes*. Since *Pterostichus* and *Harpalus* have opposite preferences regarding insect and plant (seed) food, there is no logical pattern, indicating again that they are opportunistic feeders.

Diet quality

So far, we have considered the diets of carabid beetles from the information that can be obtained from simply recording what they eat. However, a full evaluation of stomach contents can only be obtained if we know the nutritional value of each type of food for the carabid species and how the food types in the menu interact to determine the value of the complete diet. Studies taking this approach are still rather few. Comparative data are available only for species from some of the feeding guilds and for just a few food types.

Food contains four basic constituents: energy, nutrients, toxins and indigestibles. The value (or quality) of a diet or a type of food may depend on the proportion of each of these constituents. But it cannot be evaluated by measuring them, e.g. in a chemical analysis. The same type of food may be high quality to one species and of no value to another, depending on the physiological demands (specializations) of the consumer. Therefore, diet/food value is best defined by its potential contribution to the fitness of the consumer. This definition implies that the value of a food type may vary with the developmental phases of the consumer and depend on what other food it has eaten. This is not a weakness, merely a reflection of the complexity of food quality.

In practice, food value is measured by comparing the performance of individuals when the food type is added to the diet compared to the same diet without the food type. In the simplest case, the comparison is between individuals given one food type and starved controls. Such experiments are especially relevant when the food type is of low quality: if performance is reduced in a diet treatment compared to a starvation treatment, this indicates the presence of a toxic substance (cf. Toft and Wise, 1999a). More often, the comparisons are between diets with various numbers of elements, e.g. between differently mixed diets or mixed vs. monotypic diets. Evaluation of a food type should preferably include several comparisons, e.g. monotypic diet of test food vs. monotypic diet of known high-quality food type ('comparison food'); comparison food vs. mixed diet of comparison + test foods; same mixed diet with vs. without test food, etc. To our knowledge, there are still no experiments analysing the contribution of a food type to the value of a naturally mixed diet. Logistic constraints will usually prevent such experiments since natural diets may be very diverse. *A priori* one would expect the importance of single food types to decrease the more varied the total diet, so that any single element may be deleted without negative effects. However, in no case of a highly polyphagous species have we any idea of how many of the dietary elements are high quality (essential) and how many are low quality (supplementary) food types, or even if carabids are capable of detecting such differences prior to feeding.

Fitness related life-history parameters can function as indicators of diet quality (Toft, 1996). Total life-cycle fitness is not necessarily the best experimental choice because dietary demands may change during ontogeny and several gradations below zero total fitness are possible. For example, we would assign a higher value to a food type that allows a species to develop through one or two larval instars than to a food type on which the larvae die in the first instar. The life-history parameters most commonly used as quality criteria are (larval) survival, rate of development, growth rate, teneral size, longevity of imagines, and female fecundity (examples below).

A relative and partly subjective classification of food/diet quality proposed by Toft and Wise (1999a,b) will also be used in the following. 'High quality' food allows high (maximal) survival, growth rate, fecundity, etc.; 'intermediate quality' food allows submaximal performance; 'low quality' food sustains performance above zero and may contribute positively to a mixed diet; 'poor quality' food is no better than starvation or makes no improvement in a mixed diet; 'toxic' food leads to a performance below that of starved beetles (e.g. higher mortality) or reduces the value of a mixed diet. The quality assignments may result from the balance between positive effects of nutrients and negative effects of toxins and deterrents (and other factors, e.g. a hard exoskeleton). Thus, food of all quality categories may contain deterrents or toxins, probably in increasing relative amounts from high-quality to toxic foods.

Food utilization indices (Waldbauer, 1968) or equivalent co-variance analysis (Raubenheimer and Simpson, 1992; Horton and Redak, 1993) provide information about how quality is achieved, rather than indicate quality as such. If food is deterrent, low consumption rates will result in low overall quality, even if utilization efficiency is high. Food containing a toxin will usually be consumed at low rates and utilization efficiency will also be low. Food without toxins or deterrents, but of poor nutrient composition, is expected to be eaten at high rates, but with low utilization efficiency (Toft, 1996).

Considering the diversity of the diets of different species (and probably individuals), an interesting question is to what extent feeding specializations are reflected as differences in food value. Have feeding guilds evolved mainly as morphological/

behavioural (i.e. life style) adaptations that make some species more likely to encounter and acquire certain types of food/prey, which therefore make up the larger part of the diet, with no differentiation in food value? Or do these feeding specializations also have physiological/biochemical counterparts that differentially define specific essential and supplementary food types for species of different feeding guilds, or even within feeding guilds?

Value of different types of food

It can be assumed that prey types to which carabid species show specific adaptations are also high quality prey for these beetles, e.g. snails to snail specialists, Collembola to Collembola specialists, etc. The truth may not necessarily be as simple as that, however. Within each prey group, species are differently adapted to defend themselves against their predators. Thus, Digweed (1993) found varied preference of the cychrine *Scaphinotus marginatus* for different species of snails and slugs. He suggested this might be due either to differences in nutritional value or to anti-predatory defences (mucus production, shell thickness, etc.). Since food value cannot be judged from preference or consumption rate studies but only from performance experiments, the following paragraphs will deal mainly with food types for which performance results are available. *Table 3.1* summarizes the data known to us from such studies with natural prey or seeds, i.e. neglecting 'laboratory prey' like mealworms, fruit flies, etc. Included are also the results concerning the value of specific food types in mixed diets.

APHIDS

The incentive for much of the work on feeding habits of carabid beetles in agricultural habitats originated from an interest in the potential of these predators for limiting aphid populations in crops, in particular cereal fields (see Sunderland, this volume). The value of the aphids themselves is therefore of special relevance. Bilde and Toft (1994) found a low preference in *Agonum dorsale* for the cereal aphid *Rhopalosiphum padi* relative to the comparison food, *Drosophila melanogaster*. In accordance with this, the consumption capacity for the aphid was much lower than for the fruit fly, and this was the case whether tested with pure diets or a mixed diet. Even if the beetles were starved for seven days, the consumption capacity remained at the same low level. This result probably implies that the aphid consumption rate is limited by a low tolerance of the beetle to a chemical in the aphid that consequently serves as an anti-predator defence. The egg-laying rate of females kept on a pure aphid diet was much reduced, probably a composite effect of low feeding rate and low utilization efficiency. A deterrent effect of the aphid was demonstrated by Bilde and Toft (1994) by coating palatable fruit flies with aphid homogenate: *Agonum dorsale* showed a significantly lower preference for the coated flies compared to uncoated flies. So far, the presumed chemical defence of the cereal aphids has not been identified, and it is unknown whether it originates from the host plant or is synthesized by the aphids.

Bilde and Toft (1997a,b) extended the observation that aphids are of low palatability to several carabid species, and this applied partly to two additional species of cereal aphids. Capacity for aphid consumption did not exceed that for fruit fly

Table 3.1. Summary of studies determining the food value of natural insect prey or seeds [and their contribution to mixed diets] to carabid beetles. Relative value categories: High – Intermediate – Low – Poor – Toxic (defined in text).

Food/prey type	[Mixed with]	Carabid species/instar	Fitness parameter	Value	[Mixing effect]	Ref.
Seeds						
Mixed seeds 9 spp.		*Amara similata*	Fecundity	High		6
Mixed seeds 9 spp.		*Amara similata* larvae	Survival	Interm.–High		6
Mixed seeds 8 spp.		*Harpalus rufipes*	Fecundity	High		7
Mixed seeds 7 spp.		*Harpalus rufipes* larvae	Survival	Interm.		7
Poa annua		*Amara similata*	Fecundity	Low		6
Poa annua		*Amara similata* larvae I	Survival	Low		6
Poa annua		*Harpalus rufipes* larvae	Survival	High		7
Taraxacum sp.		*Amara similata*	Fecundity	Interm.		6
Taraxacum sp.		*Amara similata* larvae I	Survival	Low		6
		larvae II–III		High		
Taraxacum sp.		*Harpalus honestus* larvae	Developmental rate	High		4
Tripleurospermum inodorum		*Amara similata*	Fecundity	Interm.		6
Tripleurospermum inodorum		*Amara similata* larvae	Survival	Low		6
Tripleurospermum inodorum		*Harpalus rufipes* larvae	Survival	Interm.–High		7
Capsella bursa-pastoris		*Amara similata* larvae	Survival	High		6
Papaver rhoeas		*Amara similata* larvae	Survival	Interm.		6
Trifolium repens		*Amara similata* larvae	Survival	Low		6
Trifolium repens		*Harpalus rufipes* larvae	Survival	Interm.		7
Daucus/Umbelliferae		*Harpalus honestus* larvae	Developmental rate	High		4
Umbelliferae		*Ophonus ardosiacus* larvae	Survival	High		4
Gastropods						
Tandonia budapestensis		*Pterostichus melanarius*	Survival	Toxic		10
Earthworms						
2 spp.		*Abax parallelepipedus* larvae	Survival, growth rate	High (no comparisons)		9
Lumbricus terrestris		*Agonum dorsale*	Fecundity	Poor–Low		2
Insects						
Mixed insects 4 spp.	[Mixed seeds]	*Amara similata*	Fecundity	Low	[No]	6
Mixed insects 3spp.	[Mixed seeds]	*Amara similata* larvae	Survival	Low	[No]	6
Mixed insects 3 spp.	[Mixed seeds]	*Harpalus rufipes*	Fecundity	Low	[No]	7
Mixed insects 3 spp.	[Mixed seeds]	*Harpalus rufipes* larvae	Survival	Low	[No]	7
Lepidoptera						
Lymantria dispar larvae		*Calosoma sycophanta*	Fecundity	High		12

Rhopalosiphum padi	[*Poa annua* seeds]	*Amara similata*	Fecundity	Poor–Low	[No]	6
Rhopalosiphum padi		*Calathus melanocephalus*	Fecundity	Low		3
Rhopalosiphum padi	[Cat food]	*Poecilus cupreus*	Fecundity, egg size	High?	[No]	11
Rhopalosiphum padi	[Cat food]	*Bembidion lampros*	Fecundity, egg size	Interm.	[Pos.]	11
Rhopalosiphum padi	[Cat food]	*Pterostichus melanarius*	Fecundity, egg size	Low–Interm.	[No]	11
Rhopalosiphum padi		*Agonum dorsale*	Fecundity	Low		2
Rhopalosiphum padi		*Amara similata* larvae	Survival	Poor		6
Rhopalosiphum padi		*Harpalus rufipes* larvae	Survival	Low		7
Rhopalosiphum padi		*Bembidion lampros* larvae	Survival	Interm.		8
Sitobion avenae	[*Capsella bursa-pastoris* seeds]	*Amara similata* larvae	Survival	Poor	[No]	6
Sitobion avenae		*Calathus melanocephalus*	Fecundity	Low		3
Sitobion avenae		*Bembidion lampros* larvae	Survival	Interm.		8
Metopolophium dirhodum		*Amara similata*	Fecundity	Low		6
Metopolophium dirhodum		*Calathus melanocephalus*	Fecundity	Low		3
Metopolophium dirhodum		*Amara similata* larvae	Survival	Poor		6
Metopolophium dirhodum		*Bembidion lampros* larvae	Survival	Interm.		8
R.padi+S.avenae+M.dirhodum-mix		*Calathus melanocephalus*	Fecundity	Low		3
Collembola						
Orchesella cincta		*Notiophilus biguttatus*	Fecundity	High (no comparisons)		5
Orchesella cincta		*Notiophilus biguttatus* larvae	Growth rate, survival, adult size	High (no comparisons)		5
Isotoma anglicana		*Bembidion lampros*	Fecundity	Low–Interm.		1
Isotoma anglicana		*Bembidion lampros* larvae	Survival, development, adult size	Interm.–High		1
Isotoma notabilis		*Bembidion lampros*	Fecundity	Poor		1
Isotoma notabilis		*Bembidion lampros* larvae	Survival, development, adult size	Low–Interm.		1
Isotomurus prasinus		*Bembidion lampros*	Fecundity	Low–Interm.		1
Folsomia fimetaria		*Bembidion lampros*	Fecundity	Low–Interm.		1
Folsomia fimetaria		*Bembidion lampros* larvae	Survival, development, adult size	Interm.–High		1
Lepidocyrtus cyaneus		*Bembidion lampros*	Fecundity	Low–Interm.		1
Mixed animal diets						
Rhopalosiphum padi+Lumbricus terrestris+Drosophila melanogaster		*Agonum dorsale*	Fecundity	High		2

Refs.: 1 – Bilde *et al.*, 2000; 2 – Bilde and Toft, 1994; 3 – Bilde and Toft, 1999; 4 – Brandmayr, 1990; 5 – Ernsting *et al.*, 1992; 6 – Jørgensen and Toft, 1997a; 7 – Jørgensen and Toft, 1997b; 8 – Sørensen, 1996; 9 – Symondson, 1994; 10 – Symondson, 1997; 11 – Wallin *et al.*, 1992; 12 – Weseloh, 1993.

consumption in any species studied, and was frequently significantly lower. In most species, capacity for consumption of *R. padi* was lower than for *Sitobion avenae*, which was again lower than for *Metopolophium dirhodum*. Consumption capacity for the latter species was generally not different from that of fruit flies. The background for the different palatabilities of these aphids is unknown. Consumption rates of the three cereal aphids by *Bembidion lampros* were also low (Sørensen, 1996). However, *B. lampros* larvae were able to develop through all instars to adult stage on single-species diets of the three aphids, and thus seem to be more tolerant to aphids than are the adults (Sørensen, 1996). Still, the larval developmental time and the weight of the teneral adults ranked the three aphids species in the same order as found for consumption rates of other predators. One species, *Calathus melanocephalus*, showed a consumption capacity of *R. padi* as high as that for fruit flies (Bilde and Toft, 1997b). Bilde and Toft (1999) tested the hypothesis that a high tolerance to this prey would indicate a better ability, compared to other carabid species, to utilize aphids for maintaining performance. This was not the case. Single-species diets of the three cereal aphids (as above) and the 3-species mixed diet all resulted in similarly decreasing fecundity rates, indicating depletion of resources for egg laying. Control beetles held on fruit flies maintained a high egg-laying rate. Thus, tolerance to consumption of these prey showed no correspondence to their food value, neither between aphid species nor between aphids and fruit flies.

The evidence for aphids as low quality prey for carabid beetles is all based on laboratory studies in which the diets of individual beetles can be controlled. Supporting evidence from the field can be obtained indirectly by testing predictions based on the low-quality assumption. Pollet and Desender (1988) found by stomach dissections of 12 common carabid beetles that each stomach contained only one, or very few, aphids. Collembola, on the other hand, were often found in multiples in the same species. This is the predicted pattern if aphids are low-quality prey and Collembola are high-quality prey forming the bulk of the biomass consumed (see below).

Another prediction is that carabids should aggregate where there are high densities of their preferred food, but not at high densities of aphids or other low-quality prey. In seeming contradiction of this, Bryan and Wratten (1984) demonstrated in field experiments that several carabids aggregated with aphids. The aphid densities created in this study were exceptionally high. However, Monsrud and Toft (1999) hypothesized that potential high-quality prey insects would be more abundant at aphid colonies because they were attracted to honeydew secreted by the aphids. They found a significant attraction of Diptera and several species of carabid beetles to plots where honey was sprayed, compared to plots with augmented aphid numbers, as predicted. Interestingly, they found significant aggregation of carabid and staphylinid larvae where aphid numbers had been augmented. This would be expected if aphids can be better utilized by the larvae, as indicated above. However, most larvae are relatively immobile, whereas aphid populations show considerable spatial fluctuation (Winder *et al.*, 1999). The spatial response of carabids to cereal aphids is currently investigated using naturally occurring populations (J.M. Holland, pers. comm.).

Since aphids are supplementary food that cannot usually be consumed in high quantities by individual beetles, and that cannot alone support carabid populations, the densities of carabid populations, and consequently their potential impact on aphid populations, depend on the presence of high-quality alternative prey.

COLLEMBOLA

Together with Diptera, springtails may be an important group of prey supporting carabid beetle populations, at least as regards small- to medium-sized species. They occur in most ecosystems in very high numbers and often with many species adapted to different microhabitats (Hopkin, 1997). Some are agile surface-active species with a powerful furca and a good jumping ability. Others live in the soil spaces, have reduced or no furca, and rely on chemical defences against their predators. Toxic or deterrent compounds have recently been identified from species of several families (Dettner *et al.*, 1996; Messer and Dettner, 1997; Messer *et al.*, 2000) and further behavioural evidence was reviewed by Messer *et al.* (2000). It is therefore highly relevant to ask if carabids accept Collembola indiscriminately, or whether some types of Collembola are more important than others. Very few studies have been published that specifically compare the value of Collembolan species to a carabid.

Bilde *et al.* (2000) measured the egg-laying rate of *Bembidion lampros* held on single-species diets of five Collembola species, compared to fruit flies *D. melanogaster*. Four of the species were Isotomidae and one an Entomobryidae. All Collembola species turned out to be lower quality food to *B. lampros* than fruit flies. Comparing the feeding rates and the efficiency of prey utilization created a complex pattern, indicating a diversity of mechanisms were responsible for the predator–prey interactions. Some species were eaten in only low amounts, but the efficiency of transforming food into eggs was high; such species probably have strong feeding deterrents, but a good nutrient composition and no toxins. Others were eaten in rather large amounts, though fecundity was low, indicating either poor nutrient composition or the presence of a weak toxin. One species was eaten to only a very small extent, but this stopped egg production completely. Only on fruit flies were both consumption rate and conversion efficiency high. Selecting the right kind of Collembola for food may thus have strong influence on beetle fitness.

Bilde *et al.* (2000) also analysed the value of three of the Collembola species and fruit flies to *B. lampros* larvae by raising them on single-species diets. Though differences in quality were found, and fruit flies were again of highest quality, all three Collembola species (including the one that stopped egg production) allowed complete development to adult beetles of at least some individuals. This indicates that the larvae may be better adapted to cope with the presumed chemical defences of these Collembola than are the adults. It is tempting to speculate that carabid larvae may be closer coevolved with the soil-living microarthropods.

The intermediate value of the tested Collembola species to *B. lampros* is probably not fully representative of the role of Collembola to carabid beetles. First, several species have evolved as Collembola specialists. Thus, at least some springtail species are likely to be high-quality food to them (cf. Ernsting *et al.*, 1992), but studies comparing with other prey types are not available. *Loricera pilicornis* showed a higher consumption rate of *Isotoma anglicana* than of fruit flies and aphids (Bilde and Toft, 1997b). This springtail was also tested with *B. lampros* (Bilde *et al.*, 2000). It is possible that it is a high quality prey to the Collembolan specialist, but intermediate quality to generalist *B. lampros*. Collembola species of the same genus or family may differ widely in quality as food for the same predator (Toft and Wise, 1999a; Bilde *et al.*, 2000).

DIPTERA

Adult Diptera are persistently recorded at high frequencies in stomach dissection studies of generalist insectivores (Sunderland, 1975; Hengeveld, 1980a,b; Pollet and Desender, 1985, 1987, 1989). This is one argument for our use of *D. melanogaster* (fruit fly) as standard comparison food of high quality in our studies. With very few exceptions, fruit flies have turned out as the top quality food in tests with generalist insectivores (Bilde and Toft, 1994, 1997a,b, 1999). Drosophilids are also common in most kinds of habitats, including agricultural fields, and they were recorded as food for *P. melanarius* by Pollet and Desender (1985). Unfortunately, the value of other Diptera, as well as Diptera larvae and pupae, has not been systematically studied.

Sciarid midges: specific mention of Sciaridae as carabid prey is given by Pollet and Desender (1985). Larvae of sciarid midges are abundant in most types of soils and may be an important food source for small carabid larvae and small adults. Sciarid midges are low-quality food for wolf spiders (Toft and Wise, 1999a) and the available evidence indicates that the same may be true for carabids. Thus, *A. dorsale* showed a low preference for fruit flies coated with homogenate of adult sciarids compared to uncoated flies (Bilde and Toft, 1994).

LEPIDOPTERA

Calosoma sycophanta can sustain a high reproductive rate on a monotypic gypsy moth caterpillar diet (Weseloh, 1993). Most probably, this species, and larvae and pupae of several other species of moths and sawflies, are good carabid food. Large carabids thus contribute substantially to the density-dependent pupal mortality of winter moths (cf. den Boer, 1986). Greater wax moth larvae, however, were poorly utilized by *C. sycophanta* (Spieles and Horn, 1998).

DIPLOPODS

Snider (1984) found in laboratory experiments that five species of *Pterostichus* consumed large numbers of diplopods, in spite of the defensive secretions of these animals. The sizes taken depended on the size of the beetles. *Pterostichus* larvae ate large numbers of young instar diplopods. As a *Staphylinus* sp. was able to develop from hatching to imago on a sole diplopod diet, it is suggested that the carabids might do the same.

SEEDS

It is now well established that species of Harpalini, Amarini, Zabrini, Ditomini a.o. rely on seeds as their main food source (Schremmer, 1960; Thiele, 1977; Luff, 1980). For at least some species of *Harpalus* and *Amara* this holds true for both the adults and the larvae (Brandmayer, 1990; Jørgensen and Toft, 1997 a,b). Different seeds vary in food quality to these seed specialists; however, Jørgensen and Toft (1997a) found no correspondence in seed quality between adults and larvae. This may be because quality is determined mainly by seed size and hardness rather than nutritional composition; some highly nutritious seeds may be inaccessible to the small larvae.

The evolution of granivory from carnivory is understandable given the often high concentration of protein and lipids found in seeds as in carnivore food (Slansky and Panizzi, 1986).

Seeds are accepted also by species of generalist feeding guilds, e.g. *Carabus*, *Bembidion*, *Trechus*, *Calathus* and *Pterostichus* (Goldschmidt and Toft, 1997). The amounts of seed eaten by these species were much below that of *Harpalus rufipes*, even if no better food was available. This indicates a limited ability to utilize seed in the generalist carnivores and insectivores, but nothing is known about the nutritional value of seed to these carabids. Specialist Collembola feeders (*Loricera*, *Notiophilus*) did not accept seeds. Consumption of seeds by Carabidae is discussed further in Tooley and Brust (this volume).

PLANT MATERIAL

Unspecified plant material is often listed in stomach dissection studies. Dawson (1965) distinguished fragments of higher plants, pollen, fungal hyphae and spores, and diatoms. From an ecological point of view, distinction is important, partly because of the great differences in nutrient content, partly because granivory, phytopagy and detritivory have very different impacts in the ecosystem. According to Thiele (1977), *Zabrus* 'eats young leaves of the newly sprouting winter crop', and Luka *et al.* (1998) report herbivory on rape plants (young fruits, stems, flowers) by *Amara* spp., but to a limited extent. No laboratory study has evaluated green plant material, detritus or fungi as food for carabids. Goldschmidt and Toft (1997) compared the preferences of 25 species for seeds and pieces of fresh wheat leaf. Phytophagy (consumption of green leaves) occurred in most insectivorous species (except some Collembola specialists and a few of the generalist insectivores); quantitatively, phytophagy occurred in these species at a lower or same level as granivory. In the granivore *Harpalus rufipes*, phytophagy was recorded, but in even lower frequency than in the insectivores. Brandmayr (1990) suggested that consumption of leaves and fruits by carabid beetles served only to obtain water. Sota (1984) compared the effects of minced beef and various fruits on performance of *Leptocarabus kumagaii*. On the fruit diet, oviposition ceased and the females lost weight, but they stayed alive for the one-month observation period. Also, the males were able to stay alive for a long time on the fruit diet. According to Weseloh (1993), *Calosoma sycophanta* females cease egg laying if fed only grapes. Sota (1984) concluded that acceptance of plant food might enhance survival under food limited conditions.

Phytophagy, detritivory and fructivory thus occur in many carabid species as supplementary feeding activity. Their nutritional importance for growth and reproduction is probably low.

SNAILS AND SLUGS

The fact that some carabids have specialized as snail and slug feeders in both larval and adult stages would indicate a high food quality of these prey. Other large generalist carnivore species (e.g. *Carabus* spp., *Abax parallelepipedus*, *Pterostichus melanarius*) have been found to readily accept slugs and snails (Scherney, 1959, 1961; Loreau, 1983a,b; Symondson, 1994). Symondson (1993) and Symondson *et al.*

(1996) also found evidence that slugs were preferred prey for the two latter species, and Symondson *et al.* (1996) and Bohan *et al.* (2000) presented evidence for important predation on slugs. Their results indicate preferential feeding and aggregation at high slug densities. This is only indirect and unreliable evidence of a high food quality of slugs, and slugs may become abundant without a concomitant increase of large carabids. The relative food value of slugs to various generalist carnivore and insectivore carabids thus needs more detailed study. If slugs are only mediocre food, the control efficiency of the large carabids will depend on presence of high-quality alternative prey.

One species of slug was found to be toxic to *Pterostichus melanarius* that may die within a few days after eating it (Symondson, 1997). Production of mucus by slugs may reduce or prevent predation by many carabids (Symondson, in press). Little seems to be known about predation on gastropod eggs, which are often highly vulnerable to predation from carabids. In several species, the eggs contain defensive chemicals (Symondson, in press).

EARTHWORMS

Earthworms are assumed to be high-quality food for some species of the generalist carnivore guild because they are frequently found in stomach analyses, and some species have been bred on earthworms (Symondson, 1994, and references therein). Mass rearing of *Abax parallelepipedus* larvae on a sole diet of earthworms without cannibalism is evidence of nutritional sufficiency of this prey. Symondson *et al.* (2000) found that *P. melanarius* consumed more earthworms when other prey were scarce, indicating that earthworms are only acceptable but non-preferred food. Comparative performance studies are lacking, however, for these species. Earthworms were low-quality food for two species of generalist insectivores, e.g. with respect to larval growth and survival in *P. nigrita* (Ferenz, 1973, cited after Thiele, 1977), and with respect to fecundity in *Agonum dorsale* (Bilde and Toft, 1994).

OTHER PREY GROUPS

Some prey groups may be widespread supplementary food or essential food for local populations or single species. Isopods may be extremely abundant in many habitats. In spite of this, carabid predation seems to be limited (Sunderland and Sutton, 1980). By contrast, talitrid Crustacea are claimed to be the chief dietary constituent of the seaside-living *Eurynebria complanata* (King and Stabins, 1971). Where they are abundant, terrestrial caddis fly larvae may be important prey for *Pterostichus oblongopunctatus* (den Boer, 1986). Comparative studies on the food value of these prey are lacking.

Food quality in relation to dietary guilds

Several papers give simple recipes on how to raise carabid beetles in the laboratory. Thus, Goulet (1976) claimed to have bred 80 species, including representatives from all the above listed feeding guilds, with mealworms as the only food. Thiele (1977) used mealworms, minced beef or dog biscuits, while Burakowski (1986) used ant larvae and pupae. This ability to subsist on what must be regarded 'atypical' food for several of the species involved underlines the earlier statement that carabid feeding specializations are only partial. The exact degree of specialization (i.e. relative ability to utilize the main potential food types for improved fitness) has still not been elucidated for any species. Only scattered pieces of the puzzle are available.

In the insectivore *Agonum dorsale*, Bilde and Toft (1994) found a higher fecundity when fed *D. melanogaster*, compared to earthworms. Unfortunately, no similar study seems to exist for an earthworm or a slug specialist. Two generalist carnivores (*Pterostichus madidus* and *Abax parallelepipedus*) showed different slug antigen decay rates in the crop in spite of no difference in the rate with which slug food was processed through the crop (Symondson and Liddell, 1993). This may indicate a different ability to digest slug protein and thus possibly reflect a different degree of adaptation to slug predation.

The most extensive comparative information is available on species of the granivorous guild. Cornic (1973) reported a higher proportional intake of insect food during reproductive periods in *Harpalus affinis* and *H. rufipes*, suggesting a nutritional benefit of insectivory for egg development. These results were not confirmed in the laboratory tests of Jørgensen and Toft (1997a,b), who found substantially higher fecundity in *Amara similata* and *H. rufipes* on mixed-seed diets than on mixed-insect diets. In none of the species did the combined diet of mixed seeds + mixed insects improve fecundity, compared to the reference diets. Thus, insects as supplementary food were of no value. Allen and Hagley (1982) were able to detect antigen from Lepidoptera larvae longer in both *Harpalus affinis* and *Amara* sp. than in *P. melanarius*, indicating a slower digestion rate of insect material in the granivorous than in the insectivorous species. Larvae of *A. similata* were even more dependent on seeds than the adults since they were unable to reach maturity on an insect diet (Jørgensen and Toft, 1997a). *H. rufipes* larvae, on the other hand, were better able to utilize insects for development; seeds, insects, and the combined diet gave the same success (Jørgensen and Toft, 1997b).

In conclusion, though the value of supplementary foods may vary between species of the same feeding guild, the scattered evidence supports the notion that supplementary foods are usually of lower quality than essential foods. This may not sound very surprising. However, it may help to resolve the conflict between the seemingly specialized mouthpart morphology of many species (see Ingerson-Mahar, this volume) and their 'opportunistic polyphagy', as indicated by stomach dissections. The food quality data support the conclusion of limited polyphagy, as do the morphological evidence.

As supplementary food types may sometimes turn out to have very low nutritional value as measured in fitness tests, the question arises as to why they are so often recorded in stomach contents. A good example is the consumption of aphids in cereal fields. Sunderland *et al.* (1987) found aphids in the stomachs of 81 taxa of generalist predators, but probably they are low-quality food to all of them (cf. Bilde and Toft, 1994, 1997a,b, 1999, 2000; Toft, 1995). Ease of capture, and the need to learn (if possible) the nutritional value or deterrence of a prey type in order to reject it, may be one explanation. Another possible answer may be that supplementary foods are of sufficient value to keep the beetles alive, even if insufficient for active growth and reproduction. Acceptance may thus have survival value, especially in periods of food scarcity. Extended survival on a pure aphid diet was found in nymphs of a linyphiid spider, though the aphid contributed nothing to growth (Toft, 1995). Another possibility is that the aphids in some dietary combinations do contribute significantly to predator fitness and therefore are not generally rejected (Bilde and Toft, 1994). Bilde and Toft (2000) found some evidence in a spider that aphid consumption improved fitness if the remaining diet was of relatively low nutrient quality, but not if it was high quality.

Physiological and biochemical correlates of dietary specialization

The digestive system of carabid beetles comprises all the main types of digestive enzymes, including proteases, lipase/esterases, chitinase, cellulase, amylase, laminarinase, and oligosaccharidases (Forsythe, 1982b; Jaspar-Versali and Jeuniaux, 1987; Jaspar-Versali *et al.*, 1987; Terra *et al.*, 1996). Both fluid and mixed feeders regurgitate mid-gut enzymes for initial pre-oral digestion. In all carabids, the main digestive activity is concentrated in the crop, though again accomplished by mid-gut secretions. Terra *et al.* (1996) concluded that insect digestive enzymes are quite similar, in spite of differences in diet composition, since 'most species possess the full array of hydrolases'. This generalization may be true at a superior level, but should possibly be modified in the details. Vaje *et al.* (1984) found high intra- and especially inter-specific variation in proteases in *Carabus* spp., though the functional significance of this diversity is unknown. Jaspar-Versali and Jeuniaux (1987) demonstrated differences in the activity of several digestive enzymes among species of different feeding guilds. They found much higher cellulase activity in *Abax parallelepipedus* and *P. melanarius* (both feeding frequently on vegetable matter) than in five more strictly carnivorous *Carabus* spp. These results are suggestive of a fitness-enhancing effect of vegetable consumption in some species which still has to be demonstrated. *Carabus*, on the other hand, had higher levels of protease activity. The two groups showed equal chitinase activity; for both, insects form an important part of the diet. A similar study including granivore species would be highly welcome.

Food preferences

Food preferences as they are measured in standard laboratory experiments, are always a composite phenomenon, reflecting factors relating to discovery of the prey (activity), ease of capture (e.g. size, escape ability), ease of handling (e.g. size, hardness), and food value (e.g. palatability, nutrient content) (Mitchell, 1963; Pollet and Desender, 1986; Bilde and Toft, 1994). Wheater (1988) found a positive correlation between gape distance of carabid species and median prey size. Pearson and Mury (1979) found similar relationships in tiger beetles. Hardness imposes costs of handling, but heavily sclerotized prey also wears the mandibles more than soft prey, with possible negative consequences for foraging efficiency and reproduction (Houston, 1981; Wallin, 1988). The same may be true for seeds. Hard seeds may pose mechanical difficulties for small larvae of granivore species and thus be out of the list of potential food, in spite of being high-quality food for larger larvae (Jørgensen and Toft, 1997a).

Pollet and Desender (1987) found that predation on several insect groups was positively correlated with prey density, but negatively correlated with prey activity. Ernsting and Jansen (1978) showed how prey mobility is important for the visual hunter *Notiophilus biguttatus*. A particularly instructive experiment was performed by Lang and Gsödl (2001), who compared the preferences of *Poecilus cupreus* for crickets, *Drosophila* and the cereal aphid *Rhopalosiphum padi*, when these prey were offered live and dead, respectively. When prey was offered alive, the order of preference was aphid > *Drosophila* > cricket. When offered dead, preferences were reversed. Obviously, with dead prey, the beetles selected according to food value, while ease of capture determined preference order of live prey. This may also explain the findings of Kielty *et al.* (1999), who found preference of *P. cupreus* and *P. melanarius* for cereal aphids over entomobryid Collembola in live prey experiments.

Carabid beetles of all feeding guilds prefer highly nutritious foods: animal matter or seeds. However, the requirements for specific nutrients have not been determined (Lövei and Sunderland, 1996). R. Wedfeldt and M. Kirkegaard (unpublished) used completely artificial agar-based diets to test for ability to select food with different protein:carbohydrate ratios (80:20 or 20:80). They found non-random selection for the high-protein food in two species, *Agonum dorsale* and *P. melanarius*.

Hengeveld (1980a) discussed food specializations in Carabidae, stating that 'a specialist is not simply a species with fewer diet components, but one that selects its prey'. Given the dramatic differences in quality of different potential prey types and the different consumption rates of various prey, it is clear that all carabids must show some degree of selectivity. We prefer to think of specialist and generalists as animals that select before and after capture, respectively. Specialists probably respond to a more restricted range of cues (may have an innate 'search image') characteristic of their preferred food. Generalists initially accept a very wide range of foods but become more selective as a result of experience. This increased selectivity is most likely to be based on the development of aversions to low-quality food (cf. Toft, 1997; Toft and Wise, 1999a,b). The different mechanisms of selectivity will result in a wider prey spectrum for the generalists because they need to test all potential prey types before they become selective.

Acquired aversions may develop to sub-optimal prey whether the inferiority is due to skewed nutrient balance or chemical defences (Bernays, 1993), but are not always a safeguard against toxin consumption. Herbivores or detritivores with chemical defences may have sequestered the toxin from their own food. These toxins in turn reduce the food value of the prey to their predators. In extreme cases, this may have detrimental effects on populations of carabid beetles. Den Boer (1986) described how the fecundity of *Calathus melanocephalus* was reduced during outbreaks of the heather beetle (a chrysomelid beetle). The carabid population fluctuated in a delayed density-dependent manner. Presumably, when the toxin containing larvae of the heather beetle became very numerous, the carabids ate more than they could tolerate, and the populations crashed immediately after an outbreak.

In future agricultural systems, carabids may face prey populations developing on genetically modified crops with artificial pest defences incorporated. Like natural defences, these may create tritrophic effects by interfering with the predator–herbivore interactions higher up in the food chain, whether this is due to sequestration or just to altered chemical composition of the herbivore. Jørgensen and Lövei (1999) simulated this situation by incorporating a proteinase-inhibitor into the food of a caterpillar, and subsequently offered the caterpillar to a carabid, *Harpalus affinis*. They found a significantly reduced feeding rate on the modified food compared to controls.

Dietary mixing

Generalist consumers have often been found to respond positively to offerings of mixed diet compared to diets of a single food type (Waldbauer and Friedman, 1991). It is generally assumed that the benefit obtained is due either to a higher diversity of nutrients in the mixed diet, or to the better possibilities for self-selection of the optimal nutrient balance with a diverse availability. So far, there are very few studies testing these effects for carabid beetles, representing only few of the feeding guilds. In a spider, Toft and Wise (1999a) found that no or negative effects of prey mixing were more frequent than positive effects, depending on the quality of the diet constituents, and similar results may pertain to carabids (cf. *Table 3.1*). So far, mostly neutral and positive mixing effects have been published, and no case of negative mixing effects has been reported from carabid beetles. Wallin *et al.* (1992) found both egg size and rate of egg laying in *Bembidion lampros* to be greater when fed a mixed diet of aphids and cat food, compared to the monotypic diets. In two other species, no differences were found. In *Agonum dorsale*, fecundity was greatly increased by the mixed diet (fruit flies, aphids and earthworm) compared to single-prey diets of each constituent (Bilde and Toft, 1994). In the granivore *Amara similata*, a mixed-seed diet improved female fecundity but not larval development compared to the best single-seed diet (Jørgensen and Toft, 1997). Mixing of seeds and insects, however, made no improvement compared to the mixed-seed diet. In *H. rufipes*, the mixed-seed diet gave only the same rate of larval development as the best of its elements (Jørgensen and Toft, 1997b).

Effects of diet quality may not only influence the consumer itself, they may also be transmitted to the consumer's offspring as maternal effects. Thus, under identical conditions, larvae of *Amara similata* whose mothers were kept on a monotonous seed diet survived better than larvae whose mothers had been kept on the higher-quality, mixed-seed diet. Thus, high value to the female corresponded to low performance of the young. Most likely, females on the monotypic diets laid fewer but larger eggs than females on the high-quality, mixed diets (cf. Wallin *et al.*, 1992).

If dietary mixing is advantageous, it means that the single food types alone are not high-quality. It can be hypothesized that the essential foods to which members of a feeding guild are particularly adapted are nutritionally more complete than supplementary foods. If this holds true, it can be predicted that diets should be more mixed (have more complementary constituents in order to compose a complete diet) if it consists of supplementary food types than if it consists of essential foods. An alternative hypothesis may be that supplementary foods contain more defensive deterrents and toxins than essential foods, both being equally variable with respect to nutrient composition. In this case, all foods potentially may complement each other to produce the same benefit, but inclusion of toxic food may interfere with the predator's utilization of essential food and thus prevent it from obtaining a benefit. At this stage, it is not possible to evaluate these alternatives as far as carabid beetles are concerned. Results from spiders are in best agreement with the second hypothesis (Toft and Wise, 1999ab).

Food limitation

Evidence that carabid beetles are limited by food availability in many habitats, including agricultural fields, comes from different sources. Granivorous species have repeatedly been demonstrated to respond positively to increased weed density in agricultural fields (de Snoo *et al.*, 1995; Holland *et al.*, 1999), indicating that populations are limited by seed availability (or they may be seeking out the environmental conditions provided by the weeds). For the insectivore *Agonum dorsale*, Bilde and Toft (1998) used the degree of hunger of field-caught animals to compare seasonal variations in food limitation. This index is better than direct measurement of prey availability because it also accounts for seasonal variations in food demand.

Food limitation in the early spring when carabids invade agricultural fields from surrounding uncultivated habitats may be serious. Bilde and Toft (1998) estimated an average degree of hunger equivalent to 2–3 weeks of starvation at 20°C for *A. dorsale* in a winter wheat field in May. This is likely to have consequences for the reproductive rate of this spring breeding population and thus for its ability to affect immigrating aphids later in the season (cf. Sunderland, this volume). Since aphids are low-quality food, aphid predation is unlikely to increase significantly, even if the carabids aggregate where the aphids are plentiful (cf. Monsrud and Toft, 1999). Therefore, the efficacy of predator limitation of aphids can only be enhanced by finding ways to increase the amount of alternative food available to the predators, and in this way increase the number of predators in the field.

Exploitative competition

In line with the conclusions of Hengeveld (1980b), Dennison and Hodkinson (1983) stressed the high degree of dietary overlap among coexisting carabid and staphylinid species, implying potential competition for food also among species with different specializations. However, consideration of food value is also a necessity before potential competition can be deduced from food overlap. Our results (Jørgensen and Toft, 1997a,b) indicate that seed feeders (*Amara*, *Harpalus*) will not compete with generalist insectivores, in spite of high dietary overlap with these, because the insect part of the diet has little or no importance for their fitness, and the seed part of the diet probably has little importance for the fitness of insectivores. Reduced availability of supplementary food will therefore have little importance. Also, if supplementary prey is of low quality, individuals are not likely to aggregate where they occur in high densities, even if they are tolerant of consuming a lot. Only exploitative reduction of high-quality essential food types is likely to have potential competitive effects. We therefore believe that consideration of food value should be more explicitly incorporated in the definition of species' food niches.

Field experiments aimed at demonstrating competition between carabid species have concentrated on coexisting congenerics with not only a high dietary overlap, but presumably also very similar ranges of essential prey where it is thus most likely to occur. Lenski (1982, 1984) found evidence for improved foraging success and interspecific competition between two *Carabus* species following density manipulations. In another similar case, there was no evidence for exploitative food competition, in spite of significant food limitation of the involved species (Juliano, 1986).

Food value and pest limitation

Carabids have often been mentioned as potentially important predators on several agricultural or forest pests, e.g. aphids (Sunderland, 1975; Sunderland *et al.*, 1987), cabbage root fly eggs (Hughes, 1959; Coaker and Williams, 1963), wheat midges (Floate *et al.*, 1990), wireworms (Fox and MacLellan, 1956), codling moth larvae (Hagley *et al.*, 1982), cutworms (Frank, 1971), slugs (Symondson *et al.*, 1996; Bohan *et al.*, 2000) and gipsy moths (Weseloh, 1985). In most of these cases, natural control is exerted by the community of carabid beetles and other generalist predator species that have not specialized in preying on the pest. The probable exception is that of forest caterpillars, preyed upon by *Calosoma sycophanta* (Weseloh, 1985). Here, the prey seems to be a high-quality exclusive food for the predator. In the other cases, the pest is just one of a wide range of prey taken. Often, consumption of the pest is probably of little importance to predator fitness because it forms just a small fraction of its food intake (fly eggs), in some cases it may even be supplementary low-quality food for the beetles (aphids). In the latter situation, the feeding specializations of the predator may be more or less irrelevant for their impact on the pest. Even the granivorous species prey frequently on pest insects (e.g. aphids: Sunderland *et al.*, 1987; codling moth larvae: Hagley *et al.*, 1982), though this is of little benefit to them (Jørgensen and Toft, 1997a,b). If the beetles or their larvae are numerous, supplementary predation may anyway have a significant effect on the pest, even if there is no aggregating response, and individual feeding rate on the pest is low. In order for this to be true, the pest must have a specially vulnerable phase in the population cycle, as is the case with aphids, where even a numerically low predation rate at the aphid immigration period have consequences for subsequent population growth (see further Sunderland, this volume).

A comparable example was given by Lukasiewicz (1996), who pointed out that *P. melanarius*, in spite of a low frequency of earthworm predation, might be more important for reduction of earthworms than the earthworm specialist, *Carabus granulatus*, because of its high population densities. Granivore carabids may also have an impact on the seed population of plants whose seeds are highly preferred. Kjellson (1985) estimated that *Harpalus fuliginosus* removed 65% of the seed pool of *Carex pilulifera*. Despite this, effects on the plant population are doubtful.

Conclusion

Let us 'summarize' by listing some of the main open questions revealed by this review. The answers to some of these questions are very basic to our understanding of how carabids function in the biological communities.

- Which prey are essential high-quality food for the generalist insectivorous carabids in the field? Answering this question requires studies that combine quantitative research of field consumption with laboratory determinations of food value of potentially important prey types. We expect that studies are needed at the level of prey species (rather than families or orders).
- Is the essential prey of specialists also essential prey of corresponding generalists? For example, are the Collembola, to which the microarthropod specialists are particularly adapted, also high-quality food for the generalist insectivores?

Likewise, are the high-quality slugs of the mollusc specialists also high-quality food for the large generalist carnivores?

- How are specialist carabids adapted to overcome nutrient imbalances due to a one-sided diet? The evolution of diet specialization may be costly in terms of a trade-off in growth rate, fecundity or size due to nutritional imbalance. Can the performance of specialized predators be significantly improved by dietary mixing? Are the essential prey of specialists particularly rich in nutrients? Or have specialists adapted to cope with poor nutrient composition?
- Do generalist feeders have larger quantities of essential prey available than the specialists? Do generalist feeders have a wider range of essential prey species than the specialists, or are they generalists because they accept more intermediate- and low-quality food?
- What are the physiological mechanisms of feeding specialization? To what extent does it involve resistance against specific chemical constituents of prey, or depend on the compliment of digestive enzymes?
- What are the behavioural mechanisms of food selectivity for generalists and specialists? Do specialists initially accept a more narrow variety of food types than generalists? Do specialists develop aversions to unsuitable prey earlier than generalists?
- To what extent are carabids able to select a nutritionally optimal diet, or to compose an optimal mix of prey types containing deterrents or toxins? Can the degree of dietary mixing be behaviourally modified according to the nutritional condition of the individual?
- What is the role of non-essential food in the diets of (generalist) carabids? Some prey types (e.g. aphids) may contribute to the fitness of a carabid, though only small amounts are consumed; others may be useless, in spite of a substantial consumption rate. The same prey may be valuable to some carabid species and without value to others, even if the latter consumes the most.
- Is food limited both in terms of quantity and quality in cultivated fields, and more so in conventionally managed than in organic fields?
- The ecological role of carabid larvae. The need for more detailed studies of larval feeding habits has already been stressed. We have presented evidence that, at least in *Bembidion lampros*, several types of prey that are low quality to adults (e.g. aphids and some Collembola) may allow complete development of young, even as monotypic diets. These findings are likely to mean that larvae less easily develop aversions against these prey, and thus may be better able than the adults to respond functionally and numerically to increased densities; the numerical response has also been confirmed. Thus, larvae may not only have a very different role than the adults in the detritus-based food web of the soil, they may also be far more important as aphid antagonists than recognized so far.

Acknowledgements

We are indebted to the students and colleagues whose work in our laboratory has contributed much to our understanding of the nutritional aspects of predator feeding: Jørgen Axelsen, Jørgen B. Beck, Claus Borg, Henrik Goldschmidt, Helene Bracht Jørgensen, Mette Kirkegaard, Peter D. Kruse, Bente Marcussen, David Mayntz,

Carsten Monsrud, Flemming Sørensen, Rikke Wedfeldt, and several student assistants. Helpful comments on this manuscript were provided by Bill Symondson. The original work referred to was funded by grants from the Danish Environmental Research Program to the Centre for Agricultural Biodiversity, and from the Danish Research Centre for Organic Farming.

References

ALLEN, W.R. AND HAGLEY, E.A.C. (1982). Evaluation of immuno-electrosmophoresis on cellulose polyacetate for assessing predation of Lepidoptera (Tortricidae) by Coleoptera (Carabidae) species. *The Canadian Entomologist* **114**, 1047–1058.

ANDERSON, J.M. (1972). Food and feeding of *Notiophilus biguttatus* F. (Coleoptera: Carabidae). *Revue d'Ecologie et de Biologie du Sol* **9**, 177–184.

BAUER, T. (1981). Prey capture and structure of the visual space of an insect that hunts by sight on the litter layer (*Notiophilus biguttatus* F., Carabidae, Coleoptera). *Behavioural Ecology and Sociobiology* **8**, 91–97.

BAUER, T. (1982). Prey-capture in a ground-beetle larva. *Animal Behaviour* **30**, 203–208.

BAUER, T. (1985). Beetles which use a setal trap to hunt springtails: the hunting strategy and apparatus of *Leistus* (Coleoptera, Carabidae). *Pedobiologia* **28**, 275–287.

BAUER, T. AND KREDLER, M. (1988). Adhesive mouthparts in a ground beetle larva (Coleoptera, Carabidae, *Loricera pilicornis* F.) and their function during predation. *Zoologischer Anzeiger* **221**, 145–156.

BERNAYS, E.A. (1993). Aversion learning and feeding. In: *Insect learning*. Eds. D.R. Papaj and A.C. Lewis, pp 1–17. New York: Chapman & Hall.

BILDE, T. AND TOFT, S. (1994). Prey preference and egg production of the carabid beetle *Agonum dorsale*. *Entomologia Experimentalis et Applicata* **73**, 151–156.

BILDE, T. AND TOFT, S. (1997a). Limited predation capacity by generalist arthropod predators on the cereal aphid, *Rhopalosiphum padi*. *Organic Agriculture and Horticulture* **20**, 143–150.

BILDE, T. AND TOFT, S. (1997b). Consumption by carabid beetles of three cereal aphid species relative to other prey types. *Entomophaga* **42**, 21–32.

BILDE, T. AND TOFT, S. (1998). Quantifying food limitation of arthropod predators in the field. *Oecologia* **115**, 54–58.

BILDE, T. AND TOFT, S. (1999). Prey consumption and fecundity of the carabid beetle *Calathus melanocephalus* on diets of three cereal aphids: high consumption rates of low-quality prey. *Pedobiologia* **43**, 422–429.

BILDE, T. AND TOFT, S. (2000). Evaluation of prey for the spider *Dicymbium brevisetosum* Locket (Araneae: Linyphiidae) in single-species and mixed-species diets. *Ekológia (Bratislava)* **19** Suppl. 3, 9–18.

BILDE, T., AXELSEN, J.A. AND TOFT, S. (2000). The value of Collembola from agricultural soils as food for a generalist predator. *Journal of Applied Ecology* **37**, 672–683.

BOHAN, D.A., BOHAN, A.C., SYMONDSON, W.O.C., WILTSHIRE, C.W. AND HUGHES, L. (2000). Spatial dynamics of predation by carabid beetles on slugs. *Journal of Animal Ecology* **69**, 367–379.

BRANDMAYR, T.Z. (1990). Spermophagous (seed eating) ground beetles: first comparison of the diet and ecology of the Harpaline genera *Harpalus* and *Ophonus* (Col., Carabidae). In: *The role of ground beetles in ecological and environmental studies*. Ed. N.E. Stork, pp 307–316. Andover: Intercept.

BRYAN, K.M. AND WRATTEN, S.D. (1984). The responses of polyphagous predators to prey spatial heterogeneity: aggregation by carabid and staphylinid beetles to their aphid prey. *Ecological Entomology* **9**, 251–259.

BURAKOWSKI, B. (1986). The life-cycle and food preference of *Agonum quadripunctatum* (De Geer). In: *Feeding behaviour and accessibility of food for carabid beetles*. Eds. P.J. den Boer, L. Grüm and J. Szyszko, pp 35–39. Warsaw: Warsaw Agricultural University Press.

COAKER, T.H. AND WILLIAMS, D.A. (1963). The importance of some Carabidae and Staphylinidae as predators of the cabbage root fly, *Erioischia brassicae* (Bouché). *Entomologia Experimentalis et Applicata* **6**, 156–164.

CORNIC, J.F. (1973). Étude du régime alimentaire de trois espèces de carabiques et de ses variations en verger de pommiers. *Annales de la Societé Entomologique de France (N.S.)* **9**, 69–87.

DAJOZ, R. (1987). Le régimes alimentaire des coléoptères Carabidae. *Cahiers des Naturalistes, N.S.* **43**, 61–96.

DAVIES, M.J. (1953). The contents of the crops of some British carabid beetles. *Entomologist's Monthly Magazine* **89**, 18–23.

DAWSON, N. (1965). A comparative study of the ecology of eight species of fenland Carabidae (Coleoptera). *Journal of Animal Ecology* **34**, 299–314.

DEN BOER, P.J. (1986). Facts, hypotheses and models on the part played by food in the dynamics of carabid populations. In: *Feeding behaviour and accessibility of food for carabid beetles.* Eds. P.J. Den Boer, L. Grüm and J. Szyszko, pp 81–96. Warsaw: Warsaw Agricultural University Press.

DENNISON, D.F. AND HODKINSON, I.D. (1983). Structure of the predatory beetle community in a woodland soil ecosystem. I. Prey selection. *Pedobiologia* **25**, 109–115.

DE SNOO, G.R., VAN DER POLL, R.J. AND DE LEEUW, J. (1995). Carabids in sprayed and unsprayed crop edges of winter wheat, sugar beet and potatoes. In: *Arthropod natural enemies in arable land. I Density, spatial heterogeneity and dispersal.* Eds. S. Toft and W. Riedel, pp 199–211. Århus: Aarhus University Press.

DETTNER, K., SCHEUERLEIN, A., FABIAN, K., SCHULZ, S. AND FRANCKE, W. (1996). Chemical defence in the giant springtail Tetrodontophora bielanensis (Waga) (Insecta: Collembola). *Journal of Chemical Ecology* **22**, 1051–1074.

DIGWEED, S.C. (1993). Selection of terrestrial gastropod prey by cychrine and pterostichine ground beetles (Coleoptera: Carabidae). *The Canadian Entomologist* **125**, 463–472.

DREISIG, H. (1981). The rate of predation and its temperature dependence in a tiger beetle, *Cicindela hybrida*. *Oikos* **36**, 196–202.

ERNSTING, G. AND JANSEN, J.W. (1978). Interspecific and intraspecific selection by the predator *Notiophilus biguttatus* F. (Carabidae) concerning two collembolan prey species. *Oecologia* **33**, 173–183.

ERNSTING, G., JAGER, J.C., VAN DER MEER, J. AND SLOB, W. (1985). Locomotory activity of a visually hunting carabid beetle in response to non-visual prey stimuli. *Entomologia Experimentalis et Applicata* **38**, 41–47.

ERNSTING, G., ISAAKS, J.A. AND BERG, M.P. (1992). Life cycle and food availability indices in *Notiophilus biguttatus* (Coleoptera, Carabidae). *Ecological Entomology* **17**, 33–42.

EVANS, M.E.G. AND FORSYTHE, T.G. (1984). A comparison of adaptation to running, pushing and burrowing in some adult Coleoptera: especially Carabidae. *Journal of Zoology, London* **202**, 513–534.

EVANS, M.E.G. AND FORSYTHE, T.G. (1985). Feeding mechanisms, and their variation in form, of some adult ground-beetles (Coleoptera: Caraboidea). *Journal of Zoology, London* **206**, 113–143.

FLOATE, K.D., DOANE, J.F. AND GILLOTT, C. (1990). Carabid predators of the wheat midge (Diptera, Cecidomyiidae) in Sakatchewan. *Environmental Entomology* **19**, 1503–1511.

FORBES, S.A. (1883). The food relations of the Carabidae and Coccinellidae. *Illinois State Laboratory, Natural History Bulletin* **1**, 33–64.

FORSYTHE, T.G. (1982a). Feeding mechanisms of certain ground beetles (Coleoptera: Carabidae). *The Coleopterist's Bulletin* **36**, 26–73.

FORSYTHE, T.G. (1982b). Qualitative analyses of certain enzymes of the oral defence fluids of *Pterostichus madidus* (F.), including a list of other carabid beetles which produce oral defence fluids. *Entomologist's Monthly Magazine* **118**, 1–5.

FORSYTHE, T.G. (1983). Mouthparts and feeding of certain ground beetles (Coleoptera: Carabidae). *Zoological Journal of the Linnean Society* **79**, 319–376.

FORSYTHE, T.G. (1987). The relationship between body form and habit in some Carabidae (Coleoptera). *Journal of Zoology, London* **211**, 643–666.

FORSYTHE, T.G. (1991). Feeding and locomotory functions in relation to body form in five species of ground beetle (Coleoptera: Carabidae). *Journal of Zoology, London* **223**, 233–363.

FOX, C.J.S. AND MACLELLAN, C.R. (1956). Some Carabidae and Staphylinidae shown to feed on a wireworm, *Agriotes sputator* (L.), by the Precipitin test. *The Canadian Entomologist* **88**, 228–231.

FRANK, J.H. (1971). Carabidae (Coleoptera) as predators of the red-backed cutworm (Lepidoptera: Noctuidae) in Central Alberta. *The Canadian Entomologist* **103**, 1039–1044.

GOLDSCHMIDT, H. AND TOFT, S. (1997). Variable degrees of granivory and phytophagy in insectivorous carabid beetles. *Pedobiologia* **41**, 521–525.

GREENE, A. (1975). Biology of the five species of Cychrini (Coleoptera: Carabidae) in the steppe region of Southeastern Washington. *Melanderia* **19**, 1–41.

GOULET, H. (1976). A method for rearing ground beetles (Coleoptera: Carabidae). *The Coleopterist's Bulletin* **30**, 33–36.

HAGLEY, E.A.C., HOLLIDAY, N.J. AND BARBER, D.R. (1982). Laboratory studies of the food preferences of some orchard carabids (Coleoptera: Carabidae). *The Canadian Entomologist* **114**, 431–437.

HENGEVELD, R. (1980a). Qualitative and quantitative aspects of the food of ground beetles (Coleoptera, Carabidae): a review. *Netherland Journal of Zoology* **30**, 555–563.

HENGEVELD, R. (1980b). Polyphagy, oligophagy, and food specialization in ground beetles (Coleoptera, Carabidae). *Netherland Journal of Zoology* **30**, 564–584.

HENGEVELD, R. (1980c). Food specialization in ground beetles: an ecological or a phylogenetic process? (Coleoptera, Carabidae). *Netherland Journal of Zoology* **30**, 585–594.

HENGEVELD, R. (1981). The evolutionary relevance of feeding habits of ground beetles (Coleoptera: Carabidae). *Entomologica Scandinavica Supplement* **15**, 305–315.

HINTZPETER, U. AND BAUER, T. (1986). The antennal trap of the ground beetle *Loricera pilicornis*: a specialization for feeding on Collembola. *Journal of Zoology, London* **208**, 615–630.

HOLLAND, J.M., PERRY, J.N. AND WINDER, L. (1999). The within-field spatial and temporal distribution of arthropods in winter wheat. *Bulletin of Entomological Research* **89**, 499–513.

HOPKIN, S.P. (1997). *Biology of the springtails.* Oxford: Oxford University Press.

HORTON, D.R. AND REDAK, R.A. (1993). Further comments on analysis of covariance in insect dietary studies. *Entomologia Experimentalis et Applicata* **69**, 263–275.

HOUSTON, W.W.K. (1981). The life cycle and age of *Carabus glabratus* Paykull and *C. problematicus* Herbst (Col.: Carabidae) on moorland in Northern England. *Ecological Entomology* **6**, 263–271.

HUGHES, R.D. (1959). The natural mortality of *Erioischia brassicae* (Bouché) during the egg stage of the first generation. *Journal of Animal Ecology* **28**, 343–357.

JASPAR-VERSALI, M.F. AND JEUNIAUX, C. (1987). L'équipement enzymatique digestif des Carabes envisagé sous l'angle de leur nich écologique. *Revue d'Écologie et de Biologie du Sol* **24**, 541–547.

JASPAR-VERSALI, M.F., GOFFINET, G. AND JEUNIAUX, C. (1987). The digestive system of adult carabid beetles: an ultrastructural and histoenzymological study. *Acta Phytopathologica Entomologica Hungariae* **22**, 375–382.

JØRGENSEN, H.B. AND LÖVEI, G.L. (1999). Tri-trophic effect on predator feeding: consumption by the carabid *Harpalus affinis* of *Helothis armigera* caterpillars fed on proteinase inhibitor-containing diet. *Entomologia Experimentalis et Applicata* **93**, 113–116.

JØRGENSEN, H.B. AND TOFT, S. (1997a). Role of granivory and insectivory in the life cycle of the carabid beetle *Amara similata*. *Ecological Entomology* **22**, 7–15.

JØRGENSEN, H.B. AND TOFT, S. (1997b). Food preference, diet dependent fecundity and larval development in *Harpalus rufipes* (Coleoptera: Carabidae). *Pedobiologia* **41**, 307–315.

JULIANO, S.A. (1986). A test for competition for food among adult *Brachinus* spp. (Coleoptera: Carabidae). *Ecology* **67**, 1655–1664.

KIELTY, J.P., ALLEN-WILLIAMS, L.J., UNDERWOOD, N. AND EASTWOOD, E.A. (1996).

Behavioural responses of three species of ground beetle (Coleoptera: Carabidae) to olfactory cues associated with prey and habitat. *Journal of Insect Behaviour* **9**, 237–250.

KIELTY, J.P., ALLEN-WILLIAMS, L.J. AND UNDERWOOD, N. (1999). Prey preferences of six species of Carabidae (Coleoptera) and one Lycosidae (Araneae) commonly found in UK arable crop fields. *Journal of Applied Entomology* **123**, 193–200.

KING, P.E. AND STABINS, V. (1971). Aspects of the biology of a strand-living beetle, *Eurynebria complanata* (L.). *Journal of Natural History* **5**, 17–28.

KJELLSON, G. (1985). Seed fate in a population of *Carex pilulifera* L. II. Seed predation and its consequences for dispersal and seed bank. *Oecologia* **67**, 424–429.

KOEHLER, H. (1976). Nahrungsspektrum und Nahrungskonnex von *Pterostichus oblongopunctatus* (F.) und *Pterostichus metallicus* (F.) (Coleoptera, Carabidae). *Verhandlungen der Gesellschaft für Ökologie, Göttingen* **1976**, 103–111.

LANG, A. AND GSÖDL, S. (2001). Prey vulnerability and active predator choice as determinants of prey selection: a carabid beetle and its aphid prey. *Journal of Applied Entomology* **125**, 53–61.

LAROCHELLE, A. (1990). *The food of carabid beetles.* Québec: Fabreries, Supplement 5, Association des Entomologistes Amateurs du Québec.

LENSKI, R.E. (1982). Effects of forest cutting on two *Carabus* species: evidence for competition for food. *Ecology* **63**, 1211–1217.

LENSKI, R.E. (1984). Food limitation and competition: a field experiment with two *Carabus* species. *Journal of Animal Ecology* **53**, 203–216.

LINDROTH, C.H. (1949). *Die Fennoskandischen Carabidae. Eine tiergeographische Studie. I. Spezieller Teil.* Goteborg: Elanders Boktryckeri Aktiebolag.

LINDROTH, C.H. (1992). *Ground beetles (Carabidae) of Fennoscandia: a zoogeographic study: Part 1. Specific knowledge regarding the species.* Andover: Intercept.

LOREAU, M. (1983a). Le régime alimentaire de *Abax ater* Vill. (Coleoptera, Carabidae). *Acta Oecologia, Oecologia Generalis* **4**, 253–263.

LOREAU, M. (1983b). Le régime alimentaire de huit carabides (Coleoptera) communs en milieu forestier. *Acta Oecologia, Oecologia Generalis* **4**, 331–343.

LOREAU, M. (1984). Étude expérimentale de l'alimentation de *Abax ater* Villers, *Carabus problematicus* Herbst et *Cychrus attenuatus* Fabricius (Coeloptera, Carabidae). *Annales de la Societé Royale Zoologique de Belge* **114**, 227–240.

LÖVEI, G.L. AND SUNDERLAND, K.D. (1996). Ecology and behaviour of ground beetles (Coleoptera: Carabidae). *Annual Review of Entomology* **41**, 231–256.

LUFF, M.L. (1974). Adult and larval feeding habits of *Pterostichus madidus* (F.) (Coleoptera: Carabidae). *Journal of Natural History* **8**, 403–409.

LUFF, M.L. (1980). The biology of the ground beetle *Harpalus rufipes* in a strawberry field in Northumberland. *Annals of Applied Biology* **94**, 153–164.

LUFF, M.L. (1987). Biology of polyphagous ground beetles in agriculture. *Agricultural Zoology Reviews* **2**, 237–278.

LUKA, H., PFIFFNER, L. AND WYSS, E. (1998). *Amara ovata* und *A. similata* (Coleoptera, Carabidae), zwei phytophage Laufkäferarten in Rapsfeldern. *Mitteilungen der Schweizerischen Entomologischen Gesellschaft* **71**, 125–131.

LUKASIEWICZ, J. (1996). Predation by the beetle *Carabus granulatus* L. (Coleoptera, Carabidae) on soil macrofauna in grassland and drained peats. *Pedobiologia* **40**, 364–376.

MESSER, C. AND DETTNER, K. (1997). Inhaltstoffe von Collembolen – spurenanalytische Untersuchungen und Biotests. *Mitteilungen der Deutschen Gesellschafft für Allgemeine and Angewandte Entomologie* **11**, 465–468.

MESSER, C., WALTHER, J., DETTNER, K. AND SCHULZ, S. (2000). Chemical deterrents in podurid Collembola. *Pedobiologia* **44**, 310–220.

MITCHELL, B. (1963). Ecology of two carabid beetles, *Bembidion lampros* (Herbst) and *Trechus quadristriatus* (Schrank). I. Life cycle and feeding behaviour. *Journal of Animal Behaviour* **32**, 289–299.

MONSRUD, C. AND TOFT, S. (1999). The aggregative numerical response of polyphagous predators to aphids in cereal fields: attraction to what? *Annals of Applied Biology* **134**, 265–270.

PEARSON, D.L. AND MURY, E.J. (1979). Character divergence and convergence among tiger beetles (Coleoptera: Cicindelidae). *Ecology* **60**, 557–566.

POLLET, M. AND DESENDER, K. (1985). Adult and larval feeding ecology in *Pterostichus melanarius* Ill. (Coleoptera, Carabidae). *Mededelingen van de Faculteit Landbouwwetenschappen Rijksuniversiteit Gent* **50**, 581–594.

POLLET, M. AND DESENDER, K. (1986). Prey selection in carabid beetles (Col., Carabidae): are diel activity patterns of predators and prey synchronized? *Mededelingen van de Faculteit Landbouwwetenschappen Rijksuniversiteit Gent* **51**, 957–971.

POLLET, M. AND DESENDER, K. (1987). Feeding ecology of grassland-inhabiting carabid beetles (Carabidae, Coleoptera) in relation to the availability of some prey groups. *Acta Phytopathologica Entomologica Hungariae* **22**, 223–246.

POLLET, M. AND DESENDER, K. (1988). Quantification of prey uptake in pasture inhabiting carabid beetles. *Mededelingen van de Faculteit Landbouwwetenschappen Rijksuniversiteit Gent* **53**, 1119–1129.

POLLET, M. AND DESENDER, K. (1989). Prey uptake in subdominant, small to medium-sized carabid beetles from a pasture ecosystem. *Mededelingen van de Faculteit Landbouwetenschappen Rijksuniversiteit Gent* **54**, 809–822.

RAUBENHEIMER, D. AND SIMPSON, S.J. (1992). Analysis of covariance: an alternative to nutritional indices. *Entomologia Experimentalis et Applicata* **62**, 221–231.

SCHELVIS, J. AND SIEPEL, H. (1988). Larval food spectra of *Pterostichus oblongopunctatus* and *P. rhaeticus* in the field (Coleoptera: Carabidae). *Entomologia Generalis* **13**, 61–66.

SCHERNEY, F. (1959). *Unsere Laufkäfer.* Wittenberg: A. Ziemsen Verlag.

SCHERNEY, F. (1961). Beiträge zur Biologie und ökonomische Bedeutung räuberisch lebender Käferarten. *Zeitschrift für angewandte Entomologie* **48**, 163–175.

SCHREMMER, F. (1960). Beitrag zur Biologie von *Ditomus clypeatus* Rossi, eines körnersammelden Carabiden. *Zeitschrift der Arbeitsgemeinschaft Österreichischer Entomologen* **12**, 140–145.

SHAROVA, I.CH. (1981). *Life forms of carabids (Coleoptera, Carabidae).* Moscow: Nauka Publishers.

SKUHRAVÝ, V. (1959). Die Nahrung der Feldcarabiden. *Acta Societatis Entomologicae Cechosloveniae* **56**, 1–18.

SLANSKY, F. AND PANIZZI, A.R. (1986). Nutritional ecology of seed-sucking insects. In: *Nutritional ecology of insects, mites, spiders, and related invertebrates*. Eds. F. Slansky and J.G. Rodriguez, pp 283–320. New York: John Wiley & Sons.

SNIDER, R.M. (1984). Diplopoda as food for Coleoptera: laboratory experiments. *Pedobiologia* **26**, 197–204.

SØRENSEN, O.F. (1996). The effects of day length, temperature and prey type on food consumption by *Bembidion lampros* and *Tachyporus hypnorum*. Masters Thesis. Department of Zoology, Aarhus University, Århus.

SOTA, T. (1984). Long adult life span and polyphagy of a carabid beetle, *Leptocarabus kumagaii* in relation to reproduction and survival. *Researches on Population Ecology* **26**, 389–400.

SOTA, T. (1985). Activity patterns, diets and interspecific interactions of coexisting spring and autumn breeding carabids: *Carabus yaconinus* and *Leptocarabus kumagaii* (Coleoptera, Carabidae). *Ecological Entomology* **10**, 315–324.

SPENCE, J.R. AND SUTCLIFFE, J.F. (1982). Structure and function of feeding in larvae of *Nebria* (Coleoptera: Carabidae). *Canadian Journal of Zoology* **60**, 2382–2394.

SPIELES, D.J. AND HORN, D.J. (1998). The importance of prey for fecundity and behaviour in the gypsy moth (Lepidoptera: Lymantriidae) predator *Calosoma sycophanta* (Coleoptera: Carabidae). *Environmental Entomology* **27**, 458–462.

SUNDERLAND, K.D. (1975). The diet of some predatory arthropods in cereal crops. *Journal of Applied Ecology* **12**, 507–515.

SUNDERLAND, K.D. AND SUTTON, S.L. (1980). A serological study of arthropod predation on woodlice in a dune grassland ecosystem. *Journal of Animal Ecology* **49**, 987–1004.

SUNDERLAND, K.D., CROOK, N.E., STACEY, D.L. AND FULLER, B.J. (1987). A study of feeding by polyphagous predators on cereal aphids using ELISA and gut dissection. *Journal of Applied Ecology* **24**, 907–933.

SYMONDSON, W.O.C. (1993). The effects of crop development upon slug distribution and control by *Abax parallelepipedus* (Coleoptera: Carabidae). *Annals of Applied Biology* **123**, 449–457.

SYMONDSON, W.O.C. (1994). The potential of *Abax parallelepipedus* (Coleoptera: Carabidae) for mass breeding as a biological control agent against slugs. *Entomophaga* **39**, 323–333.

SYMONDSON, W.O.C. (1997). Does *Tandonia budapestensis* (Mollusca: Pulmonata) contain toxins? Evidence from feeding trials with the slug predator *Pterostichus melanarius* (Coleoptera: Carabidae). *Journal of Molluscan Studies* **63**, 541–545.

SYMONDSON, W.O.C. (in press). Coleoptera (Carabidae, Drilidae, Lampyridae and Staphylinidae). In: *Natural enemies of terrestrial molluscs*. Ed. G.M. Barker. Oxford: CAB International.

SYMONDSON, W.O.C. AND LIDDELL, J.E. (1993). Differential antigen decay rates during digestion of molluscan prey by carabid predators. *Entomologia Experimentalis et Applicata* **69**, 277–287.

SYMONDSON, W.O.C., GLEN, D.M., WILTSHIRE, C.W., LANGDON, C.J. AND LIDDELL, J.E. (1996). Effects of cultivation techniques and methods of straw disposal on predation by *Pterostichus melanarius* (Coleoptera: Carabidae) upon slugs (Gastropoda: Pulmonata) in an arable field. *Journal of Applied Ecology* **33**, 741–753.

SYMONDSON, W.O.C., GLEN, D.M., ERICKSON, M.L., LIDDELL, J.E. AND LANGDON, C.J. (2000). Do earthworms help to sustain the slug predator *Pterostichus melanarius* (Coleoptera: Carabidae) within crops? Investigations using a monoclonal antibody-based detection system. *Molecular Ecology* **9**, 1279–1292.

TERRA, W.W., FERREIRA, C., JORDAO, B.P. AND DILLON, R.J. (1996). Digestive enzymes. In: *Biology of the insect midgut*. Eds. M.J. Lehane and P.F. Billingsley, pp 153–194. London: Chapman & Hall.

THIELE, H.-U. (1977). *Carabid beetles in their environment.* Berlin: Springer-Verlag.

TOD , M.E. (1973). Notes on beetle predators on molluscs. *The Entomologist* **106**, 196–201.

TOFT, S. (1995). Value of the aphid *Rhopalosiphum padi* as food for cereal spiders. *Journal of Applied Ecology* **32**, 552–560.

TOFT, S. (1996). Indicators of prey quality for arthropod predators. In: *Arthropod natural enemies in arable land. II. Survival, reproduction and enhancement*. Eds. C.J.H. Booij and L.J.M.F. den Nijs, pp 107–116. Århus: Aarhus University Press.

TOFT, S. (1997). Acquired food aversion of a wolf spider to three cereal aphids: intra- and interspecific effects. *Entomophaga* **42**, 63–69.

TOFT, S. AND WISE, D.H. (1999a). Growth, development and survival of a generalist predator fed single- and mixed-species diets of different quality. *Oecologia* **119**, 191–197.

TOFT, S. AND WISE, D.H. (1999b). Behavioural and ecophysiological responses of a generalist predator fed single- and mixed-species diets of different quality. *Oecologia* **119**, 198–207.

TRAUTNER, J. (1984). Zur Verbreitung und Ökologie der *Dromius*-Arten (Coleoptera, Carabidae) in Württemberg. *Jahreshefte der Gesellschaft für Naturkunde in Württemberg* **139**, 211–215.

TRAUTNER, J., GEISSLER, S. AND SETTELE, J. (1988). Zur Verbreitung und Ökologie des Laufkäfers *Diachromus germanus* (Linne 1758) (Col., Carabidae). *Mitteilungen der Entomologischen Vereins Stuttgart* **23**, 86–105.

VAJE, S.V., MOSSAKOWSKI, D. AND GABEL, D. (1984). Temporal, intra- and interspecific variation of proteolytic enzymes in carabid-beetles. *Insect Biochemistry* **14**, 313–320.

WALDBAUER, G.P. (1968). The consumption and utilization of food by insects. *Advances in Insect Physiology* **5**, 229–288.

WALDBAUER, G.P. AND FRIEDMAN, S. (1991). Self-selection of optimal diets by insects. *Annual Review of Entomology* **36**, 43–63.

WALLIN, H. (1988). Mandible wear in the carabid beetle *Pterostichus melanarius* in relation to diet and borrowing behaviour. *Entomologia Experimentalis et Applicata* **48**, 43–50.

WALLIN, H., CHIVERTON, P.A., EKBOM, B.S. AND BORG, A. (1992). Diet, fecundity and egg size in some polyphagous predatory carabid beetles. *Entomologia Experimentalis et Applicata* **65**, 129–140.

WESELOH, R.M. (1985). Changes in population size, dispersal behaviour, and reproduction of *Calosoma sycophanta* (Coleoptera: Carabidae), associated with changes in gipsy moth, *Lymantria dispar* (Lepidoptera: Lymantriidae), abundance. *Environmental Entomology* **14**, 370–377.

WESELOH, R.M. (1993). Adult feeding affects fecundity of the predator *Calosoma sycophanta* (Col.: Carabidae). *Entomophaga* **38**, 435–439.

WHEATER, C.P. (1988). Predator-prey size relationships in some Pterostichini (Coleoptera: Carabidae). *The Coleopterist's Bulletin* **42**, 237–240.

WHEATER, C.P. (1989). Prey detection by some predatory Coleoptera (Carabidae and Staphylinidae). *Journal of Zoology, London* **218**, 171–185.

WINDER, L., PERRY, J.N. AND HOLLAND, J.M. (1999). The spatial and temporal distribution of the grain aphid *Sitobion avenae* in winter wheat. *Entomologia Experimentalis et Applicata* **93**, 275–288.

ZHAVORONKOVA, T.N. (1969). Certain structural peculiarities of the Carabids in relation to their feeding habits. *Entomological Review* **48**, 462–471.

4
Relating Diet and Morphology in Adult Carabid Beetles

JOSEPH INGERSON-MAHAR

243 Blake Hall, 93 Lipman Drive, Rutgers University, New Brunswick, NJ 08901, USA

Introduction

For over 100 years entomologists have been interested in the feeding habits of carabid beetles because of their predatory habits, and especially in forest and agricultural crops where they can be significant predators of insect pests. The predatory *Calosoma sycophanta* was one of the first insects used in biological control of gypsy moth, *Lymantria dispar*, in the United States (Burgess, 1911). Other species have gained attention for their roles in suppressing pest populations in field and forage crops (Rivard, 1964; Frank, 1971; Fox, 1974; Sunderland, 1975; House and All, 1981; Loughridge and Luff, 1983; Barney and Pass, 1986a; Sunderland *et al.*, 1987; Asteraki, 1993; Holland *et al.*, 1996), orchards (Cornic, 1973; Hagely *et al.*, 1982) and vegetable crops (Davies, 1963; Baker and Dunning, 1975; Dunning *et al.*, 1975; Grafius and Warner, 1989).

For nearly as long as carabids have been regarded as predators, it has been recognized that certain species also consumed plant material, and a few species could even be regarded as pests themselves (Webster, 1900). Forbes (1883) demonstrated that many common species had consumed plant material. Because of their abundance in most terrestrial habitats, it became clear that carabids must be greatly influencing the habitats in which they occur, either as carnivores, herbivores or omnivores and scavengers. Aside from taxonomic studies, the bulk of carabid literature regards their role in ecosystems (Mitchell, 1963; Rivard, 1965; Penney, 1966; Burakowski, 1967; Johnson and Cameron, 1969; Anderson, 1972; Calkins and Kirk, 1974; Allen, 1979; Altieri and Whitcomb, 1979; Barney and Pass, 1986a; Clark *et al*, 1997, as examples).

There are different ways of approaching the question of feeding habits relative to morphology. One method would be to collect wild specimens and, through dissection, discover what they have been feeding upon, then correlating this with particular morphological characteristics. A second method would be to hold the beetles under laboratory conditions and determine the range of material that individual beetles consume with a given morphology. Thirdly, field observations of feeding habits can be made.

Abbreviations: ELISA, Enzyme-Linked ImmunoSorbent Assays

The Agroecology of Carabid Beetles

In-depth morphological studies focusing upon body form and structure, and feeding mechanisms have answered many questions on prey selection and have provided interesting insights on how carabids procure food. Such studies have raised the question as to whether there are defining morphological characteristics that would allow researchers to assess feeding habits without extensive and labour-intensive feeding studies. For example, tiger beetles of the Cicindelinae have a characteristic body form and mouthparts that allow the capture of active insect prey, regardless of the habitat in which they are found.

Nearly all discussion on morphology in this chapter will be devoted to the adults. Relatively little information exists addressing larval biology and feeding habits due to secretive habits, difficulty in rearing larvae, and because for many carabid species the larvae remain unidentified. Exceptions to this include the larvae of *Calosoma*, which feed on lepidopterous caterpillars (Allen, 1979), and *Lebia* species, which are parasitic on Chrysomelidae (Dempster *et al.*, 1959). Depending upon the species, larval food includes both invertebrates and seeds; however, to my knowledge there are no studies specifically investigating morphological adaptations for larval feeding habits.

In the following sections arguments will be made for and against particular morphological structures for predicting feeding habits, and gut dissection as a means of determining feeding habits. It is likely that the morphology of the mouthparts and digestive system will be more useful in determining dietary preferences than the whole body form and locomotion. A carabid must have suitable morphological characteristics that allow the beetle to survive in a habitat meeting the demands of finding shelter, adequate moisture and food. The same characteristics that allow a beetle to run, burrow or fly, be diurnal or nocturnal, must also bring the beetle into contact with its prey, but they may not be the same ones which process the prey.

As will be discussed, so far, no single morphological condition currently exists that answers the question for all species: what do carabids eat?

BODY FORM

Anyone studying the family Carabidae for the first time would be struck by the similarity of body form regardless of the geographic origin of species or the habitat in which they occur. Thiele (1977), Bauer and Kredler (1993) and Evans (1990) have commented on this superficial similarity, but noted that despite the similarity there are species-specific differences in morphology that adapt various taxa to the special demands of their respective niches. Many of these morphological differences are subtle and their importance might not be recognized without detailed study.

Many species of *Calosoma* and *Carabus* have strong, deep bodies that help enable them to live in situations with 'strong environmental resistance' (Forsythe, 1987). Their large size and strength helps them to subdue large prey typifying Thiele's (1977) body form classification of 'procerization'.

The Cychrini and other Carabini that have evolved a slender forebody characterize Thiele's classification of 'cychrization'. This body form enables them to either feed on snails withdrawn into their shells, to feed on injured snails with broken shells (Digweed, 1993), or to extract other prey from tight places (Noonan, 1967). The slender structure of the forebody allows for extensive twisting of the body to aid in prey capture (Thiele, 1977). Similarly, ground beetles that live in confining or restricted situations, such as soil crevices under stones or under bark have a tendency to be narrower in width and shallower in depth. The pronotum is often the same width as the elytra to reduce resistance (Forsythe, 1987). Examples would include *Scarites subterraneus*, living in the soil, and *Cymindis platicollis*, a Nearctic species, which is frequently found under bark scales of white oak trees, *Quercus alba* (Mahar, 1979). Thiele identifies this form as 'abacization' after the body form of the genus *Abax*.

Body size and shape are significant aspects that may greatly influence the mode of life of a species and the habitat(s) it will colonize, but the question here is whether body size and shape influence feeding preferences.

Wedge pushing

One distinctive feature that has helped carabids be successful is the ability to burrow into the habitat substrate. This behaviour is made possible in part because of their dorsoventrally wedge-shaped bodies (Evans, 1977). The mandibles and head form the point of the wedge that deepens through the prothorax and abdomen. A key structure allowing carabids to use this wedge-shaped form is the metatrochanter, a kidney-shaped structure on the metathoracic legs that houses a complex of muscles. The articulation and pushing strength provided by the metatrochanter allows carabids to oscillate the hind body, thereby widening the aperture of the crevice they are moving through (Evans, 1977). This ability is useful in gathering food and in escaping predators or adverse environmental conditions.

The degree of wedge-shaped body forms range from the Carabini, *Calosoma* and *Carabus*, for example, with large, deep forebodies and even thicker abdomens, to the Lebiini, with genera like *Lebia* and *Dromius*, which are medium- to small-sized beetles that tend to be very flattened. The depth of body depends upon the particular life style of the beetle in question, e.g. in forest litter, dry open ground, under bark or on plants. In the laboratory, the author has observed *Harpalus pensylvanicus* creating shallow excavations by wedge pushing, where they will retire to consume food. However, it seems that the wedge-pushing ability is more of a trait for finding shelter rather than a means of identifying potential prey choices.

Relating the forebody to lifestyle

Large, bulging compound eyes appear to be an adaptation for detection and capture of active prey (Bauer and Kredler, 1993), correlating well for a few species, a small percentage of the family Carabidae. Diurnal ground beetles with large bulging compound eyes, such as *Cicindela*, *Elaphrus* and *Asaphidion*, have narrowed pronota that are less than the width of the head and elytra. This is seen as an adaptation that allows the greatest field of vision for these beetles and may have additional benefits in making flight easier to either pursue prey or to escape from predators (Forsythe, 1987).

Notiophilus is an exception in which the pronotum is as wide as the elytra. Flight is not important in prey capture and, when disturbed, *Notiophilus* adults prefer to burrow into soil litter. Studies by Bauer (1977) indicate that the large compound eyes are forward directed, allowing binocular vision to enable *Notiophilus* to track prey movements. Unfortunately, however, much variation exists in the degree of eye protrusion across genera and species, so that it is unlikely that this condition will permit unequivocal conclusions about diet without further study.

LOCOMOTION

Length of legs and pushing

Evans (1986) describes ground beetles as having the adaptations for both running and for pushing. Long, thin legs with primitive and relatively long five-segmented tarsi aid in running, while the ventral coxal articulations, protarsi with a special feathering mechanism and large femoral flanges over the tibial articulations aid in pushing. Evans (1986) and Forsythe (1983) agree that carabid locomotion is a continuum combining speed and pushing abilities from rapid runners/weak pushers to slow runners/strong pushers. The rapid runners, or sprinters, are characterized by long legs and relatively small metatrochanters, which contain small femoral rotator muscles (*Figure 4.1*). *Cicindela* is the extreme example of these features, but species in the Nebriini, Notiophilini, Loricerini, Elaphrini and Graphipterini also have these characteristics (Evans, 1990).

Power pushers is a category of species that demonstrate slow speed but strong pushing abilities. The legs of species in this group are relatively short, which gains

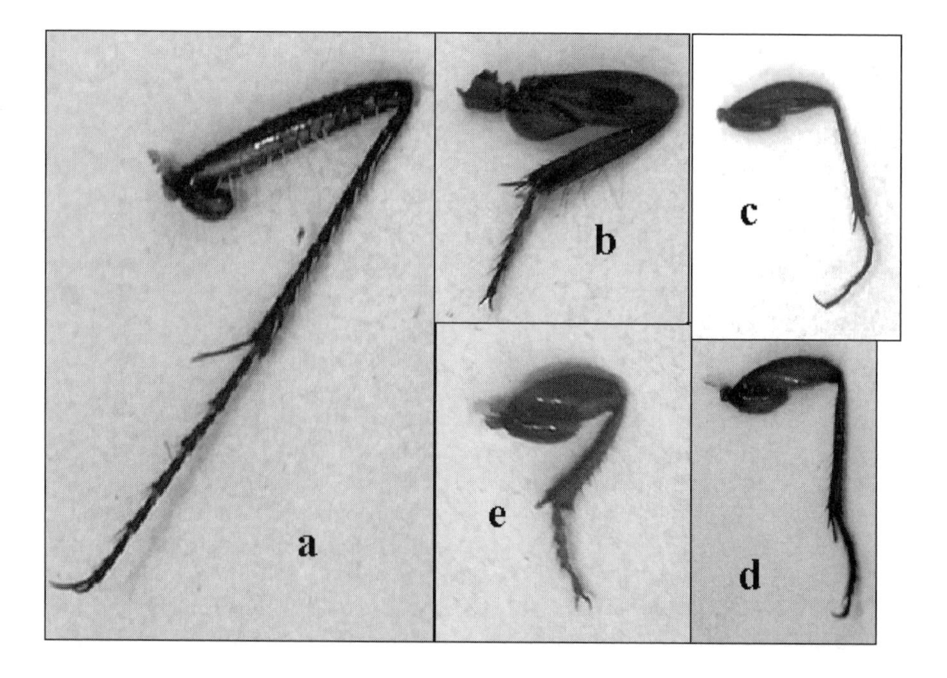

Figure 4.1. Comparison of the left metatarsal legs with trochanters (anterior view) of five species of carabids, including Cicindela. a = *Cicindela punctulata*, rapid runner; b = *Scarites subterraneus*, power pusher; c = *Pterostichus chalcites*, intermediate runner/pusher; d = *Harpalus pensylvanicus*, intermediate runner/pusher; e = *Cratacanthus dubius*, intermediate runner/pusher, or power pusher? [Photographs by author]

mechanical advantage in the strength of their movement, as illustrated by Scaritine species. Examination of the metatrochanters show that these may not be as kidney-shaped as in other groups, but rather more triangular (Forsythe, 1983) (*Figure 4.1*).

The middle group of this continuum holds the majority of species, which is that of the medium wedge-pushers. These beetles possess medium speed and medium pushing ability, which allows them to occupy many diverse habitats. They are mostly represented by the Harpalinae, which includes most of the carabid tribes. The Carabini also belong to this group, even though they outwardly appear to belong to the rapid running group. Their gait is medium, despite their long legs but, because of mechanical advantage in muscle attachment, they are relatively strong horizontal pushers (Evans, 1990).

How these morphological differences relate to feeding habits is not clear and there may not be any correlation. Speed can be seen as a useful attribute for capturing active prey, but lack of speed does not necessarily imply a non-carnivorous diet. *Scarites* species are good examples of predatory, but slow moving, species. Some species of *Pterostichus, Synuchus, Agonum,* and *Bembidion* are also rapid runners but will feed on plant material (Johnson and Cameron, 1969; Manley, 1971).

Flight

There are four flight conditions in carabids: 1) those species that are winged and capable of flight; 2) those that are winged but do not fly (Den Boer, 1969); 3) those that are dimorphic; and 4) those incapable of flight either because they lack functional flight muscles (Lindroth, 1949; Forsythe, 1987) or have reduced, brachypterous wings. Carabids capable of flight undoubtedly use this ability to find food, to disperse and to find mates.

The presence or absence of wings or the musculature needed for flight and their morphological ramifications have been examined by Forsythe (1987). Although the proportions and alignment of the sternal and episternal sclerites change depending upon the presence or absence of wings and flight muscles, Forsythe did not find any consistent pattern linking wings with feeding habits that could be applied across the Carabidae. Three species illustrate this lack of correlation. *Calosoma sycophanta* is an example of a predatory carabid that uses flight to find outbreaks of caterpillars (Lindroth, 1969). Mahar (1979) observed that *Cymindis limbata* was a common nocturnal flyer in the canopy of northern oak forests of Michigan, feeding upon live and dead insect prey and plant material. Lastly, *Amara plebeja*, at different seasons of the year, lost its wing muscles and re-grew them later, but was found to consume mostly plant material at all times (Van Huizen, 1977).

Conclusions on body form and locomotion

By combining both body form and locomotion, one can begin to understand how carabid beetles have come to occupy such a diversity of habitats and how particular species can be found over large geographical regions. However, strictly speaking, body form and locomotion themselves do not necessarily inform the researcher what the feeding preferences are for a particular species. In light of our foreknowledge regarding the diet of the more commonly studied species, the adaptations discussed so far make sense. A researcher who is not steeped in carabid biology and is attempting an ecological study or a biodiversity inventory, however, would be hard pressed to predict feeding preferences from body form and running speed. An example illustrates this point. In late summer and fall, two species of carabids occupy the disturbed landscape that forms the parking lot and lawn about the author's office building. *Cratacanthus dubius* fits the category of 'slow speed/strong pushing' because of its relatively short legs, robust body, and its metatrochanter is more than half the length of the hind femur. *Harpalus pensylvanicus* is larger, with a more flattened body and is a medium to fast runner with correspondingly long legs and medium pushing ability. Its metatrochanter is slightly less than half the length of the hind femur. *H. pensylvanicus* runs as quickly as *Pterostichus chalcites*, which in the laboratory prefers insect prey to plant material. Laboratory feeding trials indicate that both *H. pensylvanicus* and *C. dubius* prefer seeds to animal prey, although *H. pensylvanicus* occasionally consumes animal prey. Thus, although body form and speed may offer tantalizing suggestions as to feeding preferences, more morphological characteristics need to be examined.

MOUTHPARTS

Mandibles

The structure of the mandibles has long been used as an indicator/predictor of feeding preferences for carabids. Forbes (1883) was one of the first to relate mandible form and feeding habits, finding that species with long mandibles and little or no molar area were inclined to be predaceous. Bryson and Dillion (1941) compared the mandibles of *Agonoderus pallipes* (= *Stenolophus lecontei*) with *Calosoma scrutator* and *Harpalus caliginosus* and found *A. pallipes* to be intermediate in form. They concluded that since *C. scrutator* was a predator and *H. caliginosus* was predominately a herbivore, *A. pallipes* probably fed on both plant and animal material. In the laboratory, *A. pallipes* was cannibalistic, but in the field it had earned a reputation as an occasional pest, the seed-corn beetle.

The mandibles have two primary functions, to grasp the food (animal or plant) and to begin the process of digestion by converting the food into a manageable form for ingestion. Acorn and Ball (1991) considered the mandibles to be the primary means of mastication. Because of the restrictions on mandibular movement due to their articulation, they felt that diversification of food processing was accomplished by changes in the form and use of the mandibles. Processing of food could include: cutting, where the incisor ridges shear past each other; grinding, where a projection on one mandible fits into a depression on the opposing mandible; and crushing, where two more or less flat surfaces meet.

Forsythe (1982) and Evans and Forsythe (1985) developed three categories of feeding behaviour based in large part on mandibular form: fluid feeders, fragmentary feeders, and mixed feeders. Both the fluid and fragmentary feeders are predominantly zoophagous and are the easiest to define. The mandible length for fluid feeders is characteristically long, having a length to width ratio of approximately two to three. These species have one to three incisors and include the Carabini and Cychrini and also the Cicindelinae. Fragmentary feeders, such as *Nebria gyllenhalli* and *Loricera*

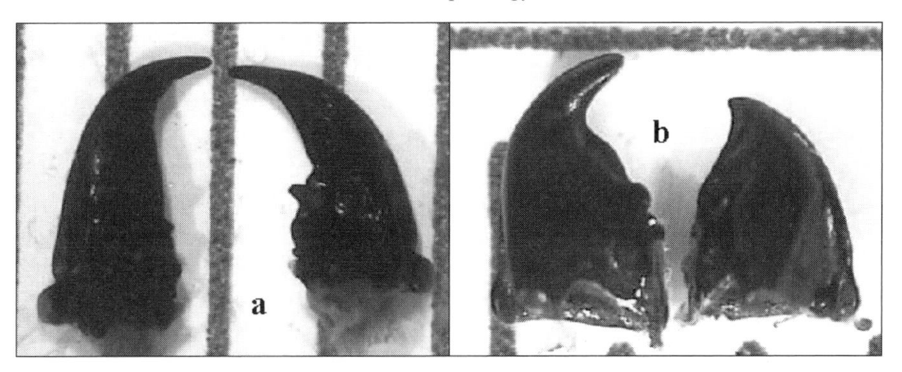

Figure 4.2. Mandibles of (a) *Broscus cephalotes* and (b) *Harpalus pensylvanicus* showing dorsal view of left and right mandibles. [Photographs by author]

pilicornis, have considerably shorter mandibles. Their respective mandibular ratio of the left mandible is 1.75 and 1.16. The mandibular width to length ratio, as it approximates 2.0, indicating they are predatory feeders that have extra-intestinal digestion.

Another character that may help identify fluid feeders is the groove and attendant setae evident on the inner margin of the ventral side of the mandibles. Both Forsythe (1982) and Sharova (1975) emphasize that the extent of mandibular setae indicates the degree of extra-intestinal digestion. The mandibles of *Carabus problematicus*, *Calsoma inquisitor* and *Cychrus caraboides* have long grooves lined with dense setae and have been shown to be fluid feeders. In other species, especially where the molar region is more developed, the setal-lined groove is much shorter and the setae distribution less dense. Sharova suggests that the latter group has only a slight filtering mechanism and, as a result, consists strictly of intra-intestinal digesters.

The third group, which contains most of the species of carabids, is the most difficult to define as they appear to use some combination of fluid feeding, semi-fluid feeding and fragmentary feeding. The actual diet ranges from predominantly zoophagous to predominantly herbivorous. Here too, however, specific characteristics of the mandibles will help define the prey preference. Longer, flatter mandibles with acute incisors characterize predaceous species, while those species with shorter and stouter, more quadrate mandibles with increased molar areas will tend toward herbivory (*Figure 4.2*). The Licinini may be an exception to these generalizations. Ball (1992) and Digweed (1993) have found the Licinini to be molluscivores. The mandibles of some species are quite dissimilar from the Cychrini, apparently because the Licinini break apart snail shells with their mandibles.

The relative degree of bluntness of the mandibles, however, should be viewed cautiously. Mandible wear has been found to correlate with the age of the beetle (Butterfield, 1996) and to be a consequence of eating hard food and burrowing by *Pterostichus melanarius* (Wallin, 1988). Relying on the acuteness of the incisors alone could be misleading in making statements about feeding behaviour.

Maxillae

Forsythe (1982) suggests that the amount of musculature available to manoeuvre the maxillae can be related to the need to hold struggling prey. By calculating the ratio of the gular width to head width, one can determine the relative proportion of muscles used to operate the maxillae and mandibles. The lower the ratio, the more likely the mandibular muscles have increased, reducing the amount of the stipital retractor muscles, implying that species with low ratios, i.e. 0.13 or less, are not likely to capture struggling prey. Live prey attempting to escape would have to be ingested as quickly as possible, causing a need for strong maxillae to hold and bring the prey to the mouth. *Cychrus caraboides*, which feeds on molluscs, has a large ratio, 0.48, indicating that the maxillae are especially important in ripping prey apart while the mandibles hold the prey.

As with any characteristic, there are exceptions to the general trend. Forsythe found in his study of the morphology-based feeding habits of 12 species that the largest gular to head width ratio belonged to *Loricera pilicornis*. However, even though the stipital retractor muscles were important, they did not account for the large gular width. Rather, *L. pilicornis* has developed a unique hunting technique in which the large setae on the underside of the head and antennae have developed into a cage for trapping Collembola. The gula has expanded, allowing a greater upper surface area for the trap.

Evans and Forsythe (1985) provide evidence that the number and arrangement of the setae on the lacinia can be used to help identify feeding behaviour. For fluid feeders, the setae appear in eight or more rows and are densely packed, forming a brush. Fragmentary feeders, since they are not concerned with the transport of digestive and prey fluids, have up to four rows of more widely spaced setae, more like a rake than a brush.

While the brush-like lacinia is present in *Carabus*, *Calosoma*, *Scarites*, *Stomis* and *Galerita*, and the rake-like lacinia is in *Nebria*, *Notiophilous*, *Leistus* and *Loricera*, this is not a consistent trait across all groups. *Elaphrus* and *Cychrus* are notable exceptions. Both are fluid feeders but have rake-like lacinia.

Mid-gular apodeme

Although not a mouthpart, the mid-gular apodeme has important relationships with the gula and the musculature that work the mandibles, maxillae and labium. Evans and Forsythe (1985) found six variations in carabid beetles in the presence and absence of the apodeme, width of the gula, and point of attachment and length of the labial muscles. Briefly, these groups are:

1. Mid-gular apodeme present (externally visible as the mid-gular sulcus) with normal gular width – includes most of the mixed feeders (also the Cicindelinae) and is the largest grouping of species.
2. Mid-gular apodeme present with gular width reduced – includes mixed feeders that are partly or largely herbivorous – this condition suggests smaller stipital muscles, implying that the beetle does not regularly consume active, struggling prey requiring strong maxillary action.
3. Mid-gular apodeme reduced or absent with flattened head and short labial muscles – largely found in the Lebiamorphs.
4. Mid-gular apodeme absent with elongate labial muscles – includes fragmentary and a few mixed feeders.

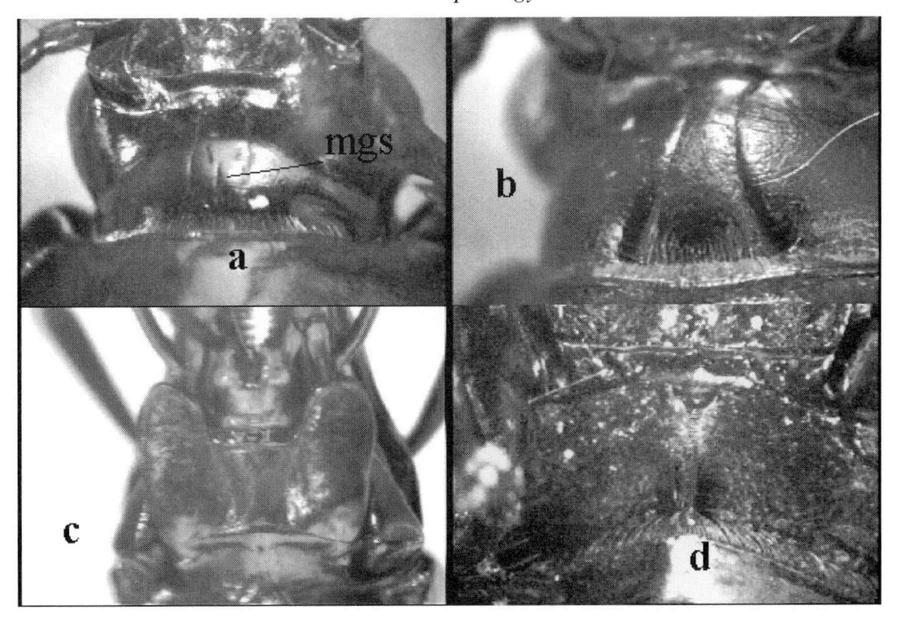

Figure 4.3. Four examples of the presence or absence of the gula and mid-gular apodeme (ventral view of head with mouthparts at the top of picture). a = *Pterostichus chalcites* with gula, mid-gular apodeme, and mid-gular sulcus (mgs); b = *Calosoma calidum* with gula but lacking mid-gular apodeme; c = *Scaphinotus* sp. same as b; d = *Scarites subterraneus* with narrow gular and lacking mid-gular apodeme. [Photographs by author]

5. Mid-gular apodeme absent with very short labial muscles and a submental apodeme sometimes present – represents fluid feeding members of the Carabitae – these species usually have well developed stipital muscles.
6. Mid-gular apodeme absent and gular width reduced – primarily the Scaratini (*Figure 4.3*). Two other categories of mid-gular apodeme development refer to non-carabid adephageans. Clearly, there are trends in feeding habits, but mixed feeders, for example, are found in three or four categories and, even though the Cicindelinae are fluid feeders, they are morphologically in a different category than the other large fluid-feeding group, the Carabitae.

Despite the thoroughness of their work on the morphology of the mouthparts and head capsules, Evans and Forsythe (1985) emphasize that the characters they present for consideration provide evidence only for the feeding mechanism and not specific prey preferences.

INTERNAL ANATOMY OF THE DIGESTIVE SYSTEM

The foregut has received the most attention in investigations of feeding habits and how prey is ultimately prepared for digestion. The mouth and oesophagus guide food to the crop, which is primarily a holding organ to further mix digestive fluids with prey material. From the crop, the food advances to the proventriculus, where it is finally prepared for entering the mid-gut. The structure of the foregut most closely related to feeding habits is the proventriculus.

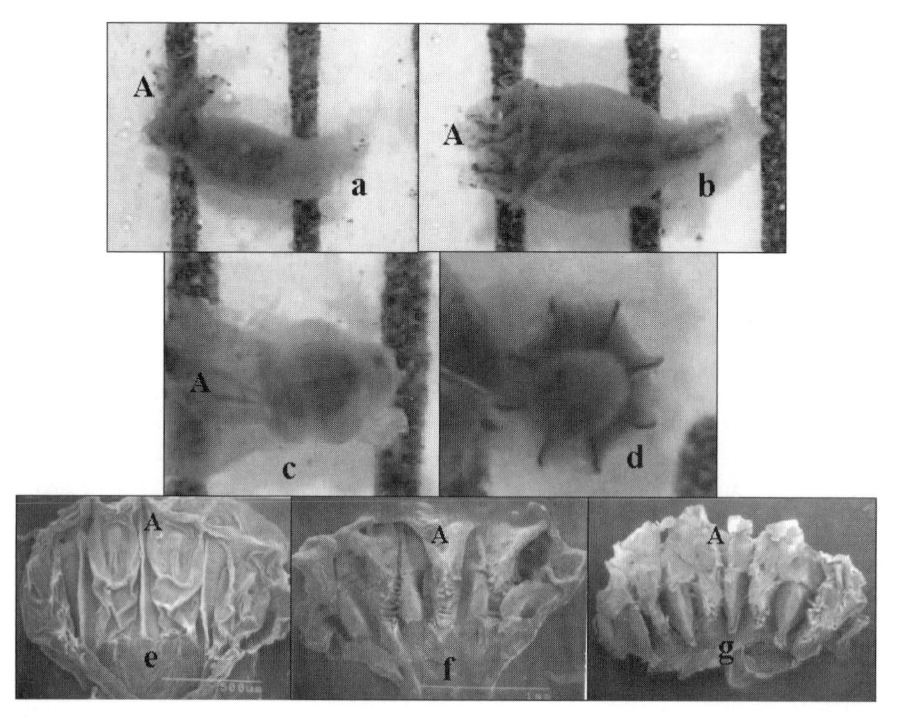

Figure 4.4. External and internal views of the proventriculus in five species of carabids. a = lateral external view of proventriculus of *Pterostichus chalcites*; b = lateral, external view of *Chlaenius tricolor*; c = lateral, external view of *Stenolophus comma* with crop attached; d = anterior view of proventriculus of *S. lineola*; e = scanning electron photograph of internal view of proventriculus of *Harpalus pensylvanicus*; f = internal view of proventriculus of *Chlaenius tricolor*; g = internal view of proventriculus of *Stenolophus lineola*; A = anterior. [Photographs by author]

The proventriculus comprises of two parts, an anterior portion with layers of circular muscles surrounding four primary lobes containing longitudinal muscle bundles, and a posterior region of varying length lacking muscles that narrows to a small aperture joining to the mid-gut (*Figure 4.4*). The anterior portion is armed internally with a variety of setae, teeth and combs that help filter or break down prey material. In between the primary lobes, four additional lobes (interlobes) may occur, developed to various degrees in different species.

In many groups, such as *Harpalus*, *Cymindis*, *Pterostichus*, *Neomyas*, *Amara*, and *Scarites*, the proventriculus is a slender structure, two to three times as long as wide, that, externally, is hardly distinguishable from the crop (Ingerson-Mahar, unpublished data). The surface of the primary lobes is covered by a variety of short setae or bristles, teeth and combs, depending upon the species. The interlobes are not developed. In other groups, the muscled portion of the proventriculus is scarcely longer than wide and quite distinct from the crop in appearance, as in *Carabus* and *Chlaenius*. Internally, the surfaces of the primary lobes are clothed in long setae in *Carabus*, but interlobes are not developed. However, the surfaces of the primary lobes in *Chlaenius* are covered in short setae or other teeth and the interlobes are produced into either blunt, tooth-like projections or into flat, highly chitinized surfaces.

In *Stenolophus comma* and *S. lineola*, the anterior portion of the proventriculus is wider than long with large, rounded, chitinous flanges protruding from it, but with a very short posterior region. Internally, the interlobes are very pronounced, with either large teeth or ridges. The surfaces of all lobes are covered by short setae and teeth.

In preparing this chapter, specimens of *Broscus cephalotes* were examined and dissected to determine the nature of the proventriculus. It was found to be intermediate between the type found in *Harpalus* and the type found in *Carabus*. That is, the anterior portion of the proventriculus was distinct from the crop because of its pronounced musculature, but overall was long and slender. Internally, the primary lobes were covered in setae but not as pronounced as in *Carabus*. This type of proventriculus fits well with the observed feeding habits of *B. cephalotes*, which is regarded as both a fluid and fragmentary feeder (Zhavoronkova, 1969; Forsythe, 1982).

The exact role of the proventriculus has been, and remains, controversial (Hengeveld, 1980c). As Hengeveld notes, the proventriculus has been 'described as a masticating organ, as a sieve, as a hoop-net and as a pump'. The proventriculus may function in all of these capacities depending upon the type of feeder. For instance, a fluid feeder that has pre-oral digestion would have different functional needs for the proventriculus than a fragmentary feeder ingesting coarse material. Cheeseman and Pritchard (1984b) disagree with Forsythe (1982) and Zhavoronkova (1969) that the proventriculus of fluid feeders acts as a filter. Cheeseman and Pritchard (1984a) demonstrated that the proventriculi of six species of fragmentary-feeding carabids acted to significantly reduce particle size of ingested insect cuticle. They further note that, since most of the digestion occurs in the crop, the most fundamental role for the proventriculus is to protect the mid-gut from large particles. Brunetti (1931) contends that the presence or absence of a proventriculus has no relation to the food consumed.

Aside from the attempt by Zhavoronkova (1969) to relate differences in the armament of the proventriculus to feeding behaviours, little has been done to relate the structure of the proventriculus to food preferences beyond noting that the proventriculi of fluid feeders contains long setae (Forsythe, 1982; Cheeseman and Pritchard, 1984b). Since the mandibles, maxillae and the proventriculus are the only structures that are responsible for the physical breakdown of food, this would seem to be a promising area of research.

CHEMORECEPTORS ON ANTENNAE AND PALPS

All insects have sensory receptors that provide environmental information to them in the form of innervated setae, peg sensilla and a variety of other types. Most work done on receptors has been with phytophagous species (Altner and Hintzpeter, 1984), with very little on the terrestrial Adephaga (Baker and Monroe, 1995; Symondson and Williams, 1997).

One only has to see a carabid beetle approaching its prey, insect or seed, to realize that the antennae are involved in sensing food. Symondson and Williams (1997) found at least four types of receptors on the last antennal segment of *Pterostichus melanarius* and *P. niger* (*Figure 4.5*). While they were not able to determine the function of these receptors, they did find differences in the structure of one of the sensors, which would provide a means of definitively identifying specimens of these closely related species.

Forsythe (1982) described how *Loricera pilicornis* used the long antennal setae to

Figure 4.5. Close-up of terminal antennal segment of *Pterostichus niger* showing chemoreceptors. [Photograph by William Symondson]

help trap prey under the head in easy reach of the mandibles. Altner and Hintzpeter (1984) hypothesized and confirmed that there was a reduction in the sensory capabilities of the antennal setae in favour of developing a trapping mechanism.

In studying the sensory receptors of the maxillary palps of *P. melanarius* and *P. niger*, Symondson and Williams (1997) compared them with *Poecilus cupreus* and *Abax parallelepipedus*, and found that all four species had similar arrangements and number of receptors. They attributed this similarity to the polyphagous feeding habits of these species. Presumably, all carabids have receptors on the maxillary and labial palps that respond to chemical constituents of food (*Figure 4.6*).

Carabid larvae probably possess palpal sensilla, as well. Mitchell (1963) detailed how larvae of *Bembidion lampros* and *Trechus quadristriatus* seem to recognize prey by touching the palps to the potential food. It was only then that the mandibles closed upon the prey. Baker and Monroe (1995) also studied the sensors of the maxillary and labial palps and galeae of the tiger beetle, *Cicindela sexguttata*. They found no significant differences between sexes in the number of sensilla and further, found that the overall number of sensilla was similar to that in other beetle families. Their conclusion was that, based on the small number of studies of Coleoptera, there seemed to be no relationship between the number of sensilla and zoophagy or phytophagy.

It seems that further study of carabid sensilla might aid in defining diagnostic characters as Symondson and Williams (1997) but, so far, there is little to suggest a correlation with feeding habits.

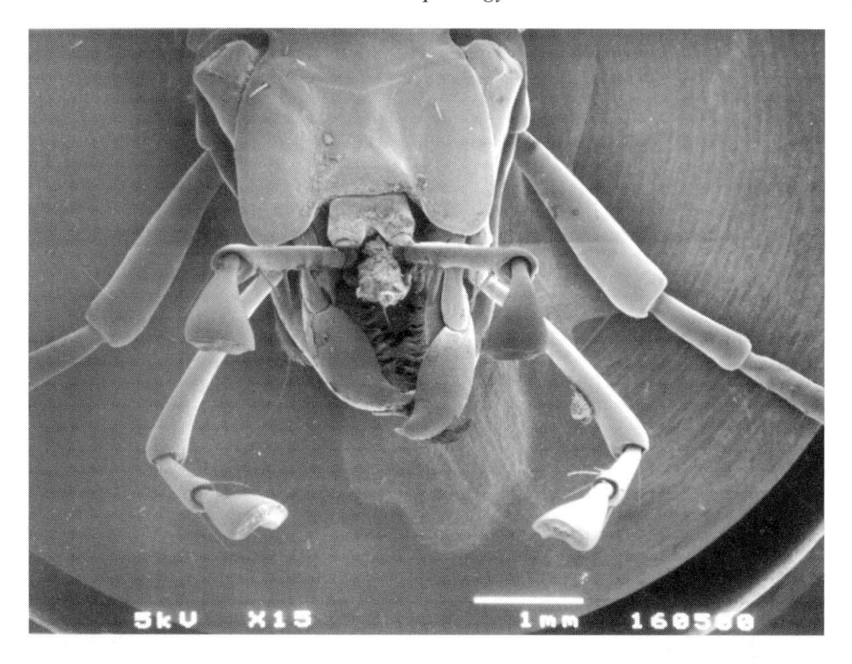

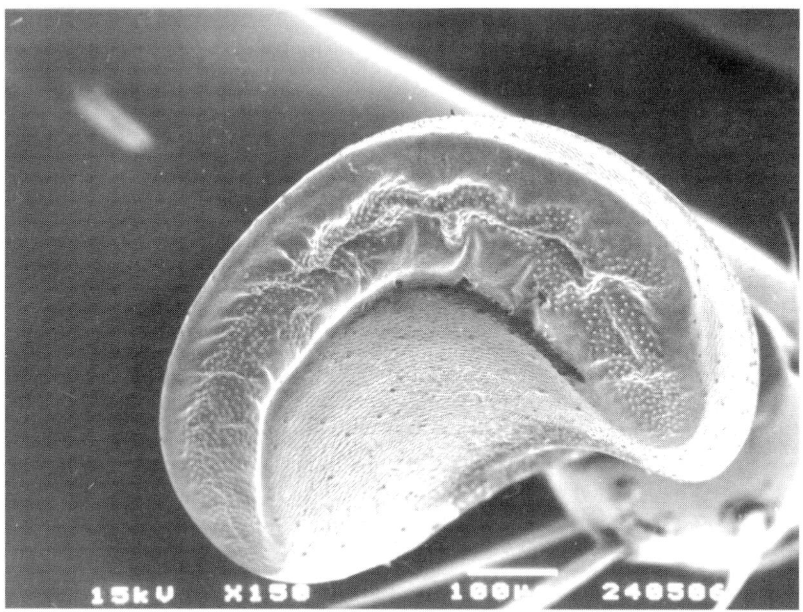

Figure 4.6. Ventral view of the mouthparts of *Scaphinotus marginalis* (a) and close-up view of the terminal segment of maxillary palpus showing chemoreceptors (b). [Photographs by William Symondson]

OBSERVATIONS OF KNOWN FEEDING HABITS

We are concerned with knowing carabid feeding habits because of: 1) their importance to agriculture and forestry; and 2) the need to know their role and interrelationships in broader ecological systems.

Defining the specific diet of carabids is a daunting task which has seen controversy through the years. Most entomologists of the late nineteenth century believed that the Carabidae were essentially predatory (Forbes, 1883). Webster (1900) shared Forbes' view that many species consumed plant material. Later, Lindroth (1949) demonstrated that even highly predaceous species sometimes ingested plant material. Johnson and Cameron (1969) further confirmed the herbivorous habits of several species. They also questioned other work on carabid feeding habits, specifically Rivard (1964, 1966), Scherny (1955) and the texts by Borror and DeLong (1966), in that these authors either dismissed or ignored phytophagous feeding habits. Currently, the pendulum has swung the other way. More recent reviews tend to dismiss *Amara* and *Harpalus*, at least, as important insect predators (Jørgensen and Toft, 1997, as an example).

Direct observations and analysis of carabid feeding abound in the literature, although reports are anecdotal and/or based on single observations. Larochelle (1990) includes feeding observations for 1054 species, which provides a range for food material for these species. More useful, although requiring careful interpretations, are laboratory studies where several specimens of one or more species are tested for feeding preferences (Shough, 1940; Davies, 1953, Johnson and Cameron, 1969; Lund and Turpin, 1977; Zetto Brandmayr, 1990;). Many researchers have analysed gut contents, including Forbes (1883), Johnson and Cameron (1969), Lindroth (1949), and Sunderland (1975). Cropland species, important in agroecosystems, were studied to establish their biology, including feeding preferences (Fox and MacLellan, 1956; Burakowski, 1967; Frank, 1971; Pausch, 1979; Grafius and Warner, 1989). Allen (1976) and Luff (1987) provided reviews of many studies conducted in the need for biological information that would define the role of carabids in agriculture and forestry in the Northern Hemisphere. An extensive review of pest control by carabids is provided by Sunderland (this volume).

There is general agreement that the Cicindelini, Carabini, Cychrini, Elaphrini, *Asaphidion*, *Loricera*, *Notiophilous* are zoophagous, borne out by feeding observations, gut analysis and morphological traits. Beyond these groups, which form a minority of species, the only general agreement is that the remaining species are mixed feeders, polyphages (generalists) and oligophages (specialists), varying in the proportions of animal and plant food consumed. Many species in this heterogenous group are subject to conflicting observations regarding consumption of plant and animal material (Johnson and Cameron, 1969; Hagley *et al.*, 1982; Barney and Pass, 1986b). Even the vilified grain pest, *Zabrus tenebroides*, has been found with insect material in its gut (Zhavoronkova, 1969). As noted by Dennison and Hodkinson (1983), even though a species has a particular morphology and feeding adaptation and related species have similar adaptations, there is still considerable overlap in food selected and consumed by morphologically dissimilar species. Toft and Bilde, in this volume, suggest feeding guilds based upon gut analysis, direct feeding observations, and morphological characteristics.

The most difficult problem is defining when a mixed-feeding carabid should be considered zoophagous or herbivorous. Zetto Brandmayr (1990) cautions that there are two types of herbivorous feeding. The first is incidental feeding by primarily zoophagous species; for example, the beetle may feed on plant material only as a source of moisture. The second type of feeding is where the members of the species ingest plant material as food, an intentional, perhaps integral, part of its diet. When one sees carabid beetles in the field feeding on plant material, it may be extremely difficult to distinguish whether it is incidental or integral feeding.

Prey availability changes over time and habitat (Skuhravy, 1959; Zhavoronkova, 1969; Luff, 1974) and prey selection may change over the life of the beetle (Penney, 1966). A primarily herbivorous species could become a significant insect predator at certain times of the year, as demonstrated by Barney and Pass (1986b; with *Amara cupreolata* and *A. impuncticollis* on alfalfa weevil, *Hypera postica* and *Harpalus pensylvanicus* on the parasites of the alfalfa weevil), and Loughridge and Luff (1983; with *Harpalus rufipes*) and Holopainen and Helenius (1992; with *H. rufipes* and *Amara plebeja* as aphid predators).

Other changes in the beetles themselves may influence prey selection and capture. In teneral beetles of *Pterostichus*, the mandible incisors are acute but, as the beetles age, the incisors become worn and blunt. Specimens of *Pterostichus* sometimes show considerable mandible wear (Wallin, 1988). How does this affect food acquisition? As the incisors become more blunt, there must be a reduction in the ability to catch and rend active prey. Does this force a change of diet for older beetles?

In studying the large plant- and animal-feeding group of carabid species, Forbes (1883) and Zhavoronkova (1969) may have best summarized the situation. Zhavoronkova defined three trophic groups based upon gut analysis: 1) fluid feeders, which were nearly entirely zoophagous; 2) fragmentary or mixed feeders that were predominantly zoophagous and; 3) fragmentary or mixed feeders that were predominantly herbivorous. These trophic levels fit very well with Forbes' assessment of carabid feeding habits, 'But as animal food is usually less abundant and less generally distributed than vegetable, it would not, at first, be to the advantage of any that they should become exclusively dependent upon the former; their interests would be best served by such modifications of structure and habit as would enable them to draw upon one of the other store, according to circumstances.' We can expect the Carabidae as a whole to include zoophagous, herbivorous and mixed feeders. All subsequent studies have done little more than confirm this impression.

Gut dissection

Perhaps the best way for an ecologist to determine the trophic role for adults of particular carabid species encountered in an ecosystem would be to accurately determine what has been consumed by field observations backed by either gut dissection and analysis, radioactive tagging, or employing other biochemical testing techniques. As mentioned above, many researchers have relied upon gut analysis to help determine the feeding habits of field collected specimens, including Forbes (1883), Lindroth (1949), Davies (1953), Penney (1966), Luff (1974), Sunderland (1975), Hengeveld (1980a), Chiverton (1984), Sunderland *et al.* (1987), Holopainen and Helenius (1992). Gut analysis of field-collected specimens allows a researcher to know immediately what prey has been consumed prior to the beetle's capture without the artificiality of laboratory feedings, where conditions and offered prey are often dissimilar to the beetles' habitats. Gut dissection is an inexpensive technique requiring little specialized equipment and has been used by some researchers, Johnson and Cameron (1969) for example, to complement laboratory feeding trials, and Sunderland *et al.* (1987) to complement and compare with enzyme-linked immunosorbent assays (ELISA) tests. Gut analysis, however, has limitations.

1. It is not useful with fluid feeders, including perhaps most carabid larvae.
2. It may underestimate the predation rate on particular prey species (Sunderland *et al.*, 1987) due to the duration of time that hard body parts are held in the gut.
3. Cannot determine whether live prey was consumed or that the beetle was scavenging.
4. Whether the food found in the gut was secondary, that is, consumed by the prey.
5. Dissection is time-consuming, meticulous work (assuming that it would take about 3 minutes to prepare a specimen for examination, a collection of 2000 beetles would require 100 hours of preparation time).
6. Not every beetle will have food in its foregut.
7. Some beetles will regurgitate or defecate consumed food upon capture (Sunderland *et al.*, in press).
8. The results are dependent upon the expertise of the researcher in their ability to determine the nature of the food remains (Lövei and Sunderland, 1996).

Other factors that need to be considered in gut analysis studies would be differences in feeding between sexes of the same species, sexual maturity, prey availability, prey preference, rate of food evacuation, method of collection of field specimens and how they are stored (Sunderland and Vickerman, 1980).

Gut dissection and analysis may be of little use on carabid larvae. In addition to the difficulty of collecting large numbers of larvae from the field, their prey preferences may be difficult to interpret because of their mode of feeding, which is similar to fluid feeders. In the case of seed-eating larvae, the majority of their food may be vegetative. Luff (1974) investigated the larval feeding habits of *Pterostichus madidus* and found few hard fragments in the gut.

MICROSCOPIC IDENTIFICATION OF REMAINS

Most papers reporting gut analysis studies record similar dietary constituents, especially when multiple species are examined. Zhavoronkova (1969) listed five general categories of food materials readily visible when using a microscope, which appear in most gut dissection studies: a) pulverized insect remains; b) parts of vegetative plant tissue with large cells and distinct cell walls, e.g. pollen grains; c) gruel-like mass containing starch granules, possibly remnants of seeds; d) homogeneous jelly-like contents, most often yellowish with large fat droplets; e) homogeneous turbid fluid or suspension found in zoophagous species. Other categories of contents that can be added to Zhavoronkova's list include cuticle fragments and chaetae from earthworms (Luff, 1974; Sunderland, 1975); whole animals in the crops, usually of mites and Collembola (Penney, 1966; Luff, 1974); and flower of Graminae and anthers, body parts of Chilopoda, body parts of Araneae, body parts of Enchytraeidae; molluscan radula (Sunderland *et al.*, 1995).

Hengeveld (1980b) provided one of the most detailed and complete descriptions of arthropod parts found in gut dissections, including legs, parts of compound eyes, and sclerites.

1. Clear chitinous remains with clear pieces represented aphids, carabid and staphylinid larvae, and mites, lepidopterous cuticle was patchy and partially pigmented.
2. Pigmented chitin ranged from opaque to black with: a) Diptera being opaque; b) adult carabid and staphylinid beetles dark brown and transparent on the edges; c) bug chitin less pigmented and often merging into lighter background or sometimes sharply delimited; d) ant chitin was usually amber; e) spider chitin was nearly transparent with little pigment
3. Various hairs with: a) spider hairs, long and flexible and when still attached to chitin had shorter bristles interspersed with no discernible pattern; b) fly bristles entirely black, long and stiff, often in recognizable patterns; c) most larvae had thin hairs in clear patterns; d) usually aphids, ants, some bugs and beetles had short hairs interspersed with occasional long hairs.
4. Eyes with: a) dipterans had compound eyes with bristles between closely packed facets; b) aphids, ants, bugs and beetles had more loosely packed facets that were circular and not hexagonal; c) ants had a chitinous diaphragm-like structure under the layer of lens.
5. Lepidopteran wing scales.

Inanimate items that were sometimes found by Hengeveld included sand particles, and crops were sometimes empty.

Not everything found in the gut can be accurately identified. Fluid feeding predators seldom include hard tissue in their crops and determining prey based on visual inspection of the crop is not possible. Green (1956) could not tell from particles in the guts of *Bembidion laterale* what had been consumed because the arthropod cuticle was so pulverized. To determine the prey, Green had to observe the beetles feeding in the field. Unless distinctive hard parts of the exoskeleton are present, it may not be possible to identify the remains. To increase the chance of identifying body parts, reference slides can be made from gut dissections of laboratory held beetles that have been fed specific prey species (Sunderland *et al.*, in press).

QUALITATIVE ASPECTS

Obviously, gut dissection and analysis are not necessarily new techniques for determining feeding habits. Over 100 years ago, Forbes (1883) used this method to begin to understand the role carabids play in nature. Despite more recent functional morphology studies (Forsythe, 1982, 1987; Evans and Forsythe, 1985; Evans, 1990), from which we better understand how carabids go about their business of feeding, we still remain with large numbers of species that cannot be classified according to feeding habits. This fact is well illustrated by the reviews of Hengeveld (1980a) and Lövei and Sunderland (1996).

One reason for this frustration is that of low numbers of specimens used in feeding studies, as pointed out by Hengeveld (1980b) and Lövei and Sunderland (1996). Number of dissected specimens ranged from 120 *Notiophilus biguttatus* by Anderson (1972) to approximately 12 000 specimens of 16 species of carabids, three species of staphylinids and one species of earwig by Sunderland and Vickerman (1980). Relatively few studies involve large numbers of specimens and these have included small numbers of species when compared to the total number of existing species (although the number of species studied may be sufficient for the respective habitat). Prey may also change over time and space, and this must be taken into account when obtaining specimens for dissection. Skuhravy (1959), Penney (1966) and Luff (1974), among others, found shifts in consumed prey over time. Even when the investigated species are predominantly zoophagous, a shift occurs according to prey availability and preference. Quality of food must also be addressed, as discussed by Toft and Bilde (this volume) because prey differ in their nutrient potential. Spatial differences also occurred. In his results, Skuhravy (1959) considered *Harpalus aeneus* to be harmful because of its plant feeding habits in Europe, yet Hagely *et al.* (1982) found *H. aeneus* to be a potential biological control agent for orchard insect pests in North America.

Most carabid beetles have broad prey ranges, as described by the lists of foodstuffs from gut dissections compiled by Sunderland (1975) and Hengeveld (1980b). Hengeveld (1980c) suggested that the Harpalinae had broader prey ranges than the Carabinae. Nevertheless, the list of food items often narrows when the researcher focuses on one species, such as Penney (1966) who listed eight categories of food items for *Nebria brevicollis*. In contrast, Luff (1974) found twice as many categories of listed food items in gut dissections of the more polyphagous *Pterostichus madidus*. A comparison with the list of food items that Sunderland (1975) found in the crops of seven carabid species shows that, collectively, the seven species had about as broad a prey range as *P. madidus* (*Table 4.1*).

Given the known feeding preferences, is there enough information available for researchers to manipulate carabids to influence or affect our existence? Lövei and Sunderland (1996) discussed early biological control efforts with *Calosoma*

Table 4.1. A comparison of the food materials found in two gut analysis studies of a single carabid species, *Nebria brevicollis* (Penney, 1966), and *Pterostichus madidus* (Luff, 1974) with a gut analysis study of multiple carabid species.

	Penney (1966)	Luff (1974)	Sunderland (1975)
# of species	1	1	7
# of dissections	305	305	807
Liquid food			x
Collembola (whole or part)	x	x?	x
Mites (whole or part)	x	x?	
Fly larvae (whole or part)	x	x	
Beetle larvae		x	x
Lepidoptera larvae		x	
Adult body parts			
Homoptera	x	x	x
Coleoptera	x	x	x
Diptera	x	x	x
Lepidoptera		x	
Spiders (whole or part)	x	x	x
Myriapods		x	
Molluscs		x	
Earthworms (cuticle)	x	x	x
Fungal material (conidia, hyphae, spores)		x?	x
Plant material		x	x

? – may be secondary food items

sycophanta on gypsy moth, *Lymantria dispar*, in North America. Entomologists have since recognized the potential of other carabid species as biological control agents (Dunning *et al*., 1975; Hagely *et al*., 1982; Grafius and Warner, 1989, as examples) and have looked at habitat manipulation and crop management as ways to encourage carabid populations (reviewed by Holland and Luff, 2000; Lee and Landis, this volume). Beginning with Johnson and Cameron (1969) in their hallmark research on phytophagous carabids employing both laboratory feeding trials and gut dissections, more investigations of the biological control of weeds have been conducted. While there are laboratory and field feeding studies on seed preferences (Lund and Turpin, 1977; Brust and House, 1988, for example) there are no gut dissection studies to my knowledge that specifically address seed consumption by field-collected carabids. As Lövei and Sunderland (1996) state, the amount of adult carabid predation is probably over estimated.

QUALITATIVE ASPECTS

Which portion of the gut is to be examined and how the information from gut dissection is used depends upon the objective(s) and attentiveness of the respective researcher. Examination of the entire gut, including the crop, proventriculus, mid-gut or mesenteron and rectum, provides the most complete information of prey consumed, percentage of individuals consuming specific prey, rates of food passage, and so forth.

As noted above, the most basic step is to be able to identify the nature of the remains (Lövei and Sunderland, 1996), which requires the researcher to be generally familiar with the possible range of food materials. Details regarding the physical identification of food materials from gut dissections are given in Hengeveld (1980b) and Sunderland *et al*. (1995). Recognition of fragments is especially important for quantifying the number of organisms consumed as multiple meals, defined by Sunderland *et al*. (1995) as 'cases where more than one prey of the same type was present and where more than one food category was found in one gut (excluding fluid, quartz, and amorphous)'. For example, in the case of a carabid feeding on Collembola, it may be difficult to count the pieces of sclerites and construct how many specimens were consumed. If the number of ingested mandibles, furculae, antennae, or other significant parts, can be ascertained, then the number of specimens consumed can be estimated. Familiarity with the prey is also crucial in understanding the biology of the carabids being studied. In an extensive ecological study of grass tussocks, gut dissection of predatory beetles including carabids and staphylinids helped Luff (1966) to complete the biology and seasonal habitat preferences of the beetles.

Once the hard food materials are identified, researchers should be cautious as to how these materials are interpreted. The rate of food passage through the gut varies with the species (Dixon and McKinlay, 1992), between individual beetles of a species (Bogya and Mols, 1995), and also depends upon the surrounding environment (Pollet and Desender, 1990). Pollet and Desender (1990) found that temperature has an important effect on the rate of food passage through the crop and mid-gut. In their laboratory experiments with three temperature levels, the warmer the ambient temperature, the more rapid the food passage. They also found that the crop provided the least reliable estimate for rate of food passage since it emptied out relatively quickly.

In a feeding preference study of 25 predatory beetles from woodlands, Dennison and Hodkinson (1983) found that, even though ELISA tests were not positive for prey antigens, prey fragments remained in the gut. Therefore, the mere presence of food fragments does not necessarily imply the food was consumed immediately prior to the beetle's capture. Pollet and Desender (1990) agree, suggesting that in *Pterostichus melanarius*, prey remains could indicate meals consumed four days earlier. However, Sunderland *et al.* (1987) reported rapid inactivation of prey antigens in certain beetles that had fed upon aphids in laboratory conditions, thus further confusing the issue of rates of digestion of soft tissue and of food passage.

One of the most common uses of gut dissection has been to establish the percentage of individuals within a population feeding on particular prey (Dawson, 1965; Penney, 1966; Sunderland, 1975; Hengeveld, 1980b). In addition to developing percentages for prey consumed, other studies have further considered the difference in consumption by male and female beetles and the stage of sexual maturity (Cornic, 1973; Sunderland, 1975; Chiverton, 1984).

Perhaps one of the more useful approaches using gut dissection and analysis is to examine beetles caught through a season or multiple growing seasons to establish which prey has been consumed over time at different levels of availability. Prey preference studies are particularly interesting if the researcher's goal is to establish whether a particular carabid is a useful predator of a crop pest. This has been especially true in the elucidation of the role of carabids and other polyphagous predators in regulating aphid populations in cereal grains in Great Britain (Sunderland, 1975; Sunderland and Vickerman, 1980; Loughridge and Luff, 1983; Holland *et al.*, 1996). One assumption from these studies was that generalist predators have an impact on low population numbers of aphids. However, with gut dissections, whether prey have been killed or scavenged cannot be ascertained and, consequently, their impact on population development. In addition, prey abundance does not always imply greater procurement of that prey by carabids. By estimating the populations of possible prey (mites and Collembola) in her study of *Nebria brevicollis*, Penney (1966) found that, though mites were more abundant and were eaten according to availability, Collembola were the preferred prey. Ease of capture may also influence the amount of prey consumed, even though it is preferred. Dawson (1965) further contends that the habitat that the beetle is found in greatly influences prey selection. In the analysis of the three primary carabid species in her study, the differences in prey consumed might have been due to habitat selection; that acceptable, Diptera prey existed in the litter, but soil-inhabiting carabids would not encounter the prey as much as litter-dwelling carabids. In cereal crops, the synchronization of aphid and predator life cycles influences the proportion of aphids contributing to the total prey consumed (Sunderland and Vickerman, 1980). The relative value of carabids in pest control is discussed more fully by Sunderland (this volume).

Other information on diet can also be gained from gut dissection. In some cases, optimal prey size can be determined from hard food materials. *N. brevicollis* was presumed to have an upper limit of prey size of 4 mm (Penney, 1966). Dawson (1965), in her study of eight carabid species, made relative measurements of prey size but did not include the actual prey size in her work. Such information may have been irrelevant, however, as she concluded that most of these species scavenged. As a result, the dead or dying prey may have been larger than the beetles would have otherwise attacked as live, active prey. As always, exceptions and unusual conditions exist. Pollet and Desender (1990) found that food fragments could be misleading, citing that a single earthworm has different sized chaetae on its cuticle. Thus, relying on the size of the chaetae could lead to erroneous assumptions about prey size and even number of prey consumed.

The accuracy of results from gut dissection and analysis depend upon the thoroughness of the researcher. However, when compared to other feeding analysis techniques such as ELISA, results are generally similar. Sunderland *et al.* (1987) compared gut dissection results with ELISA and found that negative responses agreed between the two techniques; however, the positive responses did not correlate well. It would seem that both gut analysis and other biochemical testing are needed to get a more complete picture of feeding habits.

Conclusion

In reviewing morphological traits, there seems to be no one trait that can be applied across the Carabidae that will indicate specific feeding habits. Certain external traits, such as bulbous compound eyes and narrowed pronota, appear to indicate highly predaceous species, but these represent only a small portion of the total species in the family. Further study of chemoreceptors on the palps and antennae may yield interesting insights of dietary preferences.

Internally, only the morphological structures of the proventriculus are likely to be useful in helping to decipher feeding habits. However, the proventriculus has not been studied sufficiently for there to be a consensus of opinion on its role. Since there is a range of size, amount of musculature and differences in internal armament among species, more careful study of the proventriculus may be useful. Currently, however, relying on morphological characteristics only, we find that most carabid species remain mixed feeders, that is, feeding on both plant and animal food.

Gut dissection and analysis provides a glimpse of what carabids are feeding on in the field if we are able to recognize the hard materials found in the various parts of the gut. This technique has been quite helpful because researchers have found that carabids are able to adapt to different food sources through the year, in different habitats, and at different stages of development. Still, gut dissection does not give us the complete picture of feeding habits. By considering multiple morphological characteristics, the results of gut dissection coupled with biochemical analysis of gut contents, described in the following chapter by Symondson, and direct laboratory and field feeding observations, we can get a relatively complete picture of feeding habits. Larochelle (1990) provides feeding information for 1054 species and states that the food preferences are well known for about 50 species. Considering the wealth of species in the Carabidae, there remains much work to do.

Acknowledgements

I would like to thank John Holland for this opportunity and Bill Symondson for his help and encouragement in the writing of this chapter, Jason Zhang, Rutgers University, for his help in producing the illustrations, and Martin Luff for providing me with specimens of *Broscus cephalotes*.

References

ACORN, J.H. AND BALL, G.E. (1991). The mandibles of some adult ground beetles: structure, function, and the evolution of herbivory (Coleoptera: Carabidae). *Canadian Journal of Zoology* **69**, 638–650.

ALLEN, R.T. (1979). The occurrence and importance of ground beetles in agricultural and surrounding habitats. In: *Carabid beetles: their evolution, natural history, and classification*. Ed. T.L. Erwin, pp 485–507. The Hague: Junk.

ALTIERI, M.A. AND WHITCOMB, W.H. (1979). Predaceous arthropods associated with Mexican tea in north Florida. *The Florida Entomologist* **62**, 175–182.

ALTNER, H. AND HINTZPETER, U. (1984). Reduction of sensory cells in antennal sensilla correlated with changes in feeding behaviour in the beetle *Loricera pilicornis* (Insecta, Coleoptera, Carabidae). *Zoomorphology* **104**, 171–179.

ANDERSON, J.M. (1972). Food and feeding of *Notiophilus biguttatus* F. (Coleoptera: Carabidae). *Revue d'Ecologie et de Biologie Sol* **2**, 177–184.

ASTERAKI, E.J. (1993). The potential of carabid beetles to control slugs in grass/clover swards. *Entomophaga* **38**, 193–198.

BAKER, A.N. AND DUNNING, R.A. (1975). Some effects of soil type and crop density on the activity and abundance of the epigeic fauna, particularly Carabidae, in sugar-beet fields. *Journal of Applied Ecology* **12**, 809–818.

BAKER, G.T. AND MONROE, W.A. (1995). Sensory receptors on the adult labial and maxillary palpi and galea of *Cicindela sexguttata* (Coleoptera: Cicindelidae). *Journal of Morphology* **226**, 25–31.

BALL, G. (1992). The tribe Licinini (Coleoptera: Carabidae): A review of the genus-groups and of the species of selected genera. *Journal of the New York Entomological Society* **100**, 325–380.

BARNEY, R.J. AND PASS, B.C. (1986a). Ground beetle (Coleoptera: Carabidae) populations in Kentucky alfalfa and influence of tillage. *Journal of Economic Entomology* **79**, 511–517.

BARNEY, R.J. AND PASS, B.C. (1986b). Foraging behaviour and feeding preference of ground beetles (Coleoptera: Carabidae) in Kentucky alfalfa. *Journal of Economic Entomology* **79**, 1334–1337.

BAUER, T. (1977). The relevance of the brightness to visual acuity, predation, and activity of visually hunting ground beetles (Coleoptera, Carabidae). *Oecologia* (Berl.) **30**, 63–73.

BAUER, T. AND KREDLER, M. (1993). Morphology of the compound eyes as an indicator of life-style in carabid beetles. *Canadian Journal of Zoology* **71**, 799–810.

BOGYA, S. AND MOLS, P.J.M. (1995). Ingestion, gut emptying and respiration rates of clubionid spiders (Araneae: Clubionidae) occurring in orchards. *Acta Phytopathologica et Entomologica Hungarica* **30**, 291–299.

BORROR, D.J. AND DELONG, D.M. (1966). *An introduction to the study of insects*. New York: Holt, Rhinehart & Winston.

BRUNETTI, B. (1931). Ricerche sul proventricolo degli insetti. Morfologia del proventricolo. Fomazioni chitinose. Rapporto fra la struttura delle formazioini chitinose ed il regime alimentare. *Atti della Societa Toscana di Scienze Naturali Residente in Pisa Memorie* **41**, 110–146.

BRUST, G.E. AND HOUSE, G.J. (1988). Weed seed destruction by arthropods and rodents in low-input soybean agroecosystems. *American Journal of Alternative Agriculture* **3**, 19–25.

BRYSON, H.R. AND DILLION, G.F. (1941). Observations on the morphology of the corn seed beetle (*Agonoderus pallipes* Fab., Carabidae). *Annals of the Entomological Society of America* **32**, 43–50.

BURAKOWSKI, B. (1967). Biology, ecology and distribution of *Amara pseudocommunis* Burak (Coleoptera, Carabidae). *Annales Zoologici* **24**, 485–523.

BURGESS, A.F. (1911). *Calosoma sycophanta*: its life history, behavior, and successful colonization in New England. *United States Department of Agriculture, Bulletin* **101**, 1–94.

BUTTERFIELD, J. (1996). Carabid life-cycle strategies and climate change: a study on an altitude transect. *Ecological Entomology* **21**, 9–16.

CALKINS, C.O. AND KIRK, V.M. (1974). Temporal and spatial distribution of *Pasimachus elongatus* (Coleoptera: Carabidae) a predator of false wireworm. *Annals of the Entomological Society of America* **67**, 913–914.

CHEESEMAN, M.T. AND PRITCHARD, G. (1984a). Proventricular trituration in adult carabid beetles (Coleoptera: Carabidae). *Journal of Insect Physiology* **30**, 203–209.

CHEESEMAN, M.T. AND PRITCHARD, G. (1984b). Spatial organization of digestive processes in an adult carabid beetle, *Scaphinotus marginatus* (Coleoptera: Carabidae). *Canadian Journal of Zoology* **62**, 1200–1203.

CHIVERTON, P. (1984). Pitfall-trap catches of the carabid beetle *Pterostichus melanarius*, in relation to gut contents and prey densities, in insecticide treated and untreated spring barley. *Entomologia Experimentalis et Applicata* **36**, 23–30.

CLARK, M.S., GAGE, S.H. AND SPENCE, J.R. (1997). Habitats and management associated with common ground beetles (Coleoptera: Carabidae) in a Michigan agricultural landscape. *Environmental Entomology* **26**, 519–527.

CORNIC, J.F. (1973). Etude du régime alimentaire di trios espèces de carabiques et de sez variations en verger de pommiers. *Annales de la Societe Entomologique de France* **9**, 69–87.

DAVIES, M.J. (1953). The contents of the crops of some British carabid beetles. *Entomologists Monthly Magazine* **95**, 25–28.

DAVIES, T.G. (1963). Observations of the ground beetle fauna of brassica crops. *Plant Pathology* **12**, 9–11.

DAWSON, N. (1965). A comparative study of the ecology of eight species of fenland Carabidae (Coleoptera). *Journal of Animal Ecology* **34**, 299–314.

DEMPSTER, J.P., RICHARDS, O.W. AND WALOFF, N. (1959). Carabidae as predators on the pupal stage of the chrysomelid beetle, *Phytodecta olivacea* (Forester). *Oikos* **10**, 65–70.

DEN BOER, P.J. (1969). On the significance of dispersal power for populations of carabid-beetles (Coleoptera: Carabidae). *Oecologia* **4**, 1–27.

DENNISON, D.F. AND HODKINSON, I.D. (1983). Structure of the predatory beetle community in a woodland soil ecosystem. I. Prey selection. *Pedobiologia* **25**, 109–115.

DIGWEED, S.C. (1993). Selection of terrestrial gastropod prey by Cychrine and Pterostichine ground beetles (Coleoptera: Carabidae). *Canadian Entomologist* **125**, 463–472.

DIXON, P.L. AND MCKINLAY, R.G. (1992). Pitfall trap catches of and aphid predation by *Pterostichus melanarius* and *Pterostichus madidus* in insecticide treated and untreated potatoes. *Entomologia Experimentalis et Applicata* **64**, 63–72.

DUNNING, R.A., BAKER, A.N. AND WINDLEY, R.F. (1975). Carabids in sugar beet crops and their possible role as aphid predators. *Annals of Applied Biology* **80**, 125–128.

EVANS, M.E.G. (1977). Locomotion in the Coleoptera Adephaga, especially Carabidae. *Journal of Zoology* **181**, 189–226.

EVANS, M.E.G. (1986). Carabid locomotor habits and adaptations In: *Carabid beetles: their adaptations and dynamics: XVIIth International Congress.* Ed. P.J. Den Boer, pp 59–77. Stuttgart: Gustav Fisher.

EVANS, M.E.G. (1990). Habits or habitats: Do carabid locomotor adaptations reflect habitats or lifestyles? In: *The role of ground beetles in ecological and environmental studies*. Ed. N.E. Stork, pp 295–305. Andover: Intercept.

EVANS, M.E.G. AND FORSYTHE, T.G. (1985). Feeding mechanisms, and their variation in form, of some adult ground-beetles (Coleoptera: Carabidae). *Journal of Zoology* **206**, 113–143.

FORBES, S.A. (1883). The food relations of the Carabidae and Coccinellidae. *Illinois State Laboratory of Natural History, Bulletin* **1**, 33–64.

FORSYTHE, T.G. (1982). Feeding mechanisms of certain ground beetles (Coleoptera: Carabidae). *Coleopterists Bulletin* **36**, 26–73.

FORSYTHE, T.G. (1983). Locomotion in ground beetles (Coleoptera, Carabidae) An interpretation of leg structure in functional terms. *Journal of Zoology, London* **200**, 493–507.

FORSYTHE, T.G. (1987). The relationship between body form and habit in some Carabidae (Coleoptera). *Journal of Zoology, London* **211**, 643–666.

FOX, C.J.S. (1974). The activity of ground beetles (Col.: Carabidae) in hayfields and their margins in Nova Scotia. *Phtyoprotection* **55**, 99–102.

FOX, C.J.S. AND MACLELLAN, C.R. (1956). Some Carabidae and Staphylinidae shown to feed on wireworms – *Agriotes sputator* (L.) by the precipitin test. *Canadian Entomologist* **88**, 228–231.

FRANK, J.H. (1971). Carabidae (Coleoptera) as predators of the red-backed cutworm (Lepidoptera: Notuidae) in central Alberta. *The Canadian Entomologist* **103**, 1039–1044.

GRAFIUS, E. AND WARNER, F.W. (1989). Predation by *Bembidion quadrimaculatum* (Coleoptera: Carabidae) on *Delia antiqua* (Diptera: Anthomyiidae). *Environmental Entomology* **18**, 1056–1059.

GREEN, J. (1956). The mouthparts of *Eurynebria complanata* (L.) and *Bembidion laterale* (Sam.) (Col., Carabidae). *Entomologists Monthly Magazine* **92**, 110–113.

HAGELY, E.A.C., HOLLIDAY, N.J. AND BARBER, D.R. (1982). Laboratory studies of the food preferences of some orchard carabids (Coleoptera: Carabidae). *Canadian. Entomologist* **114**, 431–437.

HENGEVELD, R. (1980a). Qualitative and quantitative aspects of the food of ground beetles (Coleoptera, Carabidae): A review. *Netherlands Journal of Zoology* **30**, 555–563.

HENGEVELD, R. (1980b). Polyphagy, oligophagy and food specialization in ground beetles (Coleoptera, Carabidae). *Netherlands Journal of Zoology* **30**, 564–584.

HENGEVELD, R. (1980c). Food specialization in ground beetles (Coleoptera, Carabidae): An ecological or a phylogenetic process? *Netherlands Journal of Zoology* **30**, 585–594.

HOLLAND, J.M. AND LUFF, M.L. (2000). The effects of agricultural practices on Carabidae in temperate agroecosystems. *Integrated Pest Management Reviews* **5**, 109–129.

HOLLAND, J.M., THOMAS, S.R. AND HEWITT, A. (1996). Some effects of the polyphagous predators on an outbreak of cereal aphid (*Sitobion avenae* F.) and orange wheat blossom midge (*Sitodoplosis mosellana* Géhin). *Agriculture, Ecosystems and Environment* **59**, 181–190.

HOLOPAINEN, J.K. AND HELENIUS, J. (1992). Gut contents of ground beetles (Col., Carabidae) and activity of these and other epigeal predators during an outbreak of *Rhopalosiphum padi* (Hom., Aphididae). *Acta Agriculturae Scandinavia, Section B. Soil and Plant Science* **42**, 57–61.

HOUSE, G.J. AND ALL, J.N. (1981). Carabid beetles in soybean agroecosystems. *Environmental Entomology* **10**, 194–196.

JOHNSON, N.E. AND CAMERON, R.S. (1969). Phytophagous ground beetles. *Annals of the Entomological Society of America* **62**, 909–914.

JØRGENSEN, H.B. AND TOFT, S. (1997). Role of granivory and insectivory in the life cycle of the carabid beetle *Amara similata*. *Ecological Entomology* **22**, 7–15.

LAROCHELLE, A. (1990). *The food of carabid beetles (Coleoptera: Carabidae, including Cicindelinae). Québec: Fabreries Supplement* 5, Association des Entomologistes Amateurs du Québec.

LINDROTH, C.H. (1949). *Die Fennoskandischen Carabidae. Eine tiergeographische Studie. I. Spezieller Teil.* Goteborg: Elanders Boktryckeri Aktiebolag.

LINDROTH, C.H. (1969). *The ground beetles (Carabidae excl. Cicindelidae) of Canada and Alaska. 1–6.* Opuscula Entomologica. Stockholm: Lund.

LOUGHRIDGE, A.H. AND LUFF, M.L. (1983). Aphid predation by *Harpalus rufipes* (DeGeer) (Coleoptera: Carabidae) in the laboratory and field. *Journal of Applied Ecology* **20**, 451–462.

LÖVEI, G.L. AND SUNDERLAND, K. (1996). Ecology and behaviour of ground beetles (Coleoptera: Carabidae). *Annual Review of Entomology* **41**, 231–256.

LUFF, M.L. (1966). The abundance and diversity of the beetle fauna of grass tussocks. *Journal of Animal Ecology* **35**, 189–208.

LUFF, M.L. (1974). Adult and larval feeding habits of *Pterostichus madidus* (F.) (Coleoptera: Carabidae). *Journal of Natural History* **8**, 403–409.

LUFF, M.L. (1987). Biology of polyphagous ground beetles in agriculture. In: *Biology and population dynamics of invertebrate crop pests.* Ed. G.E. Russell, pp 209–250. Andover: Intercept.

LUND, R.D. AND TURPIN, F.T. (1977). Carabid damage to weed seeds found in Indiana cornfields. *Environmental Entomology* **6**, 695–698.

MAHAR, J. (1979). *The life histories of* Pinacodera platicollis *Dejean and* P. limbata *Say in northern hardwood forests of Michigan.* Masters Thesis: Michigan State University.

MANLEY, G.V. (1971). A seed-cacheing carabid (Coleoptera). *Annals of the Entomological Society of America* **6**, 1474–1475.

MITCHELL, B. (1963). Ecology of two carabid beetles, *Bembidion lampros* (Herbst) and *Trechus quadristriatus* (Schrank) I. Life cycles and feeding behaviour. *Journal of Animal Ecology* **32**, 289–299.

NOONAN, G. (1967). Observations on the ecology and feeding habits of adult *Scaphinotus punctatus* LeConte. *The Pan-Pacific Entomologist* **43**, 21–23.

PAUSCH, R.D. (1979). Observations on the biology of the seed corn beetles, *Stenolophus comma* and *Stenolophus lecontei. Annals of the Entomological Society of America* **72**, 24–28.

PENNEY, M.M. (1966). Studies on certain aspects of the ecology of *Nebria brevicollis* (F.) (Coleoptera, Carabidae). *Journal of Animal Ecology* **35**, 505–512.

POLLET, M. AND DESENDER, K. (1990). Investigating the food passage in *Pterostichus melanarius* (Coleoptera: Carabidae): An attempt to explain its feeding behaviour. *Mededelingen van de Fakulteit Landbouwwetenschappen Rijksuniversiteit Gent* **55**, 527–540.

RIVARD, I. (1964). Carabid beetles (Coleoptera: Carabidae) from agricultural lands near Belleville, Ontario. *Canadian Entomologist* **96**, 517–520.

RIVARD, I. (1965). Dispersal of ground beetles (Coleoptera: Carabidae) on soil surface. *Canadian Journal of Zoology* **43**, 465–473.

RIVARD, I. (1966). Ground beetles (Coleoptera: Carabidae) in relation to agricultural crops. *Canadian Entomologist* **98**, 189–195.

SCHERNY, F. (1955). Untersuchungen über Vorkommen und Wirtschaftliche Bedeutung räuberisch lebender Käfer in Feldkulturen. *Zeitschrift Pflanzenbau Pflanzenschutz* **6**, 49–73.

SHAROVA, J.K. (1975). Evolution of imaginal life forms of ground beetles. *Zoological Zhurinal* **54**, 49–67. [In Russian]

SHOUGH, W.W. (1940). The feeding of ground beetles. *American Midland Naturalist* **24**, 336–344.

SKUHRAVY, V. (1959). [Diet of field carabids] *Casopis Ceskoslov Spolecnosti Entomologick'e* **56**, 1–19.

SUNDERLAND, K.D. (1975). The diet of some predatory arthropods in cereal crops. *Journal of Applied Ecology* **12**, 507–515.

SUNDERLAND, K.D. AND VICKERMAN, G.P. (1980). Aphid feeding by some polyphagous predators in relation to aphid density in cereal fields. *Journal of Applied Ecology* **17**, 389–396.

SUNDERLAND, K.D., CROOK, N.E., STACEY, D.L. AND FULLER, B.J. (1987). A study of feeding by polyphagous predators on cereal aphids using ELISA and gut dissection. *Journal of Applied Ecology* **24**, 907–933.

SUNDERLAND, K.D., LÖVEI, C.L. AND FENLON, J. (1995). Diets and reproductive phenologies of the introduced ground beetles *Harpalus affinis* and *Clivina australasiae* (Coleoptera: Carabidae) in New Zealand. *Australian Journal of Zoology* **43**, 39–50.

SUNDERLAND, K.D., SYMONDSON, W.O.C. AND POWELL, W. (in press). Populations and Communities. In: *Insect natural enemies* (2nd edition). Ed. M. Jervis. London: Chapman and Hall.

SYMONDSON, W.O.C. AND WILLIAMS, I.B. (1997). Low-vacuum electron microscopy of carabid chemoreceptors: a new tool for the identification of live and valuable museum specimens. *Entomologia Experimentalis et Applicata* **85**, 75–82.

THIELE, H.-U. (1977). *Carabid beetles in their environments.* New York: Springer-Verlag.

VAN HUIZEN, T.H.P. (1977). The significance of flight activity in the life cycle of *Amara plebeja* Gyll. (Coleoptera, Carabidae). *Oecologia* **29**, 27–41.

WALLIN, H. (1988). Mandible wear in the carabid beetle *Ptersotichus melanarius* in relation to diet and burrowing behaviour. *Entomologia Experimentalis et Applicata* **48**, 43–50.

WEBSTER, F.M. (1900). *Harpalus caliginosus* as a strawberry pest, with notes on other phytophagous Carabidae. *Canadian Entomologist* **32**, 265–271.

ZETTO BRANDMAYR, T. (1990). Spermaphagous (seed eating) ground beetles: first comparison of the diet and ecology of the Harpaline genera *Harpalus* and *Ophonus* (Col.; Carabidae). In: *The role of ground beetles in ecological and environmental studies.* Ed. N.E. Stork, pp 307–316. Andover: Intercept.

ZHAVORONKOVA, T.N. (1969). Certain structural peculiarities of the Carabidae (Coleoptera) in relation to their feeding habits. *Entomological Review* (Eng. Transl.) **48**, 462–471.

5
Diagnostic Techniques for Determining Carabid Diets

WILLIAM O.C. SYMONDSON

Cardiff School of Biosciences, Cardiff University, PO Box 915, Cardiff CF10 3TL, UK

Introduction

There is now a large, but highly heterogeneous, body of literature dealing with the diet of carabids beetles (major reviews include Hengeveld, 1980a,b; Larochelle, 1990; Symondson *et al.*, 2000; Toft and Bilde, this volume; Sunderland, this volume). Much of this work depends upon the results of laboratory feeding trials which, though of some value in helping to determine which prey are accepted and which the beetles are capable of killing, tell us very little about prey choice in the field where a broad range of prey are potentially available. Indeed, such trials may actually be misleading; there are reports of prey consumed in the laboratory that were apparently not consumed in the field where the prey was readily available (e.g. Rothschild, 1966). Direct observation of predation in the field is usually impractical because many carabids are nocturnal and/or live amongst dense vegetation. Even if a predation event is occasionally observed, there is no way of determining whether this was a rare or common event. The act of observing can itself affect the behaviour of the predator or prey in unpredictable ways and hence bias any attempt to quantify such behaviour. Which prey are actually killed and consumed depends on the morphology of the beetle, its prey detection systems, prey size and species preferences, the temporal and spatial distribution of predator and prey, competing densities of different prey, switching behaviour, prey defence mechanisms and escape strategies, prey toxins, functional responses, predation vs. scavenging, and many other interacting factors.

This chapter seeks to review techniques for analysing dietary preferences in the field using one of a range of chemical, biochemical, immunological or molecular techniques. Approaches that seek to predict carabid diet from morphology, or use microscopic examination of the gut contents, are covered elsewhere (Ingerson-Mahar, this volume). The review has been restricted to approaches that have been used in carabid research, even though alternative strategies have been applied in research into the diets of other polyphagous groups such as Staphylinidae, Arachnida

Abbreviations: ELISA, Enzyme-Linked ImmunoSorbent Assays; PCR, Polymerase Chain Reaction

The Agroecology of Carabid Beetles

and predatory Hemiptera. However, where new methodologies that show great promise have been developed, which could be usefully applied in carabid research, these are briefly described.

Chemical and radioactive labelling

Potential predators can be identified by marking prey with a chemical, releasing the marked prey in the field, and then sampling predators in the vicinity for the presence of that chemical after some time has elapsed. Although clearly some disturbance of the system under study is inevitable, such experiments could provide valuable information on predation in the field where the structure of the vegetation, microclimatic conditions and the presence of competing prey provide more realistic conditions than could ever be achieved in the laboratory. Such marking techniques provide a rapid, inexpensive, generally low-tech approach, without the need for specialized facilities or equipment.

A recent example of this was the release of New Zealand flatworms, *Artioposthia triangulata*, stained with neutral red, and the subsequent recovery of a carabid larva (unidentified) containing the dye (Gibson *et al.*, 1997). This serendipitous observation led to laboratory studies that confirmed that carabid larvae would feed on this alien predator of earthworms. Fluorescent dyes have been used in a similar way to mark prey, and the dye subsequently detected in the guts of polyphagous predators (earwigs) (Hawkes, 1972), although this approach does not yet appear to have been used in carabid research.

An alternative approach is to mark prey with a rare element that is unlikely to be present in the ecosystem under study. Although this technique has been widely utilized in population studies involving mark–release–recapture, it can also be applied to work on trophic interactions. For example, Johnson and Reeves (1995) fed a diet enriched with rubidium chloride to the larvae of the gypsy moth, *Lymantria dispar* (Lepidoptera: Lymantriidae). The label was subsequently detected in *Carabus nemoralis* that had fed on fourth instar *L. dispar*. Quantitative assessment of the number of prey consumed in the field would be difficult because, although rubidium concentration in the predator was positively correlated with numbers of prey consumed, it was negatively related to time since feeding.

Prey may also be marked with rare isotopes of common elements. Nienstedt and Poehling (1995, 2000) investigated the potential of ^{15}N as a marker in predation studies on spiders and carabids. The aphids *Sitobion avenae* were allowed to feed on oat seedlings treated with ^{15}N, with the ^{15}N content subsequently determined by an element analyser coupled to a mass spectrometer. A content of 10% ^{15}N was found to be sufficient to mark the aphids, and the isotope concentration was found to decrease exponentially with time once the aphids were transferred to non-enriched plants. The label appeared to have no adverse effects upon the survival or fertility of the aphids. The label was detected in the carabid *Agonum dorsale*.

A method has been devised for marking prey with mammalian immunoglobulin and then detecting these antibodies amongst the gut contents of predators using enzyme-linked immunosorbent assays (see section below on immunological techniques) (Hagler and Durand, 1994). This simple technique avoids the lengthy and expensive process of producing specific antibodies and was found to be particularly effective for detecting predation by chewing predators that consume the marked cuticle. As yet, this potentially useful approach has not been used for investigating predation by carabids.

Apart from the risks of disturbance to the system, techniques that depend upon external application of markers (fluorescent dyes, polyclonal antisera, rare isotopes) have the drawback that the prey may shed their skin as they change instars. The marker may be cleaned off by the insect itself, and will rub off over time. Even where markers are incorporated in the diet (rubidium chloride), there may be significant loss of the label within a few days once the diet containing that label is removed (Johnson and Reeves, 1995), especially where the prey has efficient mechanisms for excreting that element from its system. To overcome these problems, researchers try to minimize the time between release of the labelled prey and collection of potential predators. This may mean the released prey do not have time to re-establish themselves, either in terms of spatial separation or into their favoured microhabitats. This possibly makes them more vulnerable to predation and other density-dependent mortality factors such as parasitism and disease.

Radioactive markers were once popular (Grant and Shepard, 1985) but are now little used, partly because they are expensive and require specialized handling facilities, but mainly because they could pose a risk to the user and to the environment. Hamon *et al.* (1990) used the radioactive isotope ^{32}P to investigate predation by carabids on adult and larval *Sitonia lineatus*, a weevil pest of field beans. Test plants were induced to take up the isotope in the laboratory and weevils allowed to feed on the radioactive beans. The weevils were then introduced to field cages that were designed to allow entry by epigeal carabids. The latter were subsequently caught in gutter traps and radioactivity measured using both Geiger and scintillation counters. Adult and larval *S. lineatus* were shown to be consumed by carabids, such as *Pterostichus madidus*. The test proved to be essentially qualitative, as no correlation could be found between numbers of radioactive prey consumed in the laboratory and radioactivity of the predator. Ernsting and Joosse (1974) studied predation in the laboratory by carabids on Collembola that were allowed to feed on a ^{32}P-labelled diet, and found this to be an effective approach. Many similar field experiments have been performed elsewhere, but few involved carabids.

Protein electrophoresis

Electrophoresis was, until recently, extensively used in taxonomic studies of insects, particularly for the separation of closely related species (now almost entirely displaced by DNA techniques). At its simplest, electrophoresis is the movement of ions within an electric field, positively charged ions travelling towards the cathode and negatively charged ions towards the anode. Electrophoresis can be used to bring antibodies and antigens together, termed immunoelectrophoresis (see section on precipitin techniques), but this section is concerned mainly with the separation of proteins within a gel and, particularly, the investigation of enzyme systems. Detailed reviews of electrophoretic techniques in insect research can be found in Menken and Ulenberg (1987), Loxdale and Den Hollander (1989), Loxdale (1994), Powell *et al.* (1996) and Symondson and Hemingway (1997). As electrophoresis has been little used to investigate predation by carabids, and the electrophoretic approach is now largely redundant, readers are asked to refer to these texts for further details.

The potential for using protein electrophoresis in predation studies has been reviewed by Murray *et al.* (1989) and Solomon *et al.* (1996). Lövei (1986) conducted a laboratory trial to investigate the suitability of electrophoresis for studying predation by carabids. He used a polyacrylamide gradient slab gel and stained for esterase activity. The experimental prey were pupae of the corn borer, *Ostrinia nubilalis*, which were successfully detected after consumption by the carabids *Agonum dorsale*, *Brachinus explodens* and *Poecilus cupreus*. Esterase isoenzyme patterns of the prey could be distinguished from those of the predator and another lepidopteran prey, *L. dispar*.

Schelvis and Siepl (1988) studied feeding by larvae of *Pterostichus oblongopunctatus* and *P. rhaeticus*, following very similar electrophoretic protocols to Lövei (1986). Gradient gels were stained for esterase activity. Isozyme banding patterns for the predators and potential prey were prepared and compared with patterns for predators collected from the field. Prey were identified in 14 *P. oblongopunctatus* (Coleoptera and Diptera larvae, Collembola, Acari, Cilopoda and Oligochaeta) and nine *P. rhaeticus* (Coleoptera and Diptera larvae, Collembola).

Walrant and Loreau (1995) conducted a field study comparing the ability of isoenzyme electrophoresis with gut content examination under the microscope to detect prey remains. As in the studies above, gradient gels were used and stained for esterase activity. Fifty-seven reference gels were prepared containing isoenzyme patterns for a broad range of potential prey. Ninety predators, the carabid *Abax parallelepipedus*, were tested by both methods and results using the two systems were often different but broadly comparable. Where the remains of several prey were mixed together in the gut of a predator, interpretation of the complex banding patterns often proved to be impossible. Electrophoresis was recommended for studying predators with a narrower prey range.

Just such a predator was studied by Paill (2000), using a similar approach, in a two-year field study to look specifically at predation by *Carabus violaceus* adults and larvae on the slug, *Arion lusitanicus*. Electrophoresis was performed using iso-electric focusing on mini-polyacrylamide gels and stained for esterases, following the methods of Backeljau *et al.* (1994). Approximately 50% of adult *C. violaceus* were shown to have fed on *A. lusitanicus,* while most larvae with food remains in their guts also tested positive. The proportion of beetles giving a positive result showed an inverse relationship with slug size, neatly demonstrating a possible preference by *C. violaceus*, in the field, for smaller slugs.

Immunological techniques

An excellent recent history of the use of immunological techniques to detect predation by polyphagous predators can be found in Greenstone (1996), with reviews of the use of these approaches in Boreham and Ohiagu (1978), Boreham (1979), Sunderland (1988), Powell *et al.* (1996), and Symondson and Hemingway (1997). Not all of these techniques appear to have been used in carabid research and indeed, as can be seen from *Table 5.1*, most research has been conducted using either some form of precipitin test or enzyme-linked immunosorbent assays (ELISA). Although precipitin tests dominated until the mid-1980s, they have now been entirely displaced by the far more sensitive ELISAs. The intention here, therefore, is to discuss precipitin test very briefly, and readers may refer in particular to Boreham and Ohiagu (1978) and Boreham (1979) for further details.

RAISING AND CHARACTERIZING ANTIBODIES

Detailed descriptions of antibody production against target prey in predation studies may be found in Greenstone (1996) and Symondson and Hemingway (1997). The brief review below has been related, where possible, to studies of predation by carabids.

Polyclonal antisera

Whatever the chosen assay format, antibodies to the specific target prey must first be generated. The principles behind this process are identical to those involved with immunization against disease. A vertebrate, usually a rabbit, is injected with proteins (or other biological materials) from the target prey. These 'antigens' may be a whole-body extract of soluble proteins (Ashby, 1974; Lund and Turpin, 1977; Dennison and Hodkinson, 1983; Crook and Sunderland, 1984; Chiverton, 1987; Cameron and Reeves, 1990; Symondson and Liddell, 1996a), haemolymph from the target prey (Symondson and Liddell, 1993a) or protein fractions isolated from the prey by electrophoresis (Fichter and Stephen, 1979; Miller, 1979, 1981). Reducing the number of different protein or other materials used as antigens will, in theory, result in a reduction in the number of cross-reactions that are likely to occur. The antigens generate an immune response in the rabbit, culminating in the production of antibodies in the blood, mainly of the IgM isotype. A second immunization, approximately one month later, triggers a much greater immune response and the production of higher quantities of antibodies, especially of the more useful IgG isotype. A blood sample can be taken at this point and the serum separated from the cellular components. The serum can be used directly in immunoassays, or after further antibody purification steps. Detailed protocols for the production of polyclonal antisera can be found in Tijssen (1985), Catty and Raykundalis (1988), Hudson and Hay (1989) and Liddell and Cryer (1991), and in many of the papers relevant to carabid research listed in *Table 5.1*.

The polyclonal antiserum (so called because it contains a mixture of antibodies with a range of specificities) can then be tested against the target prey to assess sensitivity and specificity. Because the antiserum contains so many different antibodies, the chances of one or more of these cross-reacting with a site on a protein from a non-target species is high. Where such cross-reactions occur, these can sometimes be eliminated, either by passing the antiserum through a column containing the cross-reacting species (all cross-reacting antibodies are absorbed onto the column while specific antibodies pass through) (Schoof *et al.*, 1986) or by absorption. The latter simply involves mixing the antiserum and the liquid fraction of an homogenate of the cross-reacting species (e.g. Dempster, 1960, 1971; Symondson and Liddell, 1993a). All cross-reacting antibodies bind to their respective antigens and are precipitated out of solution. Both techniques result in a more specific antiserum but one with a lower

Table 5.1. Field experiments to determine consumption of target prey by carabids using immunological techniques.

Carabid species that tested positive	Technique	Prey detected	Reference
Agonum mülleri *Amara* spp. *Clivina fossor* *Harpalus* spp. *Pterostichus* spp. Larvae (unidentified)	Precipitin	*Agriotes sputator* larvae (Coleoptera: Elateridae)	Fox and MacLellan, 1956
Agonum dorsale *A. viduum* *Amara familiaris* *Bembidion lampros* *B. quadrimaculatum* *C. fossor* *Harpalus affinis* (= aeneus) *H. rufipes* *Pterostichus madidus* *P. melanarius* *Nebria brevicollis* *Trechus quadristriatus* *T. obtusus*	Precipitin	*Delia radicum* (Diptera: Anthomiidae)	Coaker and Williams, 1963
Numerous species (all negative)	Precipitin	*Conomelus anceps* (Homoptera: Delphacidae)	Rothschild, 1966
Abax parallelepipedus *H. affinis (= aeneus)* *H. rufipes* *P. melanarius* *T. quadristriatus*	Precipitin	*Pieris rapae* (Lepidopera: Pieridae)	Dempster, 1967
A. parallelepipedus *P. madidus* *P. melanarius*	Precipitin	*Operophtera brumata* pupae (Lepidoptera: Hydriomenidae)	Frank, 1967
A. parallelepipedus *Leistus* sp. *Pterostichus niger*	Precipitin	*Sitona regensteinensis* (Coleoptera: Curculionidae)	Danthanarayana, 1969
P. madidus	Precipitin	Woodlice (Isopoda: Oniscidae)	Sutton, 1970
Metabletus foveatus *Amara* sp. *Harpalus rufitarsus*	Precipitin	*Tyria jacobaeae* (Lepidoptera: Actiidae)	Dempster, 1971

Amara torrida	Precipitin	*Euxoa ochrogaster* (Lepidoptera: Noctuidae)	Frank, 1971
A. avida			
Carabus taedatus			
Calasoma calidum			
Pterostichus lucublandus			
P. adstrictus			
Harpalus amputatus			
Calathus micropterus	Precipitin	Mollusca	Tod, 1973
C. fuscipes			
C. melanocephalus			
C. piceus			
Carabus violaceus			
C. catenulatus			
C. nemoralis			
Cychrus caraboides			
Nebria brevicollis			
N. gyllenhalli			
Pterostichus anthracinus			
P. melanarius			
P. madidus			
P. niger			
Metaglymma sp.	Precipitin	*Pieris rapae* (Lepidopera: Pieridae)	Ashby, 1974
Pterostichus chalcites	Precipitin	*Agrotis ipsilon* (Lepidoptera: Noctuidae)	Lund and Turpin, 1977
P. lucublandus			
Badister bipustulatus	Precipitin	*Philoscia muscorum, Armadillium vulgare, Porcellio scaber, Oniscus asellus* (Isopoda: Oniscidae)	Sunderland and Sutton, 1980
C. melanocephalus			
Demetrias atricapillus			
Harpalus tardus			
Harpalus sp. Larvae			
Calosoma sayi	Precipitin	*Heliothis zea* (Lepidoptera: Noctuidae)	Lesiewicz *et al.*, 1982
Carabus vinctus			
Chlaenius erythropus			
Galeritula janus			
Harpalus caliginosus			
H. pensylvanicus			
Megacephala carolina			
M. virginica			
P. chalcites			
Scarites subterraneus			

continued

Table 5.1. contd.

Carabid species that tested positive	Technique	Prey detected	Reference
A. parallelepipedus abcdghik	Precipitin	Enchytraeidae a	Dennison and Hodkinson, 1983
Agonum assimile befghij		Nematoda b	
Amara plebeja ai		Lumbricidae c	
Bradycellus verbasci bfhk		Diplopoda d	
Calathus piceus abcdefghij		Isopoda e	
Carabus problematicus abcdghik		Spiders f	
C. caraboides cfjk		Mites g	
Loricera pilicornis bghj		Collembola h	
N. brevicollis abcefghik		Diptera i	
Notiophilus biguttatus bfghk		Carrion j	
Pterostichus macer bcgi		Fungi k	
P. melanarius fhik		(Letters are codes for	
P. niger bcfghij		positive species)	
(Letters indicate positive for antiserum			
against marked groups)			
Unidentified adults	Precipitin and	*Musca vetustissima*	Calver *et al.*, 1986
	immunoelectro-osmophoresis	(Diptera: Muscidae)	
A. dorsale	ELISA	*Rhopalosiphum padi*	Chiverton, 1987
B. lampros		(Homoptera: Aphididae)	
B. quadrimaculatum			
C. melanocephalus			
H. affinis (= aeneus)			
H. rufipes			
Pterostichus cupreus			
P. melanarius			
P. niger			
Synuchus nivalis			
T. quadristriatus			
Trechus secalis			
Agonum dorsale	ELISA	*Sitobion aevnae*, *Metopolophium*	Sunderland *et al.*, 1987
Amara aenea		*dirhodum* and *Rhopalosiphum padi*	
A. familiaris		(combined antigen) (Homoptera: Aphididae)	
A. plebeja			
B. lampros			
B. obtusum			

L. pilicornis			
N. brevicollis			
Notiophilus bigutatus			
P. melanarius			
Gnathaphanus sp.	ELISA	*Pieris rapae* (Lepidopera: Pieridae)	Kapuge *et al.*, 1987
Mecyclothorax ambiguus			
Rhytisternus miser			
R. liopleurus			
Notonomus gravis			
Anomotarus crudelis			
Geoscaotus sp.			
A. dorsale	ELISA	Aphids (Homoptera: Aphididae)	Hance and Renier, 1987
Asaphidion flavipes			
H. rufipes			
L. pilicornis			
P. melanarius			
P. niger			
Carabus arvensis	Precipitin	Lumbricidae, Mollusca, Copeognatha, Chilopoda, Opiliones, Arachnida, Enchytraeidae, Elateridae, Tipulidae, Isopoda.	Gruntal and Sergeyeva, 1989
C. coriaceus			
C. glabratus			
C. hortensus			
Cychrus caraboides			
(positive for all antisera against all prey groups)			
Amara aenea	Immunoelectrophoresis	*Rhagoletis pomonella* (Diptera: Tephritidae)	Allen and Hagley, 1990
A. familiaris			
Bembidion rapidus			
B. quadrimaculatum			
Diplochaeila impressifrons			
H. affinis (= aeneus)			
P. lucublandus			
P. melanarius			
Agonum placidum	Immunoelectrophoresis	*Sitodiplosis mosellana* (Diptera: Cecidomyiidae)	Floate *et al.*, 1990
Amara carinata			
A. farcta			
Bembidion nitidum			
B. nudipenne			
B. obscurellum			
B. quadrimaculatum			
B. rupicola			

continued

Table 5.1. contd.

Carabid species that tested positive	Technique	Prey detected	Reference
B. sordidum			
B. timidum			
Pterostichus adstrictus			
P. corvus			
P. femoralis			
P. lucublandis			
Unidentified carabid larvae			
Carabus frigidum	ELISA	Lymantria dispar (Lepidoptera: Lymantriidae)	DuDevoir and Reeves, 1990
C. nemoralis			
Chlaenius tricolor			
Cymindis cribricollis			
Dicaelus dilatatus			
D. politus			
Harpalus herbivagus			
H. indigens			
H. rufipes			
H. viduus			
Myas cyanescens			
Notiophilus aeneus			
Patrobus longicornis			
Pinacoderus limbata			
P. platicollis			
Platynus decentis			
Pterostichus adoxus			
P. coracinus			
P. lucublandus			
P. melanarius			
P. mutus			
P. pensylvanicus			
P. stygicus			
Sphaeroderus canadensis			
S. lecontei			
Synchus impunctatus			
Calosoma frigidum	ELISA	*Lymantria dispar* larvae (Lepidoptera: Lymantriidae)	Cameron and Reeves, 1990
C. wilcoxi			

Chlaenius emarginatus			
Dicaelus politus			
D. teter			
Ducaelus dilatatus			
Harpalus spadiceus			
M. cyanescens			
Platynus decentis			
P. hypolithus			
Pterostichus adoxus			
P. diligenda			
P. honestus			
P. madidus			
P. lucublandus			
P. pensylvanicus			
P. mutus			
P. stygicus			
P. coracinus			
P. lacrymosa			
P. moesta			
P. rostrata			
P. sculptus			
Scaphinotus viduus			
Sphaeroderus canadensis			
Synchus impunctatus			
A parallelepipedus	ELISA	Mollusca	Symondson and Liddell, 1993d
P. madidus			
P. melanarius	ELISA	Mollusca	Symondson *et al.*, 1996
B. lampros (e)	ELISA	*Otiorhynchus sulcatus* (Coleoptera:	Crook and Solomon, 1997
Calathus fuscipes (a)	(+ monoclonal antibodies)	Curculionidae)	
Carabus violaceus (a)			
H. rufipes (l,a)			
N. biguttatus (e,l)			
P. madidus (l,a)			
(Consumed eggs (e), adults (a) or larvae (l))			
P. melanarius	ELISA (+ monoclonal antibodies)	Lumbricidae	Symondson *et al.*, 2000
P. melanarius	ELISA (+ monoclonal antibodies)	Mollusca	Bohan *et al.*, 2000

antibody concentration. Unfortunately, in a great many published studies, little, if any, information is given on tests for cross-reactivity; where extensive tests are not reported, the results must be considered to be no more than provisional. Tests are also essential to ensure that the antiserum can detect prey after consumption by the predator, by no means always the case. For example, Mollett and Armbrust (1978) raised an antiserum against the alfalfa weevil, *Hypera postica*, that detected proteins that were almost immediately degraded, and thus no longer detectable after consumption by carabids. All of the field studies in *Table 5.1* (except the last three) involved the use of polyclonal antibodies.

Monoclonal antibodies

Monoclonal antibodies are produced by 'hybridoma' cells which are created in the laboratory by fusing antibody-producing B lymphocytes with myeloma cells. The end product is a limitless supply of identical antibodies (a monoclone) that will bind with a single epitope, or antigenic determinant, on a target protein (or other biological material). As long as that epitope is not found on other proteins, for example on closely related prey species, then you have a highly specific 'magic bullet' for the diagnosis of predation.

Detailed descriptions of the making of monoclonal antibodies can be found in specialized texts (e.g. Goding, 1986; Liddell and Cryer, 1991), while their technical application to predation studies has been reviewed in Greenstone (1996) and Symondson and Hemingway (1997). The basic methods and principles behind this technology were first revealed by Köhler and Milstein (1975). They showed that B lymphocytes could be immortalized by fusing their genetic material with that from myeloma cells. This allowed the resulting hybrid cells to be grown indefinitely *in vitro*, thus permitting the selective culture of useful cell lines generating specific antibody.

The process starts in a similar manner to the making of polyclonal antisera. A mouse or rat is immunized, usually twice, followed by a third immunization a few days before the mouse is killed and the spleen removed. The antibody-producing cells are extracted from the spleen and fused, usually using polyethylene glycol, with myeloma cells. The resulting mixture of lymphocytes, hybridomas and myelomas are grown in a selective medium that kills the myelomas but allows the hybrid cells to survive (the lymphocytes die of their own accord *in vitro*). Sub-cultures of hybridomas are then screened, using enzyme-linked immunosorbent assays (see below) to determine whether the antibodies they are producing bind to the antigen (protein from the target prey). Where they are shown to do so, the cells are cloned, usually by limiting dilution (Liddell and Cryer, 1991) or alternatively, where the equipment is available, using a fluorescence-activated cell sorter (Symondson *et al.*, 1999a). The aim in both cases is to create cultures from single cells (monoclonal cell lines). These cultures are grown on until they can, in turn, be screened. Until recently, clones shown to be producing useful antibodies were generally propagated in mice (Liddell and Cryer, 1991). Changes in the law, at least in the European Union, now require hybridomas to be grown in, and antibodies harvested from, *in vitro* systems, either simply in culture flasks or in bioreactors.

Once sufficient antibody is available, characterization experiments can be conducted to determine the specificity of the clone and, equally important, its ability to detect prey remains within predators such as carabids. Wherever possible, clones are selected that bind with common epitopes on proteins that are well distributed within the body of the prey, ensuring strong ELISA reactions whichever part of the prey is consumed. Specificity testing involves screening not only closely related prey species but also a representative range of taxa from potential field sites where predation will be studied. For example, antibodies were developed to detect predation on a range of different species of slugs by the carabid *Pterostichus melanarius* (Symondson and Liddell, 1993a, 1996b; Symondson *et al.*, 1997, 1999b). In each case, the antibodies were tested first against other species of slug. The different clones were shown to be specific at a range of taxonomic levels, being genus (Symondson and Liddell, 1993b), species aggregate (Symondson *et al.*, 1999b) or species specific (Symondson and Liddell, 1996b, Symondson *et al.*, 1997). Other antibodies proved to be mollusc-specific (Symondson *et al.*, 1995), and were thus useful for detecting predation on slugs and snails as a group (Bohan *et al.,* 2000). In all these studies, however, the antibodies were further tested against a wide range of other invertebrates, including a range of insect orders plus earthworms, spiders, woodlice, millipedes, centipedes, flatworms and anything else that might turn up in the gut of a carabid during a field study. Although cross-reactions are less common than with polyclonal antisera, when they do occur there is nothing that can be done about it and a new antibody must be selected.

Monoclonal antibodies will only be useful if they can detect their target prey remains, extracted from the guts of predators, for a significant period following consumption. Frequently, antibodies are created that are highly specific but are found to bind to highly labile sites that are immediately denatured by digestive enzymes. Fortunately, prey remains in the crops of carabids are usually detectable by monoclonal antibodies for longer periods than in many other predators, possibly because this part of the gut is primarily for storage rather than digestion. Detection periods and antigen decay rates are measured by feeding carabids on the target prey, then killing and freezing batches of around ten individuals at appropriate time intervals over the next few days (e.g. Symondson and Liddell, 1995). All of the beetles are then tested by ELISA and quantities, or concentrations, of remaining undigested antigen regressed against time. Detection periods using *P. melanarius* can be as long as 65 h (Symondson *et al.*, 2000), but can be much shorter (Symondson and Liddell, 1993b), depending upon the antibody and target. However, detection periods, whether using polyclonal or monoclonal antibodies, will also vary considerably between predator species (Sunderland and Sutton, 1980; Sopp and Sunderland, 1989; Hagler and Naranjo, 1997) and, therefore, each antibody–predator–prey system must be separately characterized and corrected for temperature. A comparison of detection periods for two carabids in closely related genera, *Pterostichus madidus* and *Abax parallelepipedus*, showed that slug prey could be detected in the former for 2.5 times as long as in the latter (Symondson and Liddell, 1993c). Whether the carabid is starved after feeding on the prey, or fed on alternative prey (that does not react with the antibody), will also affect detection periods. Slug remains in *P. melanarius* could be detected, using a monoclonal antibody, for significantly longer when the beetles were subsequently fed on earthworm (Symondson and Liddell, 1995). Finally, detection periods can be affected by sex; the remains of the slug *Arion distinctus* could be detected for 30%

longer in male *P. melanarius* than in females (Symondson *et al.*, 1999b). All of these variables (equally applicable to polyclonal antisera) can affect the proportion of individuals that will test positive in the field, therefore making such characterization experiments essential.

As the hybridomas can be frozen in liquid nitrogen and revived to produce more antibody whenever required, monoclonal antibodies can be generated as a uniform, highly specific reagent in limitless supply. This is a major advantage over polyclonal antisera; once a supply of the former has run out, a new antiserum must be made. As no two polyclonal antisera have the same properties, it is difficult to obtain comparable results over time or between laboratories. However, the main advantage of monoclonal antibodies is their specificity. As reported above, this may be at a range of taxonomic levels but, in addition, it can result in specificity against a particular stage, such as the eggs of slugs (Symondson *et al.,* 1995) or whiteflies (Hagler *et al.*, 1993), or even a single larval instar (Greenstone and Morgan, 1989). Crook and Solomon (1996) and Crook *et al.* (1996) raised a panel of monoclonal antibodies against the vine weevil, *Otiorhynchus sulcatus*, including antibodies specific for the eggs and adults, specifically to study predation by carabids. Such specificity has never been achieved using polyclonal antisera, and would also be difficult (but not impossible) using molecular techniques (see section on molecular detection systems). The main disadvantage of monoclonal antibodies is their cost, both in terms of labour and dedicated tissue-culture facilities. It can easily take a year to make and characterize a new monoclonal antibody with the required properties. This generally means that the monoclonal antibody approach is effective where the aim is to survey a range of predators to quantify predation on a target pest, but cannot be contemplated for analysing the prey range of a particular species of carabid. It may be possible to increase the rate at which new antibodies can be generated through the production of recombinant antibodies or antibody engineering (Liddell and Symondson, 1996).

ASSAY SYSTEMS

Antibodies have been used in a great many ingenious assay formats to track predator–prey interactions, and these have been extensively reviewed elsewhere. In carabid research, however, just two systems have been widely used, precipitin tests in various guises and ELISA. As the former seems to have been entirely displaced by the latter, precipitin tests, as techniques, are largely of historical interest. However, field results obtained by precipitin tests are clearly valid and of interest to carabid ecologists (*Table 5.1*). Although precipitin tests may be 1000+ times less sensitive than ELISA, this simply means that they probably underestimated the importance of the trophic links they have revealed.

As with all post-mortem systems of gut analysis, immunoassays cannot distinguish between predation and scavenging. In most cases, our main interest is in predation, either in purely ecological studies or, more particularly, in applied experiments to quantify the ability of carabids to control crop pests. Immunoassays, however, simply tell us whether biological material from the target prey has been consumed, and additional laboratory trials are necessary to demonstrate whether the predators are capable of capturing and killing the prey. This may not always be a problem. Symondson *et al.* (2000) studied consumption of earthworms in the field by

P. melanarius in order to determine whether the prey provided the beetles with a valuable non-target food resource that could help to sustain the predators when pests (such as slugs and aphids) were at low density. In this instance, it did not matter whether the earthworms were killed or scavenged. Scavenging is widespread amongst the Carabidae and, just as a lion would preferentially consume a dead zebra rather than expend energy and expose itself to the risk of killing a live one, so most carabids will exploit dead prey when available. Controlled experiments under semi-field conditions can show whether the predator is potentially able to reduce prey populations. Secondary predation could also, in theory, lead to false trophic links where, for example, a carabid had consumed another predator with the target pest in its gut. Recent studies using both spiders and carabids have shown that this type of error is of minimal significance, at least using monoclonal antibodies (Harwood *et al.*, 2001). This is mainly because antigens usually decay exponentially in the guts of predators during digestion. Thus, in secondary predation the prey has been subjected to two periods of maximum digestion rate in succession and little, if any, antigen will be detectable after a short period (< 1 h) in the gut of the second predator. An excellent review of the errors and difficulties involved with attempts to equate immunoassay results with predation can be found in Sunderland (1996). A range of statistical models have been developed in attempts to calculate predation in the field from immunoassay results, and are reviewed in Sopp *et al.* (1992) and Mills (1997).

Precipitin tests

Precipitin tests rely on the same principle as that described above for removing cross-reacting antibodies from polyclonal antisera. Because IgG antibodies are 'Y' shaped and are thus bivalent, having binding sites (paratopes) at the end of each arm, they can form matrices or visible precipitates when combined with their target antigens. The test is performed by homogenizing the gut contents (or whole body) of the predator, centrifuging to remove solid material then combining the supernatant with the appropriate antiserum. If antigen is present, a precipitate will form that is proportional to the amount of antigen in the sample. The test is intrinsically insensitive because enough precipitate must form to be visible.

A range of techniques has been used to bring antibodies and antigens together (Boreham and Ohiagu, 1978). The simplest is the capillary ring test in which antiserum is overlaid by a diluted predator sample in a capillary tube and a precipitate, at the boundary between the two, indicated a positive reaction. This approach was used to measure predation by carabids (amongst other predators) on cabbage root fly, *Erioischia brassicae* (Coaker and Williams, 1963), winter moth pupae, *Operophtera brumata* (Frank, 1967), the larvae of *Pieris rapae* (Dempster, 1967; Ashby, 1974), the broom weevil, *Sitona regensteinensis* (Danthanarayana, 1969), red-backed cutworms, *Euxoa ochrogaster* (Frank, 1971), cinnabar moth larvae, *Tyria jacobaeae* (Dempster, 1971) and woodlice (Sunderland and Sutton, 1980), while Dennison and Hodkinson (1983) used this technique to quantify predation by carabids on a wide range of invertebrates in a woodland soil ecosystem. A variant on this is the Modified Oakley–Fulthorpe test in which a band of agar is used to separate the antiserum from the predator sample in the tube, and antibodies and antigens allowed to diffuse through this over several days. Precipitate again forms where the two meet in a positive reaction. A widely-used technique in carabid research was the Ouchterlony, or double diffusion, test. A glass slide or petri dish is coated with a layer of agarose or agar, and antiserum and predator samples allowed to diffuse towards each other from wells cut into the gel. Precipitate again forms at the boundaries, and the visibility of this can be increased using protein stains. The technique was used to investigate predation by carabids on slugs (Tod, 1973), black cutworm larvae, *Agrotis ipsilon* (Lund and Turpin, 1977), corn earworms, *Heliothis zea* (Lesiewicz *et al.*, 1982) and bush flies, *Musca vetustissima* (Calver *et al.*, 1986). Single Radial Diffusion was used to try to quantify antigens in samples. This test was similar to the Ouchterlony test, but only predator samples were added to wells and the antiserum was incorporated in the gel. The width of the band of precipitate that formed was equivalent to the quantity of antigen in the sample (McIver, 1981). A faster, more sensitive variant on precipitin tests was cross-over electrophoresis or immunoelectroosmophoresis. Most conveniently, it was found that antibodies migrate, in an electric field, towards the cathode, the opposite direction to most other proteins, which migrate towards the anode. This allowed antibodies and antigens to be rapidly brought together within a gel, and the precipitates visualized using protein stains. The technique was used by Allen and Hagley (1982, 1990) to measure predation on *Heliothis zea*, and by Calver *et al.* (1986) and Floate *et al.* (1990) to identify carabid predators of bush flies, *M. vetustissima,* and the wheat midge, *Sitodiplosis mosellana*, respectively (*Table 5.1*). A number of variants of this approach that have been used in predation studies are described by Sunderland (1988). Healy and Cross (1975), amongst other authors, found immunoelectroosmophoresis to be superior to other precipitin techniques in comparative trials in terms of sensitivity, speed, clarity of results and economy of materials.

Enzyme-linked immunosorbent assays (ELISA)

Since it was first used by Ragsdale *et al.* (1981) to study predation in the field, ELISA has rapidly supplanted precipitin tests as the method of choice for gut content analyses. The test is relatively simple and rapid to perform but, most importantly, is considerably more sensitive than any precipitin technique and can detect less than 1 ng of target prey protein when conditions have been properly optimized (Symondson and Liddell, 1993a).

A number of different ELISA formats are possible but those most commonly used in predation studies involving carabids have been the indirect and double-antibody sandwich ELISAs. The assays have been described and compared by Voller *et al.* (1979), Powell *et al.* (1996) and Greenstone (1996), while detailed protocols are given in Symondson and Hemingway (1997). The assays are performed in 96-well microtitration plates made of polystyrene. In the indirect assay, predator gut samples, or whole predators, are first homogenized and centrifuged. The supernatants from these samples may then be further diluted and pipetted into wells on the plate. Prey proteins then bind to the polystyrene. Unbound material is then washed away and specific antiserum, or an appropriate monoclonal antibody, is added. Where target antigens are present, the antibodies will bind to them and become fixed, via the prey proteins, to the polystyrene. Again, unbound material is washed away and a second antibody is added. This second antibody is usually a commercial preparation in which the antibody has been conjugated with an enzyme. The antibody is designed to bind to other antibodies, specifically to antibodies from the species of mammal used to prepare your polyclonal antiserum or monoclonal antibodies (i.e. usually anti-rabbit or anti-mouse). Unbound second antibody is flushed away and an enzyme substrate added. This substrate will turn from a colourless to a coloured product in the presence of the enzyme, the degree of colour change in unit time being proportional to the quantity of prey antigen in the sample. The reaction is stopped using sulphuric acid and the optical density of the wells measured using an ELISA plate spectrophotometer. If a dilution series of prey antigens is included on each ELISA plate, quantities of prey antigen in test samples can be calculated using regression analysis (e.g. Lövei *et al.,* 1985; Chiverton, 1987; Symondson *et al.*, 1996). The enzymes used in ELISA are usually horseradish peroxidase or alkaline phosphatase. In the double-antibody sandwich ELISA, specific antibodies (polyclonal or monoclonal), rather than gut sample proteins, are first bound to the plate. As before, unbound material is washed away between each step. Diluted gut samples are then added and prey antigens, where present, are captured by the antibody bound to the plate. A second antibody is then added. This may be the same as the first, but to which an enzyme has been conjugated. This antibody binds to the antigen and the assay completed as for indirect assays. Other formats for double-antibody sandwich ELISA are possible, but those described are the ones mainly used in carabid research. The double-antibody sandwich is considered to be possibly more specific in that each antigen molecule must have two sites to which the antibodies can bind for the assay to work. However, as long as rigorous tests are performed to detect cross-reactivity, the indirect assay has been shown over a number of studies to be more sensitive and is now used increasingly (Greenstone, 1996).

The double-antibody sandwich has, for example, been used to study predation by carabids on aphids (Crook and Sunderland, 1984; Chiverton, 1987; Lövei *et al.*, 1987, 1990; Sunderland *et al.*, 1987; Sopp and Sunderland, 1989; Sopp *et al.,* 1992), gypsy moth larvae, *Lymantria dispar* (Cameron and Reeves, 1990), and the corn borer, *Ostrinia nubilalis* (Lövei *et al.*, 1985). However, the indirect ELISA has also been used extensively in carabid research to study predation on aphids (Hance and Rossignol, 1983; Hance and Renier, 1987; Symondson *et al.*, 1999a), *Pieris rapae* larvae (Kapuge *et al.*, 1987), New Zealand flatworms, *Artioposthia triangulata* (Symondson and Liddell, 1996a), earthworms (Symondson *et al.*, 2000), and slugs (e.g. Symondson and Liddell, 1993d; Symondson *et al.*, 1996; Bohan *et al.*, 2000).

Although dilution series are frequently included on ELISA plates, this is usually simply used to assist with the setting of positive/negative thresholds (discussed in Greenstone, 1996; Symondson and Hemingway, 1997). However, Sopp and Sunderland (1989) used a dilution series to convert absorbance values for test samples into biomass of aphid consumed, while Sopp *et al.* (1992) used such quantitative data to generate an improved model for estimating prey consumption by carabids and other predators. In two cases only has dilution series data been used to quantify antibody-recognizable prey proteins in the guts of individual predators collected from the field, and hence attempt to monitor the changing proportions and quantities of a target prey consumed over time and between treatments. Both of these studies involved the carabid, *P. melanarius*, in one case feeding on slugs (Symondson *et al.*,

1996) and in the other on earthworms (Symondson *et al.*, 2000). A full description of the techniques, calculations and rationale of this approach can be found in Symondson *et al.* (2000). Foregut samples from beetles that have eaten the target prey will contain a mixture of antibody-recognizable and unrecognizable prey proteins, the latter increasing as digestion proceeds, while partly digested material will, at the same time, flow out of the foregut into the rest of the digestive tract. Only undigested, antibody-recognizable prey remains in the foregut from one or more previous meals can be quantified. The same quantity of antigen may, therefore, be found in the foregut of a beetle that has consumed a small quantity of target prey recently and another that consumed a large quantity many hours ago. However, if the mean concentration of target prey remains in the foreguts of a number of beetles of the same species from two field treatments are compared statistically, and a significant difference is found despite individual variability, then it can be inferred that this prey forms a greater proportion of the diet of one of the groups. Similar comparisons may be possible over time if adequate consideration is given to the effects of changes in temperature.

A variant on ELISA is the immunodot assay in which direct or indirect assays are performed on nylon or nitrocellulose membranes rather than microtitration plates (Stuart and Greenstone, 1990; Greenstone and Trowell, 1994; Hagler *et al.*, 1995). The enzyme substrates used in these assays deposit a coloured particulate product on the membrane, which is generally assessed by eye as simply positive or negative, although can be quantified (Hagler *et al.*, 1995). The assay does not appear to have been used yet in carabid research but has considerable potential, not least because it avoids the considerable expense of an ELISA plate spectrophotometer.

An excellent review of the use of serological gut analysis can be found in Greenstone (1996). Although such assays can tell you whether a carabid has recently consumed the target prey, and even how much undenatured prey protein is present in the carabid gut, translation of such information into estimates of predation in the field is no simple task (Sunderland, 1996). A number of models have been constructed in an effort to do this, and have been recently reviewed in Mills (1997). However, detection depends upon a whole suite of factors, including differential digestion rates in different predator species (Symondson and Liddell, 1993c), temperature (Sopp and Sunderland, 1989; Hagler and Naranjo, 1997), the quantity of target and presence of non-target remains in the gut (Symondson and Liddell, 1995), the sex of the predator (Symondson *et al.*, 1999b), changing metabolic rates in response to starvation, the physiological state of the predator, and many other factors. Extensive laboratory experiments are ideally required to try to quantify these variables where they may be relevant to a particular study. There are many field studies which, unfortunately, make simple assumptions about which predators are ‘better’ than others as potential natural control agents, based solely upon crude percentages of the predator population that test positive. Even relatively closely-related carabids can have very different digestion rates. Symondson and Liddell (1993c) found that remains of the slug, *Deroceras reticulatum*, could be detected for 35 h in the carabid *A. parallelepipedus* but for 96 h in *P. madidus*. This appeared to contradict earlier assumptions that larger species, within a taxonomic group, have longer detection periods than smaller species (Sunderland and Sutton, 1980; Sopp and Sunderland, 1989), underlining the need for thorough testing in every case where species are compared.

Molecular detection systems

Rather than trying to detect and identify prey proteins from the guts of predators, the possibility has been discussed for some years of, instead, trying to detect undigested prey DNA (e.g. Sunderland, 1988). The argument against this has always been that DNA was thought likely to break down more rapidly, during digestion, than some of the more refractory proteins targeted by antibodies. It has only been very recently that these preconceptions have been challenged through practical experimentation, some of which has involved carabids as model predators (Zaidi *et al.*, 1999). Although in its infancy, this technology has already shown itself to have enormous potential for the future of predation studies, and there is little doubt that it will shortly become the method of choice. A more extensive review of molecular techniques for studying predation in vertebrates and invertebrates can be found in Symondson (2002).

The main impetus behind research into DNA-based detection systems has been to find a technique that is at least as specific and sensitive as monoclonal antibodies and ELISA, but which can be used to study the prey range of a predator, rather than predation by a range of predators on one, or a few, target prey. It can often be important to determine which non-pest prey, for example, can help to sustain key natural enemies within a cropping system, or which prey may be chosen by a predator in preference to target pests. Although monoclonal antibodies could, in theory, answer such questions, the costs would be prohibitive. Creating a new monoclonal antibody with the required specificity can take up to a year, and thus attempting to generate antibodies against the 50+ different prey species that might be encountered by a carabid in the field is clearly impractical. Monoclonal antibody facilities are also expensive to set up and run, whereas most universities and research establishments have molecular biology facilities. Last, but by no means least, DNA techniques avoid the use of vertebrates in the experimental process.

Until recently, DNA techniques had only been used to detect trophic interactions involving parasitism (Greenstone and Edwards, 1998; Zhu and Greenstone, 1999) or bloodmeals consumed by haematophagous insects (Coulson *et al.*, 1990; Tobolewski *et al.*, 1992; Gokool *et al.*, 1993; Lord *et al.*, 1998; Boakye *et al.*, 1999). The latter studies, however, indicated that DNA could survive in the insect gut for a reasonable period and be identified using standard molecular biology techniques.

Initial attempts to detect DNA extracted from the contents of insect predator guts largely failed. A possible reason for this was that longer DNA sequences were rapidly degraded during digestion (e.g. ~ 900 bp, Johanowicz and Hoy, 1996). It was also possible that, following even a short period in a predator gut, single-copy genomic DNA was difficult to detect, even with PCR (polymerase chain reaction) amplification. Agusti *et al.* (1999) developed PCR primers for the detection of *Helicoverpa armigera* (Lepidoptera: Noctuidae) DNA within predators (Miridae and Anthocoridae). The targets were sequences isolated from a randomly amplified polymorphic DNA (RAPD) band for which primers were made that amplified 1100, 600 and 254 bp sequences. The longest sequence, 1100 bp, could not be detected within gut samples, even immediately after consumption by the predator. The 600 pb sequence could be amplified from 50% of predators immediately after consumption, but not after 4 h in the predator gut. Detection of the 254 bp sequence was best, and was amplified from 75% of predators immediately after feeding, dropping to 45% of predators after 4 h digestion. In all cases, the predators had eaten 10 *H. armigera* eggs. At such a high level of prey ingestion, all predators would have been expected to have tested positive using antibodies. Very similar results were reported by Agusti *et al.* (2000), where the targets were RAPD-derived sequences from the whitefly, *Trialeurodes vaporariorum*, and the predators mirid bugs. A 2100 bp sequence was undetectable following prey consumption by the predator, but a 310 bp sequence was amplified, with 80% successful detection immediately after feeding, dropping to 60% after four hours. Again, 10 prey (whitefly nymphs) were consumed by each predator.

Zaidi *et al.* (1999) studied both the effect of sequence length on detection of an experimental prey within carabids, *P. cupreus*, and, most critically, the effect of targeting multiple-copy DNA sequences. The target sequences were esterase genes with 40–50-fold replication within two strains of insecticide resistant mosquitoes, *Culex quinquefasciatus*. A range of primers was used to amplify sequences of different sizes. It was soon found that, using primers that amplified sequences > 800 bp, < 50% of predators that had eaten mosquitoes tested positive after 2 h. However, when primers were used that amplified a short sequence of 148 bp, the target sequence was detected in 100% of predators feeding on one of the mosquito strains after 28 h digestion in the guts of the carabid predators. Whether the beetles had eaten one mosquito or six, digested for 0 h or 28 h, the prey were equally detectable. Clearly, DNA survives well in the foreguts of at least this species of carabid and, as long as short sequences of multiple-copy genes are targeted, molecular detection is a viable system for future predation studies.

Zaidi *et al.* (1999) speculated that an obvious future target for molecular detection systems would be mitochondrial genes which, with many hundreds of copies per cell, would provide a range of multiple-copy gene sequences, many of which are already known to be taxonomically useful (Simon *et al.*, 1994). This has recently been confirmed by Chen *et al.* (2000), who targeted mitochondrial cytochrome oxidase II gene sequences ranging from 77 to 386 bp. Primer pairs were designed to amplify specifically DNA from six species of cereal aphid. Predators were fed on a single target aphid, *Rhopalosiphum maidis*, then on five aphids of the related *Rhopalosiphum padi*. A 198 bp fragment, specific for *R. maidis*, could be amplified from 50% of predators after approximately 4 h (from the coccinelid, *Hippodamia convergens*) and 9 h (from the chrysopid, *Chrysoperla plorabunda*).

A major advantage of the molecular approach is that, once useful target sequences have been published, these can be immediately used by other workers. It has been shown, for example, that the primers developed by Chen *et al.* (2000) are equally useful for the detection of aphids within the carabid, *P. melanarius* (C.S. Dodd, W.O.C. Symondson and M.W. Bruford, unpublished data).

Current recommendations and the future

Labelling systems can be useful for very short-term studies but may be affected by the disturbance to the system that is intrinsic to this approach. Of the immunological approaches, only ELISA on microtitration plates or, as immunodot assays on membranes, can now be recommended. The assays are rapid, easy to perform, highly sensitive and, when used in combination with monoclonal antibodies, highly specific. Fully characterized monoclonal antibodies are, in addition, far less likely than polyclonal antisera to give false positives caused by cross-reactivity. Although monoclonal antibodies are initially expensive to create, once the clones that produce them have been isolated, limitless supplies of antibody can be generated inexpensively, with consistent properties. This is important because it allows results to be compared over time and between laboratories using the same clone. Once samples have been prepared, it is possible to test several hundred predator gut extracts over a 36 h period.

Once a new monoclonal antibody has been made, it is clearly important, given the time, skill and expense that goes into creating it, that it should be made available to others, and most workers are prepared to collaborate in various ways to see their clones used more widely. Most of the antibodies mentioned in this chapter so far have been created specifically for use as probes to test carabid gut samples, but *Table 5.2* gives a more complete list, including antibodies designed for use with other predators. There is no reason to believe that antibodies that can detect prey remains in coccinelids or arachnids, for example, would be any less useful for testing carabid foregut. The combination of ELISA and monoclonal antibodies is by far the best method currently employed for analysing gut samples from field-collected predators, where resources permit.

Table 5.2. Monoclonal antibodies that have been used to analyse the gut contents of polyphagous predators, or could be used for such. Where possible, a paper is quoted that gives characterization information.

Target	Reference
Lepidoptera: Noctuidae	
Helicoverpa spp. larvae	Lenz and Greenstone, 1988
Helicoverpa spp. eggs	Goodman *et al.*, 1997
Helicoverpa plus *Heliothis* spp. larvae	Greenstone *et al.*, 1991
Helicoverpa plus *Heliothis* spp. eggs	Greenstone and Trowell, 1994
Lepidoptera: Gelechiidae	
Pectinophora gossypiella eggs	Hagler *et al.*, 1994
Hemiptera: Miridae	
Lygus hesperus	Hagler *et al.*, 1992
Mollusca: Pulmonata	
General, slug adults	Symondson *et al.*, 1995
Arionidae	Symondson and Liddell, 1993b
Deroceras reticulatum adults	Symondson and Liddell, 1996b
Arion hortensis adults	Symondson *et al.*, 1999b
Tandonia budapestensis adults	Symondson *et al.*, 1997
General, slug eggs	Mendis et al., 1996
Derceras spp. eggs	Mendis *et al.*, 1996
Arion ater eggs	Mendis *et al.*, 1996
Annelida: Lumbricidae	
General, lumbricid earthworms	Symondson *et al.*, 2000
Homoptera: Aphididae	
General, aphids	Symondson *et al.*, 1999a
Homoptera: Aleyrodidae	
Trialeurodes vaporariorum	Symondson *et al.*, 1999c
Bemisia tabaci	Symondson *et al.*, 1999c
B. tabaci eggs	Hagler *et al.*, 1993
Homoptera: Delphacidae	
Sogatella furcifera eggs and adults	Pang *et al.*, 2001
Coleoptera: Curculionidae	
Otiorhynchus sulcatus	Crook and Solomon, 1996
Coleoptera: Chrysomelidae	
Cassida rubiginosa	Bacher *et al.*, 1999

In the near future, however, DNA techniques may well take over that mantle. As yet, there have been too few studies to know whether molecular techniques are as reliable as monoclonal antibody technology, nor have there been any studies involving analysis of field-collected predators. Now that we know that DNA techniques offer a practical approach with many advantages, technical improvements are likely to follow rapidly. We do not yet, for example, have a rapid assay system equivalent to ELISA which will allow large numbers of field samples to be rapidly analysed, although possible solutions to this problem are under development. Nor has anyone, to date, tried using quantitative PCR techniques that will allow us to measure the quantities of detectable prey in carabid gut samples, although possible approaches, such as TaqMan-PCR (e.g. Chiang *et al.,* 1999; Haugland *et al.*, 1999), are available from other branches of molecular biology. All of the technologies described in this chapter have their advantages and disadvantages but, in general, new systems displaced the old because they were more accurate, or could be used to address new questions. DNA approaches score highly on both of these criteria.

References

AGUSTÍ, N., DE VICENTE, M.C. AND GABARRA, R. (1999). Development of sequence amplified characterized region (SCAR) markers of *Helicoverpa armigera*: a new polymerase chain reaction-based technique for predator gut analysis. *Molecular Ecology* **8**, 1467–1474.

AGUSTÍ, N., DE VICENTE, M.C. AND GABARRA, R. (2000). Developing SCAR markers to study predation on *Trialeuropes vaporariorum. Insect Molecular Biology* **9**, 263–268.

ALLEN, W.R. AND HAGLEY, E.A.C. (1982). Evaluation of immunoelectroosmophoresis on cellulose polyacetate for assessing predation of Lepidoptera (Tortricidae) by Coleoptera (Carabidae) species. *Canadian Entomologist* **114**, 1047–1054.

ALLEN, W.R. AND HAGLEY, A.C. (1990). Epigeal arthropod as predators of mature larvae and pupae of apple maggot (Diptera: Tephritidae). *Environmental Entomology* **19**, 309–312.

ASHBY, J.W. (1974). A study of arthropod predation of *Pieris rapae* L. using serological and exclusion techniques. *Journal of Applied Ecology* **11**, 419–425.

BACHER, S., SCHENK, D. AND IMBODEN. H. (1999). A monoclonal antibody to the shield beetle *Cassida rubiginosa* (Coleoptera, Chrysomelidae): a tool for predator gut analysis. *Biological Control* **16**, 299–309.

BACKELJAU, T., BREUGELMANS, K., LEIRS, H., RODRIGUES, T. AND SHERBAKOV, D. (1994). Application of isoelectric focusing in molluscan systematics. *Nautilus (Supplement)* **2**, 156–167.

BOAKYE, D.A., TANG, J., TRUC, P., MERRIWEATHER, A. AND UNNASCH, T.R. (1999). Identification of bloodmeals in haematophagous Diptera by cytochrome B heteroduplex analysis. *Medical and Veterinary Entomology* **13**, 282–287.

BOHAN, D.A., BOHAN, A.C., GLEN, D.M., SYMONDSON, W.O.C., WILTSHIRE, C.W. AND HUGHES, L. (2000). Spatial dynamics of predation by carabid beetles on slugs. *Journal of Animal Ecology* **69**, 367–379.

BOREHAM, P.F.L. (1979). Recent developments in serological methods for predator–prey studies. *Entomological Society of America Miscellaneous Publication* **11**, 17–23.

BOREHAM, P.F.L. AND OHIAGU, C.E. (1978). The use of serology in evaluating invertebrate prey–predator relationships: a review. *Bulletin of Entomological Research* **68**, 171–194.

CALVER, M.C., MATHIESSEN, J.N., HALL, G.P., BRADLEY, J.S. AND LILLYWHITE, J.H. (1986). Immunological determination of predators of the bush fly, *Musca vetustissima* (Diptera: Muscidae) in south-western Australia. *Bulletin of Entomological Research* **76**, 133–139.

CAMERON, E.A. AND REEVES, R.M. (1990). Carabidae (Coleoptera) associated with Gypsy moth, *Lymantria dispar* (L.) (Lepidoptera: Lymantriidae), populations, subjected to *Bacillus thuringiensis* Berliner treatments in Pennsylvania. *Canadian Entomologist* **122**, 123–129.

CATTY, D. AND RAYKUNDALIS, C. (1988). Production and quality control of polyclonal antibodies. In: *Antibodies: a practical approach*, Vol. 1. Ed. D. Catty, pp 19–79. Oxford: IRL Press.

CHEN, Y., GILES, K.L., PAYTON, M.E. AND GREENSTONE, M.H. (2000). Identifying key cereal aphid predators by molecular gut analysis. *Molecular Ecology* **9**, 1887–1898.

CHIANG, P.W., WEI, W.L., GIBSON, K., BODMER, R. AND KURNIT, D.M. (1999). A fluorescent quantitative PCR approach to map gene deletions in the *Drosophila* genome. *Genetics* **153**, 1313–1316.

CHIVERTON, P.A. (1987). Predation of *Rhopalosiphum padi* (Homoptera: Aphididae) by polyphagous predatory arthropods during the aphids' pre-peak period in spring barley. *Annals of Applied Biology* **111**, 257–269.

COAKER, T.H. AND WILLIAMS, D.A. (1963). The importance of some Carabidae and Staphylinidae as predators of the cabbage root fly, *Erioischia brassicae* (Bouché). *Entomologia Experimentalis et Applicata* **6**, 156–164.

COULSON, R.M.R., CURTIS, C.F., READY, P.D., HILL, N. AND SMITH, D.F. (1990). Amplification and analysis of human DNA present in mosquito bloodmeals. *Medical and Veterinary Entomology* **4**, 357–366.

CROOK, A.M.E. AND SOLOMON, M.G. (1996). Detection of predation on vine weevil by natural enemies using immunological techniques. *Mitteilung aus der Biologischen Bundesanstalt für Land und Forstwirtschaft* **316**, 86–90.

CROOK, A.M.E. AND SOLOMON, M.G. (1997). Predators of vine weevil in soft fruit plantations. *Proceedings of the ADAS/HRI/EMRA Soft Fruit Conference, New Developments in the Soft Fruit Industry, Ashford, UK*, pp 83–87.

CROOK, A.M.E., KEANE, G. AND SOLOMON, M.G. (1996). Production and selection of monoclonal antibodies for use in detecting predation on vine weevil, *Otiorhynchus sulcatus* (Coleoptera: Curculionidae). *BCPC Symposium Proceedings 65, Diagnostics in Crop Production*, pp 287–292. Farnham, UK: British Crop Protection Council.

CROOK, N.E. AND SUNDERLAND, K.D. (1984). Detection of aphid remains in predatory insects and spiders by ELISA. *Annals of Applied Biology* **105**, 413–422.

DANTHANARAYANA, W. (1969). Population dynamics of the broom weevil *Sitona regensteinensis* (Hbst.) on broom. *Journal of Animal Ecology* **38**, 1–18.

DEMPSTER, J.P. (1960). A quantitative study of the predators on the eggs and larvae of the broom beetle *Phytodecta olivacea* Forster using the precipitin test. *Journal of Animal Ecology* **29**, 149–167.

DEMPSTER, J.P. (1967). The control of *Pieris rapae* with DDT. 1. The natural mortality of the young stages of *Pieris*. *Journal of Applied Ecology* **4**, 485–500.

DEMPSTER, J.P. (1971). The population ecology of the cinnebar moth, *Tyria jacobaeae* L. (Lepidoptera, Arctiidae). *Oecologia* **7**, 26–67.

DENNISON, D.F. AND HODKINSON, I.D. (1983). Structure of the predatory beetle community in a woodland soil ecosystem. 1. Prey selection. *Pedobiologia* **25**, 109–115.

DUDEVOIR, D.S. AND REEVES, M.R. (1990). Feeding activity of carabid beetles and spiders on gypsy moth larvae (Lepidoptera: Lymantriidae) at high density prey populations. *Journal of Entomological Science* **25**, 341–356.

ERNSTING, G. AND JOOSSE, E.N.G. (1974). Predation on two species of surface dwelling Collembola. A study with radio-isotope labelled prey. *Pedobiologia* **14**, 222–231.

FICHTER, B.L. AND STEPHEN, W.P. (1979). Selection and use of host-specific antigens. *Miscellaneous Publications of the Entomological Society of America* **11**, 25–33.

FLOATE, K.D., DOANNE, J.F. AND GILLOTT, C. (1990). Carabid predators of the wheat midge (Diptera: Cecidomyiidae) in Saskatchewan. *Environmental Entomology* **19**, 1503–1511.

FOX, C.J.S. AND MACLELLAN, C.R. (1956). Some Carabidae and Staphylinidae shown to feed on wireworm, *Agriotes sputator* (L.), by the precipitin test. *Canadian Entomologist* **88**, 228–231.

FRANK, J.H. (1967). A serological method used in the investigation of the predators of the pupal stage of the winter moth, *Operophtera brumata* (L.) (Hydromeniidae). *Quaestiones Entomologicae* **3**, 95–105.

FRANK, J.H. (1971). Carabidae (Coleoptera) as predators of the red-backed cutworm (Lepidoptera: Noctuidae) in central Alberta. *Canadian Entomologist* **113**, 1039–1044.

GIBSON, P.H., COSENS, D. AND BUCHANAN, K. (1997). A chance field observation and pilot laboratory studies of predation of the New Zealand flatworm by the larvae and adults of carabid and staphylinid beetles. *Annals of Applied Biology* **130**, 581–585.

GODING, J.W. (1986). *Monoclonal antibodies: principles and practice.* San Diego: Academic Press.

GOKOOL, S., CURTIS, C.F. AND SMITH, D.F. (1993). Analysis of mosquito bloodmeals by DNA profiling. *Medical and Veterinary Entomology* **7**, 208–216.

GOODMAN, C.L., GREENSTONE, M.H. AND STUART, M.K. (1997). Monoclonal antibodies to vitellins of bollworm and tobacco budworm (Lepidoptera: Noctuidae): biochemical and ecological implications. *Annals of the Entomological Society of America* **90**, 83–90.

GRANT, J.F. AND SHEPARD, M. (1985). Techniques for evaluating predators for control of insect pests. *Journal of Agricultural Entomology* **2**, 99–116.

GREENSTONE, M.H. (1996). Serological analysis of arthropod predation: past, present and future. In: *The ecology of agricultural pests: biochemical approaches.* Eds. W.O.C. Symondson and J.E. Liddell, pp 265–300. London: Chapman & Hall.

GREENSTONE, M.H. AND EDWARDS, M.J. (1998). DNA hybridisation probe for endoparasitism by *Microplites croceipes* (Hymenoptera: Braconidae). *Annals of the Entomological Society of America* **91**, 415–421.

GREENSTONE, M.H. AND MORGAN, C.E. (1989). Predation on *Heliothis zea*: an instar-specific ELISA for stomach analysis. *Annals of the Entomological Society of America* **84**, 457–464.

GREENSTONE, M.H. AND TROWELL, S.C. (1994). Arthropod predation: a simplified immunodot format for predator gut analysis. *Annals of the Entomological Society of America* **87**, 214–217.

GREENSTONE, M.H., STUART, M.K. AND HAUNERLAND, N.H. (1991). Using monoclonal antibodies for phylogenetic analysis: an example from the Heliothinae (Lepidoptera: Noctuidae). *Annals of the Entomological Society of America* **84**, 457–464.

GRUNTAL, S.Y. AND SERGEYEVA, T.K. (1989). Food relations characteristics of the beetles of the genera *Carabus* and *Cychrus. Zoologisch Zhurnal* **58**, 45–51.

HAGLER, J.R. AND DURAND, C.M. (1994). A new method for immunologically marking prey and its use in predation studies. *Entomophaga* **39**, 257–265.

HAGLER, J.R. AND NARANJO, S.E. (1997). Measuring the sensitivity of an indirect predator gut content ELISA: detectability of prey remains in relation to predator species, temperature, time and meal size. *Biological Control* **9**, 112–119.

HAGLER, J.H., COHEN, A.C., BRADLEY-DUNLOP, D. AND ENRIQUEZ, F.J. (1992). Field evaluation of predation on *Lygus hesperus* using a species- and stage-specific monoclonal antibody. *Environmental Entomology* **21**, 896–900.

HAGLER, J.R., BROWER, A.G., TU, Z., BYRNE, D.N., BRADLEY-DUNLOP, D. AND ENRIQUEZ, F.J. (1993). Development of a monoclonal antibody to detect predation of the sweet potato whitefly, *Bemisia tabaci. Entomologia Experimentalis et Applicata* **68**, 231–236.

HAGLER, J.R., NARANJO, S.E., BRADLEY-DUNLOP, D., ENRIQUEZ, F.J. AND HENNBERRY, T.J. (1994). A monoclonal antibody to pink bollworm (Lepidoptera: Gelechiidae) egg antigen: a tool for predator gut analysis. *Annals of the Entomological Society of America* **87**, 85–90.

HAGLER, J.R., BUCHMANN, S.L. AND HAGLER, D.A. (1995). A simple method to quantify dot blots for predator gut content analysis. *Journal of Entomological Science* **30**, 95–98.

HAMON, N., BARDNER, R., ALLEN-WILLIAMS, L. AND LEE, J.B. (1990). Carabid populations in field beans and their effect on the population dynamics of *Sitona lineatus* (L.). *Annals of Applied Biology* **117**, 51–62.

HANCE, T. AND RENIER, L. (1987). An ELISA technique for the study of the food of carabids. *Acta Phytopathologica et Entomologicae Hungaricae* **22**, 363–368.

HANCE, T. AND ROSSIGNOL, P. (1983). Essai de quantification de la prédation des Carabidae par le test ELISA. *Mededelingen van de Fakulteit Landbouwwetenschappen Rijksuniversiteit Gent* **48**, 475–485.

HARWOOD, J.D., PHILLIPS, S.W., SUNDERLAND, K.D. AND SYMONDSON, W.O.C. (2001). Secondary predation: quantification of food chain errors in an aphid–spider–carabid system using monoclonal antibodies. *Molecular Ecology* **10**, 2049–2057.

HAUGLAND, R.A., VESPER, S.J. AND WYMER, L.J. (1999). Quantitative measurement of *Stachybotrys chartarum* conidia using real time detection of PCR products with the TaqMan (TM) fluorogenic probe system. *Molecular and Cellular Probes* **13**, 329–340.

HAWKES, R.B. (1972). A fluorescent dye technique for marking insect eggs in predation studies. *Journal of Economic Entomology* **65**, 1477–1478.

HEALY, J.A. AND CROSS, T.F. (1975). Immunoelectroosmophoresis for serological identification of predators of the sheep tick *Ixodes ricinus*. *Oikos* **26**, 97–101.

HENGEVELD, R. (1980a). Polyphagy, oligophagy and food specialization in ground beetles (Coleoptera, Carabidae). *Netherlands Journal of Zoology* **30**, 564–584.

HENGEVELD, R. (1980b). Qualitative and quantitative aspects of the food of ground beetles (Coleoptera, Carabidae): a review. *Netherlands Journal of Zoology* **30**, 555–563.

HUDSON, L. AND HAY, F.C. (1989). *Practical immunology*. London: Blackwell.

JOHANOWICZ, D.L. AND HOY, M.A. (1996). *Wolbachia* in a predator–prey system: 16S ribosomal DNA analysis of two phytoseiids (Acari: Phytoseiidae) and their prey (Acari: Tetranychidae). *Annals of the Entomological Society of America* **89**, 435–441.

JOHNSON, P.C. AND REEVES, R.M. (1995). Incorporation of the biological marker rubidium in gypsy moth (Lepidoptera: Lymantriidae) and its transfer to the predator *Carabus nemoralis* (Coleoptera: Carabidae). *Environmental Entomology* **24**, 46–51.

KAPUGE, S.H., DANTHANARAYANA, W. AND HOOGENRAAD, N. (1987). Immunological investigation of prey–predator relationships for *Pieris brassicae* (L.) (Lepidoptera: Pieridae). *Bulletin of Entomological Research* **77**, 247–254.

KÖHLER, G. AND MILSTEIN, C. (1975). Continuous culture of fused cells secreting antibody of predefined specificity. *Nature* **256**, 495–497.

LAROCHELLE, A. (1990). The food of carabid beetles (Coleoptera: Carabidae, including Cicindelinae). *Fabreries*, Supplement **5**, 1–132.

LENZ, C.J. AND GREENSTONE, M.H. (1988). Production of a monoclonal antibody to the arylphorin of *Heliothis zea*. *Archives of Insect Biochemistry and Physiology* **9**, 167–177.

LESIEWICZ, D.S., LESIEWICZ, J.R., BRADLEY, J.R. AND VAN DUYN, J.W. (1982). Serological determination of carabid (Coleoptera: Carabidae) predation of corn earworm (Lepidoptera: Noctuidae) in field corn. *Environmental Entomology* **11**, 1183–1186.

LIDDELL, J.E. AND CRYER, A. (1991). *A practical guide to monoclonal antibodies*. Chichester: Wiley.

LIDDELL, J.E. AND SYMONDSON, W.O.C. (1996). The potential of combinatorial gene libraries in pest–predator relationship studies. In: *The ecology of agricultural pests: biochemical approaches*. Eds. W.O.C. Symondson and J.E. Liddell, pp 347–366. London: Chapman & Hall.

LORD, W.D., DIZINNO, J.A., WILSON, M.R., BUDOWLE, B., TAPLIN, D. AND MEINKING, T.L. (1998). Isolation, amplification and sequencing of human mitochondrial DNA obtained from human crab louse, *Pthirus pubis* (L.) blood meals. *Journal of Forensic Sciences* **43**, 1097–1100.

LÖVEI, G.L. (1986). The use of biochemical methods in the study of carabid feeding: the potential of isoenzyme analysis and ELISA. In: *Feeding behaviour and accessibility of food for carabid beetles*. Eds. P.J. den Boer, L. Grum and J. Szyszko, pp 21–27. Warsaw: Warsaw Agricultural University Press.

LÖVEI, G.L., MONOSTORI, E. AND ANDÓ, I. (1985). Digestion rate in relation to starvation in the larva of a carabid predator, *Poecilus cupreus*. *Entomologia Experimentalis et Applicata* **37**, 123–127.

LÖVEI, G.L., SOPP, P.I. AND SUNDERLAND, K.D. (1987). The effect of mixed feeding on the digestion of the carabid *Bembidion lampros*. *Acta Phytopathologica et Entomologica Hungarica* **22**, 403–407.

LÖVEI, G.L., SOPP, P.I. AND SUNDERLAND, K.D. (1990). Digestion rate in relation to alternative feeding in three species of polyphagous predators. *Ecological Entomology* **15**, 293–300.

LOXDALE, H.D. (1994). Isozyme and protein profiles of insects of agricultural and horticultural importance. In: *The identification and characterisation of pest organisms*. Ed. D.L. Hawksworth, pp 337–375. Wallingford: CAB International.

LOXDALE, H.D. AND DEN HOLLANDER, J. (Eds.) (1989). *Electrophoretic studies on agricultural pests.* Systematics Association Special volume **39**. Oxford: Clarendon Press.

LUND, R.D. AND TURPIN, F.T. (1977). Serological investigation of black cutworm larval consumption by ground beetles. *Annals of the Entomological Society of America* **70**, 322–324.

MCIVER, J.D. (1981). An examination of the utility of the precipitin test for evaluation of arthropod predator–prey relationships. *Canadian Entomologist* **113**, 213–222.

MENDIS, V.W., BOWEN, I.D., LIDDELL, J.E. AND SYMONDSON, W.O.C. (1996). Monoclonal antibodies against *Deroceras reticulatum* and *Arion ater* eggs for use in predation studies. In: *Slug and snail pests in agriculture.* I. BCPC Symposium Proceedings No. **66.** Ed. I. Henderson, pp 99–106. Farnham, UK: British Crop Protection Council.

MENKEN, S.B.J. AND ULENBERG, S.A. (1987). Biochemical characters in agricultural entomology. *Agricultural Zoology Reviews* **2**, 305–360.

MILLER, M.C. (1979). Preparatory immunodiffusion for production of specific anti-adult southern pine beetle serum. *Annals of the Entomological Society of America* **72**, 820–825.

MILLER, M.C. (1981). Evaluation of enzyme-linked immunosorbent assay of narrow- and broad-spectrum anti-adult southern pine beetle serum. *Annals of the Entomological Society of America* **74**, 279–282.

MILLS, N. (1997). Techniques to evaluate the efficacy of natural enemies. In: *Methods in ecological and agricultural entomology.* Eds. D.R. Dent and M.P. Walton, pp 271–291. Wallingford: CAB International.

MOLLET, J.A. AND ARMBRUST, E.J. (1978). Age specific serological identification of adult stages of alfalfa weevil, *Hypera postica. Annals of the Entomological Society of America* **71**, 207–211.

MURRAY, R.A., SOLOMON, M.G. AND FITZGERALD, J.D. (1989). The use of electrophoresis for determining patterns of predation in arthropods. In: *Electrophoretic studies on agricultural pests.* Eds. H.D. Loxdale and J. den Hollander, pp 467–483. Oxford: Clarendon Press.

NIENSTEDT, K. AND POEHLING, H.M. (1995). Labelling aphids with ^{15}N – an appropriate method to quantify the predation of polyphagous predators? *Mitteilungen der Deutschen Gesellschaft fur Allgemeine und Angewandte Entomologie* **10**, 227–230.

NIENSTEDT, K. AND POEHLING, H.M. (2000). ^{15}N-marked aphids for predation studies under field conditions. *Entomologia Experimentalis et Applicata* **94**, 319–323.

PAILL, W. (2000). Slugs as prey for larvae and imagines of *Carabus violaceus* (Coleoptera: Carabidae). In: *Natural history and applied ecology of carabid beetles.* Eds. P. Brandmayr, G.A. Lövei, A. Casale, A. Vigna-Taglianti and T. Zetto, pp 221–227. Moscow: Pensoft Publishers.

PANG, B., CHENG, J., CHEN, Z. AND LI, D. (2001). Development of monoclonal antibodies to the whitebacked planthopper *Sogatella furcifera* (Horath) and their characterization. *Acta Entomologica Sinica* **44**, 21–26.

POWELL, W., WALTON, M.P. AND JERVIS, M.A. (1996). Populations and communities. In: *Insect natural enemies, practical approaches to their study and evaluation.* Eds. M.A. Jervis and N.A.C. Kidd, pp 223–292. London: Chapman & Hall.

RAGSDALE, D.W., LARSON, A.D. AND NEWSOM, L.D. (1981). Quantitative assessment of the predators of *Nezara viridula* eggs and nymphs within a soybean agroecosystem using an ELISA. *Environmental Entomology* **10**, 402–405.

ROTHSCHILD, G.H.L. (1966). A study of a natural population of *Conomelus anceps* (Germar) (Homoptera: Delphacidae) including observations on predation using the precipitin test. *Journal of Animal Ecology* **35**, 413–434.

SCHELVIS, J. AND SIEPEL, H. (1988). Larval food spectra of *Pterostichus oblongopunctatus* and *P. rhaeticus* in the field (Coleoptera: Carabidae). *Entomologia Generalis* **13**, 61–66.

SCHOOF, D.D., PALCHICK, S. AND TEMPELIS, C.H. (1986). Evaluation of predator–prey relationships using an enzyme immunoassay. *Annals of the Entomological Society of America* **79**, 91–95.

SIMON, C., FRATI, F., BECKENBACH, A., CRESPI, B., LIU, H. AND FLOOK, P. (1994). Evolution, weighting and phylogenetic utility of mitochondrial gene sequences and a compilation of conserved polymerase chain reaction primers. *Annals of the Entomological Society of America* **87**, 651–701.

SOLOMON, M.G., FITZGERALD, J.D. AND MURRAY, R.A. (1996). Electrophoretic approaches to predator–prey interactions. In: *The ecology of agricultural pests: biochemical approaches.* Eds. W.O.C. Symondson and J.E. Liddell, pp 457–468. London: Chapman & Hall.

SOPP, P.I. AND SUNDERLAND, K.D. (1989). Some factors affecting the detection period of aphid remains in predators using ELISA. *Entomologia Experimentalis et Applicata* **51**, 11–20.

SOPP, P.I., SUNDERLAND, K.D., FENLON, J.S. AND WRATTEN, S.D. (1992). An improved quantitative method for estimating invertebrate predation in the field using an enzyme-linked immunosorbent assay. *Journal of Applied Ecology* **29**, 295–302.

STUART, M.K. AND GREENSTONE M.H. (1990). Beyond ELISA: a rapid, sensitive, specific immunodot assay for identification of predator stomach contents. *Annals of the Entomological Society of America* **83**, 1101–1107.

SUNDERLAND, K.D. (1988). Quantitative methods of detecting invertebrate predation occurring in the field. *Annals of Applied Biology* **112**, 201–224.

SUNDERLAND, K.D. (1996). Progress in quantifying predation using antibody techniques. In: *The ecology of agricultural pests: biochemical approaches.* Eds. W.O.C. Symondson and J.E. Liddell, pp 419–455. London: Chapman & Hall.

SUNDERLAND, K.D. AND SUTTON, S.L. (1980). A serological study of arthropod predation on woodlice in a dune grassland ecosystem. *Journal of Animal Ecology* **49**, 987–1004.

SUNDERLAND, K.D., CROOK, N.E., STACEY, D.L. AND FULLER, B.J. (1987). A study of feeding by polyphagous predators on cereal aphids using ELISA and gut dissection. *Journal of Applied Ecology* **24**, 907–933.

SUTTON, S.L. (1970). Predation on woodlice: an investigation using the precipitin test. *Entomologia Experimentalis et Applicata* **13**, 279–285.

SYMONDSON, W.O.C. (2002). Molecular identification of prey in predator diets. *Molecular Ecology* (in press).

SYMONDSON, W.O.C. AND HEMINGWAY, J. (1997). Biochemical and molecular techniques. In: *Methods in ecological and agricultural entomology.* Eds. D.R. Dent and M.P. Walton, pp 293–350. Oxford: CAB International.

SYMONDSON, W.O.C. AND LIDDELL, J.E. (1993a). The development and characterisation of an anti-haemolymph antiserum for the detection of mollusc remains within carabid beetles. *Biocontrol Science and Technology* **3**, 261–275.

SYMONDSON, W.O.C. AND LIDDELL, J.E. (1993b). A monoclonal antibody for the detection of arionid slug remains in carabid predators. *Biological Control* **3**, 207–214.

SYMONDSON, W.O.C. AND LIDDELL, J.E. (1993c). Differential antigen decay rates during digestion of molluscan prey by carabid predators. *Entomologia Experimentalis et Applicata* **69**, 277–287.

SYMONDSON, W.O.C. AND LIDDELL, J.E. (1993d). The detection of predation by *Abax parallelepipedus* and *Pterostichus madidus* (Coleoptera: Carabidae) on Mollusca using a quantitative ELISA. *Bulletin of Entomological Research* **83**, 641–647.

SYMONDSON, W.O.C. AND LIDDELL, J.E. (1995). Decay rates for slug antigens within the carabid predator *Pterostichus melanarius* monitored with a monoclonal antibody. *Entomolgia Experimentalis et Applicata* **75**, 245–250.

SYMONDSON, W.O.C. AND LIDDELL, J.E. (1996a). Immunological approaches to the detection of predation upon New Zealand Flatworms (Tricladida: Terricola): problems caused by shared epitopes with slugs (Mollusca: Pulmonata). *International Journal of Pest Management* **42**, 95–99.

SYMONDSON, W.O.C. AND LIDDELL, J.E. (1996b). A species-specific monoclonal antibody system for detecting the remains of field slugs, *Deroceras reticulatum* (Müller) (Mollusca: Pulmonata), in carabid beetles (Coleoptera: Carabidae). *Biocontrol Science and Technology* **6**, 91–99.

SYMONDSON, W.O.C., MENDIS, V.W. AND LIDDELL, J.E. (1995). Monoclonal antibodies for the identification of slugs and their eggs. *EPPO Bulletin* **25**, 377–382.

SYMONDSON, W.O.C., GLEN, D.M., WILTSHIRE, C.W., LANGDON, C.J. AND LIDDELL, J.E. (1996). Effects of cultivation techniques and methods of straw disposal on predation by

Pterostichus melanarius (Coleoptera: Carabidae) upon slugs (Gastropoda: Pulmonata) in an arable field. *Journal of Applied Ecology* **33**, 741–753.

SYMONDSON, W.O.C., ERICKSON, M.L. AND LIDDELL, J.E. (1997). Species-specific detection of predation by Coleoptera on the milacid slug *Tandonia budapestensis* (Mollusca: Pulmonata). *Biocontrol Science and Technology* **7**, 457–465.

SYMONDSON, W.O.C., ERICKSON, M.L., LIDDELL, J.E. AND JAYAWARDENA, K.G.I. (1999a). Amplified detection, using a monoclonal antibody, of an aphid-specific epitope exposed during digestion in the gut of a predator. *Insect Biochemistry and Molecular Biology* **29**, 873–882.

SYMONDSON, W.O.C., ERICKSON, M.L. AND LIDDELL, J.E. (1999b). Development of a monoclonal antibody for the detection and quantification of predation on slugs within the *Arion hortensis* agg. (Mollusca: Pulmonata). *Biological Control* **16**, 274–282.

SYMONDSON, W.O.C., GASULL, T. AND LIDDELL, J.E. (1999c). Rapid identification of adult whiteflies in plant consignments using monoclonal antibodies. *Annals of Applied Biology* **134**, 271–276.

SYMONDSON, W.O.C., GLEN, D.M., ERICKSON, M.L., LIDDELL, J.E. AND LANGDON, C.J. (2000). Do earthworms help to sustain the slug predator *Pterostichus melanarius* (Coleoptera: Carabidae) within crops? Investigations using a monoclonal antibody-based detection system. *Molecular Ecology* **9**, 1279–1292.

TIJSSEN, P. (1985). *Practice and theory of enzyme immunoassays*. Oxford: Elsevier.

TOBOLEWSKI, J., KALISZEWSKI, M.J., COLWELL, R.K. AND OLIVER, J.H. (1992). Detection and identification of mammalian DNA from the gut of museum specimens of ticks. *Journal of Medical Entomology* **29**, 1049–1051.

TOD, M.E. (1973). Notes on beetle predators of molluscs. *The Entomologist* **106**, 196–201.

VOLLER, A., BIDWELL, D.E. AND BARTLETT, A. (1979). *The Enzyme-Linked Immunosorbent Assay (ELISA)*. Guernsey: Dynatech Europe.

WALRANT, A. AND LOREAU, M. (1995). Comparison of iso-enzyme electrophoresis and gut content examination for determining the natural diets of the groundbeetle species *Abax ater* (Coleoptera: Carabidae). *Entomologia Generalis* **19**, 253–259.

ZAIDI, R.H., JAAL, Z., HAWKES, N.J., HEMINGWAY, J. AND SYMONDSON, W.O.C. (1999). Can the detection of prey DNA amongst the gut contents of invertebrate predators provide a new technique for quantifying predation in the field? *Molecular Ecology* **8**, 2081–2088.

ZHU, Y.-C. AND GREENSTONE, M.H. (1999). Polymerase chain reaction techniques for distinguishing three species and two strains of *Aphelinus* (Hymenoptera: Aphelinidae) from *Diuraphis noxia* and *Schizaphis graminum* (Homoptera: Aphididae). *Annals of the Entomological Society of America* **92**, 71–79.

6
Invertebrate Pest Control by Carabids

KEITH D. SUNDERLAND

Department of Entomological Sciences, Horticulture Research International, Wellesbourne, Warwickshire CV35 9EF, UK

Introduction

There is a growing global interest in biological control of pests because it is realized that practical problems (such as pest resistance and pollution of the environment) render unsustainable those agricultural systems that rely entirely on inputs of chemical pesticides (Pimentel, 1995). Carabids are one component of the multifarious natural enemy assemblages, which form the raw material out of which effective biocontrol systems must be forged. The Carabidae are present worldwide and occur in many habitats, including our crops and adjacent areas (Thiele, 1977; Lövei and Sunderland, 1996). The family contains species that range from specialist to generalist in their feeding habits (Hengeveld, 1980). Because they are found in crops (annual and perennial, including orchards and forest plantations), it is logically possible that they may feed, to some extent, on pest invertebrates that attack crop plants. Post-mortem examination of beetles collected from crop fields provides evidence of pest consumption (but this cannot be interpreted rigorously as predation), and observation of feeding behaviour in laboratory and field indicates which species are physically and behaviourally capable of killing pests. Beyond this, to measure whether (and to what extent) carabids can have an impact on pest populations that is economically significant for farmers, it is usually necessary to carry out manipulative experiments in the field. This review aims to examine the evidence for consumption and predation of pests by carabids and for carabid-induced modification of the population dynamics of pest species. Since carabids rarely occur alone, pest control by predator assemblages that include carabids as a numerically significant component is also reviewed.

This review extends beyond the earlier information reviewed by Basedow *et al.* (1976), Allen (1979), Luff (1987), Sunderland (1988), Larochelle (1990) [records for 1054 species up to 1977], Kromp (1999), and Symondson *et al.* (2002). Additional references on carabid predation of pest Mollusca can be found in Symondson (in press), Symondson (this volume) and Port *et al.* (2001). The effects of pest populations on the fitness of carabids will not be considered here, but the topic is reviewed by Toft

Abbreviations: ELISA, Enzyme-Linked ImmunoSorbent Assays

The Agroecology of Carabid Beetles

and Bilde (thisvolume). The minor role of carabids as pests (Thiele, 1977; Luff, 1987) is also not included. Natural enemies can affect the behaviour, distribution and abundance of pests indirectly (i.e. by mechanisms other than direct predation, parasitism and pathogenesis). For example, the mere presence of natural enemies may reduce the feeding rate of pests (Snyder and Wise, 2000), cause them to alter their spatial distribution (Stamp, 1997), or even induce morphological changes (Weisser *et al*., 1999). These topics are little researched with respect to Carabidae, and so will not be reviewed here.

Tiger beetles (supertribe Cicindelitae) are included in this review as part of the family Carabidae. The carabid nomenclature used here has been checked and updated by Dr R.L. Davidson (Carnegie Museum of Natural History, Pittsburgh, USA) using recent sources, such as Lorenz (1998a,b).

Evidence that carabids consume pests (but without proof of predation)

The guts of field-collected beetles can be examined, in various ways, to determine if they have been feeding on pests. The use of such post-mortem techniques (including dissection, labelling, electrophoretic, immunological and molecular methods) for determining aspects of the diet of carabids is reviewed by Symondson (this volume) and Sunderland *et al*. (in press). Some examples (mainly from the last two decades) of pest consumption by carabids are listed in *Table 6.1*. Overall, carabid adults were found to have consumed slugs (e.g. *Deroceras reticulatum*), snails (e.g. *Oxychilus* sp.), chrysomelid beetles (e.g. *Leptinotarsa decemlineata*), wireworms (e.g. *Agriotes sputator*), weevils (e.g. *Otiorhynchus sulcatus*), root flies (e.g. *Delia radicum*), apple maggots (e.g. *Rhagoletis pomonella*), blossom midges (e.g. *Sitodiplosis mosellana*), leatherjackets (Tipulidae), wheat bugs (e.g. *Eurygaster integriceps*), aphids (e.g. *Aphis pomi*), sawflies (e.g. *Neodiprion sertifer*), caterpillars (e.g. *Cydia pomonella*), cutworms (e.g. *Agrotis ipsilon*) and termites (e.g. *Odontotermes wallonensis*). Most of the studies summarized in *Table 6.1* employed examination, by gut dissection or serological methods, of beetles collected from cereals and forest habitats. This would appear to reflect the main focus of research in recent decades. Given this bias, it is worth noting that 30% of the 182 beetle species listed had consumed Coleoptera, 28% Lepidoptera, 16% Homoptera and 13% Diptera. Species that were clearly polyphagous (in that they consumed pests belonging to three or more orders) were *Abax parallelepipedus* [*Abax ater*], *Anchomenus dorsalis* [*Agonum dorsale*], *Amara aenea*, *Bembidion lampros*, *Bembidion properans*, *Bembidion quadrimaculatum*, *Clivina fossor*, *Diplocheila impressicollis* [*Diplocheila impressifrons*], *Harpalus affinis* [*Harpalus aeneus*], *Harpalus rufipes*, *Nebria brevicollis*, *Trechus quadristriatus* and eight pterostichine species, *Poecilus cupreus* [*Pterostichus cupreus*], *Poecilus lucublandus* [*Pterostichus lucublandus*], *Pterostichus aethiops*, *Pterostichus madidus*, *Pterostichus melanarius, Pterostichus niger, Pterostichus nigrita* and *Pterostichus oblongopunctatus*. *Pterostichus* were especially polyphagous, with one species (*P. melanarius*) consuming at least fourteen pest species belonging to five orders. The converse cannot be argued (i.e. that other species in *Table 6.1* are monophagous or oligophagous) because the results may reflect research choices by humans as much as prey choices by beetles. A few records were found for carabid larvae, and they had consumed slugs, aphids, wireworms, blossom midges and the New Zealand flatworm (*Table 6.1*).

Table 6.1. Examples of pest consumption by carabids (predation not proven). Where carabid names have been updated (by Dr R.L. Davidson, Carnegie Museum of Natural History), the original name used in the cited publication(s) is also shown [in square brackets]. Junior synonyms of current names are shown (in round brackets).

Carabid species	Pest species (reference in brackets; see key)
Adults	
Abax parallelus	MOLLUSCA: slugs (41)
Abax parallelepipedus [Abax parallelopipedus; Abax ater]	MOLLUSCA: slugs and snails (10, 40); slugs (20); *Arion* sp. (50); HOMOPTERA: aphids (40); LEPIDOPTERA: *Operophtera brumata* (33); *Pieris rapae* (32)
Acupalpus meridianus	COLEOPTERA: *Leptinotarsa decemlineata* (53)
Agonum duftschmidi	COLEOPTERA: *Leptinotarsa decemlineata* (53)
Agonum ericeti	HYMENOPTERA: *Neodiprion sertifer* (8)
Agonum gracilipes	COLEOPTERA: *Leptinotarsa decemlineata* (53)
Agonum lugens	COLEOPTERA: *Leptinotarsa decemlineata* (53)
Agonum marginatum	COLEOPTERA: *Leptinotarsa decemlineata* (53)
Agonum muelleri	COLEOPTERA: Elateridae (31); HOMOPTERA: cereal aphids (11)
Agonum placidum	DIPTERA: *Sitodiplosis mosellana* (9)
Agonum viduum	DIPTERA: *Delia radicum* (16)
Agonum viridicupreum	COLEOPTERA: *Leptinotarsa decemlineata* (53)
Amara aenea	COLEOPTERA: *Leptinotarsa decemlineata* (53); DIPTERA: *Rhagoletis pomonella* (2); HOMOPTERA: *Aphis pomi* (15); cereal aphids (28, 29) aphids (55); LEPIDOPTERA: *Cydia pomonella* (1)
Amara aulica [*Curtonotus aulicus*]	MOLLUSCA: slugs (20); COLEOPTERA: *Leptinotarsa decemlineata* (53)
Amara avida	LEPIDOPTERA: *Euxoa ochrogaster* (34)
Amara carinata	DIPTERA: *Sitodiplosis mosellana* (9)
Amara familiaris	DIPTERA: *Delia radicum* (16), *Rhagoletis pomonella* (2); HOMOPTERA: *Aphis pomi* (15); cereal aphids (27, 28, 29)
Amara farcta	DIPTERA: *Sitodiplosis mosellana* (9)
Amara lunicollis	MOLLUSCA: slugs (20)
Amara ovata	COLEOPTERA: *Leptinotarsa decemlineata* (53)
Amara plebeja	HOMOPTERA: *Rhopalosiphum padi* (18); cereal aphids (28, 29)
Amara similata	MOLLUSCA: slugs (20); COLEOPTERA: *Leptinotarsa decemlineata* (53)
Amara torrida	LEPIDOPTERA: *Euxoa ochrogaster* (34)
Anchomenus dorsalis [Agonum dorsale]	COLEOPTERA: *Gastrophysa polygoni* (46); *Leptinotarsa decemlineata* (53); DIPTERA: *Delia radicum* (16); HOMOPTERA: *Rhopalosiphum padi* (37); cereal aphids (3, 11, 23, 27, 28, 29, 48)
Anisodactylus binotatus	COLEOPTERA: *Leptinotarsa decemlineata* (53)
Anisodactylus nemorivagus	COLEOPTERA: *Leptinotarsa decemlineata* (53)
Anisodactylus sanctaecrucis	LEPIDOPTERA: *Cydia pomonella* (1)
Anisodactylus signatus	COLEOPTERA: *Leptinotarsa decemlineata* (53)
Anomotarus crudelis	LEPIDOPTERA: *Pieris rapae* (38)
Asaphidion flavipes	HOMOPTERA: cereal aphids (11, 23, 28, 29); aphids (41)
Bembidion biguttatum	COLEOPTERA: *Leptinotarsa decemlineata* (53)

Table 6.1. continued

Carabid species	Pest species (reference in brackets; see key)
Adults (continued)	
Bembidion bruxellense	HOMOPTERA: *Rhopalosiphum padi* (18)
Bembidion guttula	HOMOPTERA: *Rhopalosiphum padi* (18)
Bembidion lampros	COLEOPTERA: *Leptinotarsa decemlineata* (53); *Otiorhynchus sulcatus* (44); DIPTERA: *Delia radicum* (16); HOMOPTERA: *Rhopalosiphum padi* (37, 47); *Sitobion avenae* (47); cereal aphids (11, 27, 28, 29, 48); aphids (55)
Bembidion lunulatum	HOMOPTERA: aphids (55)
Bembidion nitidum	DIPTERA: *Sitodiplosis mosellana* (9)
Bembidion nudipenne	DIPTERA: *Sitodiplosis mosellana* (9)
Bembidion obscurellum	DIPTERA: *Sitodiplosis mosellana* (9)
Bembidion obtusum	HOMOPTERA: *Rhopalosiphum padi* (47); *Sitobion avenae* (47); cereal aphids (11, 29, 52)
Bembidion properans	COLEOPTERA: *Leptinotarsa decemlineata* (53); DIPTERA: Tipulidae (13); HETEROPTERA: *Eurygaster integriceps* (49); HOMOPTERA: *Rhopalosiphum padi* (18); aphids (13, 55)
Bembidion quadrimaculatum	COLEOPTERA: *Leptinotarsa decemlineata* (53); DIPTERA: *Delia radicum* (16), *Rhagoletis pomonella*, (2) *Sitodiplosis mosellana* (9); HOMOPTERA: *Aphis pomi* (15); *Rhopalosiphum padi* (18, 37); LEPIDOPTERA: *Cydia pomonella* (1)
Bembidion rapidum	DIPTERA: *Rhagoletis pomonella* (2)
Bembidion rupicola	DIPTERA: *Sitodiplosis mosellana* (9)
Bembidion sordidum	DIPTERA: *Sitodiplosis mosellana* (9)
Bembidion tetracolum	HOMOPTERA: cereal aphids (11)
Bembidion timidum	DIPTERA: *Sitodiplosis mosellana* (9)
Bembidion varium	COLEOPTERA: *Leptinotarsa decemlineata* (53)
Brachinus crepitans	COLEOPTERA: *Leptinotarsa decemlineata* (53); HETEROPTERA: *Eurygaster integriceps* (49)
Brachinus ejaculans	HETEROPTERA: *Eurygaster integriceps* (49)
Brachinus elegans (Brachinus ganglbaueri)	COLEOPTERA: *Leptinotarsa decemlineata* (53)
Brachinus explodens	COLEOPTERA: *Leptinotarsa decemlineata* (53)
Brachinus psophia	COLEOPTERA: *Leptinotarsa decemlineata* (53); HETEROPTERA: *Eurygaster integriceps* (49)
Brachinus sclopeta	HETEROPTERA: *Eurygaster integriceps* (49)
Broscus cephalotes	COLEOPTERA: *Leptinotarsa decemlineata* (53); HETEROPTERA: *Eurygaster integriceps* (49)
Calathus erratus	COLEOPTERA: *Leptinotarsa decemlineata* (53)
Calathus fuscipes	COLEOPTERA: *Leptinotarsa decemlineata* (53); *Otiorhynchus sulcatus* (44); HOMOPTERA: cereal aphids (27, 28)
Calathus melanocephalus	COLEOPTERA: *Leptinotarsa decemlineata* (53); HOMOPTERA: *Rhopalosiphum padi* (37)
Calosoma calidum	LEPIDOPTERA: *Euxoa ochrogaster* (34)
Calosoma frigidum	LEPIDOPTERA: *Lymantria dispar* (6, 7)
Calosoma sayi	LEPIDOPTERA: *Heliothis zea* (36)

Table 6.1. continued

Carabid species	Pest species (reference in brackets; see key)
Adults (continued)	
Calosoma sycophanta	LEPIDOPTERA: *Lymantria dispar* (25)
Carabus arcensis (Carabus arvensis)	COLEOPTERA: *Leptinotarsa decemlineata* (53)
Carabus cancellatus	COLEOPTERA: *Leptinotarsa decemlineata* (53)
Carabus euchromus sacheri [Carabus obsoletus]	COLEOPTERA: *Leptinotarsa decemlineata* (53)
Carabus goryi (= limbatus) [Carabus limbatus]	LEPIDOPTERA: *Lymantria dispar* (7)
Carabus granulatus	MOLLUSCA: *Arion lusitanicus* (54); COLEOPTERA: *Leptinotarsa decemlineata* (53)
Carabus nemoralis	LEPIDOPTERA: *Lymantria dispar* (6)
Carabus perrini transcaucasicus (Carabus campestris)	HETEROPTERA: *Eurygaster integriceps* (49)
Carabus rothi hampei [Carabus hampei]	COLEOPTERA: *Leptinotarsa decemlineata* (5, 53)
Carabus taedatus	LEPIDOPTERA: *Euxoa ochrogaster* (34)
Carabus ullrichii [*Carabus ullrichi*]	COLEOPTERA: *Leptinotarsa decemlineata* (53)
Carabus violaceus	MOLLUSCA: slugs (20); *Arion lusitanicus* (54); COLEOPTERA: *Leptinotarsa decemlineata* (53); *Otiorhynchus sulcatus* (44)
Carabus vinctus	LEPIDOPTERA: *Heliothis zea* (36)
Carabus wilcoxi	LEPIDOPTERA: *Lymantria dispar* (7)
Carabus zawadszkyii [*Carabus zawadszkyi*]	COLEOPTERA: *Leptinotarsa decemlineata* (53)
Chlaenius aeneocephalus	HETEROPTERA: *Eurygaster integriceps* (49)
Chlaenius emarginatus	LEPIDOPTERA: *Lymantria dispar* (7)
Chlaenius erythropus	LEPIDOPTERA: *Heliothis zea* (36)
Chlaenius festivus	COLEOPTERA: *Leptinotarsa decemlineata* (53)
Chlaenius nigricornis	COLEOPTERA: *Leptinotarsa decemlineata* (53)
Chlaenius spoliatus	COLEOPTERA: *Leptinotarsa decemlineata* (53)
Chlaenius tricolor	LEPIDOPTERA: *Lymantria dispar* (6)
Cicindela germanica	COLEOPTERA: *Leptinotarsa decemlineata* (53)
Clivina australasiae	HOMOPTERA: aphids (30)
Clivina collaris	COLEOPTERA: *Leptinotarsa decemlineata* (53)
Clivina fossor	COLEOPTERA: Elateridae (31); *Leptinotarsa decemlineata* (53); DIPTERA: *Delia radicum* (16); HOMOPTERA: *Rhopalosiphum padi* (18, 37); aphids (13, 55);
Clivina impressefrons [*Clivina impressifrons*]	LEPIDOPTERA: *Cydia pomonella* (1)
Cychrus caraboides	MOLLUSCA: slugs (20)
Cymindis cingulata	COLEOPTERA: *Leptinotarsa decemlineata* (53)
Cymindis cribricollis	LEPIDOPTERA: *Lymantria dispar* (6)
Cymindis limbatus [Pinacodera limbata]	LEPIDOPTERA: *Lymantria dispar* (6)

Table 6.1. continued

Carabid species	Pest species (reference in brackets; see key)
Adults (continued)	
Cymindis platicollis [*Pinacodera platicollis*]	LEPIDOPTERA: *Lymantria dispar* (6)
Demetrias atricapillus	HOMOPTERA: cereal aphids (27, 28, 29)
Dicaelus dilatatus	LEPIDOPTERA: *Lymantria dispar* (6, 7)
Dicaelus politus	LEPIDOPTERA: *Lymantria dispar* (6, 7)
Dicaelus teter	LEPIDOPTERA: *Lymantria dispar* (7)
Diplocheila impressicollis [*Diplocheila impressifrons*]	DIPTERA: *Rhagoletis pomonella* (2); HOMOPTERA: *Aphis pomi* (15); LEPIDOPTERA: *Cydia pomonella* (1)
Galerita janus [*Galeritula janus*]	LEPIDOPTERA: *Heliothis zea* (36)
Gynandromorphus etruscus	HETEROPTERA: *Eurygaster integriceps* (49)
Harpalus affinis [*Harpalus aeneus*]	COLEOPTERA: *Leptinotarsa decemlineata* (53); DIPTERA: *Delia radicum* (16); *Rhagoletis pomonella* (2); HOMOPTERA: *Aphis pomi* (15); *Rhopalosiphum padi* (37); cereal aphids (28, 29); aphids (30); LEPIDOPTERA: *Cydia pomonella* (1); *Operophtera brumata* (42); *Pieris rapae* (32)
Harpalus amputatus	LEPIDOPTERA: *Euxoa ochrogaster* (34)
Harpalus caliginosus	LEPIDOPTERA: *Heliothis zea* (36)
Harpalus distinguendus	COLEOPTERA: *Leptinotarsa decemlineata* (53); HETEROPTERA: *Eurygaster integriceps* (49)
Harpalus griseus [*Pseudoophonus griseus*]	COLEOPTERA: *Leptinotarsa decemlineata* (53)
Harpalus herbivagus	HOMOPTERA: *Aphis pomi* (15); LEPIDOPTERA: *Lymantria dispar* (6)
Harpalus honestus	COLEOPTERA: *Leptinotarsa decemlineata* (53)
Harpalus hospes [*Ophonus hospes*]	HETEROPTERA: *Eurygaster integriceps* (49)
Harpalus indigens	LEPIDOPTERA: *Lymantria dispar* (6)
Harpalus latus	MOLLUSCA: slugs (20)
Harpalus pensylvanicus [*Harpalus pennsylvanicus*]	LEPIDOPTERA: *Heliothis zea* (36)
Harpalus providens (=viduus) [*Harpalus viduus*]	LEPIDOPTERA: *Lymantria dispar* (6)
Harpalus rubripes	COLEOPTERA: *Leptinotarsa decemlineata* (53)
Harpalus rufipes [*Ophonus rufipes*; *Pseudophonus rufipes*]	MOLLUSCA: slugs (20, 42); COLEOPTERA: *Leptinotarsa decemlineata* (53); *Otiorhynchus sulcatus* (44); DIPTERA: *Delia radicum* (16); HETEROPTERA: *Eurygaster integriceps* (49); HOMOPTERA: *Rhopalosiphum padi* (18, 37); cereal aphids (23, 27, 28, 29); aphids (21, 42); LEPIDOPTERA: *Lymantria dispar* (6); *Pieris rapae* (32)
Harpalus spadiceus	LEPIDOPTERA: *Lymantria dispar* (7)
Harpalus tardus	COLEOPTERA: *Leptinotarsa decemlineata* (53)
Leistus rufomarginatus	HOMOPTERA: aphids (41)
Loricera pilicornis	HOMOPTERA: cereal aphids (11, 23, 28, 29, 48); aphids (55)
Mecyclothorax ambiguus	LEPIDOPTERA: *Pieris rapae* (38)
Megacephala carolina	LEPIDOPTERA: *Heliothis zea* (36)
Megacephala virginica	LEPIDOPTERA: *Heliothis zea* (36)

Table 6.1. continued

Carabid species	Pest species (reference in brackets; see key)
Adults (continued)	
Microlestes minutulus	HETEROPTERA: *Eurygaster integriceps* (49)
Myas cyanescens	LEPIDOPTERA: *Lymantria dispar* (6, 7)
Nebria brevicollis	MOLLUSCA: slugs (20); COLEOPTERA: *Gastrophysa polygoni* (46); DIPTERA: *Delia radicum* (16); Tipulidae adults, larvae (55); HOMOPTERA: cereal aphids (11, 27, 28, 29); aphids (41, 55)
Notiophilus aeneus	LEPIDOPTERA: *Lymantria dispar* (6)
Notiophilus biguttatus	COLEOPTERA: *Otiorhynchus sulcatus* (44); HOMOPTERA: *Rhopalosiphum padi* (47); *Sitobion avenae* (47); cereal aphids (27, 28, 29); aphids (41)
Notonomus gravis	LEPIDOPTERA: *Pieris rapae* (38)
Omphra pilosa	ISOPTERA: Termitidae (26)
Patrobus longicornis	HOMOPTERA: *Aphis pomi* (15); LEPIDOPTERA: *Lymantria dispar* (6)
Platynus assimilis	COLEOPTERA: *Leptinotarsa decemlineata* (53)
Platynus decentis	LEPIDOPTERA: *Lymantria dispar* (6, 7)
Platynus hypolithos [*Platynus hypolithus*]	LEPIDOPTERA: *Lymantria dispar* (7)
Platynus parmarginatus [*Platynus paramarginatum*]	LEPIDOPTERA: *Lymantria dispar* (7)
Poecilus chalcites [*Pterostichus chalcites*]	COLEOPTERA: *Agrotis ipsilon* (35); LEPIDOPTERA: *Cydia pomonella* (1); *Heliothis zea* (36)
Poecilis corvus [*Pterostichus corvus*]	DIPTERA: *Sitodiplosis mosellana* (9)
Poecilis crenuliger [*Pterostichus crenuliger*]	HETEROPTERA: *Eurygaster integriceps* (49)
Poecilus cupreus [*Pterostichus cupreus*]	COLEOPTERA: *Leptinotarsa decemlineata* (5, 53); HETEROPTERA: *Eurygaster integriceps* (49); HOMOPTERA: *Acyrthosiphon pisum* (17); *Rhopalosiphum padi* (37)
Poecilus lepidus	COLEOPTERA: *Leptinotarsa decemlineata* (53)
Poecilus lucublandus [*Pterostichus lucublandus*]	COLEOPTERA: *Agrotis ipsilon* (35); DIPTERA: *Rhagoletis pomonella* (2); *Sitodiplosis mosellana* (9); HOMOPTERA: *Aphis pomi* (15); LEPIDOPTERA: *Cochylis hospes* (24); *Lymantria dispar* (6, 7)
Poecilus puncticollis [*Pterostichus puncticollis*]	HETEROPTERA: *Eurygaster integriceps* (49)
Poecilus versicolor	COLEOPTERA: *Leptinotarsa decemlineata* (53)
Pterostichus adoxus	LEPIDOPTERA: *Lymantria dispar* (6, 7)
Pterostichus adstrictus	DIPTERA: *Sitodiplosis mosellana* (9); LEPIDOPTERA: *Euxoa ochrogaster* (34)
Pterostichus aethiops	MOLLUSCA: slugs (39); COLEOPTERA: Elateridae (39); DIPTERA: Tipulidae (39)
Pterostichus anthracinus	COLEOPTERA: *Leptinotarsa decemlineata* (53)
Pterostichus chameleon [*Pterostichus chamaeleon*]	COLEOPTERA: *Leptinotarsa decemlineata* (53)
Pterostichus coracinus	LEPIDOPTERA: *Lymantria dispar* (6, 7)
Pterostichus diligendus	LEPIDOPTERA: *Lymantria dispar* (7)
Pterostichus femoralis	DIPTERA: *Sitodiplosis mosellana* (9)
Pterostichus honestus	LEPIDOPTERA: *Lymantria dispar* (7)

Table 6.1. continued

Carabid species	Pest species (reference in brackets; see key)
Adults (continued)	
Pterostichus lachrymosus [Pterostichus lacrymosus]	LEPIDOPTERA: *Lymantria dispar* (7)
Pterostichus macer	COLEOPTERA: *Leptinotarsa decemlineata* (53)
Pterostichus madidus [Feronia madida]	MOLLUSCA: slugs and snails (10); slugs (20); COLEOPTERA: *Gastrophysa polygoni* (46); *Otiorhynchus sulcatus* (44); *Sitona lineatus* (43); Elateridae (39); DIPTERA: Tipulidae (39); *Delia radicum* (16); HOMOPTERA: *Macrosiphum euphorbiae* (14); cereal aphids (22, 29); LEPIDOPTERA: *Lymantria dispar* (7); *Operophtera brumata* (33)
Pterostichus melanarius [Feronia vulgaris; Platysma vulgare]	MOLLUSCA: slugs (4, 20, 39, 42); slug eggs (12); *Arion lusitanicus* (54); COLEOPTERA: *Gastrophysa polygoni* (46); *Leptinotarsa decemlineata* (5, 53); DIPTERA: *Delia radicum* (16); *Rhagoletis pomonella* (2); Tipulidae (12); HOMOPTERA: *Acyrthosiphon pisum* (17); *Aphis pomi* (15); *Macrosiphum euphorbiae* (14); *Rhopalosiphum padi* (37); cereal aphids (3, 11, 22, 23, 27, 28, 29, 48); aphids (12, 21); LEPIDOPTERA: *Cydia pomonella* (1); *Lymantria dispar* (6); *Operophtera brumata* (33); *Pieris rapae* (32)
Pterostichus melas	COLEOPTERA: *Leptinotarsa decemlineata* (53)
Pterostichus minor	COLEOPTERA: *Leptinotarsa decemlineata* (53)
Pterostichus moestus	LEPIDOPTERA: *Lymantria dispar* (7)
Pterostichus mutus	LEPIDOPTERA: *Lymantria dispar* (6, 7)
Pterostichus niger	MOLLUSCA: slugs (20, 39); COLEOPTERA: Elateridae (39); *Leptinotarsa decemlineata* (53); DIPTERA: Tipulidae (39); HOMOPTERA: *Rhopalosiphum padi* (37); cereal aphids (23)
Pterostichus nigrita [*Pterostichus nigritus*]	MOLLUSCA: slugs (20); COLEOPTERA: *Leptinotarsa decemlineata* (53); HYMENOPTERA: *Neodiprion sertifer* (8)
Pterostichus oblongopunctatus	MOLLUSCA: slugs (39); COLEOPTERA: Elateridae (39); DIPTERA: Tipulidae (39); HOMOPTERA: aphids (41)
Pterostichus ovoideus	COLEOPTERA: *Leptinotarsa decemlineata* (53)
Pterostichus pensylvanicus	LEPIDOPTERA: *Lymantria dispar* (6, 7)
Pterostichus rostratus	LEPIDOPTERA: *Lymantria dispar* (7)
Pterostichus sculptus	LEPIDOPTERA: *Lymantria dispar* (7)
Pterostichus strenuus	COLEOPTERA: Curculionidae adult (55); *Leptinotarsa decemlineata* (53); HOMOPTERA: aphids (55)
Pterostichus stygicus	LEPIDOPTERA: *Lymantria dispar* (6, 7)
Pterostichus vernalis	COLEOPTERA: *Leptinotarsa decemlineata* (53); HOMOPTERA: aphids (55)
Rhytisternus liopleurus	LEPIDOPTERA: *Pieris rapae* (38)
Rhytisternus miser	LEPIDOPTERA: *Pieris rapae* (38)
Scaphinotus viduus	LEPIDOPTERA: *Lymantria dispar* (7)
Scarites subterraneus	HOMOPTERA: *Aphis pomi* (15); LEPIDOPTERA: *Heliothis zea* (36)
Sphaeroderus canadensis	LEPIDOPTERA: *Lymantria dispar* (6, 7)
Sphaeroderus stenostomus lecontei [*Sphaeroderus lecontei*]	LEPIDOPTERA: *Lymantria dispar* (6)
Stomis pumicatus	COLEOPTERA: *Leptinotarsa decemlineata* (53)
Synuchus impunctatus	LEPIDOPTERA: *Lymantria dispar* (6, 7)

Table 6.1. continued

Carabid species	Pest species (reference in brackets; see key)
Adults (continued)	
Synuchus vivalis (=nivalis) *[Synuchus nivalis]*	COLEOPTERA: *Leptinotarsa decemlineata* (53); HOMOPTERA: *Rhopalosiphum padi* (37); cereal aphids (28)
Trechus obtusus	DIPTERA: *Delia radicum* (16)
Trechus quadristriatus	DIPTERA: *Delia radicum* (16); HOMOPTERA: *Rhopalosiphum padi* (37, 47); *Sitobion avenae* (47); cereal aphids (29, 45); LEPIDOPTERA: *Pieris rapae* (32)
Trechus secalis	HOMOPTERA: *Rhopalosiphum padi* (37)
Carabid species (larvae)	
Carabus violaceus	MOLLUSCA: *Arion lusitanicus* (54)
Demetrias atricapillus	HOMOPTERA: cereal aphids (29)
Loricera pilicornis	HOMOPTERA: cereal aphids (29)
Nebria brevicollis	HOMOPTERA: *Rhopalosiphum padi* (47); *Sitobion avenae* (47)
Notiophilus biguttatus	HOMOPTERA: cereal aphids (29)
Pterostichus oblongopunctatus	COLEOPTERA: Elateridae (19)
Pterostichus rhaeticus	COLEOPTERA: Elateridae (19)
Trechus spp.	HOMOPTERA: cereal aphids (29)
Unidentified	TRICLADIDA: *Artioposthia triangulata* (51); COLEOPTERA: Elateridae (31); DIPTERA: *Sitodiplosis mosellana* (9); HOMOPTERA: cereal aphids (29)

Key to Table 6.1: D = dye; E = electrophoretic method; G = gut dissection; R = examination of prey remains; S = serological method; T = radiotracers.

1: Hagley and Allen (1988), S, E, apple; 2: Allen and Hagley (1990), S, E, apple; 3: Chiverton and Sotherton (1991), G, wheat; 4: Symondson *et al.* (1996), S, rape; 5: Koval (1986), S, potato; 6: DuDevoir and Reeves (1990), S, forest; 7: Cameron and Reeves (1990), S, forest; 8: Uzenbaev (1984), S, forest; 9: Floate *et al.* (1990), S, E, wheat; 10: Symondson and Liddell (1993), S, forest; 11: Janssens and De Clercq (1990), G, wheat; 12: Pollet and Desender (1985), G, pasture; 13: Pollet and Desender (1987a), G, pasture; 14: Dixon and McKinlay (1992), G, potato; 15: Hagley and Allen (1990), S, apple; 16: Coaker and Williams (1963), S, brassicas; 17: Ekbom (1994), G, peas, clover, alfalfa; 18: Holopainen and Helenius (1992), G, barley; 19: Schelvis and Siepel (1988), E, forest; 20: Ayre and Port (1996), S, oilseed rape, wheat; 21: Kabacik-Wasylik (1989), G, potato; 22: Holland and Thomas (1997a), G, wheat; 23: Hance and Renier (1987), S, wheat; 24: Bergmann and Oseto (1990), G, sunflower; 25: Fuester and Taylor (1996), R, forest; 26: Kumar and Rajagopal (1990), G, maize, orchards, forests; 27: Sunderland (1975), G, barley, wheat; 28: Sunderland and Vickerman (1980), G, barley, wheat; 29: Sunderland *et al.* (1987), S, wheat; 30: Sunderland *et al.* (1995b), G, arable; 31: Fox and MacLellan (1956), S, arable; 32: Dempster (1967), S, sprouts; 33: Frank (1967a), S, forest; 34: Frank (1971), S, barley, oats; 35: Lund and Turpin (1977), S, maize; 36: Lesiewicz *et al.* (1982), S, maize; 37: Chiverton (1987), GS, barley; 38: Kapuge *et al.* (1987), S, cabbage; 39: Sergeyeva and Gryuntal (1990), S, forest; 40: Loreau (1983a), G, forest; 41: Loreau (1983b), G, forest; 42: Cornic (1973), G, apple; 43: Hamon *et al.* (1990), T, beans; 44: Crook and Solomon (1997), S, strawberry, blackcurrant; 45: Burn (1992), S, wheat; 46: Sotherton (1982), G, wheat; 47: Sopp and Chiverton (1987), S, wheat; 48: Scheller (1984), G, barley; 49: Titova and Kupershteyn (1976), S, wheat; 50: Walrant and Loreau (1995), EG, forest; 51: Gibson *et al.* (1997), D, garden; 52: Kennedy (1994), G, wheat; 53: Koval (1999), S, potato; 54: Grimm *et al.* (2000), E, beans; 55: Pollet and Desender (1987b), G, pasture.

Post-mortem methods provide data that are only a very approximate indication of the predation potential of carabids, and the interpretation of such data must be approached with great caution (Sunderland, 1996). This is because pests that have not been killed by the beetle may, nevertheless, be consumed (by scavenging or secondary predation) and, conversely, the beetle may kill pests but contain no identifiable pest remains (due to extra-oral digestion (Vaje and Mossakowski, 1986; Cheeseman and Gillott, 1987; Cohen, 1995), wasteful killing (e.g. Constantineanu and Constantineanu, 1996) or fatal wounding). Recent quantitative data suggest that secondary predation (e.g. the carabid consumes another predator that has consumed the pest) is unlikely to account for a significant proportion of positive results in most systems (Harwood *et al.*, 2001), but scavenging remains a problem. Carabids tested by enzyme-linked immunosorbent assays (ELISA) for feeding on gypsy moth (*Lymantria dispar*) were especially likely to have eaten carrion because the moth population was being killed by a nucleopolyhedrosis virus (Cameron and Reeves, 1990). Similarly, gut dissection of carabids collected from crops sprayed with aphicides revealed remains of aphids, but many of these may have been scavenged after being killed by the chemical (Chiverton, 1984; Dixon and McKinlay, 1992). Halsall and Wratten (1988a) obtained direct video evidence in a wheat field of scavenging on dead aphids by *Bembidion lampros*, *Bembidion obtusum* and *Trechus quadristriatus*. Observations (or video surveillance) in laboratory and field can also be used to establish which pest species a given carabid species is capable of killing, and these data are reviewed in the next section.

Evidence that carabids can kill pests

Records for predation of pests by 214 species of adult Carabidae (from 110 studies) are summarized in *Table 6.2*. 64% of these studies described feeding observations in simple arenas in the laboratory, and in 36% the authors observed predation in the field (mainly in cereal and forest habitats). Overall, pest Lepidoptera were killed by 43% of carabid species, followed by Diptera at 20%, Coleoptera at 12% and Homoptera at 12%. Pests that were seen to be killed in the field included slugs (e.g *Arion ater*), landsnails (e.g. *Monacha obstructa*), rootworm adults (e.g. *Diabrotica virgifera*), eggs and larvae of chrysomelid beetles (e.g. *Leptinotarsa decemlineata*), rootfly eggs (e.g. *Delia radicum*), eggs of carrot fly (e.g. *Psila rosae*), aphids (e.g. *Aphis craccivora*), planthoppers (e.g. *Sogatella furcifera*), cutworms (e.g. *Agrotis ipsilon*), caterpillars (e.g. *Spodoptera frugiperda*), moth pupae (e.g. *Choristoneura fumiferana*), mole crickets (*Scapteriscus* spp.) and termites (*Odontotermes wallonensis*). Polyphagous species (using the arbitrary criterion of being capable of killing pests belonging to three or more orders) were found in the genera *Anchomenus* [*Agonum*] (1 species), *Clivina* (1), *Demetrias* (1), *Megacephala* (1), *Nebria* (1), *Amara* (2), *Bembidion* (2), *Cyclotrachelus* [*Evarthrus*] (2), *Harpalus* (2), *Pterostichus* (2) and *Poecilus* [*Pterostichus*] (3). It is reassuring that these results are broadly in agreement with the species distribution of polyphagy determined by post-mortem examination of field-collected beetles (previous section). *Poecilus lucublandus* [*Pterostichus lucublandus*] killed 12 pest species belonging to 4 orders (12,4). This was exceeded by *Bembidion quadrimaculatum* (14,4), *Harpalus pensylvanicus* (16,4), *Pterostichus melanarius* (16,5) and *Poecilus cupreus* [*Pterostichus cupreus*] (18,5). Thus, the pterostichine carabids are the most polyphagous species in terms of not only consumption (previous section) but also predation of pests. Only positive results are reported in *Table 6.2*. Negative results are more difficult to interpret because they might result from low sample size, inappropriate physical conditions or poor health of the beetles, rather than inability of the carabid ever to kill the pest species under investigation.

Table 6.2. Examples of predation on pests by carabids. Where carabid names have been updated (by Dr R.L. Davidson, Carnegie Museum of Natural History), the original name used in the cited publication(s) is also shown [in square brackets]. Junior synonyms of current names are shown (in round brackets).

Carabid species	Pest species (**L** = laboratory; **F** = field) (reference in brackets; see key)
Adults	
Abax parallelepipedus [*Abax parallelopipedus*]	MOLLUSCA: *Deroceras reticulatum* **L** (73), **F** (112); slugs **L** (71); LEPIDOPTERA: *Operophtera brumata* pupa **L** (63, 106)
Acupalpus dubius	DIPTERA: *Delia radicum* eggs **L** (11)
Acupalpus meridianus	DIPTERA: *Delia radicum* eggs **L** (11)
Agonum chalcomum [*Platynus chalcomus*]	LEPIDOPTERA: *Plutella xylostella* larva **L** (34)
Agonum cupreum	LEPIDOPTERA: *Euxoa ochrogaster* eggs **L** (64)
Agonum cupripennis	LEPIDOPTERA: *Pseudaletia unipuncta* larva **L** (20)
Agonum gracile	DIPTERA: *Delia radicum* eggs **L** (11)
Agonum marginatum	DIPTERA: *Delia radicum* eggs **L** (11)
Agonum moestum	DIPTERA: *Delia radicum* eggs **L** (11)
Agonum muelleri	DIPTERA: *Delia antiqua* egg, larva, pupa **L** (24); *Delia radicum* eggs **L** (11); *Drosophila melanogaster* pupa **F** (10)
Agonum octopunctatum	LEPIDOPTERA: *Agrotis ipsilon* larva **F** (21); *Ostrinia nubilalis* larva **F** (21); *Papaipema nebris* larva **F** (21); *Pseudaletia unipuncta* larva **F** (21)
Agonum placidum	LEPIDOPTERA: *Euxoa ochrogaster* eggs **L** (64)
Agonum punctiforme	LEPIDOPTERA: *Pseudaletia unipuncta* larva **L** (20); *Spodoptera frugiperda* larva **F** (22)
Agonum sexpunctatum	DIPTERA: *Drosophila melanogaster* pupa **F** (10)
Agonum thoreyi	DIPTERA: *Delia radicum* eggs **L** (11)
Amara aenea	DIPTERA: *Delia radicum* eggs **L** (11); *Drosophila melanogaster* pupa **F** (10); HOMOPTERA: *Sitobion avenae* **L** (62)
Amara apricaria	DIPTERA: *Delia floralis* eggs **L** (81); HOMOPTERA: *Rhopalosiphum padi* **L** (43); LEPIDOPTERA: *Euxoa ochrogaster* eggs **L** (64)
Amara aulica	TRICLADIDA: *Artioposthia triangulata* **L** (105)
Amara avida	LEPIDOPTERA: *Euxoa ochrogaster* eggs **L** (64)
Amara bifrons	DIPTERA: *Delia floralis* eggs **L** (81); *Delia radicum* eggs **L** (11); HOMOPTERA: *Rhopalosiphum padi* **L** (43)
Amara chalcites	LEPIDOPTERA: *Plutella xylostella* larva **L** (34)
Amara congrua	LEPIDOPTERA: *Plutella xylostella* larva **L** (34)
Amara cupreolata	COLEOPTERA: *Hypera postica* larva **L** (25); LEPIDOPTERA: *Agrotis ipsilon* larva **F** (21); *Ostrinia nubilalis* larva **F** (21); *Papaipema nebris* larva **F** (21); *Pseudaletia unipuncta* larva **F** (20, 21); *Spodoptera frugiperda* larva **F** (22)
Amara ellipsis	LEPIDOPTERA: *Euxoa ochrogaster* eggs **L** (64)
Amara eurynota	DIPTERA: *Delia radicum* eggs **L** (11)
Amara familiaris	DIPTERA: *Delia radicum* eggs **L** (11); HOMOPTERA: *Sitobion avenae* **L** (62); LEPIDOPTERA: *Pseudaletia unipuncta* larva **L** (20)
Amara fulva	DIPTERA: *Delia floralis* eggs **L** (81)
Amara impuncticollis	COLEOPTERA: *Popillia japonica* eggs **L** (27); LEPIDOPTERA: *Agrotis ipsilon* larva **F** (21); *Colias eurytheme* larva **L** (25); *Melanchra picta* larva **L** (25); *Ostrinia nubilalis* larva **F** (21); *Papaipema nebris* larva **F** (21); *Plathypena scabra* larva **L** (25); *Pseudaletia unipuncta* larva **F** (21); *Spodoptera frugiperda* larva **F** (22)

Table 6.2. continued

Carabid species	Pest species (**L** = laboratory; **F** = field) (reference in brackets; see key)
Adults (continued)	
Amara latior	LEPIDOPTERA: *Euxoa ochrogaster* eggs **L** (64)
Amara montivaga	DIPTERA: *Delia floralis* eggs **L** (81)
Amara nitens (=macronota) [*Amara macronota*]	LEPIDOPTERA: *Plutella xylostella* larva **L** (34)
Amara ovata	DIPTERA: *Delia radicum* eggs **L** (11)
Amara pallipes	COLEOPTERA: *Popillia japonica* eggs **L** (27)
Amara patruelis	LEPIDOPTERA: *Euxoa ochrogaster* eggs **L** (64)
Amara plebeja	DIPTERA: *Delia floralis* eggs **L** (81); *Delia radicum* eggs **L** (11); *Drosophila melanogaster* pupa **F** (10); HOMOPTERA: *Rhopalosiphum padi* **L** (43); *Sitobion avenae* **L** (62)
Amara quenseli	LEPIDOPTERA: *Euxoa ochrogaster* eggs **L** (64)
Amara similata	DIPTERA: *Delia radicum* eggs **L** (11)
Amara torrida	LEPIDOPTERA: *Euxoa ochrogaster* eggs **L** (64)
Amphasia sericea	LEPIDOPTERA: *Agrotis ipsilon* larva **F** (21); *Ostrinia nubilalis* larva **F** (21); *Papaipema nebris* larva **F** (21); *Pseudaletia unipuncta* larva **F** (21); *Spodoptera frugiperda* larva **F** (22)
Anchomenus dorsalis [*Agonum dorsale*]	COLEOPTERA: *Gastrophysa polygoni* eggs, larva **L** (79); DIPTERA: *Contarinia tritici* larva **L** (14); *Delia radicum* eggs **L** (11); *Drosophila melanogaster* pupa **F** (10); HOMOPTERA: *Metopolophium dirhodum* **L** (14); *Rhopalosiphum padi* **L** (14, 43, 103); *Sitobion avenae* **L** (62); LEPIDOPTERA: *Mamestra brassicae* larva **L** (17); *Ostrinia nubilalis* pupa **L** (110)
Anisodactylus californicus	LEPIDOPTERA: *Cydia pomonella* larva **L** (33)
Anisodactylus carbonarius	LEPIDOPTERA: *Pseudaletia unipuncta* larva **L** (20)
Anisodactylus punctatipennis	LEPIDOPTERA: *Plutella xylostella* larva **L** (34)
Anisodactylus sanctaecrucis [*Anisodactylus santaecrucis*]	COLEOPTERA: *Leptinotarsa decemlineata* eggs **L** (72); *Listronotus oregonensis* all stages **L** (5); larva and pupa **L** (36)
Anisodactylus signatus	LEPIDOPTERA: *Plutella xylostella* larva **L** (34)
Anomotarus crudelis	LEPIDOPTERA: *Pieris rapae* egg, larva **L** (38)
Anthracus consputus [*Acupalpus consputus*]	DIPTERA: *Delia radicum* eggs **L** (11)
Asaphidion flavipes	DIPTERA: *Delia radicum* eggs **L** (11); HOMOPTERA: *Sitobion avenae* **L** (62)
Asaphidion semilucidum	LEPIDOPTERA: *Plutella xylostella* larva **L** (34)
Bembidion andreae	DIPTERA: *Delia radicum* eggs **L** (11)
Bembidion articulatum	DIPTERA: *Delia radicum* eggs **L** (11)
Bembidion assimile	DIPTERA: *Delia radicum* eggs **L** (11)
Bembidion bandotaro	LEPIDOPTERA: *Plutella xylostella* larva **L** (34)
Bembidion bimaculatum	LEPIDOPTERA: *Euxoa ochrogaster* eggs **L** (64)
Bembidion canadianum	LEPIDOPTERA: *Euxoa ochrogaster* eggs **L** (64)
Bembidion femoratum	HOMOPTERA: *Aphis fabae* **L** (82)
Bembidion guttula	HOMOPTERA: *Metopolophium dirhodum* **L** (88); *Rhopalosiphum padi* **L** (88)
Bembidion lampros	DIPTERA: *Delia floralis* eggs **L** (81); *Delia radicum* eggs **L** (11); *Drosophila melanogaster* pupa **F** (10); HOMOPTERA: *Aphis fabae* **L** (82); *Myzus persicae* **L** (39); *Metopolophium dirhodum* **L** (88); *Rhopalosiphum padi* **L** (43, 80; 88); *Sitobion avenae* **L** (62)

Table 6.2. continued

Carabid species	Pest species (**L** = laboratory; **F** = field) (reference in brackets; see key)
Adults (continued)	
Bembidion lunatum	DIPTERA: *Delia radicum* eggs **L** (11)
Bembidion morawitzi	LEPIDOPTERA: *Plutella xylostella* larva **L** (34)
Bembidion mutatum	LEPIDOPTERA: *Euxoa ochrogaster* eggs **L** (64)
Bembidion nitidum	DIPTERA: *Delia radicum* eggs **F** (83); LEPIDOPTERA: *Euxoa ochrogaster* eggs **L** (64)
Bembidion obscurellum	LEPIDOPTERA: *Euxoa ochrogaster* eggs **L** (64)
Bembidion obtusum	DIPTERA: *Delia radicum* eggs **L** (11); HOMOPTERA: *Sitobion avenae* **L** (62)
Bembidion quadrimaculatum	COLEOPTERA: *Listronotus oregonensis* all stages **L** (5); larva and pupa **L** (36); *Popillia japonica* eggs **L** (27); DIPTERA:*Delia antiqua* egg, larva **L** (24, 31); *Delia floralis* eggs **L** (81); *Delia radicum* eggs **LF** (11); *Psila rosae* eggs **L** (84); HOMOPTERA: *Aphis fabae* **L** (82); *Sitobion avenae* **L** (62); LEPIDOPTERA: *Agrotis ipsilon* larva **F** (21); *Euxoa ochrogaster* eggs **L** (64); *Ostrinia nubilalis* larva **F** (21); *Papaipema nebris* larva **F** (21); *Pseudaletia unipuncta* larva **F** (21); *Spodoptera frugiperda* larva **F** (22)
Bembidion rupicola	LEPIDOPTERA: *Euxoa ochrogaster* eggs **L** (64)
Bembidion scapulare (=oblongum) [*Bembidion oblongum*]	DIPTERA: *Delia antiqua* egg, larva, pupa **L** (24)
Bembidion tetracolum [*Bembidion ustulatum*]	COLEOPTERA: *Otiorhynchus sulcatus* eggs **L** (67); DIPTERA: *Delia antiqua* egg, larva **L** (24); *Delia radicum* eggs **L** (11); *Drosophila melanogaster* pupa **F** (10); HOMOPTERA: *Aphis fabae* **L** (82); *Rhopalosiphum padi* **L** (43); LEPIDOPTERA: *Mamestra brassicae* eggs, larva **L** (28)
Bembidion versicolor	LEPIDOPTERA: *Euxoa ochrogaster* eggs **L** (64)
Blemus discus [*Trechus discus*]	DIPTERA: *Delia radicum* eggs **L** (11)
Brachinus explodens	LEPIDOPTERA: *Lymantria dispar* pupa **L** (110); *Ostrinia nubilalis* pupa **L** (110)
Bradycellus verbasci	DIPTERA: *Delia radicum* eggs **L** (11)
Calathus fuscipes	DIPTERA: *Delia radicum* eggs **L** (11); HOMOPTERA: *Metopolophium dirhodum* **L** (14); *Rhopalosiphum padi* **L** (14)
Calathus melanocephalus	DIPTERA: *Delia floralis* eggs **L** (81); *Delia radicum* eggs **L** (11); HOMOPTERA: *Rhopalosiphum padi* **L** (14, 43)
Calathus mollis	DIPLOPODA: *Ommatoiulus moreletii* **L** (58)
Calathus ruficollis	LEPIDOPTERA: *Cydia pomonella* larva **L** (33)
Calleida decora	COLEOPTERA: *Leptinotarsa decemlineata* eggs and larvae **L** (2); LEPIDOPTERA: *Anticarsia gemmatalis* eggs and larva **L** (26)
Calleida nilgirensis	LEPIDOPTERA: *Cydia leucostoma* larva **LF** (57, 60)
Calleida splendidula	LEPIDOPTERA: *Opisina arenosella* larva **L** (56)
Calleida viridipennis	LEPIDOPTERA: *Archips argyrospila* larva **F** (95)
Calosoma affine	LEPIDOPTERA: *Colias eurytheme* **L** (19); *Spodoptera exigua* **L** (19); *Spodoptera praefica* **L** (19)
Calosoma calidum	LEPIDOPTERA: *Plathypena scabra* pupa **L** (29)
Calosoma externum	LEPIDOPTERA: *Alabama argillacea* larvae **F** (92)
Calosoma frigidum	MOLLUSCA: *Arion ater* **LF** (66); LEPIDOPTERA: *Choristoneura fumiferana* pupa **F** (42); *Lymantria dispar* larva **L** (15)

Table 6.2. continued

Carabid species	Pest species (**L** = laboratory; **F** = field) (reference in brackets; see key)
Adults (continued)	
Calosoma himalayanum	LEPIDOPTERA: *Lymantria obfuscata* larva **F** (35, 55); *L. obfuscata* pupa **F** (55)
Calosoma maderae	LEPIDOPTERA: *Hyblaea puera* larva **L** (107); *Hyblaea puera* larva, pupa **F** (107); *Lymantria obfuscata* larva **F** (35); *Mythimna separata* larva, pupa **LF** (108)
Calosoma maximowiczi	LEPIDOPTERA: *Hyssia adusta* larva **L** (9)
Calosoma sayi	COLEOPTERA: *Diabrotica undecimpunctata* **L** (18); LEPIDOPTERA: *Alabama argillacea* larvae **F** (92); *Anticarsia gemmatalis* pupa **L** (32); *Helicoverpa zea* larva **L** (18); *Manduca sexta* larva **L** (18); *Spodoptera frugiperda*, all stages, **L** (18)
Calosoma schayeri	LEPIDOPTERA: *Heliothis* spp. larva, **LF** (102)
Calosoma scrutator	LEPIDOPTERA: *Heliothis zea* larvae and adult **F** (92)
Calosoma sycophanta	LEPIDOPTERA: *Lymantria dispar* larva **F** (50, 61)
Carabus amplipennis (=luetgensi) [*Carabus luetgensi*]	DIPLOPODA: *Ommatoiulus moreletii* **L** (58)
Carabus arcensis (Carabus arvensis)	HYMENOPTERA: *Pristiphora abietina* larva, pupa **L** (104)
Carabus auronitens	HYMENOPTERA: *Pristiphora abietina* larva **L** (104)
Carabus brandti	MOLLUSCA: *Bradybaena ravida* **L** (51)
Carabus caschmirensis [*Carabus cashmirensis*]	LEPIDOPTERA: *Lymantria obfuscata* larva, pupa **F** (55)
Carabus hortensis	HYMENOPTERA: *Pristiphora abietina* larva **L** (104)
Carabus hybridus (=impressus) [*Carabus impressus*]	MOLLUSCA: 14 species of landsnail **F** (78)
Carabus lusitanicus	DIPLOPODA: *Ommatoiulus moreletii* **L** (58)
Carabus melancholicus	DIPLOPODA: *Ommatoiulus moreletii* **L** (58)
Carabus nemoralis	LEPIDOPTERA: *Lymantria dispar* larva **L** (15)
Carabus problematicus	MOLLUSCA: slugs **L** (71); HOMOPTERA: *Macrosiphon albifrons* **L** (45)
Carabus rothi (= hampei) [*Carabus hampei*]	COLEOPTERA: *Leptinotarsa decemlineata* **LF** (3)
Carabus rugosus	DIPLOPODA: *Ommatoiulus moreletii* **L** (58)
Carabus serratus	LEPIDOPTERA: *Euxoa ochrogaster* larva **L** (64)
Carabus taedatus	LEPIDOPTERA: *Euxoa ochrogaster* larva **L** (64)
Carabus violaceus	MOLLUSCA: *Arion fasciatus* **L** (53); *Deroceras reticulatum* **L** (53); HOMOPTERA: *Brevicoryne brassicae* **L** (12); *Metopolophium dirhodum* **L** (12); *Rhopalosiphum padi* **L** (12); *Sitobion avenae* **L** (12)
Chlaenius abstersus	LEPIDOPTERA: *Plutella xylostella* larva **L** (34)
Chlaenius bimaculatus (Chlaenius rayotus)	LEPIDOPTERA: *Hyblaea puera* larva, pupa **F** (107)
Chlaenius conformis [*Pachydinodes conformis*]	LEPIDOPTERA: *Anomis flava* larva **L** (54); *Diparopsis watersi* larva **L** (54); *Earias* spp. larvae **L** (54); *Heliothis armigera* larva **L** (54); *Spodoptera littoralis* larva **L** (54); *Syllepte derogata* larva **L** (54); unidentified larvae **F** (54)
Chlaenius fasciger [*Cyaneodinodes fasciger*]	COLEOPTERA: *Mesoplatys ochroptera* eggs, larva **F** (74)

Table 6.2. continued

Carabid species	Pest species (**L** = laboratory; **F** = field) (reference in brackets; see key)
Adults (continued)	
Chlaenius goryi (=venator) [*Lissauchenius venator*]	LEPIDOPTERA: *Anomis flava* larva **L** (54); *Diparopsis watersi* larva **L** (54); *Earias* spp. larvae **L** (54); *Heliothis armigera* larva **L** (54); *Spodoptera littoralis* larva **L** (54); *Syllepte derogata* larva **L** (54); unidentified larvae **F** (54)
Chlaenius micans	LEPIDOPTERA: *Plutella xylostella* larva **L** (34)
Chlaenius pallipes	LEPIDOPTERA: *Plutella xylostella* larva **L** (34)
Chlaenius panagaeoides	HOMOPTERA: *Aphis craccivora* **LF** (108)
Chlaenius posticalis	LEPIDOPTERA: *Plutella xylostella* larva **L** (34)
Chlaenius tomentosus	COLEOPTERA: *Popillia japonica* eggs **L** (27); LEPIDOPTERA: *Agrotis ipsilon* larva **F** (21); *Ostrinia nubilalis* larva **F** (21); *Papaipema nebris* larva **F** (21); *Pseudaletia unipuncta* larva **F** (21); *Spodoptera frugiperda* larva **F** (22)
Chlaenius tricolor [*Chalaenius tricolor*]	COLEOPTERA: *Popillia japonica* eggs **L** (27); LEPIDOPTERA: *Agrotis ipsilon* larva **F** (21); *Ostrinia nubilalis* larva **F** (21); *Papaipema nebris* larva **F** (21); *Pseudaletia unipuncta* larva **F** (21); *Spodoptera frugiperda* larva **F** (22)
Chlaenius virgulifer	LEPIDOPTERA: *Plutella xylostella* larva **L** (34)
Cicindela germanica	HOMOPTERA: cereal aphids **F** (91)
Clivina fossor	COLEOPTERA: *Listronotus oregonensis* egg, larva, pupa **L** (5); DIPTERA: *Delia floralis* eggs **L** (81); HOMOPTERA: *Rhopalosiphum padi* **L** (43)
Colliuris pensylvanica [*Colliuris pennsylvanica*]	LEPIDOPTERA: *Agrotis ipsilon* larva **F** (21); *Ostrinia nubilalis* larva **F** (21); *Papaipema nebris* larva **F** (21); *Pseudaletia unipuncta* larva **F** (21); *Spodoptera frugiperda* larva **F** (22)
Craspedophorus angulatus	LEPIDOPTERA: *Hyblaea puera* larva, pupa **F** (107)
Cratacanthus dubius	COLEOPTERA: *Popillia japonica* eggs, larva **L** (27)
Cychrus attenuatus	MOLLUSCA: slugs, snails **L** (71); COLEOPTERA: Scarabaeidae larvae **L** (71)
Cychrus caraboides	MOLLUSCA: *Arion fasciatus* **L** (53); *Deroceras reticulatum* **L** (53)
Cyclotrachelus alternans [*Evarthrus alternans*]	MOLLUSCA: Limacidae **L** (87); COLEOPTERA: *Diabrotica undecimpunctata* adult **L** (87); LEPIDOPTERA: *Agrotis ipsilon* larva **L** (87); *Plathypena scabra* larva **L** (87)
Cyclotrachelus sodalis [*Evarthrus sodalis*]	COLEOPTERA: *Acalymma vittata* all stages **L** (75); *Diabrotica undecimpunctata* all stages **L** (75; 76); *Hypera postica* adult **L** (25); *Popillia japonica* larva **L** (27); *Sitona hispidulus* adult **L** (25); HETEROPTERA: *Anasa tristis* all stages **L** (75; 77); LEPIDOPTERA: *Colias eurytheme* larva **L** (25); *Melanchra picta* larva **L** (25); *Peridroma saucia* larva **L** (25); *Plathypena scabra* larva **L** (25); *Spodoptera frugiperda* larva **L** (25); *Spodoptera ornithogalli* larva **L** (25)
Cymindis platicollis [*Pinacodera platicollis*]	LEPIDOPTERA: *Lymantria dispar* larva **L** (15)
Creagris labrosa	LEPIDOPTERA: *Opisina arenosella* larva **F** (52)
Demetrias atricapillus	COLEOPTERA: *Gastrophysa polygoni* eggs, larva **L** (79); DIPTERA: *Delia radicum* eggs **L** (11); HOMOPTERA: *Sitobion avenae* **L** (62); *Rhopalosiphum padi* **L** (103)
Dicheirotrichus cognatus [*Trichocellus cognatus*]	LEPIDOPTERA: *Euxoa ochrogaster* eggs **L** (64)
Diplocheila striatopunctata	MOLLUSCA: *Arion ater* **L** (66)

Table 6.2. continued

Carabid species	Pest species (**L** = laboratory; **F** = field) (reference in brackets; see key)
Adults (continued)	
Dolichus halensis	LEPIDOPTERA: *Plutella xylostella* larva **L** (34)
Elaphropus incurvus [*Tachyura incurva; Trechus incurvus*]	DIPTERA: *Delia antiqua* egg, larva **L** (24); *Delia radicum* eggs **F** (83)
Elaphrus cupreus	DIPTERA: *Delia radicum* eggs **L** (11)
Elaphrus riparius	DIPTERA: *Delia radicum* eggs **L** (11)
Graphipterus obsoletus	LEPIDOPTERA: *Anomis flava* larva **L** (54); *Diparopsis watersi* larva **L** (54); *Earias* spp. larvae **L** (54); *Heliothis armigera* larva **L** (54); *Spodoptera littoralis* larva **L** (54); *Syllepte derogata* larva **L** (54); unidentified larvae **F** (54)
Harpalus affinis [*Harpalus aeneus*]	DIPTERA: *Delia floralis* eggs **L** (81); *Delia radicum* eggs **L** (11); HOMOPTERA: *Aphis fabae* **L** (82); *Rhopalosiphum padi* **L** (43); *Sitobion avenae* **L** (62)
Harpalus amputatus	LEPIDOPTERA: *Euxoa ochrogaster* eggs, larva **L** (64)
Harpalus caliginosus	COLEOPTERA: *Popillia japonica* eggs, larva **L** (27); LEPIDOPTERA: *Plathypena scabra* pupa **L** (29)
Harpalus chalcentus	LEPIDOPTERA: *Plutella xylostella* larva **L** (34)
Harpalus eous	LEPIDOPTERA: *Plutella xylostella* larva **L** (34)
Harpalus griseus	LEPIDOPTERA: *Plutella xylostella* larva **L** (34)
Harpalus jureceki	LEPIDOPTERA: *Plutella xylostella* larva **L** (34)
Harpalus pensylvanicus [*Harpalus pennsylvanicus*]	COLEOPTERA: *Diabrotica undecimpunctata* adult **L** (87); *Diabrotica virgifera* adult **F**, eggs, larvae, adults **L** (86); *Hypera postica* adult **L** (25); *Sitona hispidulus* adult **L** (25); HETEROPTERA: *Anasa tristis* all stages **L** (75); HOMOPTERA: *Acyrthosiphon kondoi* **L** (13); *Acyrthosiphon pisum* **L** (13); LEPIDOPTERA: *Agrotis ipsilon* larva **L** (87), **F** (21); *Colias eurytheme* larva **L** (25); *Cydia pomonella* larva **LF** (33); *Melanchra picta* larva **L** (25); *Ostrinia nubilalis* larva **F** (21, 86); *Papaipema nebris* larva **F** (21); *Plathypena scabra* larva **L** (25, 87), pupa **L** (29); *Pseudaletia unipuncta* larva **F** (21); *Spodoptera frugiperda* larva **F** (22)
Harpalus reversus (*=funerarius*) [*Harpalus funerarius*]	LEPIDOPTERA: *Euxoa ochrogaster* eggs, larva **L** (64)
Harpalus rufipes	DIPTERA: *Contarinia tritici* larva **L** (14); *Delia floralis* eggs **L** (81); *Delia radicum* eggs **L** (11); HOMOPTERA: *Aphis fabae* **L** (82); *Brevicoryne brassicae* **L** (12); *Metopolophium dirhodum* **L** (12, 14); *Myzus persicae* **L** (39); *Rhopalosiphum padi* **L** (12, 14, 43, 103); *Sitobion avenae* **L** (12, 62); LEPIDOPTERA: *Mamestra brassicae* larva **L** (17, 28)
Harpalus sinicus	LEPIDOPTERA: *Plutella xylostella* larva **L** (34)
Harpalus tardus	DIPTERA: *Delia radicum* eggs **L** (11)
Harpalus tinctulus	LEPIDOPTERA: *Plutella xylostella* larva **L** (34)
Helluo insignis	LEPIDOPTERA: *Heliothis* spp. pupa, **L** (102)
Lebia analis	COLEOPTERA: *Leptinotarsa decemlineata* eggs and larvae **LF** (2); LEPIDOPTERA: *Heliothis zea* eggs **L** (92)
Lebia grandis	COLEOPTERA: *Leptinotarsa decemlineata* larvae **F** (2); eggs and larvae **L** (69); eggs **L** (72)

Table 6.2. continued

Carabid species	Pest species (**L** = laboratory; **F** = field) (reference in brackets; see key)
Adults (continued)	
Lophoglossus tartaricus [*Pterostichus tartaricus*]	LEPIDOPTERA: *Agrotis ipsilon* larva **F** (22); *Ostrinia nubilalis* larva **F** (22); *Papaipema nebris* larva **F** (22); *Pseudaletia unipuncta* larva **F** (22); *Spodoptera frugiperda* larva **F** (22)
Loricera pilicornis	HOMOPTERA: *Rhopalosiphum padi* **L** (43); *Sitobion avenae* **L** (62)
Megacephala affinis	COLEOPTERA: *Onthophagus marginicollis* adult **L** (90); *Onthophagus landolti* adult **F** (90)
Megacephala carolina	LEPIDOPTERA: *Anticarsia gemmatalis* pupa **L** (32)
Megacephala fulgida	ORTHOPTERA: *Scapteriscus* spp. nymphs **F** (96)
Megacephala virginica	COLEOPTERA: *Popillia japonica* eggs, larva, pupa **L** (27); LEPIDOPTERA: *Anticarsia gemmatalis* pupa **L** (32); *Spodoptera frugiperda* larva **L** (27); Noctuidae adult **F** (92); ORTHOPTERA: *Scapteriscus* spp. nymphs **L** (89)
Megadromus antarcticus	MOLLUSCA: *Deroceras panormitanum* **L** (47); *Deroceras reticulatum* **L** (47)
Nebria brevicollis	COLEOPTERA: *Gastrophysa polygoni* eggs, larva **L** (79); DIPTERA: *Delia radicum* eggs **L** (23); HOMOPTERA: *Brevicoryne brassicae* **L** (12); *Metopolophium dirhodum* **L** (12; 88); *Rhopalosiphum padi* **L** (12; 88); *Sitobion avenae* **L** (12, 62); LEPIDOPTERA: *Operophtera brumata* pupa **L** (63)
Notiophilus biguttatus	HOMOPTERA: *Metopolophium dirhodum* **L** (88); *Rhopalosiphum padi* **L** (88); *Sitobion avenae* **L** (62)
Notonomus gravis	COLEOPTERA: Scarabaeidae larva **L** (68); LEPIDOPTERA: *Pieris rapae* egg, larva, pupa **L** (38); *Pieris* pupa **L** (68)
Notonomus phillippi	COLEOPTERA: Scarabaeidae larva **L** (68); LEPIDOPTERA: *Pieris* pupa **L** (68)
Olisthopus rotundatus	DIPTERA: *Delia radicum* eggs **L** (11)
Omphra pilosa	ISOPTERA: Termitidae all stages **F** (59); termite workers **LF** (108); LEPIDOPTERA: *Hyblaea puera* larva **L** (107); *Hyblaea puera* larva, pupa **F** (107)
Onypterygia fulgens	COLEOPTERA: *Leptinotarsa decemlineata* eggs, larva **L** (37)
Onypterygia tricolor	COLEOPTERA: *Leptinotarsa decemlineata* eggs, larva **L** (37)
Ophionea indica [*Casnoidea indica*]	HOMOPTERA: *Nilaparvata lugens* **L** (98); *Nilaparvata lugens* nymphs, adults **F** (108); *Sogatella furcifera* **LF** (46)
Ophionea nigrofasciata	LEPIDOPTERA: *Rivula atimeta* eggs **L** (38)
Oxylobus dekkanus	ISOPTERA: termite workers **L** (108); LEPIDOPTERA: *Hyblaea puera* larva, pupa **F** (107)
Paranchus albipes [*Agonum albipes*]	DIPTERA: *Delia radicum* eggs **L** (11)
Parena nigrolineata	LEPIDOPTERA: *Hyphantria cunea* larva **F** (100); *Opisina arenosella* larva **LF** (108)
Patrobus atrorufus	HOMOPTERA: *Myzus persicae* **L** (39)
Plochionus timidis	LEPIDOPTERA: *Archips argyrospila* larva **F** (95)
Poecilus chalcites [*Pterostichus chalcites*]	COLEOPTERA: *Diabrotica virgifera* eggs, larvae, adults **L** (85); *Leptinotarsa decemlineata* eggs and larvae **LF** (1); DIPTERA: *Delia antiqua* pupae **L** (7); LEPIDOPTERA: *Agrotis ipsilon* larva **F** (21); *Anticarsia gemmatalis* pupa **L** (32); *Ostrinia nubilalis* larva **F** (21);

Table 6.2. continued

Carabid species	Pest species (**L** = laboratory; **F** = field) (reference in brackets; see key)
Adults (continued)	
	Papaipema nebris larva **F** (21); *Pseudaletia unipuncta* larva **L** (20, 21, 70); *Spodoptera frugiperda* larva **F** (22)
Poecilus corvus [*Pterostichus corvus*]	ORTHOPTERA: *Melanoplus bivittatus* eggs **L** (40)
Poecilus cupreus [*Pterostichus cupreus*]	COLEOPTERA: *Chaetocnema breviuscula* adult **L** (4); Elateridae larva **L** (4); *Oulema melanopus* eggs **L** (109); *Tanymecus palliatus* adult and larva **L** (4); DIPTERA: *Contarinia tritici* larva **L** (14); *Delia radicum* eggs **L** (11); *Drosophila melanogaster* larva **F** (10); *Opomyza florum* pupa **L** (109); *Pegomya betae* larva and pupa **L** (4); *Phorbia securis* pupa **L** (109); HETEROPTERA: *Eurygaster maura* eggs, larva **L** (109); HOMOPTERA: *Metopolophium dirhodum* **L** (14; 88); *Myzus persicae* **L** (39); *Rhopalosiphum padi* **L** (14, 80, 88, 103); *Sitobion avenae* **L** (62, 109); LEPIDOPTERA: *Agrotis segetum* larva **L** (4); *Operophtera brumata* pupa **L** (63); *Ostrinia nubilalis* eggs **L** (109); *Ostrinia nubilalis* pupa **L** (110)
Poecilus cursitor [*Pterostichus cursitor*]	LEPIDOPTERA: *Cydia pomonella* larva **L** (33)
Poecilus lucublandus [*Pterostichus lucublandus*]	MOLLUSCA: Limacidae **L** (87); COLEOPTERA: *Leptinotarsa decemlineata* eggs **L** (72); *Listronotus oregonensis* all stages **L** (5); larva and pupa **L** (36); DIPTERA: *Delia antiqua* egg, larva **L** (24); *Delia antiqua* pupae **L** (7, 24); LEPIDOPTERA: *Agrotis ipsilon* larva **F** (21, 87); *Euxoa ochrogaster* eggs, larva **L** (64); *Ostrinia nubilalis* larva **F** (21); *Papaipema nebris* larva **F** (21); *Plathypena scabra* pupa **L** (29, 87); *Pseudaletia unipuncta* larva **F** (20, 21); *Spodoptera frugiperda* larva **F** (22)
Poecilus versicolor (=planicollis) [*Pterostichus planicollis*]	LEPIDOPTERA: *Plutella xylostella* larva **L** (34)
Pterostichus adstrictus	LEPIDOPTERA: *Choristoneura fumiferana* pupa **F** (42); *Euxoa ochrogaster* larva, pupa **L** (64)
Pterostichus brevipennis	DIPLOPODA: *Ommatoiulus moreletii* **L** (58)
Pterostichus burmeisteri	HYMENOPTERA: *Pristiphora abietina* larva **L** (104)
Pterostichus coracinus	LEPIDOPTERA: *Agrotis ipsilon* larva **F** (21); *Choristoneura fumiferana* pupa **F** (42); *Ostrinia nubilalis* larva **F** (21); *Papaipema nebris* larva **F** (21); *Pseudaletia unipuncta* larva **F** (21)
Pterostichus diligens	DIPTERA: *Delia radicum* eggs **L** (11)
Pterostichus femoralis	ORTHOPTERA: *Melanoplus bivittatus* eggs **L** (40)
Pterostichus haptoderoides	LEPIDOPTERA: *Plutella xylostella* larva **L** (34)
Pterostichus lustrans	LEPIDOPTERA: *Cydia pomonella* larva **L** (33)
Pterostichus madidus [*Feronia madida*]	COLEOPTERA: *Gastrophysa polygoni* larva, adult **L** (79); *Hylobius abietis* larva **L** (93); DIPTERA: *Delia radicum* eggs **L** (11); HOMOPTERA: *Brevicoryne brassicae* **L** (12); *Metopolophium dirhodum* **L** (12; 88); *Rhopalosiphum padi* **L** (12; 88); *Sitobion avenae* **L** (12, 62); LEPIDOPTERA: *Operophtera brumata* pupa **L** (63, 106)
Pterostichus melanarius	MOLLUSCA: *Deroceras reticulatum* **L** (6, 111); COLEOPTERA: *Gastrophysa polygoni* larva, adult **L** (79); *Leptinotarsa decemlineata* eggs **L** (72); *Listronotus oregonensis* all stages **L** (5); larva and pupa **L** (36); DIPTERA: *Contarinia tritici* larva **L** (14); *Delia antiqua* pupae **L** (7); *Delia floralis* eggs **L** (81); *Drosophila melanogaster* larva **F** (10); Tipulidae larvae **L** (8); HOMOPTERA: *Aphis fabae* **L** (82); *Brevicoryne brassicae* **L** (12); *Metopolophium dirhodum* **L** (12, 14); *Rhopalosiphum*

Table 6.2. continued

Carabid species	Pest species (**L** = laboratory; **F** = field) (reference in brackets; see key)
Adults (continued)	
	padi **L** (12, 14, 43); *Sitobion avenae* **L** (12, 62); LEPIDOPTERA: *Mamestra brassicae* larva **L** (17, 28); *Operophtera brumata* pupa **L** (63)
Pterostichus mutus	LEPIDOPTERA: *Lymantria dispar* larva **L** (15)
Pterostichus niger	MOLLUSCA: *Arion fasciatus* **L** (53); *Deroceras reticulatum* **L** (53)
Pterostichus nigrita	DIPTERA: *Delia radicum* eggs **L** (11)
Pterostichus oblongopunctatus	HYMENOPTERA: *Pristiphora abietina* larva **L** (104)
Pterostichus paulinoi [*Pterostichus paulini*]	DIPLOPODA: *Ommatoiulus moreletii* **L** (58)
Pterostichus pensylvanicus	LEPIDOPTERA: *Lymantria dispar* larva **L** (15)
Pterostichus permundus [*Abacidus permundus*]	DIPTERA: *Delia antiqua* pupae **L** (7); LEPIDOPTERA: *Agrotis ipsilon* larva **F** (21); *Ostrinia nubilalis* larva **F** (21); *Papaipema nebris* larva **F** (21); *Pseudaletia unipuncta* larva **F** (21); *Spodoptera frugiperda* larva **F** (22)
Pterostichus quadrifoveolatus	HYMENOPTERA: *Pristiphora abietina* larva **L** (104)
Pterostichus vernalis	DIPTERA: *Drosophila melanogaster* larva **F** (10)
Scaphinotus marginatus	MOLLUSCA: *Deroceras reticulatum* **L** (41)
Scarites cyclops	DIPLOPODA: *Ommatoiulus moreletii* **L** (58)
Scarites substriatus	COLEOPTERA: *Diabrotica undecimpunctata* adult **L** (87); *Popillia japonica* eggs, larva **L** (27); LEPIDOPTERA: *Agrotis ipsilon* larva **F** (87); *Plathypena scabra* pupa **L** (87); *Pseudaletia unipuncta* larva **L** (70)
Scarites subterraneus	COLEOPTERA: *Popillia japonica* eggs, larva **L** (27); LEPIDOPTERA: *Agrotis ipsilon* larva **F** (21); *Ostrinia nubilalis* larva **F** (21); *Papaipema nebris* larva **F** (21); *Pseudaletia unipuncta* larva **F** (20, 21); *Spodoptera frugiperda* larva **F** (22)
Scarites terricola	LEPIDOPTERA: *Plutella xylostella* larva **L** (34)
Simodontus australis	LEPIDOPTERA: *Perthida glyphopa* larva **F** (97)
Sphaeroderus stenostomus lecontei [Sphaeroderus lecontei]	LEPIDOPTERA: *Lymantria dispar* larva **L** (15)
Stenolophus lecontei [*Agonoderus lecontei*]	DIPTERA: *Delia radicum* eggs **F** (83)
Stenolophus mixtus	DIPTERA: *Delia radicum* eggs **L** (11)
Synuchus vivalis (=nivalis) [*Synuchus nivalis*]	DIPTERA: *Delia radicum* eggs **L** (11)
Tanystoma maculicolle	LEPIDOPTERA: *Cydia pomonella* larva **L** (33)
Trechus quadristriatus	DIPTERA: *Contarinia tritici* larva **L** (14); *Delia floralis* eggs **L** (81); *Delia radicum* eggs **L** (11); *Psila rosae* eggs **LF** (84); HOMOPTERA: *Aphis fabae* **L** (82); *Metopolophium dirhodum* **L** (14; 88); *Myzus persicae* **L** (39); *Rhopalosiphum padi* **L** (14; 88); *Sitobion avenae* **L** (62)
Trechus secalis	HOMOPTERA: *Rhopalosiphum padi* **L** (43)
Carabid species (larvae)	
Amara spp.	HOMOPTERA: *Sitobion avenae* **L** (62)
Anchomenus dorsalis [*Agonum dorsale*]	HOMOPTERA: *Sitobion avenae* **L** (62)

Table 6.2. continued

Carabid species	Pest species (**L** = laboratory; **F** = field) (reference in brackets; see key)
Larvae (continued)	
Bembidion tetacolum	HOMOPTERA: *Sitobion avenae* **L** (99)
Calleida decora	LEPIDOPTERA: *Anticarsia gemmatalis* larva **F** (30)
Calleida nilgirensis	LEPIDOPTERA: *Cydia leucostoma* larva **L** (60)
Calleida splendidula	LEPIDOPTERA: *Opisina arenosella* larva **L** (56)
Calleida viridipennis	LEPIDOPTERA: *Archips argyrospila* larva **F** (95)
Calosoma himalayanum	LEPIDOPTERA: *Lymantria obfuscata* larva **F** (35, 55); *L. obfuscata* pupa **F** (55)
Calosoma maderae	LEPIDOPTERA: *Hyblaea puera* larva, pupa **F** (107); *Mythimna separata* larva, pupa **F** (108)
Calosoma sayi	LEPIDOPTERA: *Anticarsia gemmatalis* pupa **L** (32)
Calosoma sycophanta	LEPIDOPTERA: *Lymantria dispar* larva and pupa **L** (16); all stages **F** (50)
Carabus caschmirensis [*Carabus cashmirensis*]	LEPIDOPTERA: *Lymantria obfuscata* larva, pupa **F** (55)
Carabus hybridus (= *impressus*) [*Carabus impressus*]	MOLLUSCA: *Eopolita protensa* **F** (78); *Monacha haifaensis* **F** (78); *Theba pisana* **F** (78); *Xeropicta vestalis* **F** (78)
Chlaenius bimaculatus (*Chlaenius rayotus*)	LEPIDOPTERA: *Hyblaea puera* larva, pupa **F** (107)
Chlaenius conformis [*Pachydinodes conformis*]	LEPIDOPTERA: *Anomis flava* larva **L** (54); *Diparopsis watersi* larva **L** (54); *Earias* spp. larvae **L** (54); *Heliothis armigera* larva **L** (54); *Spodoptera littoralis* larva **L** (54); *Syllepte derogata* larva **L** (54); unidentified larvae **F** (54)
Chlaenius fasciger [*Cyaneodinodes fasciger*]	COLEOPTERA: *Mesoplatys ochroptera* eggs, larva **F** (74)
Chlaenius goryi (=*venator*) [*Lissauchenius venator*]	LEPIDOPTERA: *Anomis flava* larva **L** (54); *Diparopsis watersi* larva **L** (54); *Earias* spp. larvae **L** (54); *Heliothis armigera* larva **L** (54); *Spodoptera littoralis* larva **L** (54); *Syllepte derogata* larva **L** (54); unidentified larvae **F** (54)
Chlaenius micans	LEPIDOPTERA: *Plutella xylostella* larva **L** (34)
Chlaenius posticalis	LEPIDOPTERA: *Cnaphalocrocis medinalis* larva **F** (44); *Plutella xylostella* larva **L** (34)
Craspedophorus angulatus	LEPIDOPTERA: *Hyblaea puera* larva, pupa **F** (107)
Demetrias atricapillus	HOMOPTERA: *Sitobion avenae* **L** (62)
Graphipterus obsoletus	LEPIDOPTERA: *Anomis flava* larva **L** (54); *Diparopsis watersi* larva **L** (54); *Earias* spp. larvae **L** (54); *Heliothis armigera* larva **L** (54); *Spodoptera littoralis* larva **L** (54); *Syllepte derogata* larva **L** (54); unidentified larvae **F** (54)
Harpalus pensylvanicus	COLEOPTERA: *Diabrotica virgifera* eggs, larvae **L** (86)
Harpalus spp.	COLEOPTERA: *Diabrotica undecimpunctata* larva **F** (48)
Lachnocrepis japonica	LEPIDOPTERA: *Chilo suppressalis* larva **F** (101)
Lebia analis	COLEOPTERA: *Ceratoma trifurcata* larvae, pupae **L** (92); *Diabrotica undecimpunctata* larvae, pupae **L** (92)
Nebria brevicollis	HOMOPTERA: *Sitobion avenae* **L** (62)
Notiophilus biguttatus	HOMOPTERA: *Sitobion avenae* **L** (62)
Omphra pilosa	ISOPTERA: Termitidae all stages **F** (59); termites **LF** (108); LEPIDOPTERA: *Hyblaea puera* larva, pupa **F** (107)

Table 6.2. continued

Carabid species	Pest species (**L** = laboratory; **F** = field) (reference in brackets; see key)
Larvae (continued)	
Ophionea indica	DIPTERA: *Orseolia oryzae* pupae **F** (94)
Oxylobus dekkanus	LEPIDOPTERA: *Hyblaea puera* larva, pupa **F** (107)
Parena laesipennis	LEPIDOPTERA: *Pidorus glaucopis* larva **LF** (49)
Parena nigrolineata	LEPIDOPTERA: *Opisina arenosella* larva **LF** (108)
Poecilus cupreus	HOMOPTERA: *Sitobion avenae* **L** (62)
Plochionus timidis	LEPIDOPTERA: *Archips argyrospila* larva **F** (95)
Pterostichus spp.	COLEOPTERA: *Diabrotica undecimpunctata* larva **F** (48)
Unspecified	TRICLADIDA: *Artioposthia triangulata* **L** (105); COLEOPTERA: *Popillia japonica* eggs **L** (27); LEPIDOPTERA: *Agrotis ipsilon* larva **F** (21); *Ostrinia nubilalis* larva **F** (21); *Papaipema nebris* larva **F** (21); *Pseudaletia unipuncta* larva **F** (21); *Spodoptera frugiperda* larva **F** (22)

Key to Table 6.2: O = visual observation; T = tethered prey (artificial prey, sentinel prey, placed-out prey, baits); V = video recording
1: Heimpel and Hough-Goldstein (1992), O, potato; 2: Hilbeck and Kennedy (1996), O, potato; 3: Koval (1985), O, potato; 4: Berim and Novikov (1983), O; 5: Baines *et al.* (1990), O; 6: Symondson and Liddell (1996), O; 7: Menalled *et al.* (1999), O; 8: Chapman (1994), O; 9: Gong *et al.* (1988), O; 10: Lys (1995), OVT, wheat; 11: Finch (1996), O; 12: Kielty *et al.* (1999), O; 13: Losey and Denno (1998a), O; 14: Lübke-Al Hussein and Tritsch (1994), O; 15: Johnson and Reeves (1995), O; 16: Weseloh (1988), O; 17: Vasconcelos *et al.* (1996), O; 18: Young (1984), O; 19: Wallin (1991), O; 20: Clark *et al.* (1994), O; 21: Brust *et al.* (1986a), OT, maize; 22: Brust *et al.* (1986b), OT, maize; 23: Finch and Elliott (1992), O; 24: Tomlin *et al.* (1985), O; 25: Barney and Pass (1986), O; 26: Fuller (1988), O; 27: Terry *et al.* (1993), O; 28: Johansen (1997), O; 29: Bechinski *et al.* (1983), O; 30: Godfrey *et al.* (1989), O, soybean; 31: Grafius and Warner (1989), O; 32: Lee *et al.* (1990), O; 33: Riddick and Mills (1994), OT, apple; 34: Suenaga and Hamamura (1998), O; 35: Dharmadhikari *et al.* (1985), O, forest; 36: Zhao *et al.* (1990), O; 37: Cappaert *et al.* (1991), O; 38: Van den Berg *et al.* (1992), O; 39: Landis and Van der Werf (1997), O; 40: Songa and Holliday (1997), O; 41: Digweed (1993), O; 42: Kelly and Régnière (1985), O, forest; 43: Andersen (1992), O; 44: Barrion and Litsinger (1985), O, rice; 45: Wink (1986), O; 46: Singh *et al.* (1998), O, rice; 47: Chapman *et al.* (1997), O; 48: Brust (1991), T, maize; 49: Mochizuki (1990), O, trees; 50: Constantineanu and Constantineanu (1996), O, forest; 51: Chen *et al.* (1996), O; 52: Pillai and Nair (1986), O, coconut; 53: Pakarinen (1994), O; 54: Deguine (1991), O, cotton; 55: Rishi and Shah (1985), O, forest, orchard; 56: Pillai and Nair (1990), O; 57: Selvasundaram and Muraleedharan (1987), O, tea; 58: Baker (1985), O; 59: Kumar and Rajagopal (1990), O, maize, orchard, forest; 60: Muraleedharan *et al.* (1991), O; 61: Ferrero (1985), O, forest; 62: Sunderland *et al.* (1987), O; 63: Frank (1967b), O; 64: Frank (1971), O; 65: Kapuge *et al.* (1987), O; 66: Poulin and O'Neil (1969), O; 67: Evenhuis (1983), O; 68: Horne (1992), O; 69: Hough-Goldstein *et al.* (1993), O; 70: Laub and Luna (1992), O; 71: Loreau (1984), O, 72: Hazzard *et al.* (1991), O; 73: Symondson (1989), O; 74: Sileshi *et al.* (2001), O, sesbania; 75: Snyder and Wise (1999), O; 76: Snyder and Wise (2000), O; 77: Snyder and Wise (2001), O; 78: Mienis (1988), O; 79: Sotherton (1982), O; 80: Chiverton (1988), V; 81: Andersen *et al.* (1983), O; 82: Dunning *et al.* (1975), O; 83: Wishart *et al.* (1956), O, cabbage; 84: Burn (1982), O, carrot; 85: Kirk (1975), O; 86: Kirk (1973), O, maize; 87: Best and Beegle (1977), O; 88: Mundy *et al.* (2000), O; 89: Hudson *et al.* (1988), O; 90: Young (1980), O, pasture; 91: Lövei and Szentkirályi (1984), O, maize; 92: Whitcomb and Bell (1964), O, cotton; 93: Salisbury and Leather (1998), O; 94: Kobayashi *et al.* (1995), O, rice; 95: Braun *et al.* (1990), O, forest; 96: Fowler (1987), O, pasture; 97: Mazanec (1987), O, forest; 98: Luong (1987), O; 99: Theiss and Heimbach (1993), O; 100: Habu *et al.* (1963) in Ito and Miyashita (1968), O, forest; 101: Ito *et al.* (1962), O, rice; 102: Room (1979), O, cotton; 103: Asín and Pons (1998), O; 104: Schmied and Fuhrer (1996), O; 105: Gibson *et al.*. (1997), O; 106: East (1974), O; 107: Loganathan and David (1999), O, forest; 108: Rajagopal and Kumar (1992), O, maize, forest, cowpea, coconut, rice; 109: Malschi and Mustea (1995), O; 110: Lövei (1986), O; 111: McKemey *et al.* (2001), O; 112: Symondson (1993), O, protected lettuce.

Observations of predation on pests by 36 species of carabid larvae are summarized in *Table 6.2*. In the field, larvae were observed to kill snails (e.g. *Monacha haifaensis*), beetle eggs (e.g. *Mesoplatys ochroptera*), rootworms (e.g. *Diabrotica undecimpunctata*), caterpillars (e.g. *Anticarsia gemmatalis*), stem borers (e.g. *Chilo suppressalis*), moth pupae (e.g. *Lymantria obfuscata*), rice gall midge pupae (e.g. *Orseolia oryzae*) and termites (e.g. *Odontotermes wallonensis*). Larvae of *Ophionea indica* were found to bore holes into the galls produced by the rice gall midge (*Orseolia oryzae*) and attack the midge pupa within (Kobayashi *et al.*, 1995).

Characteristic damage showed that larvae of *Calleida viridipennis* and *Plochionus timidis* had entered silken webs of the leafroller, *Archips argyrospila*, and attacked the caterpillars inside 17% of the 1772 webs examined (Braun *et al.*, 1990). Similarly, Habu *et al.* (1963) (reported in Ito and Miyashita, 1968) observed *Parena nigrolineata* feeding on larvae of *Hyphantria cunea* in its communal web on trees. There were insufficient records to attempt generalizations about the relative degree of polyphagy of different species of carabid larvae.

Laboratory-based observations of predation, especially by hungry beetles in small Petri dishes (e.g. Room, 1979; Sunderland *et al.*, 1987; Cappaert *et al.*, 1991; Gibson *et al.*, 1997), determines only whether the beetle is capable of killing the pest in question. As the degree of realism of the test system increases, from laboratory arenas containing a small number of plants (e.g. Zhao *et al.*, 1990; Suenaga and Hamamura, 1998), through larger-scale insectary experiments (Sopp *et al.*, 1992), to observations in field cages (e.g. Van den Berg *et al.*, 1992), so it becomes progressively safer to accept the results as an indication of what is likely to happen under undisturbed field conditions. Observations under semi-natural conditions created in the laboratory can enable researchers to gain valuable insights into how carabids forage for pests. For example, Chiverton (1988) filmed the responses of two carabid beetles to the aphid, *Rhopalosiphum padi*, in a laboratory arena containing spring barley plants. *Bembidion lampros* was observed to be a non-climber, but found aphids (apparently at random) walking on the soil surface. However, when *Poecilus cupreus* [*Pterostichus cupreus*] discovered an aphid colony, it climbed and searched the plant and neighbouring plants. Subsequent individuals always climbed the same plants where previous carabids had found aphids, suggesting a response to olfactory cues left by searching carabids or to aphids or aphid products. *Harpalus rufipes* and *Pterostichus melanarius* have been shown to respond to aphid alarm pheromone in a continuous-airflow olfactometer (Kielty *et al.,* 1996), and more carabids were caught in plots in a wheat field where alarm pheromone (E-beta-farnesene) was being dispensed, compared with control plots (Kirkland *et al.*, 1998). This would make aphid predation non-random and more efficient (but see under *Trophic generalists* in the next section). Laboratory experiments by Mundy *et al.* (2000) showed, however, that *P. cupreus* was less likely to climb aphid-infested plants if entomobryid Collembola were present on the ground, and that this carabid did not respond to the odour of *R. padi* when the aphids were undisturbed (and so unlikely to be emitting alarm pheromone).

In simple Petri dish tests, pests (including species and life stages that gain safety in cryptic locations in the field) are fully exposed, which can produce misleading results. For example, onion rootfly eggs (*Delia antiqua*) on the surface of Petri dishes were 70% predated by *Bembidion quadrimaculatum* compared with only 18% when eggs were buried one centimetre under soil. In onion fields, only 13–43% of onion fly eggs are oviposited on or near the soil surface (Grafius and Warner, 1989). Similarly, Zhao *et al.* (1990) reported *B. quadrimaculatum* to be an excellent predator of carrot weevil eggs (*Listronotus oregonensis*) in Petri dishes (Baines, 1987 cited in Zhao *et al.*, 1990), but this beetle failed to remove eggs from their oviposition sites on the petiole and crown of carrots. However, cryptic locations do not protect pests from all carabids. *Poecilus corvus* [*Pterostichus corvus*], for example, ate twice as many grasshopper eggs buried in soil than exposed on the soil surface (Songa and Holliday, 1997), and the larvae of *Calosoma sayi* [*Calosoma alternans sayi*] consumed pupae of *Anticarsia gemmatalis* below ground, as well as on the soil surface (Lee *et al.*, 1990). Occasionally, relationships are discovered that hold for a range of carabid species. Wheater (1988), for example, found a positive correlation between the mandible gape of carabids and the size of pest attacked, and Finch (1996) demonstrated that the number of cabbage root fly eggs (*Delia radicum*) predated was related positively to the length of carabid beetles in the range 3–10 mm.

Pest individuals can be placed out in the field so that they can be observed for long periods to determine which species of carabid will attack them, but to render this method feasible it is sometimes necessary to restrain the movement of the pest. In an apple orchard, Riddick and Mills (1994) tethered codling moth larvae (*Cydia pomonella*) around the abdominal segments using tough black cotton thread. Small carabids attacked the larvae but were often unable to kill them, whereas large carabids killed and ate a larva within thirty minutes. Godfrey *et al.* (1989) were able to observe velvetbean caterpillars (*Anticarsia gemmatalis*) placed on soybean foliage (and on the ground below) without the need to restrain their movement. The majority of small caterpillars on the foliage were consumed by larvae of the carabid, *Calleida decora*. Researchers have devised innovative solutions to the problem of observing predation on cryptic pests in the field. For example, Brust (1991) observed underground predation of larvae of southern corn rootworm (*Diabrotica undecimpunctata*) on maize roots. This was done by putting a plexiglass plate close to the roots and having a bag of soil that could be removed to allow the observer to watch (which was done at 2 h intervals for 24 h). This method showed larvae of *Harpalus* and *Pterostichus* to be significant predators of first to third instar rootworm larvae. Records of predation are sometimes gleaned as a bonus during routine inspection of crop plants to count pest numbers (e.g. Hilbeck and Kennedy, 1996), or whilst manipulating pest density in field cages (e.g. Kelly and Régnière, 1985), but they may also come from planned, intensive, long-term observations of the activity of predators (e.g. Griffiths *et al.*, 1985; Contantineanu and Contantineanu, 1996). From observations alone, however, it is rarely possible to do more than guess at the impact that carabids may be making on a pest population. Other methods, such as modelling and manipulation of carabid density in the field, are usually necessary to investigate impact on pests, and these are reviewed in the next section.

The contribution of carabids to pest control

CARABIDS ACTING ALONE

Some species of carabid have a very restricted diet, and can be considered as trophic specialists. A few such species specialize in feeding on pests. Trophic specialists and trophic generalists are considered separately (below) because the mechanisms of interaction with pest populations differ.

Trophic specialists

Specialists probably select their prey before capture, generalists after capture. Generalists may, to a limited extent, become more selective due to the development of aversions, but prey sampling is needed before aversions can develop, and thus generalists have a wider prey spectrum than specialists (Toft and Bilde, this volume). Carabid species are known that are specialist feeders on Mollusca, Isoptera, Lepidoptera and Orthoptera.

Cychrus and *Scaphinotus* have a body shape and modified mouthparts that enable them to extract snails from their shells, and some other carabid genera (e.g. *Procerus*) have mandibles that are large and strong enough to permit them to crush the shell to gain access to the snail inside. Although these beetles are known to consume a range of different foods, it seems that snails and slugs are their main prey. There is evidence that they can detect and follow mucus trails and also avoid slug defences, such as the enhanced production of especially viscous mucus. The current state of knowledge of this subject has been summarized recently by Port *et al.* (2001) and reviewed extensively by Symondson (in press). Data appear to be unavailable concerning the effectiveness of biological control by mollusc specialists acting alone. Altieri *et al.* (1982) recorded a significant level of mollusc reduction due to release of *Scaphinotus stratiopunctatus* into a commercial daisy field. Metal sheets were placed in the field as beetle refuges but these shelters attracted garter snakes, which also feed on molluscs, and the relative contribution of these two predators to mollusc control was not determined.

Omphra pilosa adults and larvae (common in maize, plantations and orchards in India) were not seen to eat anything but termites in the field. Gut dissection of adults revealed only termite remains, and the beetles could be reared on an exclusively termite diet in the laboratory. Impact on termite populations has not been investigated (Kumar and Rajagopal, 1990).

Calosoma sycophanta is capable of feeding on a range of Lepidoptera, but its abundance in the USA (imported from Europe in 1906) is closely coupled to that of a forest pest, the gypsy moth (*Lymantria dispar*). Beetles are active for only about a month when caterpillars and pupae are present, and they will only reproduce if gypsy moth is abundant. Hence, in the USA, this species is effectively a specialist predator (Weseloh *et al.*, 1995). *C. sycophanta* is a tenacious and efficient forager. Ferrero (1985) observed a single individual adult to climb a tree, catch a *L. dispar* caterpillar, fall with it to the ground and partially consume it, then re-climb the tree and repeat the process five times. Larvae of *C. sycophanta* destroyed a mean of 40% of *L. dispar* pupae in a year, but it is not known whether the beetle can prevent significant defoliation occurring (Weseloh, 1985). Releases of adult beetles increased gypsy moth pupal mortality in one year (Weseloh, 1990), but failed to do so in a later year (Weseloh *et al.*, 1995). Lepidoptera specialists are not effective at preventing all outbreaks of a pest species, but can make a considerable contribution to curtailing them and to reducing the likelihood of recurrence in the following year. For example, *Calosoma maximowiczi* has a strong positive numerical response to the beech-defoliating caterpillar, *Quadricalcarifera punctatella*, in Japan. No beetles were caught when caterpillar density was low, but beetles increased greatly in outbreak years as a result of aggregative and reproductive numerical responses. Predation rates (in the laboratory), taken together with relative densities of carabids and caterpillars in the field, suggested that the carabid makes an important contribution to suppressing pest density during an outbreak (Kamata and Igarashi, 1995).

Adults of *Megacephala fulgida* are attracted to synthesized songs of mole crickets (*Scapteriscus* spp.) at sound traps in Brazil and Uruguay (Guido and Fowler, 1988).

This tiger beetle is found commonly in mole cricket burrows, has been seen to dig out and eat nymphal mole crickets, and displays numerical and functional responses to its prey (Fowler, 1987). Mole cricket populations are sparser in South America than in USA, where they are an exotic pest (Guido and Fowler, 1988). It is possible that natural enemies, such as *M. fulgida,* inhibit *Scapteriscus* spp. population increase in South America.

Trophic generalists

Superfluous killing (also called wasteful killing), i.e. where more pests are killed than are consumed, is a valuable trait for a natural enemy to have. It can occur in generalist carabids, e.g. *Trechus quadristriatus* attacking cecidomyiid larvae, *Dasineura brassicae*, in the laboratory (Wolf-Schwerin, 1993, reported in Ekschmitt *et al.,* 1997), but data on its incidence in the field appear to be lacking. It is unlikely to occur at low pest densities, yet it is at low pest density early in the season (when predator to prey ratios are high) that generalist predators are most likely to make an impact on pest populations. Their ability to subsist on early-season alternative prey, before the pest arrives, is a positive feature of generalist carabids in relation to their suitability for pest control (Edwards *et al.*, 1979; Helenius, 1990; Laub and Luna, 1992; Landis and Van der Werf, 1997).

Although trophic generalists are less likely than trophic specialists to detect pests at a distance, they may, nevertheless, have a reduced emigration rate from randomly-found patches of high pest density compared with areas of low pest density. This would result in an aggregative numerical response to pest density (by a process analagous to that described for ladybirds by Ives *et al.*, 1993) that might improve their efficiency as biocontrol agents. This matter has been investigated in relation to the response of carabids to aphid population density and distribution in wheat. Bryan and Wratten (1984) caught more carabids in artificially-created patches of high aphid density, and Halsall and Wratten (1988a,b) used video monitoring to show that four carabid species entered aphid-rich patches more frequently than patches with lower aphid density. Monsrud and Toft (1999) found, however, that carabids (especially *Anchomenus dorsalis* [*Agonum dorsale*]) were not attracted directly to aphids in the field, but rather to honeydew produced by aphids (which they simulated by spraying honey solution) or to alternative prey types (such as Diptera) that were attracted to honeydew. Furthermore, Winder *et al.* (1997) modelled foraging by *A. dorsalis* and concluded that this beetle would not detect patches of the cereal aphid *Sitobion avenae* at aphid densities of 0.1–5.0 aphids per shoot. In the Bryan and Wratten (1984) experiment, artificial aphid infestations were at densities up to 707 times higher than controls, and Winder *et al.* (1997) suggested that patch responses by carabids under these conditions are artificial and not representative of realistic field situations. Available data suggest, therefore, that carabids do not respond to aphid density *per se*, but may aggregate in areas where honeydew and honeydew-feeding Diptera are available. This would increase, indirectly, their encounter rate with falling and dislodged aphids, and stimulate plant climbing by some species of carabid (see earlier section 'Evidence that carabids can kill pests'). Toft and Bilde (this volume) have shown that aphids are supplementary food that cannot usually be consumed in high quantities by individual beetles, and that cannot alone support carabid populations.

Table 6.3. Examples of studies where impact on pest populations by carabids (alone) has been investigated. In the 'Impact type' column, the maximum reduction and pest growth stage (for holometabolous species) affected are shown. An asterisk in the 'Authors' column denotes that differences between treated and control plots were tested statistically. Where carabid names have been updated (by Dr R.L. Davidson, Carnegie Museum of Natural History), the original name used in the cited publication(s) is also shown [in square brackets].

Pest species	Crop	Carabid species	Method (see key)	Impact type (see key)	Authors (see key)
MOLLUSCA					
Deroceras panormitanum	Lettuce	*Megadromus antarcticus*	B, D	N (50%)	6*
Deroceras reticulatum	Lettuce	*Abax parallelepipedus*	A, B in polyhouse	N (78%), D	10*
Deroceras reticulatum	Grass/clover	*Abax parallelepipedus* *Pterostichus madidus*	B, D in greenhouse	N (100%), Y	4*
Deroceras reticulatum	Grass	*Plocamostethus planiusculus*, *Holcapsis mucronata* *[Holeaspis mucranata]*, *Ctenognathus bidens*	C, D	N (74%)	5*
COLEOPTERA					
Sitona lineatus	Beans	*Pterostichus madidus* and others	C, radiotracers, calculation	N (adults 24%) N (larvae 11%)	8
DIPTERA					
Delia antiqua	Onion	*Bembidion quadrimaculatum*	C, D	N (larvae 57%)	14*
Delia radicum	Cabbage	*Bembidion lampros* & others	B, D	N (eggs 52%)	20
Sitodiplosis mosellana	Wheat	14 species (see paper)	serology, calculation	N (larvae 11%)	12
HOMOPTERA					
Aphis fabae	Sugar beet	*Asaphidion flavipes*, *Anchomenus dorsalis* *[Agonum dorsale]*, *Pterostichus melanarius*	C, D	N (up to 100%)	9
Aphididae (unidentified)	Barley, ley	*Poecilus cupreus* *[Pterostichus cupreus]*	B, D	N (NS)	19*

continued

Table 6.3. continued

LEPIDOPTERA					
Anticarsia gemmatalis	Soybean	*Calleida decora*	C, D	N (larvae 58%)	13*
Cydia pomonella	Apple	*Agonum punctiforme, Calathus ruficollis, Harpalus pensylvanicus*	tethered prey, calculation	N (larvae 60%)	17
Geometridae	Orchard	*Calosoma inquisitor, Calosoma denticolle [Calosoma denticolor]*	pupal examination	N (pupae 20%)	3
Lymantria dispar	Forest	*Calosoma sycophanta*	pupal examination	N (pupae 40%)	1
Lymantria dispar	Forest	*Calosoma sycophanta*	A, pupal examination	N (larvae 12%) N (pupae 35%)	18*
Lymantria dispar	Forest	*Calosoma sycophanta* larva	pupal examination	N (pupae 31%)	7
Pidorus glaucopis	Amenity trees	*Parena laesipennis* larva	field counts and correlation	N (larvae)	2
Plathypena scabra	Soybean	*Harpalus pensylvanicus [Harpalus pennsylvanicus], H. caliginosus, Poecilus lucublandus [Pterostichus lucublandus]*	B, C pupal examination	N (pupae 5%)	16
Rivula atimeta	Rice	*Ophionea nigrofasciata*	C	No reduction (eggs)	11
Spodoptera litura	Tomato	*Megadromus antarcticus*	A, C in greenhouse	N (larvae 49%) D (larvae 52%)	15*

Key toTable 6.3: *Method*: A = augmentation of carabid density (above natural field density); B = barrier-enclosed plots; C = field cages; D = manipulation of carabid density; *Impact type* (and maximum percentage change compared to controls): N = pest numbers reduced; D = plant damage reduced; Y = yield increased; NS = no signi ficant difference; *Authors*: 1: Weseloh (1985); 2: Mochizuki (1990); 3: Lapa and Tkachev (1992); 4: Asteraki (1993); 5: Barker (1991); 6: Chapman *et al.* (1997); 7: Fuester *et al.* (1997); 8: Hamon *et al.* (1990); 9: Hance (1987); 10: Symondson (1993); 11: Van den Berg *et al.* (1992); 12: Floate *et al.* (1990); 13: Fuller (1988); 14: Grafius and Warner (1989); 15: Hodge *et al.* (1999); 16: Bechinski *et al.* (1983); 17: Riddick and Mills (1994); 18: Weseloh (1990); 19: Bommarco (1999); 20: Ryan and Ryan (1973).

The densities of carabid populations, and consequently their potential impact on aphid populations, depend on the presence of high-quality alternative prey, such as Diptera. Aggregation in honeydew-contaminated patches could, therefore, provide carabids with the Diptera prey that would enable them to maintain a steady (if low-level) predation pressure on aphid populations. There appears to be a lack of data from realistic manipulative field studies on the contribution that carabid populations (alone) make to aphid control. Significant aphid reductions have been recorded in sealed field cages in sugar beet (Hance, 1987) and small barriered plots in alfalfa (Losey and Denno, 1998a), which were cleared of all invertebrates and then seeded with known numbers of aphids and carabids. However, in these experimental designs the movements of predator and prey were constrained and the predator had no access to alternative prey. Losey and Denno (1998a) also recorded aphid suppression in open field plots which they attributed mostly to the carabid *Harpalus pensylvanicus*, but this assessment was based on pitfall captures, which are known to give a misleading impression of the relative abundance of various predator groups (Sunderland *et al.*, 1995a).

Carabid impact on pests is sometimes inferred from predator–pest associations. For example, Speight and Lawton (1976) found a significant positive relationship between the number of freeze-killed *Drosophila* pupae (artificial prey placed on the ground) consumed in ten areas of variable weediness in a wheat field and the number of ground beetles caught in pitfalls in these ten areas. Chiverton (1986) manipulated abundance of all predators in barriered plots in spring barley and found significant negative relationships between peak levels of the aphid, *Rhopalosiphum padi*, and abundance (in pitfall traps) of *Bembidion quadrimaculatum* and other *Bembidion* species. Similarly, Menalled *et al.* (1999) reported a significant positive relationship between the disappearance rate of freeze-killed pupae of onion fly, *Delia antiqua*, placed on the ground in maize, and abundance of the four commonest carabid species (*Pterostichus melanarius*, *Poecilus chalcites* [*Pterostichus chalcites*], *Poecilus lucublandus* [*Pterostichus lucublandus*], *Pterostichus permundus*) in plots where predator density was manipulated. Carabids (two species of *Ophionea*) were, however, a non-significant term in a multiple linear regression for rice leaffolder (*Cnaphalocrocis medinalis* and *Marasmia patnalis*) egg disappearance in irrigated rice fields (De Kraker *et al.*, 1999). Carabids have a patchy distribution in wheat fields and the position of these patches changes dynamically through time, within a season (Holland *et al.*, 1999, 2000). Bohan *et al.* (2000) found that the spatial distribution of *Pterostichus melanarius* in a field of wheat was dynamically associated with the distribution of slugs (seven species, but predominantly *Deroceras reticulatum* and *Arion intermedius*). In June, there was a significant positive relationship between the abundance of beetles that had been feeding on slugs (determined by ELISA) and the abundance of slugs, but this relationship became negative in July. The authors found no evidence that these correlations could be explained by slugs and carabids reacting to common factors (related to crop or environment), and they considered it likely that carabids aggregated to patches of high slug density and then reduced slug abundance in those patches to a significant degree by predation. Furthermore, later studies (Symondson *et al.*, 2002) showed that growth of the *P. melanarius* population was strongly related to slug density in the previous year. Using the method of Spatial Analysis by Distance Indices (SADIE), it has been demonstrated (Winder *et al.*, 2001) that *P. melanarius* responds to patches of the aphid *Metopolophium dirhodum* in wheat in a manner analogous to that described above for slugs, i.e. the beetles appear to be dynamically exploiting high-density patches of aphids. Significant negative regression relationships were obtained between the density of *P. melanarius* and the rate of increase of the aphids, *M. dirhodum* and *Sitobion avenae*, during the early stages of population development. It is surprising that this extremely polyphagous beetle (see first two sections) should respond in this way to single prey species, but perhaps the explanation is that: a) in regions where slugs are abundant they may dominate the biomass of prey available to *P. melanarius* in fields; and b) in high-density patches of aphids there will be an abundance not only of aphids, but also of honeydew-feeding prey (Monsrud and Toft, 1999) and generalist predators (which are also consumed by *P. melanarius* as a top predator – Sunderland, 1975; Pollet and Desender, 1985). Other generalist carabid species may also aggregate in prey-rich patches. Warner *et al.* (2000), using SADIE, reported significant positive associations between the abundance of brassica pod midge (*Dasineura brassicae*), falling from winter rape plants, and pitfall catches of adult *Harpalus rufipes* and *Anchomenus dorsalis* [*Agonum dorsale*]. There were also spatial associations between larval cabbage stem flea beetle (*Psylliodes chrysocephala*) and *Pterostichus madidus* and *Trechus quadristriatus* (Warner, 2001). However, in these studies, no gut analyses or feeding trials were reported, and it remains possible that the carabids, midge larvae and flea beetle larvae were spatially associated for some reason other than a trophic relationship. The methods of biochemical analysis of predator diet (Symondson, this volume; Sunderland *et al.*, in press) and of statistical analysis of spatial distributions (e.g. SADIE, Perry, 1998) that have become available recently, hold considerable promise for elucidating how generalist predators forage for pests. After these methods have been applied to a wider range of predator–pest systems, generalizations about how the efficiency of pest control relates to predator foraging behaviour may become possible.

Some authors have attempted to calculate, from pest consumption and predator density data, the effect that carabids have on pest populations. For example, Hamon *et al.* (1990) used radiotracers to label adults and larvae of the weevil, *Sitona lineatus*, in beans, and recorded the proportion of carabid adults that gained the label. Mark–recapture was used to estimate the density of three dominant carabid species in the field. Based on these data, plus some assumptions about carabid response to *Sitona* density, it was estimated that carabids killed 24% of newly-emerged *Sitona* in one year. This is a valuable first estimate, but since the prey were disturbed by the labelling process and were presented to carabids in the field at high density in an open cage, the true predation rate in an undisturbed system could be different. Floate *et al.* (1990) used a serological method to detect antigens of wheat midge (*Sitodiplosis mosellana*) in carabids, then estimated minimum and maximum daily consumption rates using appropriate equations (Sunderland and Sutton, 1980). Based on these estimates, adult carabids were considered to have reduced numbers of midge larvae by a maximum of 11%. However, the authors point out that nocturnal predation may have been underestimated and that predation by carabid larvae (half of which were serologically positive) was not included in the estimates (Floate *et al.*, 1990).

An alternative approach to investigating the effect that carabids might have on pest populations is to integrate relevant field and laboratory data in simulation models. Scheller (1984) simulated the growth of populations of the cereal aphids, *S. avenae* and *M. dirhodum*, on spring barley during a twenty-day period in summer, and compared aphid population growth in the presence and in the absence of carabid populations. Aphid mortality rates (for a density of 62 carabids m^{-2}, which was estimated by removal trapping) were taken from functional response values derived mathematically from laboratory-determined data for meal retention rates at 16°C. The results suggested that carabids could have a large inhibitory effect on the growth of aphid populations. However, carabid density was unusually high (attributed to small plots surrounded by grassy borders) and various untested assumptions (e.g. that aphid catches by carabids are random) had to be made in order to derive functional response values. Ekbom *et al.* (1992) simulated the effects of predation by *Bembidion lampros* and *Poecilus cupreus* [*Pterostichus cupreus*] on populations of the aphid, *Rhopalosiphum padi*, in spring barley. The results suggested strongly that predation early in the season was more effective than predation acting later. Although the effects of alternative food were not simulated, the model was fairly successful at predicting *R. padi* densities in the field, on the assumption that carabid densities in the field were two to five times greater than densities used in the model. Carabid densities reported in other studies were, indeed, greater than the base densities used in the model. The simulations suggested that, given a moderate level of aphid immigration, early predation by these carabids can prevent aphid outbreaks developing (maintaining aphid populations below the economic threshold). Petersen and Holst (1997) used a metabolic pool approach to produce a preliminary model to investigate conditions for the control of *R. padi* by *B. lampros*. Early arrival of the aphid population improved the probability of successful control. The availability of alternative food for *B. lampros* increased its reproduction, but reduced its impact on the aphid population. This implies that the effectiveness of carabids may have been overestimated in the model of Ekbom *et al.* (1992), which did not consider the effects of alternative food. However, realistic modelling of the effects of alternative foods must be guided by a knowledge of real predator food preferences in the field, and this information is currently lacking for the majority of species. Winder *et al.* (1988) used a simulation model to estimate that the cereal aphid, *Sitobion avenae*, which peaked at 1.5 per shoot in a field of wheat, would have reached 17.1 per shoot in the absence of predation. When the effects of known densities of carabids (all feeding at rates observed in the laboratory) were incorporated, the aphid population peaked at only 2.6 per shoot. The authors then attempted to estimate more realistic predation rates by basing calculations on the proportion of carabids that contained aphid remains (from an earlier serological study – Sunderland *et al.,* 1987); this time, aphids peaked at 15.5 per shoot (close to the value for no predation). This exercise demonstrated that carabids, at existing densities, could control aphid populations if the carabids were to feed almost exclusively on the pest. More realistically, it will be necessary to obtain aphid control by increasing carabid density, or to benefit from the carabid contribution within an assemblage of natural enemies.

Recording the changes in pest density that result from experimental alteration of the density of carabids is a potentially powerful and convincing technique for investigating predator impact in the field. In sward boxes in the greenhouse (Asteraki, 1993), in semi-field miniplots (McKemey, 2000), in field cages (Barker, 1991), and in field enclosures (Chapman *et al.,* 1997), slug numbers were reduced significantly by carabids. Grafius and Warner (1989) confined *Bembidion quadrimaculatum*, at a range of densities, in small field enclosures containing 2–4 onion plants infested with eggs of onion fly (*Delia antiqua*). Surviving *D. antiqua* larvae were counted seven days later. There was a significant negative correlation between onion fly survival and *B. quadrimaculatum* density. All these experiments provide a useful first indication of the likelihood of biological control by carabids, but the results have to be interpreted with caution because predator foraging was constrained and alternative prey were absent or at reduced levels compared with the open field. Ryan and Ryan (1973) used pitfall traps to reduce predator abundance (95% were carabids) in a large (7 m × 21 m) barriered cabbage plot. Densities of the eggs of cabbage root fly (*Delia radicum*) were found to be 41% higher in the reduced-carabid plot compared with a control plot containing normal field densities of carabids, but, unfortunately, there was no treatment replication in this study. Bommarco (1999) added adult *Poecilus cupreus* [*Pterostichus cupreus*] to 6.3 m^2 barriered plots of ley and spring barley to give initial densities of approximately 0, 2 and 10 beetles m^{-2}. Densities of aphids, thrips and weevils did not respond significantly in relation to carabid density. However, the experiment was of short duration because beetle densities in manipulated plots converged after fifteen days (due to mortality and escapes). So, although there have been a number of manipulative field studies that have indicated a likely effect of carabids, acting alone, in supressing pest populations, definitive studies (e.g. employing large replicated plots, with normal field levels of alternative prey, where beetles can forage normally) appear to be lacking.

The above strictures about alternative prey do not apply in the case of augmentative biocontrol in protected crops. Symondson (1993) added *Abax parallelepipedus* to plots of lettuce in a polyhouse. Mortality of the slug, *Deroceras reticulatum*, was increased significantly in plots where beetles were added, compared with control plots. This effect was greatest for female beetles and small plants (the beetles did not seem able to capture slugs efficiently within large plants). *A. parallelepipedus* is relatively easy to rear and will survive for nine months in polyhouses. The economics of rear and release are uncertain, however, because, although the beetle is not cannibalistic, there is a high level of mortality at the pre-pupal stage; the life cycle from egg to adult can take 110 days, and undersized adults may be produced (Symondson, 1994).

Results of a selection of studies where the impact of carabids (acting alone) on pest populations has been investigated are summarized in *Table 6.3*. In approximately half of these studies, treated and control plots were replicated and differences were tested statistically. Nearly all studies were confined to a single pest growth stage and the effects of carabids on the overall population dynamics of the pest were not determined. Effects on crop damage and yield were investigated in only three studies (Asteraki, 1993; Symondson, 1993; Hodge *et al*., 1999), and, in these, carabids were found to reduce, significantly, crop damage by slugs and caterpillars. Examination of the table suggests that slug control by carabids is a promising option, but otherwise, to date, there is little firm evidence for a strong economic advantage to agriculture attributable to carabids, by themselves, suppressing pest populations. However, if the carabid component to pest restraint were to be combined with components from a range of other natural enemies, then perhaps the value of carabids, as part of a biocontrol package, would become evident. This possibility is reviewed in the next section.

CARABIDS ACTING WITH OTHER NATURAL ENEMIES

Trophic specialists

In a few studies, it has been demonstrated that specialist carabids make more impact on pest populations than do other categories of predator. Examination of pupae and

Table 6.4. Examples of studies where impact on a pest population by a natural enemy complex, including carabids, has been investigated. In the 'Impact type' column the maximum reduction and pest growth stage affected (for holometabolous species) are shown. An asterisk in the 'Authors' column denotes that differences between treated and control plots were tested statistically.

Pest species	Crop	Predators in addition to carabids	Method (see key)	Impact type (see key)	Authors (see key)
MOLLUSCA					
Limax maximus , *Helix aspersa*	Daisy	Garter snakes	A, refuges	N (50%)	5*
Deroceras reticulatum, Arion intermedius	Grass	Starlings	C, D	N (57%)	6*
Deroceras reticulatum	Wheat	Spiders, staphylinids	B, D	N (89%)	24*
COLEOPTERA					
Diabrotica undecimpunctata	Squash	Lycosids	B, D	N (adults 53%)Y (200%)	21*
Gastrophysa polygoni	Wheat	Spiders, staphylinids	B, C, D	N (eggs 49%, larvae 61%)	25*
Popillia japonica	Turfgrass	Spiders, staphylinids, ants	Fate of placed out eggs monitored	N (eggs 51%)	20*
DIPTERA					
Dasineura brassicae	Brassicas	Spiders, staphylinids	B, D	N (adults 12%)	10
Delia coarctata	Not given	Not given	B, D	N (pupae 29% NS)	34*
Delia radicum	Cabbage	Not given	B, D	N (eggs 80%), D (78%)	1*
Delia radicum	Cabbage	Staphylinids	B, D	N (eggs, pupae NS) D (NS)	33*
Delia radicum	Cauliflower	Staphylinids	B, D	N (eggs 25%, larvae 29% NS, pupae 23% NS), D (41%),Y (60% NS)	31*
Psila rosae	Carrots	Spiders, staphylinids, harvestmen	B, D	D (NS)	19*
Sitodiolosis mosellana	Wheat	Spiders, staphylinids	B, D	N (larvae NS), Y (NS)	9*
Sitodiolosis mosellana	Wheat	Spiders, staphylinids	B, D	N (adults 84%)	10
Sitodiolosis mosellana	Wheat	Spiders, staphylinids	B, D	N (pupae in soil 31%), Y (NS)	35*
HETEROPTERA					
Anasa tristis	Squash	Lycosids, nabids	B, D	N (67%), Y (133%)	22*
HOMOPTERA					
Aphis nasturtii	Potato	Spiders, staphylinids	A, B, D	N (49%)	7*
Diuraphis noxia	Wheat, barley	Spiders, ladybirds, hoverflies, nabids, lacewings, parasitoids	C	N (95%)	38
Macrosiphum euphorbiae	Potato	Spiders, staphylinids	A, B, D	N (80%)	7*
Myzus persicae	Potato	Spiders, staphylinids	A, B, D	N (88%)	7*
Myzus persicae	Sugar beet	Spiders, staphylinids, cantharids etc	B, D	D (35% virus reduced)	17*

Table 6.4. continued

Rhopalosiphum padi	Barley	Spiders, staphylinids, coccinellids	B, D	N (84%)	26
Rhopalosiphum padi	Barley	Spiders, staphylinids	B, D	N (58%)	27*
Rhopalosiphum padi	Maize	Spiders	A, C, D	N (77% NS)	32*
Rhopalosiphum padi, Sitobion avenae	Wheat	Spiders, staphylinids	B, D	N (50%)	18
Sitobion avenae	Wheat	Spiders, staphylinids	B, D	N (NS), Y (NS)	9*
Sitobion avenae	Wheat	Spiders, staphylinids	B, D	N (28%), Y (NS)	11*
Sitobion avenae	Wheat	Spiders, staphylinids	B, D	N (NS), Y (NS)	16*
Sitobion avenae	Wheat	Spiders, staphylinids	B, D	N (63%)	23*
Cereal aphids	Wheat	Spiders, staphylinids	B, D	N (86%)	3
Cereal aphids	Wheat	Spiders, staphylinids	B, D	N (62%)	28
Cereal aphids	Wheat	Spiders, staphylinids, earwigs, mites	B, D	N (77%)	4
Cicadellidae	Maize	Spiders	A, C, D	N (82%)	32*
LEPIDOPTERA					
Agrotis ipsilon	Maize	Spiders, staphylinids, harvestmen	B, D, placed out pest	D (43%)	29*
Cochylis hospes	Sunflower	Staphylinids, cantharids	B, placed out pest	N (larvae + pupae 44%)	13
Diatraea saccharalis	Sorghum	Ants, spiders, bugs, earwigs	I	N (larvae 60%), D (44%), Y (22%)	37*
Diatraea saccharalis	Sugarcane	Ants, spiders, bugs, earwigs	I	N (larvae 45%), D (72%), Y (19%)	37*
Hemileuca oliviae	Pasture	Ants, mantids, spiders, gryllids	B, C, D	N (larvae 42%)	36*
Lymantria dispar	Oak woods	Ants, harvestmen, pentatomids, spiders	Pupal examination	N (pupae 36%)	15
Operophtera brumata	Oak woods	Shrews	Pupal examination	N (pupae 45%)	39
Ostrinia nubilalis	Maize	Ants, spiders, fungus beetles, birds	C, placed out pest	N (larvae in stems 40%)	30*
Pseudaletia unipuncta	Maize	Spiders, lycosids, thomisids, harvestmen	Correlation	N (small larvae)	12
Pseudaletia unipuncta	Maize	Spiders, ants, staphylinids	B, D	D (74%)	14*
Spodoptera frugiperda	Turfgrass	Spiders, staphylinids, ants	Pupal examination	N (pupae NS)	20*
Thanatarctia imparilis	Mulberry	Ants, spiders, parasitoids	Correlation	N (larvae 90%)	2
THYSANOPTERA					
Cereal thrips	Maize	Spiders	A, C, D	N (60%)	32*

Key to Table 6.4: *Method*: A = augmentation of carabid density (above natural field density); B = barrier-enclosed plots; C = field cages; D = manipulation of carabid density; I = insecticidal check method; *Impact type*: (maximum percentage change compared to controls): N = pest numbers reduced; D = plant damage reduced; Y = yield increased; NS = no significant difference; *Authors*: 1: Armstrong *et al.* (1998); 2: Hondo and Morimoto (1997); 3: Edwards *et al.* (1979); 4: Sunderland *et al.* (1980); 5: Altieri *et al.* (1982); 6: Barker (1991); 7: Boiteau (1986); 8: Helenius (1990); 9: Holland *et al.* (1996); 10: Basedow (1975); 11: Holland and Thomas (1997b); 12: Laub and Luna (1992); 13: Bergmann and Oseto (1990); 14: Clark *et al.* (1994); 15: Fuester and Taylor (1996); 16: Holland and Thomas (1997a); 17: Landis and Van der Werf (1997); 18: Lübke-Al Hussein and Triltsch (1994); 19: Rämert and Ekbom (1996); 20: Terry *et al.* (1993); 21: Snyder and Wise (1999); 22: Snyder and Wise (2001); 23: Winder (1990); 24: Burn (1992); 25: Sotherton (1982); 26: Chiverton (1986); 27: Gravesen and Toft (1987); 28: De Clercq and Pietraszko (1983); 29: Brust *et al.* (1985); 30: Coll and Bottrell (1992); 31: Wright *et al.* (1960); 32: Lang *et al.* (1999); 33: Tukahirwa and Coaker (1982); 34: Ryan (1975); 35: Holland and Thomas (1996); 36: Shaw *et al.* (1987); 37: Fuller and Reagan (1988); 38: Mohamed *et al.* (2000); 39: East (1974).

pupal remains of gypsy moth (*Lymantria dispar*) indicated invertebrate predation levels up to 36% in oak woodland. Most attacked pupae had large jagged wounds characteristic of predation by larvae of the carabid, *Calosoma sycophanta,* and there was a significant positive correlation between the incidence of predated pupae and abundance of these larvae. *C. sycophanta* was considered to be the most important mortality agent of gypsy moth pupae (Fuester and Taylor, 1996). *Parena perforata* is a specialist carabid, attacking communal larvae of the Mulberry Tiger Moth (*Thanatarctica imparilis*). Predation by generalist feeders (ants and spiders) was restricted to small nests of caterpillars and was not an important cause of mortality at high pest density. *P. perforata* adults, however, cut through the silk of nests with their mandibles and attacked the caterpillars, even in high-density nests. *P. perforata* were found to be highly mobile and voracious (capable of consuming a hundred caterpillars per day in the laboratory) and acted most strongly during the pest decline phase (Hondo and Morimoto, 1997).

Trophic generalists

When carabids (and especially generalist carabids) are part of large multi-species complexes of natural enemies, a wide range of interactions between the different groups of natural enemies can occur. Interactions within complexes of natural enemies can have significant effects on the success of biological control (Ferguson and Stiling, 1996; Roy and Pell, 2000). This topic is treated in detail in Sunderland *et al.* (1997), but some examples of positive and negative interactions (in relation to biocontrol efficiency), involving carabids, are given below. Positive interactions can occur in relation to foraging location. For example, in laboratory trials, the aphid, *Acyrthosiphon pisum*, dropped from alfalfa plants as an escape response to ladybirds (*Coccinella 7-punctata*) foraging on the plant. This made the aphids available to the ground-foraging carabid, *Harpalus pensylvanicus*, which otherwise caused virtually no aphid mortality. Aphid mortality in the joint presence of both predator species was significantly greater than the sum of the aphid mortalities in the presence of each predator separately. Thus, these predators inflicted synergistic mortality on the aphid, at least under laboratory conditions (Losey and Denno, 1998b). Carabids can also interact positively with the pathogens of pests, by dissemination of the pathogens. The carabid, *Pterostichus madidus,* consumed significantly more dead-infected than dead-uninfected aphids (*Acyrthosiphon pisum* attacked by *Erynia neoaphidis*) (Roy *et al.*, 1998), and carabids that had fed on baculovirus-infected cabbage moth (*Mamestra brassicae*) caused a significant degree of transmission of the virus to healthy caterpillars in a cabbage field (Vasconcelos *et al.*, 1996). Similarly, *Harpalus rufipes* and *Harpalus affinis* [*Harpalus aeneus*] dispersed the caterpillar pathogen, *Bacillus thuringiensis* var. *kurstaki*, by at least 135 m in a cabbage field (Pedersen *et al.*, 1995). Young and Hamm (1986) considered that the behaviour and digestive biology of *Calosoma sayi* suited it well for the dissemination of pathogens, such as baculoviruses, to fall armyworm, *Spodoptera frugiperda*. Negative interactions are also possible, and carabids have been found sometimes to hinder biological control through intraguild predation, or by inflicting damage on parasitoid populations. Snyder and Ives (2001), for example, carried out laboratory experiments and field cage experiments to investigate interactions between aphids (*A. pisum*) on alfalfa, the carabid, *Pterostichus melanarius*, and the parasitoid wasp, *Aphidius ervi*. When alfalfa plants were short, the carabid controlled aphid populations, but when plants were tall, the carabid disrupted biological control by predating selectively on parasitized aphids and preventing aphid control by the parasitoid. Generalizations cannot yet be made safely about the biocontrol outcomes of such community interactions, which are dependent on the exact combinations of organisms involved. For example, in an aphid–parasitoid–ladybird system on cotton, Colfer and Rosenheim (2001) found that, although ladybirds ate mummified aphids, this did not significantly inhibit biological control of the aphid pest. Snyder and Wise (2001) carried out field manipulations of the abundance of carabids and lycosid spiders (separately and combined) in fenced vegetable plots and recorded the effects on pests and crop yield. Lycosids reduced populations of striped cucumber beetle (*Acalymma vittata*), significantly, on cucumber in spring, which resulted in significant increases in yield; this effect was reinforced by the addition of carabids. In the summer, carabids reduced numbers of squash bug (*Anasa tristis*), significantly, on squash, which increased squash yield significantly. Lycosids, however, caused an increase in squash bug populations, which cancelled out the positive effect of the carabids. Lycosids were considered to have produced this effect by strong intraguild predation of nabid bugs (which are important predators of squash bugs) because field populations of nabids were significantly reduced in the presence of lycosids, but little affected by carabids. Thus, it is clear that there can be complex interactions between different groups of generalist predators, in one season, working additively to control pests, but in another season producing interference effects detrimental to pest control. This degree of complexity is likely to be the norm for field crops, and there remains much to be learnt about how best to manage communities of natural enemies (including carabids) to optimize the biological control of pests.

The impact (or absence of impact) of assemblages of generalist predators (including carabids) on pest populations has been investigated directly by experimental manipulations of predator density in the field. In all examples described below, the complex of natural enemies that was manipulated included carabid beetles. Levels of damage to carrots caused by carrot fly (*Psila rosae*) were unaffected by experimental alterations of the abundance of generalist predators in the field (Rämert and Ekbom, 1996), but, in sugar beet, early-season predation by a complex of predators reduced significantly both the population density of the aphid, *Myzus persicae,* and the damage it caused by vectoring beet yellows virus (Landis and Van der Werf, 1997). Fuller and Reagan (1988) used the insecticidal check method to reduce predator populations in plots of sorghum and sugarcane. They found that numbers of sugarcane borer (*Diatraea saccharalis*) larvae and damage levels were increased significantly, and yields reduced significantly, in plots with reduced-predator populations, compared to plots with normal field populations of predators. The economic injury level was exceeded in reduced-predator plots, but was not exceeded in plots with normal predator densities. The authors did not, however, discuss the possibility of pest increase being stimulated directly by the insecticide (i.e. hormoligosis – sublethal quantities of a stressing agent may stimulate growth or fecundity directly (Luckey, 1968), or indirectly (trophobiosis), via changes in the physiology of the host plant (Brandenburg and Kennedy, 1987; Kidd and Rummel, 1997)). A number of studies have also been carried out in maize. Menalled *et al.* (1999) manipulated predator density in this crop using various types of barrier and demonstrated significantly higher levels of removal of sentinel prey (freeze-killed pupae of onion fly, *Delia antiqua*, placed on the ground) in plots where predator density was augmented. Although spiders and ladybirds were also present in the field, carabids were the most abundant group of predators (in pitfalls). In a study by Brust *et al.* (1985), damage to maize plants by black cutworm larvae (*Agrotis ipsilon*) was found to be significantly greater within barriered plots, where the density of a complex of predators (including the carabid genera *Amara*, *Amphasia*, *Poecilus* [*Pterostichus*], and *Pterostichus* [*Abacidus*]) had been reduced experimentally. Coll and Bottrell (1992) compared survival of corn borer larvae (*Ostrinia nubilalis*) in caged and uncaged maize stems and found that predators (believed to be non-sucking predators, including carabids, ants and spiders) caused a significant reduction of survival by about 40%. Other, similar, manipulative field studies have been carried out in European cereals (wheat, barley and oats). Helenius (1990) used barriers and ingress and egress trenches to manipulate the abundance of predators in oats. Peak densities of the aphid, *Rhopalosiphum padi,* were always higher, and grain yields 19–22% lower, where predators were reduced. Similarly, Chiverton (1986) and Gravesen and Toft (1987) recorded *R. padi* densities in spring barley two to six times greater in predator-reduced plots than in control plots. In contrast, Holland *et al.* (1996), in winter wheat, found no consistent significant increase in aphid (*Sitobion avenae*) and blossom midge (*Sitodiplosis mosellana*) densities, and no decrease in grain yield, in plots where predators were reduced substantially. This may have been because predator (carabids, staphylinids and spiders) activity was reduced and aphid reproduction increased rapidly during a period of hot, dry weather at a late crop development stage. In a later study (Holland and Thomas, 1997b), predators did reduce this aphid by 31% and also reduced pupae of *S. mosellana* in the soil in July by 31% (Holland and Thomas, 1996). Basedow (1975) used cages rather than open enclosures (so redistribution of flying pests between treatments was suppressed) and estimated that predators reduced *S. mosellana* by 84%. Burn (1992) found that, in some years, reduction of predators inside barriered plots of wheat allowed the development of significantly higher populations of *S. avenae* and *Metopolophium dirhodum*. Burn (1992) also showed that the slug, *Deroceras reticulatum*, was increased significantly by the exclusion of predators in eight out of fifteen trials over a period of five years. Winder (1990) recorded significant increases in density of *S. avenae* inside barriered plots where predators were reduced. He also showed (by the use of aphid fall-off and climb-back traps) that aphid fall-off rates were the same in all plots, but aphid climb-back rates were highest where predator densities were lowest, suggesting that at least a part of the overall effect of predators on aphid density was due to live aphids being predated on the ground. Duffield *et al.* (1996) used similar methods, and, from a regression of aphid reclimb rate against predator abundance (in pitfall traps), they estimated that 86% of cereal aphids would reclimb wheat plants in the absence of ground predators. Approximately 15% of aphids reclimbed in control plots containing normal field populations of predators. From gut dissections, Griffiths *et al.* (1985) showed that aphids formed a high proportion of the food of the carabid, *Anchomenus dorsalis* [*Agonum dorsale*], in July. However, during many hours of field observation (by day and night), they never saw this beetle eating aphids on wheat plants, although it foraged frequently on the ground. Sunderland *et al.* (1988) showed that 51 out of 57 species of carabid were confined to the ground zone of UK cereals. After three years of trapping, Lövei and Szentkirályi (1984) caught only ten carabid beetles (of four species) from the stalks of maize plants in Hungary. It seems likely, therefore, that most of the contribution that carabids make to aphid control in European cereals results from foraging on the ground. This would be consistent with what is known for aphid control in alfalfa in the USA (Losey and Denno, 1998b – see above).

Summarized results of 39 studies, where the impact of a natural enemy complex (including carabids) on pest populations has been investigated, are given in *Table 6.4*. Treated and control plots were replicated and differences were tested statistically in nearly three quarters of the investigations listed. Significant reductions of the density of dipterous pests by predators were recorded in only a third of cases, but other pests (molluscs, beetles, aphids, bugs, caterpillars and thrips) were reduced significantly in 87% of cases. Significant reductions of crop damage and/or increases in yield, attributable to predation of pests, were recorded in relation to damage caused by beetles, bugs and caterpillars. Predators reduced damage caused by flies and aphids to a significant degree in some cases, but not in others (*Table 6.4*). The contribution that carabids made to the biological control of pests was not quantified for the majority of studies in *Table 6.4*.

Conclusions

This review has shown that, although there is ample evidence that carabids consume pests and are capable of killing a wide range of pest types, there are relatively few rigorous demonstrations that carabids can depress pest populations to a degree that is economically beneficial for farmers. Circumstantial evidence for the value of carabids as biocontrol agents comes from correlations, calculations and simulation modelling, but the most convincing demonstration of their role is an increase in pest numbers and crop damage following experimental reduction of predator density in the field. Carabids that are trophic specialists have a valuable role in pest control in a limited number of cases (mainly for lepidopterous pests), but the majority of carabid species are trophic generalists. There is little evidence that trophic generalist carabids can, by themselves, make a significant impact on pest populations (with the possible exception of slugs). There is more evidence that, when carabids form part of an assemblage of generalist predators, the assemblage as a whole can often reduce pests to a significant degree. The relative contribution that carabids make to the overall pest control success of the assemblage has rarely been determined, but studies such as that of Snyder and Wise (2001), where groups of predators are manipulated separately and in combination, suggest that the contribution of any one group of predator will vary according to crop, pest and season, and that interactions between groups of predators will have a significant influence on the pest control outcome. This means that, in the context of conservation biocontrol (which is relevant to a much larger proportion of agriculture than is classical or augmentative biological control), we should study and learn how to manage assemblages of natural enemies, rather than just one group of predators, such as the Carabidae.

Acknowledgements

I am very grateful to Dr R.L. Davidson (Carnegie Museum of Natural History, Pittsburgh, USA) for checking and updating the names of carabids in the tables, to Dr W.O.C. Symondson (Cardiff University, UK), Dr P. Chiverton (Sweden University of Agricultural Sciences – SLU) and Prof. G.M. Tatchell (Horticulture Research International – HRI, UK) for helpful comments on the manuscript, to Lidija Kravar-Garde (HRI) for Russian translation, and to the HRI library staff for locating obscure papers. The work was funded by the UK Department for Environment, Food & Rural Affairs.

References

ALLEN, R.T. (1979). The occurrence and importance of ground beetles in agricultural and surrounding habitats. In: *Carabid beetles: their evolution, natural history and classification.* Eds. T.L. Erwin, G.E. Ball, D.L. Whitehead and A.L. Halpern, pp 485–505. The Hague: Dr W. Junk.

ALLEN, W.R. AND HAGLEY, E.A.C. (1990). Epigeal arthropods as predators of mature larvae and pupae of the apple maggot (Diptera: Tephritidae). *Environmental Entomology* **19**, 309–312.

ALTIERI, M.A., HAGEN, K.S., TRUJILLO, J. AND CALTAGIRONE, L.E. (1982). Biological control of *Limax maximus* and *Helix aspersa* by indigenous predators in a daisy field in central coastal California. *Acta Oecologica* **3**, 387–390.

ANDERSEN, A. (1992). Predation by selected carabid and staphylinid species on the aphid *Rhopalosiphum padi* in laboratory and semifield experiments. *Norwegian Journal of Agricultural Sciences* **6**, 265–273.

ANDERSEN, A., HANSEN, A.G., RYDLAND, N. AND ØYRE, G. (1983). Carabidae and Staphylinidae (Col.) as predators of eggs of the turnip root fly *Delia floralis* Fallén (Diptera, Anthomyiidae) in cage experiments. *Zeitschrift für angewandte Entomologie* **95**, 499–506.

ARMSTRONG, G., MFUGALE, O.B.J. AND CHAPMAN, P.A. (1998). Intercropping for pest control: the role of predators. *1998 Brighton Crop Protection Conference – Pests and Diseases*, British Crop Protection Council, Farnham, UK, **2**, pp 607–612.

ASÍN, L. AND PONS, X. (1998). Aphid predators in maize fields. *Bulletin OILB/SROP* **21(8)**, 163–170.

ASTERAKI, E.J. (1993). The potential of carabid beetles to control slugs in grass/clover swards. *Entomophaga* **38**, 193–198.

AYRE, K. AND PORT, G.R. (1996). Carabid beetles recorded feeding on slugs in arable fields using ELISA. *Slug and snail pests in agriculture. Proceedings of a Symposium, University of Kent 1996, British Crop Protection Council* **66**, pp 411–418.

BAINES, D., STEWART, R. AND BOIVIN, G. (1990). Consumption of carrot weevil (Coleoptera: Curculionidae) by five species of carabids (Coleoptera: Carabidae) abundant in carrot fields in southwestern Quebec. *Environmental Entomology* **19**, 1146–1149.

BAKER, G.H. (1985). Predators of *Ommatoiulus moreletii* (Lucas) (Diplopoda: Iulidae) in Portugal and Australia. *Journal of the Australian Entomological Society* **24**, 247–252.

BARKER, G.M. (1991). Biology of slugs (Agriolimacidae and Arionidae: Mollusca) in New Zealand hill country pastures. *Oecologia* **85**, 581–595.

BARNEY, R.J. AND PASS, B.C. (1986). Foraging behaviour and feeding preference of ground beetles (Coleoptera: Carabidae) in Kentucky alfalfa. *Journal of Economic Entomology* **79**, 1334–1337.

BARRION, A.T. AND LITSINGER, J.A. (1985). *Chlaenius* spp. (Coleoptera: Carabidae), a leaffolder (LF) predator. *International Rice Research Newsletter* **10**, 21–22.

BASEDOW, T. (1975). Predaceous arthropods in agriculture, their influence upon the insect pests, and how to spare them while using insecticides. *Semaine d'Étude Agriculture et Hygiène des Plantes*, 1975, 311–323.

BASEDOW, T., BORG, A., DE CLERCQ, R., NIJVELDT, W. AND SCHERNEY, F. (1976).

Untersuchungen über das Vorkommen der Laufkäfer [Col.: Carabidae] auf Europäischen Getreidefeldern. *Entomophaga* **21**, 59–72.

BECHINSKI, E.J., BECHINSKI, J.F. AND PEDIGO, L.P. (1983). Survivorship of experimental green cloverworm (Lepidoptera: Noctuidae) pupal cohorts in soybeans. *Environmental Entomology* **12**, 662–668.

BERGMANN, D.J. AND OSETO, C.Y. (1990). Life tables of the Banded Sunflower Moth (Lepidoptera: Tortricidae) in the Northern Great Plains. *Environmental Entomology* **19**, 1418–1421.

BERIM, N.G. AND NOVIKOV, N.V. (1983). Feeding specialisation of ground beetles. *Zashchita Rastenii* **7**, 18.

BEST, R. AND BEEGLE, C.C. (1977). Food preferences of five species of carabids commonly found in Iowa cornfields. *Environmental Entomology* **6**, 9–12.

BOHAN, D.A., BOHAN, A.C., GLEN, D.M., SYMONDSON, W.O.C., WILTSHIRE, C.W. AND HUGHES, L. (2000). Spatial dynamics of predation by carabid beetles on slugs. *Journal of Animal Ecology* **69**, 367–379.

BOITEAU, G. (1986). Native predators and the control of potato aphids. *Canadian Entomologist* **118**, 1177–1183.

BOMMARCO, R. (1999). Feeding, reproduction and community impact of a predatory carabid in two agricultural habitats. *Oikos* **87**, 89–96.

BRANDENBURG, R.L. AND KENNEDY, G.G. (1987). Ecological and agricultural considerations in the management of twospotted spider mite (*Tetranychus urticae* Koch). *Agricultural Zoology Reviews* **2**, 185–236.

BRAUN, D.M., GOYER, R.A. AND LENHARD, G.J. (1990). Biology and mortality agents of the fruit tree leafroller (Lepidoptera: Tortricidae), on bald cypress in Louisiana. *Journal of Entomological Science* **25**, 176–184.

BRUST, G.E. (1991). A method for observing below-ground pest–predator interactions in corn agroecosystems. *Journal of Entomological Sciences* **26**, 1–8.

BRUST, G.E., STINNER, B.R. AND MCCARTNEY, D.A. (1985). Tillage and soil insecticide effects on predator-black cutworm (Lepidoptera: Noctuidae) interactions in corn agroecosystems. *Journal of Economic Entomology* **78**, 1389–1392.

BRUST, G.E., STINNER, B.R. AND MCCARTNEY, D.A. (1986a). Predation by soil inhabiting arthropods in intercropped and monoculture agroecosystems. *Agriculture, Ecosystems and Environment* **18**, 145–154.

BRUST, G.E., STINNER, B.R. AND MCCARTNEY, D.A. (1986b). Predator activity and predation in corn agroecosystems. *Environmental Entomology* **15**, 1017–1021.

BRYAN, K.M. AND WRATTEN, S.D. (1984). The responses of polyphagous predators to prey spatial heterogeneity: aggregation by carabid and staphylinid beetles to their cereal aphid prey. *Ecological Entomology* **9**, 251–259.

BURN, A.J. (1982). The role of predator searching efficiency in carrot fly egg loss. *Annals of Applied Biology* **101**, 154–159.

BURN, A.J. (1992). Interactions between cereal pests and their predators and parasites. In: *Pesticides, cereal farming and the environment.* Eds. P. Grieg-Smith, G. Frampton and A. Hardy, pp 110–131. London: HMSO.

CAMERON, E.A. AND REEVES, R.M. (1990). Carabidae (Coleoptera) associated with gypsy moth, *Lymantria dispar* (L.) (Lepidoptera: Lymantriidae), populations subjected to *Bacillus thuringiensis* Berliner treatments in Pennsylvania. *Canadian Entomologist* **122**, 123–129.

CAPPAERT, D.L., DRUMMOND, F.A. AND LOGAN, P.A. (1991). Incidence of natural enemies of the Colorado Potato Beetle, *Leptinotarsa decemlineata* [Coleoptera: Chysomelidae] on a native host in Mexico. *Entomophaga* **36**, 369–378.

CHAPMAN, P.A. (1994). Control of leatherjackets by natural enemies: the potential role of the ground beetle *Pterostichus melanarius. 1994 Brighton Crop Protection Conference – Pests and Disease*s, BCPC, Farnham, UK, pp 933–934.

CHAPMAN, R.B., SIMEONIDIS, A.S. AND SMITH, J.T. (1997). Evaluation of metallic green ground beetle as a predator of slugs. *Proceedings of the 50th New Zealand Plant Protection Conference*, New Zealand Plant Protection Society, Rotorua, New Zealand, pp 51–55.

CHEESEMAN, M.T. AND GILLOTT, C. (1987). Organization of protein digestion in adult *Calosoma calidum* (Coleoptera: Carabidae). *Journal of Insect Physiology* **33**, 1–8.

CHEN, D.N., ZHANG, G.Q., XY, W.X., LIU, Y.H., CHENG, X.M. AND WU, J.Q. (1996). A study of complementary techniques for snail control. *1996 BCPC Symposium Proceedings 'Slug and snail pests in agriculture'*, British Crop Protection Council, Surrey, UK, **66**, 425–432.

CHIVERTON, P.A. (1984). Pitfall-trap catches of the carabid beetle *Pterostichus melanarius*, in relation to gut contents and prey densities, in insecticide treated and untreated spring barley. *Entomologia Experimentalis et Applicata* **36**, 23–30.

CHIVERTON, P.A. (1986). Predator density manipulation and its effects on populations of *Rhopalosiphum padi* (Hom.: Aphididae) in spring barley. *Annals of Applied Biology* **109**, 49–60.

CHIVERTON, P.A. (1987). Predation of *Rhopalosiphum padi* (Homoptera: Aphididae) by polyphagous predatory arthropods during the aphids' pre-peak period in spring barley. *Annals of Applied Biology* **111**, 257–269.

CHIVERTON, P.A. (1988). Searching behaviour and cereal aphid consumption by *Bembidion lampros* and *Pterostichus cupreus*, in relation to temperature and prey density. *Entomologia Experimentalis et Applicata* **47**, 173–182.

CHIVERTON, P.A. AND SOTHERTON, N.W. (1991). The effects on beneficial arthropods of the exclusion of herbicides from cereal crop edges. *Journal of Applied Ecology* **28**, 1027–1039.

CLARK, M.S., LUNA, J.M., STONE, N.D. AND YOUNGMAN, R.R. (1994). Generalist predator consumption of armyworm (Lepidoptera: Noctuidae) and effect of predator removal on damage in no-till corn. *Environmental Entomology* **23**, 617–622.

COAKER, T.H. AND WILLIAMS, D.A. (1963). The importance of some Carabidae and Staphylinidae as predators of the cabbage root fly, *Erioischia brassicae* (Bouché). *Entomologia Experimentalis et Applicata* **6**, 156–164.

COHEN, A.C. (1995). Extra-oral digestion in predaceous terrestrial Arthropoda. *Annual Review of Entomology* **40**, 85–103.

COLFER, R.G. AND ROSENHEIM, J.A. (2001). Predation on immature parasitoids and its impact on aphid suppression. *Oecologia* **126**, 292–304.

COLL, M. AND BOTTRELL, D.G. (1992). Mortality of European corn borer larvae by natural enemies in different corn microhabitats. *Biological Control* **2**, 95–103.

CONSTANTINEANU, I. AND CONSTANTINEANU, R. (1996). Contributions of entomophagous predator insects in limiting the outbreak of *Lymantria dispar* L. (Lep., Lymantriidae). *Revue Roumaine de Biologie, Serie de Biologie Animale* **41**, 69–77.

CORNIC, J.F. (1973). Étude du régime alimentaire de trois espèces de carabiques et de ses variations en verger de pommiers. *Annales Société Entomologique de France* **9**, 69–87.

CROOK, A.M.E. AND SOLOMON, M.G. (1997). Predators of vine weevil in soft fruit plantations. *Proceedings of the ADAS/HRI/EMRA Soft Fruit Conference, New Developments in the Soft Fruit Industry, Ashford, UK*, pp 83–87.

DE CLERCQ, R. AND PIETRASZKO, R. (1983). Epigeal arthropods in relation to predation of cereal aphids. In: *Aphid antagonists.* Ed. R. Cavalloro, pp 88–92. Rotterdam: A.A. Balkema.

DEGUINE, J.P. (1991). Observations on carabid predators of lepidopteran cotton pests in North Cameroon. *Coton et Fibres Tropicales* **46**, 249–255.

DE KRAKER, J., VAN HUIS, A., VAN LENTEREN, J.C., HEONG, K.L. AND RABBINGE, R. (1999). Egg mortality of rice leaffolders *Cnaphalocrocis medinalis* and *Marasmia patnalis* in irrigated rice fields. *BioControl* **44**, 449–471.

DEMPSTER, J.P. (1967). The control of *Pieris rapae* with DDT. 1. The natural mortality of the young stages of *Pieris. Journal of Applied Ecology* **4**, 485–500.

DHARMADHIKARI, P.R., RAMASESHIAH, G. AND ACHAN, P.D. (1985). Survey of *Lymantria obfuscata* and its natural enemies in India. *Entomophaga* **30,** 399–408.

DIGWEED, S.C. (1993). Selection of terrestrial gastropod prey by Cychrine and Pterostichine ground beetles (Coleoptera: Carabidae). *Canadian Entomologist* **125,** 463–472.

DIXON, P.L. AND MCKINLAY, R.G. (1992). Pitfall trap catches of and aphid predation by *Pterostichus melanarius* and *Pterostichus madidus* in insecticide treated and untreated potatoes. *Entomologia Experimentalis et Applicata* **64**, 63–72.

DUDEVOIR, D.S. AND REEVES, R.M. (1990). Feeding activity of carabid beetles and spiders on gypsy moth larvae (Lepidoptera: Lymantriidae) at high-density prey populations. *Journal of Entomological Science* **25,** 341–356.

DUFFIELD, S.J., JEPSON, P.C., WRATTEN, S.D. AND SOTHERTON, N.W. (1996). Spatial changes in invertebrate predation rate in winter wheat following treatment with dimethoate. *Entomologia Experimentalis et Applicata* **78**, 9–17.

DUNNING, R.A., BAKER, A.N. AND WINDLEY, R.F. (1975). Carabids in sugar beet crops and their possible role as aphid predators. *Annals of Applied Biology* **80,** 125–128.

EAST, R. (1974). Predation on the soil-dwelling stages of the winter moth at Wytham Woods, Berkshire. *Journal of Animal Ecology* **43**, 611–626.

EDWARDS, C.A., SUNDERLAND, K.D. AND GEORGE, K.S. (1979). Studies on polyphagous predators of cereal aphids. *Journal of Applied Ecology* **16**, 811–823.

EKBOM, B. (1994). Arthropod predators of the pea aphid, *Acythosiphon pisum* Harr. (Hom., Aphididae) in peas (*Pisum sativum* L.), clover (*Trifolium pratense* L.) and alfalfa (*Medicago sativa* L.). *Journal of Applied Entomology* **117**, 469–476.

EKBOM, B.S., WIKTELIUS, S. AND CHIVERTON, P.A. (1992). Can polyphagous predators control the bird cherry-oat aphid (*Rhopalosiphum padi*) in spring cereals? *Entomologia Experimentalis et Applicata* **65**, 215–223.

EKSCHMITT, K., WOLTERS, V. AND WEBER, M. (1997). Spiders, carabids and staphylinids: the ecological potential of predatory macroarthropods. In: *Fauna in soil ecosystems.* Ed. G. Benckiser, pp 307–362. New York: Marcel Dekker.

EVENHUIS, H.H. (1983). Role of carabids in the natural control of the black vine weevil, *Otiorhynchus sulcatus. Mitteilungen der Deutschen Gessellschaft für Allgemeine und Angewandte Entomologie* **4**, 83–85.

FERGUSON, K.I. AND STILING, P. (1996). Non-additive effects of multiple natural enemies on aphid populations. *Oecologia* **108**, 375–379.

FERRERO, F. (1985). A precious forest auxiliary insect: *Calosoma sycophanta. Phytoma* **370**, 28.

FINCH, S. (1996). Effect of beetle size on predation of cabbage root fly eggs by ground beetles. *Entomologia Experimentalis et Applicata* **81**, 199–206.

FINCH, S. AND ELLIOTT, M.S. (1992). Predation of cabbage root fly eggs by Carabidae. *Bulletin OILB/SROP* **15** (4), 176–183.

FLOATE, K.D., DOANE, J.F. AND GILLOTT, C. (1990). Carabid predators of the wheat midge (Diptera: Cecidomyiidae) in Saskatchewan. *Environmental Entomology* **19**, 1503–1511.

FOWLER, H.G. (1987). Predatory behaviour of *Megacephala fulgida* (Coleoptera: Cicindelidae). *The Coleopterists Bulletin* **41**, 407–408.

FOX, C.J.S. AND MACLELLAN, C.R. (1956). Some Carabidae and Staphylinidae shown to feed on a wireworm, *Agriotes sputator* (L.), by the precipitin test. *Canadian Entomologist* **88,** 228–231.

FRANK, J.H. (1967a). A serological method used in the investigation of the predators of the pupal stage of the winter moth, *Operophtera brumata* (L.) (Hydriomenidae). *Quaestiones Entomologicae* **3**, 95–105.

FRANK, J.H. (1967b). The insect predators of the pupal stage of the winter moth, *Operophtera brumata* (L.) (Lepidoptera: Hydriomenidae). *Journal of Animal Ecology* **36**, 375–389.

FRANK, J.H. (1971). Carabidae (Coleoptera) as predators of the red-backed cutworm (Lepidoptera: Noctuidae) in central Alberta. *Canadian Entomologist* **103**, 1039–1044.

FUESTER, R.W. AND TAYLOR, P.B. (1996). Differential mortality in male and female Gypsy Moth (Lepidoptera: Lymantriidae) pupae by invertebrate natural enemies and other factors. *Environmental Entomology* **25,** 536–547.

FUESTER, R.W., SANDRIDGE, P.T., DILL, N.H., MCLAUGHLIN, J.M., TAYLOR, P.B., SIGMOND, J.O.D. AND NEWLON, C.J. (1997). Apparent fate of Gypsy Moth (Lepidoptera: Lymantriidae) pupae stung by the introduced parasite *Coccygomimus disparis* (Hymenoptera: Ichneumonidae). *Environmental Entomology* **26**, 1442–1451.

FULLER, B.W. (1988). Predation by *Calleida decora* (F.) (Coleoptera: Carabidae) on velvetbean caterpillar (Lepidoptera: Noctuidae) in soybean. *Journal of Economic Entomology* **81**, 127–129.

FULLER, B.W. AND REAGAN, T.E. (1988). Comparative predation of the sugarcane borer (Lepidoptera: Pyralidae) on sweet sorghum and sugarcane. *Journal of Economic Entomology* **81**, 713–717.

GIBSON, P.H., COSENS, D. AND BUCHANAN, K. (1997). A chance field observation and pilot laboratory studies of predation of the New Zealand flatworm by the larvae and adults of carabid and staphylinid beetles. *Annals of Applied Biology* **130**, 581–585.

GODFREY, K.E., WHITCOMB, W.H. AND STIMAC, J.L. (1989). Arthropod predators of velvetbean caterpillar, *Anticarsia gemmatalis* Hübner (Lepidoptera: Noctuidae), eggs and larvae. *Environmental Entomology* **18**, 118–123.

GONG, O.J., ZHANG, H.M., LI, Y.G. AND CHEN, S.L. (1988). A study on *Hyssia adusta* Draudt. *Acta Entomologica Sinica* **31**, 280–286.

GRAFIUS, E. AND WARNER, F.W. (1989). Predation by *Bembidion quadrimaculatum* (Coleoptera: Carabidae) on *Delia antiqua* (Diptera: Anthomyiidae). *Environmental Entomology* **18**, 1056–1059.

GRAVESEN, E. AND TOFT, S. (1987). Grass fields as reservoirs for polyphagous predators (Arthropoda) of aphids (Homopt., Aphididae). *Journal of Applied Entomology* **104**, 461–473.

GRIFFITHS, E., WRATTEN, S.D. AND VICKERMAN, G.P. (1985). Foraging by the carabid *Agonum dorsale* in the field. *Ecological Entomology* **10**, 181–189.

GRIMM, B., PAILL, W. AND KAISER, H. (2000). The 'Spanish slug': autecology, predators and wild plants as food plants. *Förderungsdienst* **48**, 11–16.

GUIDO, A.S. AND FOWLER, H.G. (1988). *Megacephala fulgida* (Coleoptera: Cicindelidae): a phonotactically orienting predator of *Scapteriscus* mole crickets (Orthoptera: Gryllotalpidae). *Cicindela* **20**, 51–52.

HAGLEY, E.A.C. AND ALLEN, W.R. (1988). Ground beetles (Coleoptera: Carabidae) as predators of the Codling Moth, *Cydia pomonella* (L.) (Lepidoptera: Tortricidae). *Canadian Entomologist* **120**, 917–925.

HAGLEY, E.A.C. AND ALLEN, W.R. (1990).The green apple aphid, *Aphis pomi* Degeer (Homoptera: Aphididae), as prey of polyphagous arthropod predators in Ontario. *Canadian Entomologist* **122**, 1221–1228.

HALSALL, N.B. AND WRATTEN, S.D. (1988a). Video recording of aphid predation by Carabidae in a wheat crop. *1988 Brighton Crop Protection Conference – Pests and Diseases*, British Crop Protection Council, Farnham, UK, **3**, pp 1047–1052.

HALSALL, N.B. AND WRATTEN, S.D. (1988b). Video recordings of aphid predation in a wheat crop. *Aspects of Applied Biology* **17**, 277–280.

HAMON, N., BARDNER, R., ALLEN-WILLIAMS, L. AND LEE, J.B. (1990). Carabid populations in field beans and their effect on the population dynamics of *Sitona lineatus* (L.). *Annals of Applied Biology* **117**, 51–62.

HANCE, T. (1987). Predation impact of carabids at different population densities on *Aphis fabae* development in sugar beet. *Pedobiologia* **30**, 251–262.

HANCE, T. AND RENIER, L. (1987). An ELISA technique for the study of the food of carabids. *Acta Phytopathologica et Entomologica Hungarica* **22**, 363–368.

HARWOOD, J.D., PHILLIPS, S.W., SUNDERLAND, K.D. AND SYMONDSON, W.O.C. (2001). Secondary predation: quantification of food chain errors in an aphid–spider–carabid system using monoclonal antibodies. *Molecular Ecology* **10**, 2049–2057.

HAZZARD, R.V., FERRO, D.N., VAN DRIESCHE, R.G. AND TUTTLE, A.F. (1991). Mortality of eggs of Colorado Potato Beetle (Coleoptera: Chrysomelidae) from predation by *Coleomegilla maculata* (Coleoptera: Coccinellidae). *Environmental Entomology* **20**, 841–848.

HEIMPEL, G.E. AND HOUGH-GOLDSTEIN, J.A. (1992). A survey of arthropod predators of *Leptinotarsa decemlineata* (Say) in Delaware potato fields. *Journal of Agricultural Entomology* **9**, 137–142.

HELENIUS, J. (1990). Effect of epigeal predators on infestation by the aphid *Rhopalosiphum padi* and on grain yield of oats in monocrops and mixed intercrops. *Entomologia Experimentalis et Applicata* **54**, 225–236.

HENGEVELD, R. (1980). Polyphagy, oligophagy and food specialization in ground beetles (Coleoptera, Carabidae). *Netherlands Journal of Zoology* **30**, 564–584.

HILBECK, A. AND KENNEDY, G.G. (1996). Predators feeding on the Colorado Potato Beetle in insecticide-free plots and insecticide-treated commercial potato fields in Eastern North Carolina. *Biological Control* **6**, 273–282.

HODGE, S., WRATTEN, S., SMITH, J., THOMAS, M. AND FRAMPTON, C. (1999). The role of leaf wounding and an epigeal predator on caterpillar damage to tomato plants. *Annals of Applied Biology* **134,** 137–141.

HOLLAND, J.M. AND THOMAS, S.R. (1996). Quantifying the impact of polyphagous invertebrate predators in controlling cereal pests and in preventing quantity and quality reductions. *1996 Brighton Crop Protection Conference – Pests and Diseases*, pp 629–634.

HOLLAND, J.M. AND THOMAS, S.R. (1997a). Assessing the role of beneficial invertebrates in conventional and integrated farming systems during an outbreak of *Sitobion avenae*. *Biological Agriculture and Horticulture* **15**, 73–82.

HOLLAND, J.M. AND THOMAS, S.R. (1997b). Quantifying the impact of polyphagous invertebrate predators in controlling cereal aphids and in preventing wheat yield and quality reductions. *Annals of Applied Biology* **131**, 375–397.

HOLLAND, J.M., THOMAS, S.R. AND HEWITT, A. (1996). Some effects of polyphagous predators on an outbreak of cereal aphid (*Sitobion avenae* F.) and orange wheat blossom midge (*Sitodiplosis mosselana* Géhin). *Agriculture, Ecosystems and Environment* **59**, 181–190.

HOLLAND, J.M., PERRY, J.M. AND WINDER, L. (1999). The within-field spatial and temporal distribution of arthropods in winter wheat. *Bulletin of Entomological Research* **89**, 499–513.

HOLLAND, J.M., WINDER, L. AND PERRY, J.N. (2000). The impact of dimethoate on the spatal distribution of beneficial arthropods in winter wheat. *Annals of Applied Biology* **136**, 93–105.

HOLOPAINEN, J.K. AND HELENIUS, J. (1992). Gut contents of ground beetles (Col., Carabidae), and activity of these and other epigeal predators during an outbreak of *Rhopalosiphum padi* (Hom., Aphididae). *Acta Agriculturae Scandinavica, Section B, Soil and Plant Science* **42**, 57–61.

HONDO, M. AND MORIMOTO, N. (1997). Effect of predation by the specialist predator, *Parena perforata* Bates (Coleoptera: Carabidae) on population changes of the Mulberry Tiger Moth, *Thanatarctia imparilis* Butler (Lepidoptera: Arctiidae). *Applied Entomology and Zoology* **32**, 311–316.

HORNE, P.A. (1992). Comparative life histories of two species of *Notonomus* (Coleoptera: Carabidae) in Victoria. *Australian Journal of Zoology* **40**, 163–171.

HOUGH-GOLDSTEIN, J.A., HEIMPEL, G.E., BECHMANN, H.E. AND MASON, C.E. (1993). Arthropod natural enemies of the Colorado potato beetle. *Crop Protection* **12**, 324–334.

HUDSON, W.G., FRANK, J.H. AND CASTNER, J.L. (1988). Biological control of *Scapteriscus* spp. mole crickets (Orthoptera: Gryllotalpidae) in Florida. *Bulletin of the Entomological Society of America* **34**, 192–198.

ITO, Y. AND MIYASHITA, K. (1968). Biology of *Hyphantria cunea* Drury (Lepidoptera: Arctiidae) in Japan. V. Preliminary life tables and mortality data in urban areas. *Researches in Population Ecology* **10**, 177–209.

ITO, Y., MIYASHITA, K. AND SEKIGUCHI, K. (1962). Studies on the predators of the rice crop insect pests, using the insecticidal check method. *Japanese Journal of Ecology* **12**, 1–11.

IVES, A.R., KAREIVA, P. AND PERRY, R. (1993). Response of a predator to variation in prey density at three hierarchical scales: lady beetles feeding on aphids. *Ecology* **74**, 1929–1938.

JANSSENS, J. AND DE CLERCQ, R. (1990). Observations on Carabidae, Staphylinidae and Araneae as predators of cereal aphids in winter wheat. *Mededelingen van de Fakulteit Landbouwwetenschappen Rijksuniversiteit Gent* **55** (2b), 471–475.

JOHANSEN, N.S. (1997). Mortality of eggs, larvae and pupae and larval dispersal of the cabbage moth, *Mamestra brassicae*, in white cabbage in south-eastern Norway. *Entomologia Experimentalis et Applicata* **83,** 347–360.

JOHNSON, P.C. AND REEVES, R.M. (1995). Incorporation of the biological marker rubidium in Gypsy Moth (Lepidoptera: Lymantriidae) and its transfer to the predator *Carabus nemoralis* (Coleoptera: Carabidae). *Environmental Entomology* **24**, 46–51.

KABACIK-WASYLIK, D. (1989). The food of two Carabidae species in potato crops. *Polish Ecological Studies* **15**, 111–117.

KAMATA, N. AND IGARASHI, Y. (1995). An example of numerical response of the carabid beetle, *Calosoma maximowiczi* Morawitz (Col., Carabidae), to the beech caterpillar,

Quadricalcarifera punctatella (Motschulsky) (Lep., Notodontidae). *Journal of Applied Entomology* **119**, 139–142.

KAPUGE, S.H., DANTHANARAYANA, W. AND HOOGENRAAD, N. (1987). Immunological investigation of prey–predator relationships for *Pieris rapae* (L.) (Lepidoptera: Pieridae). *Bulletin of Entomological Research* **77**, 247–254.

KELLY, B. AND RÉGNIÈRE, J. (1985). Predation on pupae of the spruce budworm (Lepidoptera: Tortricidae) on the forest floor. *Canadian Entomologist* **117**, 33–38.

KENNEDY, T.F. (1994). The ecology of *Bembidion obtusum* (Ser.) (Coleoptera: Carabidae) in winter wheat fields in Ireland. *Proceedings of the Royal Irish Academy, Series B* **94**, 33–40.

KIDD, P.W. AND RUMMEL, D.R. (1997). Effect of insect predators and a pyrethroid insecticide on cotton aphid, *Aphis gossypii* Glover, population density. *Southwestern Entomologist* **22**, 381–393.

KIELTY, J.P., ALLEN-WILLIAMS, L.J., UNDERWOOD, N. AND EASTWOOD, E.A. (1996). Behavioral responses of three species of ground beetle (Coleoptera: Carabidae) to olfactory cues associated with prey and habitat. *Journal of Insect Behavior* **9**, 237–250.

KIELTY, J.P., ALLEN-WILLIAMS, L.J. AND UNDERWOOD, N. (1999). Prey preferences of six species of Carabidae (Coleoptera) and one Lycosidae (Araneae) commonly found in UK arable crop fields. *Journal of Applied Entomology* **123**, 193–200.

KIRK, V.M. (1973). Biology of a ground beetle, *Harpalus pensylvanicus*. *Annals of the Entomological Society of America* **66**, 513–517.

KIRK, V.M. (1975). Biology of *Pterostichus chalcites*, a ground beetle of cropland. *Annals of the Entomological Society of America* **68**, 855–858.

KIRKLAND, D.L., EVANS, K.A. AND LOLA-LUZ, T. (1998). Manipulating the behaviour of beneficial insects in cereal crops to enhance control of aphids. *1998 Brighton Crop Protection Conference – Pests and Diseases* **2**, pp 663–668.

KOBAYASHI, M., KUDAGAMAGE, C. AND NUGALIYADDE, L. (1995). Distribution of larvae of *Ophionea indica* Thunberg (Carabidae), a predator of the Rice Gall Midge, *Orseolia oryzae* (Wood-Mason) in paddy fields of Sri Lanka. *Japan Agricultural Research Quarterly* **29**, 89–93.

KOVAL, A.G. (1985). The Carpathian ground beetle – a natural enemy of the Colorado beetle. *Zashchita Rastenii* **6**, 25–26.

KOVAL, A.G. (1986). Predatory carabids as natural enemies of the Colorado potato beetle. *Zashchita Rastenii* **11**, 45–46.

KOVAL, A.G. (1999). Contribution to the knowledge of carabids (Coleoptera, Carabidae) preying on Colorado Potato Beetle in potato fields of the Transcarpathian Region. *Entomological Review* **79**, 523–532.

KROMP, B. (1999). Carabid beetles in sustainable agriculture: a review on pest control efficacy, cultivation impacts and enhancement. *Agriculture, Ecosystems and Environment* **74**, 187–228.

KUMAR, P. AND RAJAGOPAL, D. (1990). Carabid beetle, *Omphra pilosa* Klug (Coleoptera: Carabidae) a potential predator on termites. *Journal of Biological Control* **4**, 105–108.

LANDIS, D.A. AND VAN DER WERF, W. (1997). Early-season predation impacts the establishment of aphids and spread of beet yellows virus in sugar beet. *Entomophaga* **42**, 499–516.

LANG, A., FILSER, J. AND HENSCHEL, J.R. (1999). Predation by ground beetles and wolf spiders on herbivorous insects in a maize crop. *Agriculture, Ecosystems and Environment* **72**, 189–199.

LAPA, A.M. AND TKACHEV, V.M. (1992). Ecological aspects of an integrated pest management system in orchards. *Acta Phytopathologica et Entomologica Hungarica* **27**, 401–403.

LAROCHELLE, A. (1990). *The food of carabid beetles (Coleoptera: Carabidae, including Cicindelinae)*. Fabreries, Supplément 5. Québec: Association des Entomologistes Amateurs du Québec.

LAUB, C.A. AND LUNA, J.M. (1992). Winter cover crop suppression practices and natural enemies of armyworm (Lepidoptera: Noctuidae) in no-till corn. *Environmental Entomology* **21**, 41–49.

LEE, J.H., JOHNSON, S.J. AND WRIGHT, V.L. (1990). Quantitative survivorship analysis of the velvetbean caterpillar (Lepidoptera: Noctuidae) pupae in soybean fields in Louisiana. *Environmental Entomology* **19**, 978–986.

LESIEWICZ, D.S., LESIEWICZ, J.L., BRADLEY, J.R. AND VAN DUYN, J.W. (1982). Serological determination of carabid (Coleoptera: Adephaga) predation of corn earworm (Lepidoptera: Noctuidae) in field corn. *Environmental Entomology* **11**, 1183–1186.

LOGANATHAN, J. AND DAVID, P.M.M. (1999). Predator complex of the teak defoliator, *Hyblaea puera* Cramer (Lepidoptera: Hyblaeidae) in an intensively managed teak plantation at Veeravanallur, Tamil Nadu. *Entomon* **24**, 259–263.

LOREAU, M. (1983a). Le régime alimentaire de *Abax ater* Vill. (Coleoptera, Carabidae). *Acta Oecologica* **4**, 253–263.

LOREAU, M. (1983b). Le régime alimentaire de huit carabides (Coleoptera) communs en milieu forestier. *Acta Oecologica* **4**, 331–343.

LOREAU, M. (1984). Experimental study of the feeding of *Abax ater* Villers, *Carabus problematicus* Herbst and *Cychrus attenuatus* Fabricius (Coleoptera, Carabidae). *Annales Société Royale Zoologique de Belgique* **114,** 227–240.

LORENZ, W. (1998a). *Systematic list of extant ground beetles of the world (Insecta Coleoptera 'Geadephaga': Trachypachidae and Carabidae incl. Paussinae, Cicindelinae, Rhysodinae).* Tutzing, Germany, (privately published by the author, after meeting the Zoological Code's criteria), 503 pages.

LORENZ, W. (1998b). *Nomina Carabidarum: a directory of the scientific names of ground beetles (Insecta Coleoptera 'Geadephaga': Trachypachidae and Carabidae incl. Paussinae, Cicindelinae, Rhysodinae).* Tutzing, Germany, (privately published by the author, after meeting the Zoological Code's criteria), 939 pages.

LOSEY, J.E. AND DENNO, R.F. (1998a). Positive predator–predator interactions: enhanced predation rates and synergistic suppression of aphid populations. *Ecology* **79**, 2143–2152.

LOSEY, J.E. AND DENNO, R.F. (1998b). Interspecific variation in the escape responses of aphids: effect on risk of predation from foliar-foraging and ground-foraging predators. *Oecologia* **115**, 245–252.

LÖVEI, G.L. (1986). The use of the biochemical methods in the study of carabid feeding: the potential of isoenzyme analysis and ELISA. In: *Feeding behaviour and accessibility of food for carabid beetles.* Eds. P.J. Den Boer, L. Grüm and J. Szyszko, pp 21–27. Warsaw: Warsaw Agricultural University Press.

LÖVEI, G.L. AND SUNDERLAND, K.D. (1996). Ecology and behaviour of ground beetles (Coleoptera: Carabidae). *Annual Review of Entomology* **41**, 231–256.

LÖVEI, G.L. AND SZENTKIRÁLYI, F. (1984). Carabids climbing maize plants. *Zeitschrift für angewandte Entomologie* **97**, 107–110.

LÜBKE-AL HUSSEIN, M. AND TRILTSCH, H. (1994). Some aspects about polyphagous arthropods as antagonists of aphids in cereal fields. *Bulletin OILB/SROP* **17** (4), 168–178.

LUCKEY, T.D. (1968). Insecticide hormoligosis. *Journal of Economic Entomology* **61**, 7–12.

LUFF, M.L. (1987). Biology of polyphagous ground beetles in agriculture. *Agricultural Zoloogy Reviews* **2**, 237–278.

LUND, R.D. AND TURPIN, F.T. (1977). Serological investigation of black cutworm larval consumption by ground beetles. *Annals of the Entomological Society of America* **70**, 322–324.

LUONG, M.C. (1987). Predators of brown planthopper *Nilaparvata lugens* Stål (BPH) in ricefields of the Mekong Delta, Vietnam. *International Rice Research Newsletter* **12**, 31–32.

LYS, J.A. (1995). Observation of epigeic predators and predation on artificial prey in a cereal field. *Entomologia Experimentalis et Applicata* **75**, 265–272.

MALSCHI, D. AND MUSTEA, D. (1995). Protection and use of entomophagous arthropod fauna in cereals. *Romanian Agricultural Research* **4,** 93–99.

MAZANEC, Z. (1987). Natural enemies of *Perthida glyphopa* Common (Lepidoptera: Incurvariidae). *Journal of the Australian Entomological Society* **26**, 303–308.

MCKEMEY, A.R. (2000). *Integrating behavioural aspects of carabid–slug interactions with immunological data on predator population ecology*. PhD thesis, University of Cardiff, UK.

MCKEMEY, A.R., SYMONDSON, W.O.C., GLEN, D.M. AND BRAIN, P. (2001). Effects of slug size on predation by *Pterostichus melanarius* (Coleoptera: Carabidae). *Biocontrol Science and Technology* **11**, 83–93.

MENALLED, F.D., LEE, J.C. AND LANDIS, D.A. (1999). Manipulating carabid beetle abundance alters prey removal rates in corn fields. *BioControl* **43**, 441–456.

MIENIS, H.K. (1988). Additional records of predation on landsnails by the ground beetle *Carabus impressus* in Israel. *The Conchologist's Newsletter* **106**, 121–123.

MOCHIZUKI, A. (1990). A carabid predator, *Parena laesipennis* (Bates) (Coleoptera: Carabidae) attacking *Pidorus glaucopis* (Drury) (Lepidoptera: Zygaenidae). *Applied Entomology and Zoology* **25**, 319–320.

MOHAMED, A.H., LESTER, P.J. AND HOLTZER, T.O. (2000). Abundance and effects of predators and parasitoids on the Russian Wheat Aphid (Homoptera: Aphididae) under organic farming conditions in Colorado. *Environmental Entomology* **29**, 360–368.

MONSRUD, C. AND TOFT, S. (1999). The aggregative numerical response of polyphagous predators to aphids in cereal fields: attraction to what? *Annals of Applied Biology* **134**, 265–270.

MUNDY, C.A., ALLEN-WILLIAMS, L.J., UNDERWOOD, N. AND WARRINGTON, S. (2000). Prey selection and foraging behaviour by *Pterostichus cupreus* L. (Col., Carabidae) under laboratory conditions. *Journal of Applied Entomology* **124**, 349–358.

MURALEEDHARAN, N., RADHAKRISHNAN, B. AND SELVASUNDARAM, R. (1991). Observations on the life-history and feeding rates of *Calleida nilgirensis* Straneo (Coleoptera: Carabidae): a predator of the flushworm of tea. *Journal of Ecobiology* **3**, 189–192.

PAKARINEN, E. (1994). The importance of mucus as a defence against carabid beetles by the slugs *Arion fasciatus* and *Deroceras reticulatum*. *Journal of Molluscan Studies* **60**, 149–155.

PEDERSEN, J.C., DAMGAARD, P.H., EILENBERG, J. AND HANSEN, B.M. (1995). Dispersal of *Bacillus thuringiensis* var. *kustaki* in an experimental cabbage field. *Canadian Journal of Microbiology* **41**, 118–25.

PERRY, J.N. (1998). Measures of spatial pattern for counts. *Ecology* **79**, 1008–1017.

PETERSEN, M.K. AND HOLST, N. (1997). Modelling natural control of cereal aphids. II. The carabid *Bembidion lampros*. *Acta Jutlandica* **72**, 207–219.

PILLAI, G.B. AND NAIR, K.R. (1986). Additions to the natural enemy complex of the coconut caterpillar *Opisina arenosella* Wlk. *Journal of Plantation Crops* **14**, 138–140.

PILLAI, G.B. AND NAIR, K.R. (1990). On the biology of *Calleida splendidula* (F.) (Coleoptera: Carabidae), a predator of the coconut leaf eating caterpillar, *Opisina arenosella* Wlk. *Indian Coconut Journal* **20**, 14–17.

PIMENTEL, D. (1995). Ecological theory, pest problems and biologically based solutions. In: *Ecology and integrated farming systems.* Eds. D.M. Glen, M.P. Greaves and H.M. Anderson, pp 69–82. Chichester: John Wiley.

POLLET, M. AND DESENDER, K. (1985) Adult and larval feeding ecology in *Pterostichus melanarius* Ill. (Coleoptera, Carabidae). *Mededelingen van de Fakulteit Landbouwwetenschappen Rijksuniversiteit Gent* **50** (2b), 581–594.

POLLET, M. AND DESENDER, K. (1987a). The consequences of different life histories in ground beetles for their feeding ecology and impact on other pasture arthropods. *Mededelingen van de Fakulteit Landbouwwetenschappen Rijksuniversiteit Gent* **52** (2a), 179–190.

POLLET, M. AND DESENDER, K. (1987b). Feeding ecology of grassland-inhabiting carabid beetles (Carabidae, Coleoptera) in relation to the availability of some prey groups. *Acta Phytopathologica et Entomologica Hungarica* **22**, 223–246.

PORT, G.R., GLEN, D.M. AND SYMONDSON, W.O.C. (2001). Success in biological control of terrestrial molluscs. In: *Measures of success in biological control.* Eds. G. Gurr and S.D. Wratten. Dordrecht: Kluwer Academic Publishers.

POULIN, G. AND O'NEIL, L.C. (1969). Observations sur les prédateurs de la limace noire, *Arion ater* (L.) (Gastéropodes, Pulmonés, Arionidés). *Phytoprotection* **50**, 1–6.

RAJAGOPAL, D. AND KUMAR, P. (1992). Carabids (Coleoptera: Carabidae) as potential predators on major crop pests in South India. *Journal of Biological Control* **6**, 13–17.

RÄMERT, B. AND EKBOM, B. (1996). Intercropping as a management strategy against carrot rust fly (Diptera: Psilidae): a test of enemies and resource concentration hypotheses. *Environmental Entomology* **25**, 1092–1100.

RIDDICK, E.W. AND MILLS, N.J. (1994). Potential of adult carabids (Coleoptera: Carabidae) as predators of fifth-instar Codling Moth (Lepidoptera: Tortricidae) in apple orchards in California. *Environmental Entomology* **23**, 1338–1345.

RISHI, N.D. AND SHAH, K.A. (1985). Survey and bioecological studies on the natural enemy complex of Indian gypsy moth, *Lymantria obfuscata* Walker (Lepidoptera: Lymantriidae). *Journal of Entomological Research* **9**, 82–93.

ROOM, P.M. (1979). Parasites and predators of *Heliothis* spp. (Lepidoptera: Noctuidae) in cotton in the Namoi valley, New South Wales. *Journal of the Australian Entomological Society* **18**, 223–228.

ROY, H.E. AND PELL, J.K. (2000). Interactions between entomopathogenic fungi and other natural enemies: implications for biological control. *Biocontrol Science and Technology* **10**, 737–752.

ROY, H.E., PELL, J.K., CLARK, S.J. AND ALDERSON, P.G. (1998). Implications of predator foraging on aphid pathogen dynamics. *Journal of Invertebrate Pathology* **71**, 236–247.

RYAN, M.F. (1975). The natural mortality of wheat bulb fly pupae, *Leptohylemyia coarctata* (Fall.) (Dipt., Anthomyiidae). *Plant Pathology* **24**, 27–30.

RYAN, M.F. AND RYAN, J. (1973). The natural mortality of cabbage root fly eggs in peatland. *Scientific Proceedings of the Royal Dublin Society, Series B* **3**, 195–199.

SALISBURY, A.N. AND LEATHER, S.R. (1998). Migration of larvae of the large pine weevil, *Hylobius abietis* L. (Col., Curculionidae): possible predation a lesser risk than death by starvation? *Journal of Applied Entomology* **122**, 295–299.

SCHELLER, H.V. (1984). The role of ground beetles (Carabidae) as predators on early populations of cereal aphids in spring barley. *Zeitschrift für angewandte Entomologie* **97**, 451–463.

SCHELVIS, J. AND SIEPEL, H. (1988). Larval food spectra of *Pterostichus oblongopunctatus* and *P. rhaeticus* in the field (Coleoptera: Carabidae). *Entomologia Generalis* **13**, 61–66.

SCHMIED, A. AND FÜHRER, E. (1996). The impact of ground beetle species (Coleoptera: Carabidae) in spruce stands, damaged by *Pristiphora abietina* (Hymenoptera: Tenthredinidae). *Entomologia Generalis* **21**, 81–94.

SELVASUNDARAM, R. AND MURALEEDHARAN, N. (1987). Natural enemies of certain leaf folding caterpillar pests of tea in southern India. *Journal of Coffee Research* **17**, 118–119.

SERGEYEVA, T.K. AND GRYUNTAL, S.Y. (1990). Relationships of ground beetles, *Pterostichus* species, with their food resources. *Entomological Review* **69**, 58–67.

SHAW, P.B., OWENS, J.C., HUDDLESTON, E.W. AND RICHMAN, D.B. (1987). Role of arthropod predators in mortality of early instars of the range caterpillar, *Hemileuca oliviae* (Lepidoptera: Saturniidae). *Environmental Entomology* **16**, 814–820.

SILESHI, G., KENIS, M., OGOL, C.K.P.O. AND SITHANANTHAM, S. (2001). Predators of *Mesoplatys ochroptera* in sesbania-planted-fallows in Eastern Zambia. *Biocontrol* **46**, 289–310.

SINGH, H., SINGH, J. AND SADANA, G.L. (1998). Studies on the natural enemies of whitebacked planthopper, *Sogatella furcifera* (Horvath), (Delphacidae: Hemiptera). *Agricultural Science Digest* **18**, 51–53.

SNYDER, W.E. AND IVES, A.R. (2001). Generalist predators disrupt biological control by a specialist parasitoid. *Ecology* **82**, 705–716.

SNYDER, W.E. AND WISE, D.H. (1999). Predator interference and the establishment of generalist predator populations for biocontrol. *Biological Control* **15**, 283–292.

SNYDER, W.E. AND WISE, D.H. (2000). Antipredator behavior of spotted cucumber beetles (Coleoptera: Chrysomelidae) in response to predators that pose varying risks. *Environmental Entomology* **29**, 35–42.

SNYDER, W.E. AND WISE, D.H. (2001). Contrasting trophic cascades generated by a community of generalist predators. *Ecology* **82**, 1571–1583.

SONGA, J.M. AND HOLLIDAY, N.J. (1997). Laboratory studies of predation of grasshopper eggs, *Melanoplus bivittatus* (Say), by adults of two species of *Pterostichus* Bonelli (Coleoptera: Carabidae). *Canadian Entomologist* **129**, 1151–1159.

SOPP, P. AND CHIVERTON, P. (1987). Autumn predation of cereal aphids by polyphagous predators in Southern England: a 'first look' using ELISA. *Bulletin OILB/SROP* **10** (1), 103–108.

SOPP, P.I., SUNDERLAND, K.D., FENLON, J.S. AND WRATTEN, S.D. (1992). An improved quantitative method for estimating invertebrate predation in the field using an enzyme-linked immunosorbent assay (ELISA). *Journal of Applied Ecology* **29**, 295–302.

SOTHERTON, N.W. (1982). Predation of a chrysomelid beetle (*Gastrophysa polygoni*) in cereals by polyphagous predators. *Annals of Applied Biology* **101**, 196–199.

SPEIGHT, M.R. AND LAWTON, J.H. (1976). The influence of weed-cover on the mortality imposed on artificial prey by predatory ground beetles in cereal fields. *Oecologia* **23**, 211–223.

STAMP, N.E. (1997). Behavior of harassed caterpillars and consequences for host plants. *Oikos* **79**, 147–154.

SUENAGA, H. AND HAMAMURA, T. (1998). Laboratory evaluation of carabid beetles (Coleoptera: Carabidae) as predators of Diamondback Moth (Lepidoptera: Plutellidae) larvae. *Environmental Entomology* **27**, 767–772.

SUNDERLAND, K.D. (1975). The diet of some predatory arthropods in cereal crops. *Journal of Applied Ecology* **12,** 507–515.

SUNDERLAND, K.D. (1988). Carabidae and other invertebrates. In: *Aphids, their biology, natural enemies and control, Volume B.* Eds. A.K. Minks and P. Harrewijn, pp 293–310. Amsterdam: Elsevier Science Publishers.

SUNDERLAND, K.D. (1996). Progress in quantifying predation using antibody techniques. In: *The ecology of agricultural pests: biochemical approaches.* Eds W.O.C. Symondson and J.E. Liddell, pp 419–455. London: Chapman & Hall.

SUNDERLAND, K.D. AND SUTTON, S.L. (1980). A serological study of arthropod predation on woodlice in a dune grassland ecosystem. *Journal of Animal Ecology* **49**, 987–1004.

SUNDERLAND, K.D. AND VICKERMAN, G.P. (1980). Aphid feeding by some polyphagous predators in relation to aphid density in cereal fields. *Journal of Applied Ecology* **17,** 389–396.

SUNDERLAND, K.D., STACEY, D.L. AND EDWARDS, C.A. (1980). The role of polyphagous predators in limiting the increase of cereal aphids in winter wheat. *Bulletin OILB/SROP* **1980/III/4,** 85–92.

SUNDERLAND, K.D., CROOK, N.E., STACEY, D.L. AND FULLER, B.J. (1987). A study of aphid feeding by polyphagous predators on cereal aphids using ELISA and gut dissection. *Journal of Applied Ecology* **24,** 907–933.

SUNDERLAND, K.D., CHAMBERS, R.J. AND CARTER, O.C.R. (1988). Potential interactions between varietal resistance and natural enemies in the control of cereal aphids. In: *Integrated crop protection in cereals.* Eds. R. Cavalloro and K.D. Sunderland, pp 41–56. Rotterdam: A.A. Balkema.

SUNDERLAND, K.D., DE SNOO, G.R., DINTER, A., HANCE, T., HELENIUS, J., JEPSON, P., KROMP, B., LYS, J.A., SAMU, F., SOTHERTON, N.W., TOFT, S. AND ULBER, B. (1995a). Density estimation for invertebrate predators in agroecosystems. *Acta Jutlandica* **70**, 133–164.

SUNDERLAND, K.D., LÖVEI, G.L. AND FENLON, J. (1995b). Diets and reproductive phenologies of the introduced ground beetles *Harpalus affinis* and *Clivina australasiae* (Coleoptera: Carabidae) in New Zealand. *Australian Journal of Zoology* **43**, 39–50.

SUNDERLAND, K.D., AXELSEN, J.A., DROMPH, K., FREIER, B., HEMPTINNE, J-L., HOLST, N.H., MOLS, P.J.M., PETERSEN, M.K., POWELL, W., RUGGLE, P., TRILTSCH, H. AND WINDER, L. (1997). Pest control by a community of natural enemies. *Acta Jutlandica*, **72**, 271–326.

SUNDERLAND, K.D., POWELL, W. AND SYMONDSON, W.O.C. (in press) Populations and communities. Chapter 6. In: *Insects as natural enemies.* Ed. M.A. Jervis. Dordrecht: Kluwer Academic Publishers.

SYMONDSON, W.O.C. (1989). Biological control of slugs by carabids. In: *Slugs and snails in world agriculture.* Ed. I. Henderson, pp 295–300. Thornton Heath: British Crop Protection Council Monograph 41.

SYMONDSON, W.O.C. (1993). The effects of crop development upon slug distribution and control by *Abax parallelepipedus* (Coleoptera: Carabidae). *Annals of Applied Biology* **123**, 449–457.

SYMONDSON, W.O.C. (1994). The potential of *Abax parallelepipedus* (Col.: Carabidae) for mass breeding as a biological control agent against slugs. *Entomophaga* **39**, 323–333.

SYMONDSON, W.O.C. (in press). Coleoptera (Carabidae, Drilidae, Lampyridae and Staphylinidae). In: *Natural enemies of terrestrial molluscs.* Ed. G.M. Barker. Oxford: CAB International.

SYMONDSON, W.O.C. AND LIDDELL, J.E. (1993). The detection of predation by *Abax parallelepipedus* and *Pterostichus madidus* (Coleoptera: Carabidae) on Mollusca using a quantitative ELISA. *Bulletin of Entomological Research* **83**, 641–647.

SYMONDSON, W.O.C. AND LIDDELL, J.E. (1996). A species-specific monoclonal antibody system for detecting the remains of field slugs, *Deroceras reticulatum* (Miller) (Mollusca: Pulmonata), in carabid beetles (Coleoptera: Carabidae). *Biocontrol Science and Technology* **6**, 91–99.

SYMONDSON, W.O.C., SUNDERLAND, K.D. AND GREENSTONE, M.H. (2002). Can generalist predators be effective biocontrol agents? *Annual Review of Entomology* **47**, 561–594.

SYMONDSON, W.O.C., GLEN, D.M., WILTSHIRE, C.W., LANGDON, C.J. AND LIDDELL, J.E. (1996). Effects of cultivation techniques and methods of straw disposal on predation by *Pterostichus melanarius* (Coleoptera: Carabidae) upon slugs (Gastropoda: Pulmonata) in an arable field. *Journal of Applied Ecology* **33,** 741–753.

SYMONDSON, W.O.C., GLEN, D.M., IVES, A.R., LANGDON, C.J. AND WILTSHIRE, C.W. (2002). Dynamics of the relationship between a generalist predator and slugs over five years. *Ecology* **83**, 137–147.

TERRY, L.A., POTTER, D.A. AND SPICER, P.G. (1993). Insecticides affect predatory arthropods and predation on Japanese Beetle (Coleoptera: Scarabaeidae) eggs and Fall Armyworm (Lepidoptera: Noctuidae) pupae in turfgrass. *Journal of Economic Entomology* **86**, 871–878.

THEISS, S. AND HEIMBACH, U. (1993). Fütterungsversuche an Carabidenlarven als Beitrag zur Klärung ihre Biologie. *Mitteilungen der Deutschen Gesellschaft für allgemeine und angewandte Entomologie* **8**, 841–847.

THIELE, H.-U. (1977). *Carabid beetles in their environments.* Berlin: Springer-Verlag.

TITOVA, E.V. AND KUPERSHTEYN, M.L. (1976). Ground beetles (Coleoptera, Carabidae) of a wheatfield biocoenosis in the North Caucasian Steppe Zone and use of the precipitin test to evaluate their trophic links with *Eurygaster integriceps* Put. *Entomological Review* **55**, 1–8.

TOMLIN, A.D., MILLER, J.J., HARRIS, C.R. AND TOLMAN, J.H. (1985). Arthropod parasitoids and predators of the onion maggot (Diptera: Anthomyiidae) in Southwestern Ontario. *Journal of Economic Entomology* **78**, 975–981.

TUKAHIRWA, E.M. AND COAKER, T.H. (1982). Effect of mixed cropping on some insect pests of brassicas: reduced *Brevicoryne brassicae* infestations and influences on epigeal predators and the disturbance of oviposition behaviour in *Delia brassicae*. *Entomologia Experimentalis et Applicata* **32**, 129–140.

UZENBAEV, S.D. (1984). Natural enemies of pseudolarvae of the red pine sawfly (*Neodiprion sertifer*) in southern Karelia. *Zoologicheskii Zhurnal* **63**, 1012–1018.

VAJE, S. AND MOSSAKOWSKI, D. (1986). Occurrence of proteolytic enzymes in relation to accessibility of food in carabid beetles. In: *Feeding behaviour and accessibility of food for carabid beetles.* Eds. P.J. Den Boer, L. Grüm and J. Szyszko, pp 29–34. Warsaw: Warsaw Agricultural University Press.

VAN DEN BERG, H., LITSINGER, J.A., SHEPARD, B.M. AND PANTUA, P.C. (1992). Acceptance of eggs of *Rivula atimeta*, *Naranga aenescens* (Lep.: Noctuidae) and *Hydrellia philippina* (Dipt.: Ephydridae) by insect predators on rice. *Entomophaga* **37**, 21–28.

VASCONCELOS, S.D., WILLIAMS, T., HAILS, R.S. AND CORY, J.S. (1996). Prey selection and baculovirus dissemination by carabid predators of Lepidoptera. *Ecological Entomology* **21**, 98–104.

WALLIN, H. (1991). Movement patterns and foraging tactics of a caterpillar hunter inhabiting alfalfa fields. *Functional Ecology* **5**, 740–749.

WALRANT, A. AND LOREAU, M. (1995). Comparison of iso-enzyme electrophoresis and gut examination for determining the natural diet of the groundbeetle species *Abax ater* (Coleoptera: Carabidae). *Entomologia Generalis* **19**, 253–259.

WARNER, D.J. (2001). Spatio-temporal relationships between *Psylliodes chrysocephala* (cabbage stem flea beetle) larvae and predatory carabid beetles in a crop of winter oilseed rape. *Bulletin of the Royal Entomological Society* **25**, 58–60.

WARNER, D.J., ALLEN-WILLIAMS, L., FERGUSON, A. AND WILLIAMS, I.H. (2000). Pest–predator spatial relationships in winter rape: implications for integrated crop management. *Pest Management Science* **56**, 977–982.

WEISSER, W.W., BRAENDLE, C. AND MINORETTI, N. (1999). Predator-induced morphological shift in the pea aphid. *Proceedings of the Royal Society of London Series B – Biological Sciences* **266**, 1175–1181.

WESELOH, R.M. (1985). Predation by *Calosoma sycophanta* L. (Coleoptera: Carabidae): evidence for a large impact on gypsy moth, *Lymantria dispar* L. (Lepidoptera: Lymantriidae), pupae. *Canadian Entomologist* **117**, 1117–1126.

WESELOH, R.M. (1988). Prey preferences of *Calosoma sycophanta* L. (Coleoptera: Carabidae) larvae and relationship of prey consumption to predator size. *Canadian Entomologist* **120**, 873–880.

WESELOH, R.M. (1990). Experimental forest releases of *Calosoma sycophanta* (Coleoptera: Carabidae) against the Gypsy Moth. *Journal of Economic Entomology* **83**, 2229–2234.

WESELOH, R.M., BERNON, G., BUTLER, L., FUESTER, R., MCCULLOUGH, D. AND STEHR, F. (1995). Releases of *Calosoma sycophanta* (Coleoptera: Carabidae) near the edge of Gypsy Moth (Lepidoptera: Lymantriidae) distribution. *Environmental Entomology* **24**, 1713–1717.

WHEATER, C.P. (1988). Predator–prey size relationships in some Pterostichini (Coleoptera: Carabidae). *Coleopterists Bulletin* **42**, 237–240.

WHITCOMB, W.H. AND BELL, K. (1964). Predaceous insects, spiders, and mites of Arkansas cotton fields. *Bulletin of the Arkansas Agricultural Experimental Station* **690**, 1–83.

WINDER, L.H. (1990). Predation of the cereal aphid *Sitobion avenae* by polyphagous predators on the ground. *Ecological Entomology* **15**, 105–110.

WINDER, L.H., CARTER, N. AND WRATTEN, S.D. (1988). Assessing the cereal aphid control potential of ground beetles with a simulation model. *1998 Brighton Crop Protection Conference – Pests and Diseases* **3**, pp 1155–1160.

WINDER, L.H., WRATTEN, S.D. AND CARTER, N. (1997). Spatial heterogeneity and predator searching behaviour – can carabids detect patches of their aphid prey? *Acta Jutlandica* **72**, 47–62.

WINDER, L.H., ALEXANDER, C.J., HOLLAND, J.M., WOOLLEY, C. AND PERRY, J.N. (2001). Modelling the dynamic spatio-temporal response of predators to transient prey patches in the field. *Ecology Letters* **4**, 568–576.

WINK, M. (1986). Acquired toxicity – the advantages of specializing on alkaloid-rich lupins to *Macrosiphon albifrons* (Aphidae). *Naturwissenschaften* **73**, 210–212.

WISHART, G., DOANE, J.F. AND MAYBEE, G.E. (1956). Notes on beetles as predators of eggs of *Hylemya brassicae* (Bouché) (Diptera: Anthomyiidae). *Canadian Entomologist* **88**, 634–639.

WRIGHT, D.W., HUGHES, R.D. AND WORRALL, J. (1960). The effect of certain predators on the numbers of cabbage root fly (*Erioischia brassicae* (Bouché)) and on subsequent damage caused by the pest. *Annals of Applied Biology* **48**, 756–763.

YOUNG, O.P. (1980). Predation by tiger beetles (Coleoptera: Cicindelidae) on dung beetles (Coleoptera: Scarabaeidae) in Panama. *The Coleopterists Bulletin* **34**, 63–64.

YOUNG, O.P. (1984). Prey of adult *Calosoma sayi* (Coleoptera: Carabidae). *Journal of the Georgia Entomological Society* **19**, 503–507.

YOUNG, O.P. AND HAMM, J.J. (1986). Rate of food passage and fecal production in *Calosoma sayi* (Coleoptera: Carabidae). *Entomological News* **97**, 21–27.

ZHAO, D.X., BOIVIN, G. AND STEWART, R.K. (1990). Consumption of carrot weevil, *Listronotus oregonensis*, (Coleoptera: Curculionidae) by four species of carabids on host plants in the laboratory. *Entomophaga* **35**, 57–60.

7
Weed Seed Predation by Carabid Beetles

JOSEPHINE TOOLEY[1] AND GERALD E. BRUST[2]

[1]Department of Agricultural Botany, School of Plant Sciences, The University of Reading, Building 2, Earley Gate, Reading, Berkshire RG6 6AU, UK and [2]SW Purdue Agricultural Research Center, Purdue University, 4369 North Purdue Road, Vincennes, IN 47591 3043, USA

Introduction

The consumption of seeds by animals and insects can be a major factor affecting plant density, abundance and distribution. Seed production is an important stage of the life cycle of plants, contributing to the replacement of adult plants and expansion of the population size. Seed predators come from a diverse range of taxa, but include animals such as birds, mice and squirrels, as well as insects such as beetles, flies and parasitoids.

Seed predation was a term first used by Janzen (1971) to distinguish between animals that consume seeds and destroy them in the process and animals that just ingest seeds, leaving their fate undecided. Predation is normally a term relating to the capture of prey; however, in this chapter it will be used to describe the removal of potentially viable seeds from the seedbank. Seed predation can be conveniently divided into pre-seedfall and post-seedfall predation. Since carabids are predominantly ground dwellers, this chapter will deal only with post-seedfall predation. However, this does not mean that carabids are incapable of climbing plants to feed from seed heads.

Seed predation is thought to be responsible for a significant amount of seed loss, both post seed-shed and prior to seeds being incorporated into the soil seedbank. It is particularly important when investigating plant population dynamics to be able to quantify any loss of seeds and its potential impact on seedling recruitment.

One area of research where seed predators may have a beneficial effect is that of weed control within farming systems. Weed control in integrated farming systems is a combination of chemical and cultural methods. Quantifying the impact of seed predators upon populations of weeds may result in more effective methods of weed control, which take advantage of natural populations of seed predators.

There has been increasing interest over the past few years in the role of carabids in

---

Abbreviations: Plant Relative Yield (PRY)

The Agroecology of Carabid Beetles

agroecosystems. They are already used as bio-indicators of environmental change and their potential as a biological control agent for many crop pests is still being investigated, as summarized by Sunderland (this volume). Many studies have been carried out to determine the diet of carabids, with emphasis being placed on their consumption of other arthropods. Some species are known to be polyphagous, but relatively few investigations have focused on the importance of plant material in their diet. In particular, certain species are known to feed on a variety of weed seeds, although until recently the significance of this in relation to weed control was overlooked.

In this chapter, the role of carabids as seed consumers will be discussed, with particular emphasis placed on their contribution to weed control. Relatively little research has been carried out as to the impact these beetles have as seed consumers. Therefore, it has been necessary to rely, in some instances, upon data from studies using other seed predators.

Overview of seed predation

Relatively few investigations of seed predation have looked at seed loss within agricultural systems (Povey *et al.*, 1993; Cardina, 1996; Cromar *et al.*, 1999; Menalled *et al.*, 2000). Instead, most have focused upon deserts (Reichman, 1979; Inouye *et al.*, 1980; Abramsky, 1983), grasslands and old-fields (Mittelbach and Gross, 1984; Reader, 1993; Hulme, 1998; Edwards and Crawley, 1999), or woodlands (Kjellsson, 1985). Two of the most widely studied seed predators are ants and rodents (Reichman, 1979; Inouye *et al.*, 1980; Abramsky, 1983; Diaz, 1992; Edwards and Crawley, 1999). The rate of seed consumption by both of these taxa has been shown to be highly variable, although comparative studies indicate that rodents are responsible for a higher proportion of seed predation than ants. Results from Reichman (1979) indicated that ants retrieved approximately 45% of seeds they detected during a 24 h period, with rodents removing 96%. Edwards and Crawley (1999) concur, with rodents removing an average of 90–100% of seeds nightly, compared with 11–67% for ants. These studies suggest that invertebrates are less significant seed predators than vertebrates. Indeed, Hulme (1996) found that arthropods were responsible for reducing plant biomass by 9.5% in contrast to 14% by molluscs and 26% by rodents. However, Kjellsson's (1985) investigation into the predation of *Carex pilulifera* (Juncaceae) seeds discovered that the ground beetle, *Harpalus fuliginosus*, was responsible for reducing its seed pool by 65%, compared with only 21% by the mouse, *Sylvaemus flavicollis.*

Within agricultural ecosystems, particularly in Europe and North America, the role of carabid beetles as seed predators is starting to be investigated. Cardina *et al.* (1996) carried out exclosure experiments, which attributed approximately 50% of seed predation to carabids and slugs, with mice responsible for the remaining 50%. Seed loss attributable to invertebrates was found by Cromar *et al.* (1999) to be greater still, at around 80–90%. The experiments relating to carabid beetles will be discussed in more detail further on in this chapter.

Most common weed seed predators

Ground beetles have been found in a great variety of habitats, ranging from sand dunes to the snow line of mountainous regions. A number of surveys have shown that many species are common to agricultural land, as reviewed by Luff (this volume). It should therefore be relatively easy to discuss the distribution and abundance of phytophagous beetles. However, Allen (1979) points out the difficulties in comparing the results from different surveys because of the variation in crop type and number of pitfall traps used. Rivard (1966) also found that, of the 102 species captured in his three-year survey of beetles in agricultural crops, the nature of the crop greatly influenced beetle numbers.

Thiele (1977) compiled the results of 29 surveys taken across Europe from the United Kingdom to Northern Russia. This demonstrated a surprising homogeneity, with the spermophagous species, *Harpalus rufipes* and *H. aeneus*, occurring in over two-thirds of the areas investigated. Jones (1979), in her study of carabids in winter wheat crops in the UK, found that *H. rufipes* was numerous every year, together with *Pterostichus melanarius*. However, whilst there are thousands of species of carabids, only a small number have been studied as seed predators. Zetto Brandmayr (1990) states that "spermophagous carabids are found only in *Amarini*, *Zabrini* and *Harpalini*". Although the range of seed-eaters may be broader than this, it is certainly true that the majority of investigations so far have focused upon beetles from these genera, and in particular *H. rufipes* (Briggs, 1965; Luff, 1980; Jorgensen and Toft, 1997). In addition, only a few are present in sufficient numbers in agricultural systems to have much of an impact on weed biocontrol.

Ten different carabid genera have been identified as seed predators in North America (*Harpalus calignosus*, *H. erraticus*, *H. pensylvanicus* + spp.; *Chlaenius* spp.; *Agonum punctiform* + spp.; *Lebia* spp.; *Bembidion* spp.; *Amara cuproleota impuncticollis,* + spp.; *Anisodactylus mercula Germer* + spp.; *Selenophorus* spp.; *Pterostrichus* spp.; *Stenolophus* spp.). Of these, the genera *Harpalus* and *Amara* are very important seed predators in most agricultural systems (Kirk, 1972; Brust and House, 1988; Brust, 1994). *Harpalus* species, especially *H. pensylvanicus*, are thought to synchronize their breeding period (starting in late summer to early autumn) to coincide with the ripening of grass weeds. Ironically, this same genus has been found in greenhouse experiments to selectively feed on broad-leaved weed seeds when they are in mixed populations with grass seeds (Brust, 1994). This illustrates the potentially contradictory conclusions that may be made about any one particular weed seed predator. Carabids, like other predators, can switch to other prey or have preferences for particular prey species based on the specific time and environment within which the study is conducted. Therefore, while feeding trials are important in determining the potential contribution that carabid species may make to weed biocontrol, the myriad of combinations of weed seed species, seed predators and environmental conditions in agricultural fields, makes prediction of which species will be controlled almost impossible. Continued studies are needed to try to understand, in part, the contributions that carabids and other insects make as weed seed predators and weed biocontrol agents under field conditions.

Adaptations to a seed-feeding habit

As the old idea that carabid beetles are predominantly predatory has been revised, a study has been made of the adaptations exhibited by beetles to the phytophagous mode of feeding. Thiele (1977) states that phytophagy plays a substantial role in the nutrition of many carabids and the degree of specialization of phytophages can be large or small. To feed from plants, seeds or fruits, carabids face two major challenges which they must overcome; firstly, locating the plant material and secondly, utilizing it as a food source.

Adult plants may be fairly conspicuous; however, beetles commonly thought to be ground dwellers, must be able to exploit this vertical habitat. Numerous reports demonstrate the ability of carabids to climb plants to reach the inflorescences where they are found to feed. *Amara familiaris* has been observed eating the flowers of a species of chickweed (*Stellaria* spp.) (Aubrook, 1949) and *Amara aulica* is reported to be commonly found near Asteraceae, which are climbed by the beetles at night (Lindroth, 1974). *Zabrus tenebriodes* is also described as extremely skilled at climbing the stalks of cereals to devour the grains, and can be a pest.

Post-dispersal seed predators have a particularly difficult challenge in locating inconspicuous items, which can be scattered at low densities over soil or even buried. Experiments have shown that some predatory species of carabid belonging to the *Cincindelida*e can visually detect the presence of fast moving insect prey such as crickets (Wheater, 1989), but do not respond to immobile prey. Since sessile organisms such as plants do not provide these moving visual cues, it is likely that beetles have to rely more heavily upon olfactory and tactile cues. In experiments using a four-arm olfactometer, Kielty *et al.* (1996) found that *H. rufipes* and *P. melanarius* were both able to detect a crude extract of wheat using olfactory senses.

It has been observed that adults and larvae are extremely skilled at manipulating seeds and have the ability to carefully peel away the outer fruit or seed coat. Briggs (1965) discovered that *H. rufipes* tended to eat only the endosperm of strawberry seeds, the husks of which were found beneath the plants. Forsythe (1982) and Ingerson-Mahar (this volume) discuss in detail how the morphology of some carabids can be related to their feeding methods. *A. aulica* is described as having blunt incisor lobes, but possesses large crenellated molar regions suitable for crushing and breaking open seeds and then grinding them up, rather than piercing prey. In comparison with other species, there is also an increase in the size of the mandibular muscles, reflecting the power of the mandibular movements required to deal with plant material. *H. rufipes* was found to have mouthparts very similar to *A. aulica*, although it is also capable of piercing and crushing small arthropods. The dense beard of spines on the inner lobes of the proventriculus is also thought to help break up and strain vegetable matter.

Some larvae are particularly well adapted to the seed-feeding habit. The larvae of *H. rufipes* are known to be seed eating, and dig vertical burrows to which seeds are carried and stored. The third larval instar is thought to remain within the burrow and feed exclusively on the stored seeds (Luff, 1980). In experiments comparing the larval development of *H. rufipes* fed on a mixed seed or insect diet, Jorgensen and Toft (1997) found no significant difference in mortality rates between diets, although development times were 50% shorter for the mixed seed diet compared with the mixed insect diet. This suggests that, at some stage of development, seeds are of a higher nutritional value to the larvae.

The larvae of *Z. tenebrioides* also construct burrows into which they drag the leaves and stalks of cereal plants, sometimes causing much damage to fields of wheat (Bassett, 1978). The burrowing abilities of such species may protect vulnerable larvae from predators and help to maintain humidity levels, preventing water loss, while the cache of seeds ensures a high quality food source remains available for a long time. There is, however, a clear differentiation between 'true' spermatophagy where real feeding specializations are shown and the general trend of carabids to feed from rotten fruits, seedlings, etc. under certain conditions, such as when water is scarce (Zetto Brandmayr, 1990). An evolutionary pathway has been proposed which suggests that the first seed-eating carabids came from waterside habitats where large numbers of *Harpalini* and *Amarini* are still found. Whatever their origins, phytophagous beetles, and in particular seed feeders, exploit a food source which is nutritionally of very high quality, and thus co-evolution between specialist seed-eating beetles and seed-producing plants such as *Compositae* is extremely likely.

Methods of detecting predation

A large variety of methodologies have been used to try to assess seed predation by carabids in the laboratory and field. Unfortunately, the value of many qualitative studies is questionable, since the physiology and ecology of carabids can make it very difficult to quantify predation, and it is important to understand the limitations of the methods used.

Direct observation, although reliable, is extremely time consuming. However, surveillance cameras using time-lapse or continuous video recording equipment have been used to study both sessile predators and prey in the laboratory and field. Semi-automated techniques can provide the equivalent of many hours of direct observations, allowing a reliable estimate of the rate of predation. When studying nocturnal beetles, care must be taken in the choice of illumination used. Sixty watt, red tungsten bulbs have been used to film *Agonum dorsale* (Griffiths *et al.*, 1985) and also for the image analysis of olfactometer experiments with *P. melanarius, H. rufipes* and *Nebria brevicollis* (Kielty *et al.*, 1996), but Sunderland (1988) warned that generalizations cannot be made and care must be taken not to modify behaviour.

Gut analysis has frequently been used to determine the diet of carabids (e.g. Davies, 1953; Johnson and Cameron, 1969; Sunderland, 1975; Symondson and Liddell, 1993; reviewed in Symondson, this volume). The method normally employed involves removing the crop, proventriculus, mid-gut and hind gut and teasing them open on a microscope slide. Jervis and Kidd (1996) discuss the importance of killing the predators as soon as possible after collection to prevent the loss of gut contents through defecation and regurgitation. When choosing a sample of beetles for examination, care is also needed in choosing an unbiased sample; therefore, daily feeding cycles must be taken into account. Visual analysis of the contents of the digestive tracts of beetles is only suitable for those with little extra-oral digestion. Sunderland (1975) found 'unidentified plant material' in the crops of *H. rufipes*, *N. brevicollis* and *Pterostichus melanarius* (= *Feronia melanaria*). Although useful in identifying some insect material, this method has not been used successfully in the accurate identification of seed remains.

In the future, the identification of seed proteins is probably the key to discovering seed predators. Electrophoresis and serological techniques have both been used to analyse the components of predator diets (Stuart and Greenstone, 1990). If proteins specific to certain types of seeds can be identified, then antisera can be raised against them to identify seed fragments within carabid crop samples. However, this technique is able to test only a limited number of seeds that the carabid may consume, and entails a need for some preliminary feeding studies.

Until a satisfactory method of detecting seed predation is developed, data concerning seed consumption will continue to come largely from observational laboratory feeding studies and field-based seed baiting experiments using exclusion methods.

Weed seed preferences of carabid beetles

Post-seedfall predators are thought to be a major determinant of seed survival, plant species distribution patterns and plant community composition. However, predation by this group is very hard to detect and quantify, and so may often be underestimated. Carabid seed-feeding preferences are relatively easy to discover in the laboratory (Johnson and Cameron, 1969; Best and Beegle, 1977; Hagley *et al.*, 1982; Goldschmidt and Toft, 1997; Jorgensen and Toft, 1997), although caution must be used when trying to interpret results as levels of seed consumption are often much higher in the laboratory than in the field.

Goldschmidt and Toft's (1997) study of phytophagy among 24 species of insectivorous carabids found that spermatophagy was most prevalent among the Harpalinae. The carabids, *Pterostichus versicolor, P. cupreus* and *H. rufipes*, consumed, respectively, an average of 1.4, 3.6 and 8.1 seeds of *Capsella bursa-pastoris* a day. Jorgensen and Toft (1997) used cafeteria-style feeding trials where *H. rufipes* adults were given a mixture of seeds to eat over a 24 h period. Each beetle's preference for the different seed species was then calculated. In general, it was found that *Taraxacum* species and *Viola arvensis* seeds were preferred the most, and *Polygonum persicaria*, *Veronica arvensis* and *Lithospermum arvense* were the least popular.

In similar feeding trials, Tooley and Froud-Williams (1999) fed *A. aenea*, *H. rufipes* and *P. melanarius* two different mixed seed diets. Diet 1 consisted of *Chenopodium album*, *Stellaria media*, *Viola arvensis*, *Poa trivialis* and *Matricaria* species. Diet 2 was a combination of *Avena fatua*, *Alopecurus myosuroides*, *Bromus sterilis*, *Polygonum persicaria* and *Galium aparine*. *Tables 7.1* and *7.2* show the mean number of seeds from both diets consumed by each carabid species over a 24 h period. The results indicate that for diet 1, the most favoured seeds for *A. aenea*, *H. rufipes* and *P. melanarius* were *P. trivialis*, *V. arvensis* and *V. arvensis* respectively. The most preferred seeds from diet 2 were less clear since far fewer were eaten, albeit, *H. rufipes* has a significant preference for *A. fatua* and *A. myosuroides*.

Since many carabids are polyphagous feeders, it is important to look at what proportion of their diet is made up of seeds, compared with other food types such as insect and plant material. In an examination of this, Tooley (unpublished) found that seeds were still a major part of the diet for the carabids, *A. aenea*, *H. rufipes* and *P. melanarius*. When beetles were given a choice of aphids, mealworms, wheat seedlings and their favourite seeds from diets 1 and 2, the number of seeds eaten as a proportion of the total food items eaten was 39%, 62% and 31% for *A. aenea*, *H. rufipes* and *P. melanarius* respectively. The importance of seeds within the diet can be directly linked to beetle fecundity. Jorgensen and Toft (1997) demonstrated that females of *H. rufipes* laid significantly more eggs when fed upon mixed seed diets, compared with a purely insect diet.

Table 7.1. Mean number of seeds from diet 1 consumed by carabids over 48 h.

	C. album	*S. media*	*V. arvensis*	*P. trivialis*	*Matricaria* spp.
A. aenea	1.00^{ab}	4.65^{c}	0.43^{a}	5.57^{d}	1.83^{bc}
H. rufipes	13.05^{c}	6.0^{a}	38.4^{d}	7.45^{b}	11.05^{b}
P. melanarius	0.80^{a}	3.00^{c}	5.34^{d}	2.44^{c}	1.07^{b}

Different letters indicate statistically significant differences where P ³ 0.05

Table 7.2. Mean number of seeds from diet 2 consumed by carabids over 48 h.

	A. fatua	*A. myosuroides*	*B. sterilis*	*P. persicaria*	*G. aparine*
A. aenea	0.16^{b}	0.06^{ab}	0.09^{ab}	0.00^{a}	0.09^{ab}
H. rufipes	0.66^{b}	0.72^{b}	0.09^{a}	0.22^{a}	0.00^{a}
P. melanarius	1.36^{a}	0.53^{a}	0.33^{a}	0.20^{a}	0.20^{a}

Different letters indicate statistically significant differences where P ³ 0.05

In some seed baiting experiments, seed predators have been found to remove (consume) large percentages of seeds per day, up to 60–70% (Reichman, 1979). However, other studies have found more modest seed removal of 4–20% per day (Mittelbach and Gross, 1984; Brust and House, 1988). Over a five-week trial feeding period in the field, Brust and House (1988) found that carabid species removed 40% of *Ambrosia artemisiifolia*, 30% of *Amaranthus retroflexus*, 31% of *Cassia obtusifolia*, and 20% of *Datura stramonium* seeds. In laboratory studies looking at these same weed species, carabids consumed from 1.5–2 times as many seeds as in field studies. Thus, caution is needed when extrapolating the possible consumption rates from laboratory to field, and requires further investigation. Determining seed preferences for carabids within the field can be complicated because it is often difficult to distinguish between different types of invertebrate seed predation, i.e. slugs, ants and beetles, let alone preferences among carabid species. This would require constant observation, perhaps using video cameras, and is not a method that has been employed to collect quantitative data. Currently, exclosure methods using a combination of physical and chemical barriers are commonly employed to differentiate between seed predators (Mittelbach and Gross, 1984; Povey *et al.*, 1993; Cardina *et al.*, 1996; Hulme, 1996; Cromar *et al.*, 1999; Menalled *et al.*, 2000)

Factors affecting carabid seed choice

Selective feeding is thought to occur because seed predators are maximizing the return/handling time investment in finding and opening a seed (Reichman, 1977). A number of factors, including abundance, morphology and the nutritional quality of the seed, might be important in its selection (Inouye *et al.*, 1980). Since carabids feed by holding seeds between their tarsal spurs, it is likely that there is a maximum seed size upon which each beetle species is capable of feeding. Seeds which are very large or protected by glumes would be difficult to manipulate. This was seen in the case of *A. aenea* and *A. familiaris* (Tooley, unpublished), species that preferred feeding on seeds less than 3 mm in length. Interestingly, this does not result in larger carabid species selecting larger seeds, as *H. rufipes* was found to consume 58% more *Viola arvensis* seeds than *A. fatua* in feeding trials. Other morphological factors that may influence carabid seed choice are thickness of seed coat and the proportion of endosperm to seed coat. Seed properties are discussed further in Janzen (1969) and Janzen (1970). Biochemical factors that may affect selection include the protein content and toxicity of seeds (Janzen, 1969; Hendry *et al.*, 1994). Lipids are the principle reserve material in most seeds and are usually present in the form of triglycerides of fatty acids, stored in the cotyledon or the endosperm. These are important components of insect diets (Chapman, 1982).

Implications for plant populations and weed control

One of the primary means by which seed predators may affect the composition of plant communities is through selective seed feeding. If the seed predators showed no preference and consumed seed of every species in equal amounts, they would have much less of an effect on the make-up of the plant species within a system. While this effect might appear obvious, data from the relatively few studies in which this phenomenon has been studied do not indicate a simple, direct relationship between seed predators and plant populations (Brown *et al.*, 1975; Harper, 1977; Reichman, 1979; Inouye *et al.*, 1980; Anderson, 1982; Mehlhop and Scott, 1983; Carroll and Risch, 1984; Smith, 1987).

Many studies on seed predators have found that they selectively consume certain seed species, but the outcome of this selectivity is not always certain. However, there is strong evidence that any selective predation will probably reduce the reproductive ability of the selected species (Janzen, 1970; Harper, 1977). Thus, the key to the importance of carabids as seed predators, and therefore biological control of certain weed species, is both an understanding of their preference for certain weed seed species and the factors that influence this feeding preference. Both laboratory and field studies where carabids as seed predators have been examined indicate that they readily feed on *A. retroflexus*, *C. album*, *A. artemisiifolia*, *Xanthium pennsylvanicum*, *D. stramonium* and, to a lesser extent, on *Digitaria sanguinalis* and other *Digitaria* and *Panicum* species, including *P. dichotomiflorum*.

Because seed predators have the ability to feed selectively on certain species, they can influence the final plant composition of a habitat. Risch and Carroll (1986), using a mixture of seed species based on seed mass, which is probably a superior method of comparison for competition studies (Trenbath, 1974), observed that one seed predator (an ant) could alter the competitive ability for several of the seed mixtures. With seed predators present, certain weed species that were unable to compete became dominant (measured in biomass). In a temperate old-field system, Mittelbach and Gross (1984) concluded that, even though the total biomass of weed seed removed by seed predators was comparatively low, their selective feeding could determine the distribution, and possibly the abundance, of some old-field plant species. A greenhouse study by Brust (1994) demonstrated that selective feeding by carabids does influence biomass of competing weed species. In this study, an equal mass of broad-leaved and grass-weed seeds was mixed and allowed to germinate and compete for two months in the presence or absence of carabids. It was found that Plant Relative Yield (PRY) of broad-leaved weeds (*A. retroflexus* and *C. album*) was reduced by 50–60% when carabids were present. This reduction in PRY of the broad-leaved weeds indicates that their ability to compete with the grass-weed species was significantly reduced. This suggests that grass species, which would not normally compete with broad-leaved weed species, should begin to dominate systems involving large numbers of carabid weed seed predators. Indeed, this is what was observed in a low-input, no-tillage system (Brust and House, 1988). At the start of the study, *A. retroflexus* and *C. album* comprised 48% of all weed species, but after three years, they made up only 28% of the total. Grass species comprised a larger percentage of all weed species after three years (35%) than they did at the start of the study, when they only accounted for 26%.

Similar conclusions have been drawn in most seed-feeding studies: seed predators have an effect on plant species composition because of their selective feeding, but confounding factors limited the ability to predict which particular species may dominate. Factors included:

1. Percentage composition of preferred seed versus non-preferred seed, as the amount of non-preferred increased so that the preferred seed was less predated (Risch and Carroll, 1986).
2. Size of seed (Smith, 1987).
3. Whether seed predators fed selectively on a numerical or biomass dominant basis (Mittelbach and Gross, 1984).
4. Nutritive quality of seed (i.e. protein, sugar, fibre, essential oils, and tannins (Levin, 1974; Reichman, 1977; Smith, 1987).
5. Adult species dominance, i.e. the more dominant the mature species, the greater the predation of its seed (Janzen, 1971).
6. Size of seed predator, i.e. rodents and larger carabids (15–25 mm) usually fed on larger seed species, while smaller carabids (< 15 mm), ants, and crickets usually fed on smaller ones (Mittelbach and Gross, 1984; Brust, 1994). However, with ants, seed size is not always the most important factor, providing a further confounding effect.

Thus, although most studies of seed predators reached the conclusion that seed predators can reduce seed biomass and, through selective feeding, affect species composition, few of these reported the total elimination by seed predators of any particular plant species in an area. Therefore, in agricultural systems there may simply be a change in weed species hierarchy as a result of seed predators. If the objective of biological weed control is to change the plant species composition in a particular area (Harris, 1988), then carabids could be considered as biocontrol agents of weeds.

There are several scenarios where the shift in weed species dominance could be useful in agricultural systems. If the shift could be made to grass species when a broad-leaved crop (for example, soybeans) is grown, the use of a grass herbicide in this case could possibly mean less damage to the crop and better weed control. Another benefit of species shifting could be the delay in competition. Because carabids consume many weed species, even those species that become dominant are still less competitive when compared to environments where seed predators are absent (Brust, 1994). This reduction in competitive ability may be sufficient for the crop to successfully out-compete the weed.

Another potential benefit of weed control by carabids is that they can reduce the seedling biomass of individual weed species, and this maybe extended to maturity. Risch and Carroll (1986) and Brust (1994) found that early growth reduction of certain selectively predated species was followed by a reduced competitive ability in older plants. Seed predators can reduce the reproductive ability of certain weed species and, over time, reduce the population of the species by predating their seeds. Therefore, for this mechanism of weed control to be effective in agricultural systems, weed seed predators must be given time to build up their populations, which will impact on selected weed species over a period of years. However, before management plans can be devised to encourage weed-feeding carabids, more information is needed on which weed species are affected by which seed predators. Additionally, how seed predators can be manipulated and what densities are needed before they have an impact on the competitive ability of the selected weeds needs further investigation.

Impact of farming systems

Many of the studies on seed predators and their impact on plant community structure have been conducted in desert ecosystems. This is not surprising as these systems usually have fewer plant species and fewer biotic interactions. The abiotic interactions of water, soil and temperature are thought to be paramount in determining the abundance and distribution of plants in desert communities (Hastings and Turner, 1965; Mulroy and Rundel, 1977; Ehleringer *et al.*, 1979). Therefore, studies of biotic interactions, such as competition, seed dispersal, herbivory and, of course, seed predation (Campbell, 1929; Glendening, 1952; Fonteyn and Mahall, 1978; Inouye *et al.*, 1980), were thought to be more approachable in these systems than in more species-rich systems. Indeed, this may be true when we examine species-rich systems that have a high level of species redundancy, i.e. similar functions are performed by several species. If there is a loss of a few species in a complex, natural system, this loss can be compensated for by other similarly functioning species and the overall community structure is not noticeably different (Paine, 1966; Mellinger and McNaughton, 1975; McNaughton, 1977; Pimm, 1982). Therefore, in less complex systems, such as desert communities and agricultural systems, there is likely to be a greater impact by keystone species (i.e. seed predators) than in more complex systems (Risch and Carroll, 1986). This can be observed in studies in arid communities where seed removal rates are very high, reaching 70% (Soholt, 1973; Brown *et al.*, 1975; Davidson, 1977; Reichman, 1979; Inouye *et al.*, 1980). However, in more complex systems such as old-field communities, seed removal rates were much less, up to 20% (Mittelbach and Gross, 1984).

The question then arises as to what the ‘best’ type of system could be that will optimize seed predators and weed seed predation – a more complex or less complex one? The studies reviewed here have shown that there is no simple answer. Constraints are placed on any agricultural system as to the best way to manipulate it. One fairly simple and grower-accepted way to alter an agriculture system to favour seed predators is to use reduced tillage methods or no-till. This type of system has little soil disturbance and the detritus (mulch) left on the surface increases the number of detritivores, which increases the number of seed predators (House and Stinner, 1987). Moreover, most seed predators are omnivores and feed on a wide variety of material such as fungi, detritus, seeds, other predators, etc. Mulching is also often used with minimum tillage and this creates an environment that aids carabid survival (House and Parmelee, 1985; House and Stinner, 1987; Pickett and Squeirs, 1989). In this type of system, carabids are thought to play a major role in determining weed species success.

In temperate, no-till agricultural fields, carabids were most responsible for the reduction in competitive ability of broad-leaved weed species (Brust, 1994). Using endemic carabid densities found in a no-till system, it was discerned that seed predators could reduce seedling biomass of broad-leaved weeds by 50–60%, compared to where no seed predators were present. Reducing the carabid numbers by 50% did not significantly reduce the impact of seed predators on the ability of broad-leaved plants to compete. This study illustrates the possibility of using carabids encountered in most temperate agricultural systems as 'keystone' seed predators. However, more work is needed to determine the density of carabids required and which species are most important.

Similarly, losses of *Abutilon theophrasti* (Malvaceae) seed from the soil surface in Ohio ranged from 1.2–57% per day, depending on the time of year. Rates of loss were similar, irrespective of tillage system (no-till and moldboard plough). Likewise, predator numbers were similar between tillage regimes. Predation rate was described by an exponential decay function of seed dormancy, with high rates of loss at low seed densities. Both vertebrate and invertebrates were implicated in seed predation (Cardina *et al.,* 1996). Elsewhere, in Ontario, 80–90% of total losses of *Chenopodium album* (Chenopodiaceae) and *Echinochloa crus-galli* (Poaceae) were attributed to ground-dwelling invertebrates. Predation was greatest (32%) in no-till and moldboard plough treatments, but was lowest (24%) with chisel plough (Cromar *et al.*, 1999).

Another method to encourage carabids involves manipulating the ecosystem by creating refugia in the field (Van Emden, 1965). This approach has not been specifically investigated for spermophagous species, with most studies concentrating on improving pest control, as reviewed by Lee and Landis (this volume). This method would necessitate greater management by growers to be successful. The refugia can take on many different forms, but usually consists of an unmanaged legume/grass mixture. These refugia are usually only a small portion of the field. The habitats create a much more inviting area for carabids to overwinter, breed and then disperse into the adjacent field. However, several studies have shown that dispersal from these refugia occurs only over a short distance (Rivard, 1966; Thiele, 1977; Blumburg and Crossley, 1983; Thomas *et al.*, 1991), and penetration of the field was often limited (e.g. Van Emden, 1965; House and All, 1981; McPherson, *et al.*, 1981; Sotherton, 1985). Even so, there are examples of where dispersal throughout the field has been recorded, and this subject is discussed more fully in Thomas *et al.* (this volume). For biocontrol of weeds, however, more complete coverage is needed than for pest control and, consequently, a larger portion of refugia sites would be needed. Other methods of trying to manipulate (modify) the agricultural system to favour carabids are based on the same general idea of creating refugia within the field, as described by Lee and Landis (this volume). This includes intercropping, or companion planting (Brust *et al.*, 1986), or restricted use of synthetic chemicals in certain areas (Coaker, 1965; Brust *et al.*, 1986). Almost all of these methods have been investigated with the expectation of increasing natural enemies of insect pests but, since a great number of carabids are also weed seed predators, these methods could also be used to increase potential weed biocontrol.

From this brief description of carabid weed seed feeding it is obvious much more work is needed on this aspect of carabid biology. Investigations so far have established that carabids do feed selectively on seeds and that the numbers of seeds predated can be extremely high, albeit non-uniform throughout the field. However, little is understood about why certain seeds are more favoured and how these are selected, although it is likely to be a combination of physical and biochemical factors. This knowledge will enable field environments to be much more successfully manipulated to enhance natural populations of beneficial carabid beetles.

Acknowledgements

Josephine Tooley would like to acknowledge MAFF for their financial support as part of CASE studentship with The Game Conservancy and Allerton Research and Educational Trust. Thanks also go to Drs R.J. Froud-Williams and J.M. Holland for assistance with this chapter, and also Dr N.D. Boatman for help and advice with my research.

References

ABRAMSKY, Z. (1983). Experiments on seed predation by rodents and ants in the Israeli desert. *Oecologia* **57** 328–332.

ALLEN, R.T. (1979). The occurrence and importance of ground beetles in agriculture and surrounding habitats. In: *Carabid beetles: their evolution, natural history, and classification*. Eds. T.L. Erwin, G.E. Ball, D.R. Whitehead and A.L. Halpern, pp 485–505. The Hague: Junk.

ANDERSON, A. (1982). Seed removal by ants in the mallee of northwestern Victoria. In: *Ant–plant interactions in Australia*. Ed. R.C. Buckley, pp 31–43. The Hague: Junk.

AUBROOK, E.W. (1949). *Amara familiaris*. *Entomologists Monthly Magazine* **85**, 44.

BASSETT, P. (1978). Damage to winter cereals by *Zabrus tenebriodes* (Goeze) (Coleoptera: Carabidae). *Plant Pathology* **27**, 48.

BEST, R.L. AND BEEGLE, C.C. (1977). Food preferences of five species of carabids commonly found in Iowa cornfields. *Environmental Entomology* **6**, 9–12.

BLUMBURG, A.Y. AND CROSSLEY, JR., D.A. (1983). Comparison of soil surface arthropod populations in conventional tillage, no-tillage, and old-field systems. *Agro-Ecosystems* **8,** 247–253.

BRIGGS, J.B. (1965). Biology of some ground beetles (Col., Carabidae) injurious to strawberries. *Bulletin of Entomological Research* **56**, 79–93.

BROWN, J.H., BROVER, J.J., DAVIDSON, D.W. AND LIEBERMAN, G.A. (1975). A preliminary study of seed predation in desert and mountain habitats. *Ecology* **56**, 987–992.

BRUST, G.E. (1994). Carabids affect the ability of broadleaf weeds to compete. *Agriculture, Ecosystems, and Environment* **48**, 27–34.

BRUST, G.E. AND HOUSE, G.J. (1988). Weed seed destruction by arthropods and rodents in low-input soybean agroecosystems. *American Journal of Alternative Agriculture* **3,** 19–25.

BRUST, G.E., STINNER, B.R. AND MCCARTNEY, D.A. (1986). Predation by soil inhabiting arthropods in intercropped and monoculture agroecosystems. *Agriculture, Ecosystems, and Environment* **18**, 145–154.

CAMPBELL, R.S. (1929). Vegetative succession in the *Prosopis* sand dunes of southern New Mexico. *Ecology* **10**, 329–398.

CARDINA, J., NORQUAY, H.M., STINNER, B.R. AND MCCARTNEY, D.A. (1996). Post-dispersal predation of velvetleaf (*Abutilon threophrasti*) seeds. *Weed Science* **44**, 534–539.

CARROLL, C.R. AND RISCH, S.J. (1984). The dynamics of seed harvesting in early successional communities by a tropical ant, *Solenopsis geminata*. *Oecologia* **61**, 388–392.

CHAPMAN, R.F. (1982). *The insects: structure and function.* Harvard: Harvard University Press.

COAKER, T.H. (1965). Further experiments on the effect of beetle predators on the numbers of the cabbage root fly, *Erioischia brassicae* (Bouche), attacking *Brassica* crops. *Annals of Applied Biology* **56**, 7–20.

CROMAR, H.E., MURPHY, S.D. AND SWANTON, C.J. (1999). Influence of tillage and crop residue on post dispersal predation of weed seeds. *Weed Science* **47**, 184–194.

DAVIDSON, D.W. (1977). Foraging ecology and community organization in desert seed-eating ants. *Ecology* **58**, 725–737.

DAVIES, M.J. (1953). The contents of the crops of some British carabid beetles. *Entomologists Monthly Magazine* **89**, 18–23.

DIAZ, M. (1992). Spatial and temporal patterns of granivorous ant seed predation in patchy cereal crop areas of central Spain. *Oecologia* **91**, 561–568.

EDWARDS, G.R. AND CRAWLEY, M.J. (1999). Rodent seed predation and seedling recruitment in meseic grassland. *Oecologia* **118**, 288–296.

EHLERINGER, J., MOONEY, H.A. AND BERRY, J.A. (1979). Photosynthesis and microclimate of *Camissonia claviformis*, a desert winter annual. *Ecology* **60,** 280–286.

FONTEYN, P.J. AND MAHALL, B.E. (1978). Competition among desert perennials. *Nature* **275**, 544–545.

FORSYTHE, T.G. (1982). Feeding mechanisms of certain ground beetles. *The Coleopterists Bulletin* **36**, 26–73.

GLENDENING, G.E. (1952). Some quantitative data on the increase of mesquite and cactus on a desert grassland range in southern Arizona. *Ecology* **33**, 319–328.

GOLDSCHMIDT, H. AND TOFT, S. (1997). Variable degrees of granivory and phytophagy in insectivorous carabid beetles. *Pedobiologia* **41**, 521–525.

GRIFFITHS, E., WRATTEN, S.D. AND VICKERMAN, G.P. (1985). Foraging by the carabid *Agonum dorsale* in the field. *Ecological Entomology* **10**, 181–189.

HAGLEY, E.A.C., HOLLIDAY, N.J. AND BARBER, D.R. (1982). Laboratory studies of the food preferences of some carabids (Coleoptera: Carabidae). *Canadian Entomologist* **114**, 431–437.

HARPER, J.L. (1977). *Populations of plants*. London: Academic Press.

HARRIS, P. (1988). Environmental impact of weed control insects. *Bioscience* **38**, 542–548.

HASTINGS, J.R. AND TURNER, R.M. (1965). *The changing mile: an ecological study of vegetation change with time in the lower mile of an arid and semiarid region*. Tucson: University of Arizona Press.

HENDRY, G.A.F., THOMSON, K., MOSS, C.J., EDWARDS, E. AND THORPE, P.C. (1994). Seed persistence: a correlation between seed longevity in the soil and *ortho*-dihydroxyphenol concentration. *Functional Ecology* **8**, 658–664.

HOUSE, G.J. AND ALL, J.N. (1981). Carabid beetles in soybean agroecosystems. *Environmental Entomology* **10**, 194–196.

HOUSE, G.J. AND PARMELEE, R.W. (1985). Comparison of soil arthropods and earthworms from conventional and no-tillage agroecosystems. *Soil Tillage Research* **5**, 351–360.

HOUSE, G.J. AND STINNER, B.R. (1987). Arthropods in conservation tillage systems. *Entomological Society of America – Miscellaneous Publications* No. 65.

HULME, P.E. (1996). Herbivores and the performance of grassland plants: a comparison of arthropod, mollusc and rodent herbivory. *Journal of Ecology* **84**, 45–51.

HULME, P.E. (1998). Post-dispersal seed predation and seed bank persistence. *Seed Science Research* **8**, 513–519.

INOUYE, R.S., BYERS, G.S. AND BROWN, J.H. (1980). Effects of predation and competition on survivorship, fecundity and community structure of desert annuals. *Ecology* **61**, 1344–1351.

JANZEN, D.H. (1969). Seed-eaters versus seed size, number, toxicity and dispersal. *Evolution* **23**, 1–27.

JANZEN, D.J. (1970). Herbivores and the number of tree species in tropical forests. *American Naturalist* **104**, 501–528.

JANZEN, D.H. (1971). Seed predation by animals. *Annual Review of Ecology and Systematics* **2**, 465–492.

JERVIS, M. AND KIDD, N. (Eds.) (1996). *Insect natural enemies: practical approaches to their study and evaluation.* London: Chapman and Hall.

JOHNSON, N.E. AND CAMERON, R.S. (1969). Phytophagous ground beetles. *Annals of the Entomological Society of America* **62**, 909–914.

JONES, M.J. (1979). The abundance and reproductive activity of common Carabidae in a winter wheat crop. *Ecological Entomology* **4**, 31–43.

JORGENSEN, H.B. AND TOFT, S. (1997). Food preference, diet dependent fecundity and larval development in *Harpalus rufipes* (Coleoptera: Carabidae). *Pedobiologia* **41**, 307–315.

KIELTY, J.P., ALLEN-WILLIAMS, L.J., UNDERWOOD, N. AND EASTWOOD, E.A. (1996). Behavioural responses of three species of ground beetle (Coleoptera: Carabidae) to olfactory cues associated with prey and habitat. *Journal of Insect Behavior* **9**, 237–251.

KIRK, V.M. (1972). Seed-caching by larvae of two ground beetles, *Harpalus pensylvanicus* and *H. erraticus*. *Annals of the Entomological Society of America* **65**, 1426–1428.

KJELLSSON, G. (1985). Seed fate in a population of *Carex Pilulifera L.* II. Seed predation and its consequences for dispersal and seed bank. *Oecologia* **67**, 424–429.

LEVIN, D.A. (1974). The oil content of seeds: an ecological perspective. *American Naturalist* **108**, 193–206.

LINDROTH, C.H. (1974). *Handbook for the identification of British insects, Coleoptera: Carabidae, IV, part 2*. London: Royal Entomological Society of London.

LUFF, M.L. (1980). The biology of the ground beetle *Harpalus rufipes* in a strawberry field in Northumberland. *Annals of Applied Biology* **94**, 153–164.

MCNAUGHTON, S.J. (1977). Diversity and stability of ecological communities: a comment on the role of empiricisms in ecology. *American Naturalist* **111**, 515–525.

MCPHERSON, R.M., SMITH, J.C. AND ALLEN, W.A. (1981). Incidence of arthropod predators in different soybean cropping systems. *Environmental Entomology* **11**, 685–689.

MEHLHOP, P. AND SCOTT, JR., N.J. (1983). Temporal patterns seed use and availability in a guild of desert ants. *Ecological Entomology* **8**, 69–85.

MELLINGER, M.V. AND MCNAUGHTON, S.J. (1975). Structure and function of successional vascular plant communities in coastal New York. *Ecological Monographs* **45**, 161–182.

MENALLED, F.D., MARINO, P.C., RENNER, K.A. AND LANDIS, D.A. (2000). Post-dispersal weed seed predation in Michigan crop fields as a function of agricultural landscape structure. *Agriculture, Ecosystems and Environment* **77**, 193–202.

MITTELBACH, G.G. AND GROSS, K.L. (1984). Experimental studies of seed predation in old-fields. *Oecologia* **65**, 7–13.

MULROY, T.W. AND RUNDEL, P.W. (1977). Annual plants: adaptations to desert environments. *Bioscience* **27**, 109–114.

PAINE, R.T. (1966). Food web complexity and species diversity. *American Naturalist* **100**, 65–75.

PICKETT, S.T.A. AND SQUEIRS, E.R. (1989). Experimental approaches to understanding community organization and dynamics: Confirmation and surprise in Piedmont old-fields. *Ecological Society of America – Abstracts* **72**, 136.

PIMM, S.L. (1982). *Food webs*. New York: Chapman and Hall.

POVEY, F.D., SMITH, H. AND WATT, T.A. (1993). Predation of annual grass weed seeds in arable field margins. *Annals of Applied Biology* **122**, 323–328.

READER, R.J. (1993). Control of seedling emergence by ground cover and seed predation in relation to seed size for some old-field species. *Journal of Ecology* **81**, 169–175

REICHMAN, O.J. (1977). Optimization of diets through food preferences by heteromyid rodents. *Ecology* **58**, 454–457.

REICHMAN, O.J. (1979). Desert granivore foraging and its impact on seed densities and distributions. *Ecology* **60**, 1085–1092.

RISCH, S.J. AND. CARROLL, C.R. (1986). Effects of seed predation by a tropical ant on competition among weeds. *Ecology* **67**, 1319–1327.

RIVARD, I. (1966). Ground beetles (Coleoptera: Carabidae) in relation to agricultural crops. *Canadian Entomologist* **98,** 189–193.

SMITH, T.J. (1987). Seed predation in relationship to tree dominance and distribution in mangrove forests. *Ecology* **68**, 266–273.

SOHOLT, L. (1973). Consumption of primary production by a population of kangaroo rats (*Dipodomys merriami*) in the Mojave Desert. *Ecological Monographs* **43**, 357–376.

SOTHERTON, N.W. (1985). The distribution and abundance of predatory Coleoptera overwintering in field boundaries. *Annals of Applied Biology* **106**, 17–21.

STUART, M.K. AND GREENSTONE, M.H. (1990). Beyond ELISA: A rapid, sensitive immunodot assay for identification of predator stomach contents. *Annals of the Entomological Society of America* **83,** 1101–1107.

SUNDERLAND, K.D. (1975). The diet of some predatory arthropods in cereal crops. *Journal of Applied Ecology* **12**, 507–515.

SUNDERLAND, K.D. (1988). Quantitative methods for detecting invertebrate predation occurring in the field. *Annals of Applied Biology* **112**. 201–224.

SYMONDSON, W.O.C. AND LIDDELL, J.E. (1993). The detection of predation by *Abax parallelepipedus* and *Pterostichus madidus* (Coleoptera: Carabidae) on Mollusca using a quantitative ELISA. *Bulletin of Entomological Research* **83**, 641–647.

THIELE, H.-U. (1977). *Carabid beetles in their environments*. Berlin: Springer-Verlag.

THOMAS, M.B., WRATTEN, S.D. AND SOTHERTON, N.W. (1991). Creation of “island” habitats in farmland to manipulate populations of beneficial arthropods: predator densities and emigration. *Journal of Applied Ecology* **28,** 906–917.

TOOLEY, J.A. AND FROUD-WILLIAMS, R.J. (1999). Laboratory studies of weed seed predation by carabid beetles. *The 1999 Brighton Conference: Weeds* **2**, 571–572.

TRENBATH, B.R. (1974). Biomass productivity of mixtures. *Advances in Agronomy* **25,** 177–181.

VAN EMDEN, H.F. (1965). The role of uncultivated land in the biology of crop pests and beneficial insects. *Scientia Horticulturae* **17,** 121–136.

WHEATER, C.P. (1989). Prey detection by some predatory Coleoptera. *Journal of Zoology* **218**, 171–185.

ZETTO BRANDMAYR, T. (1990). Spermophagous (seed-eating) ground beetles: First comparison of the diet and ecology of the harpaline genera *Harpalus* and *Ophonus* (Col., Carabidae). In: *The role of ground beetles in ecological and environmental studies*. Ed. N.E. Stork, pp 307–316. Andover: Intercept.

8
Impact of Cultivation and Crop Husbandry Practices

THIERRY HANCE

Unité d'écologie et de biogéographie, Centre de Recherches sur la Biodiversité, Université Catholique de Louvain, Place Croix du Sud, 4–5, B-1348 Louvain-la-Neuve, Belgium

Introduction

The productivity of agricultural ecosystems is linked to the amount of anthropogenic energy input received in different forms, such as fertilization, cultivation, genetic selection, insect and weed control and manpower (Odum, 1971). Consequently, agroecosystems are frequently disturbed in their natural evolution, and for the soil fauna these disturbances may be considered as catastrophic and unpredictable events. On the other hand, because of its role in organic matter decomposition and insect pest control, soil fauna is in itself an essential component of the productivity we can expect of our crops. As they are mobile, polyphagous and diverse, carabids are good indicators of that fauna. It is thus essential to well understand the impact of husbandry practices on carabid beetles and their population development and species assemblages.

At the landscape level, agroecosystems are composed of a juxtaposition of temporary crops separated by different kinds of uncultivated habitats, some of very small size, such as paths, ditches, hedges, rows of trees or grassy strips, road borders, etc. (Petit and Usher, 1998). These adjacent patches constitute the only permanent habitats spared from crop cultivation, even though they undergo some side effects, i.e. spray drift. Husbandry results in an oversimplification of the environment as only the desired plant species is allowed to grow, and the phytophagous community that could potentially live on it is heavily suppressed (Potts and Vickerman, 1974; Pietrasko and DeClercq, 1980). Given these conditions, it is surprising to find a quite diverse carabid fauna inhabiting crop fields. In his famous book, Thiele (1977) stressed that this fauna is surprisingly homogenous. In a survey of 32 publications from all across Western Europe, he pointed out that eight species were particularly frequent in two-thirds of the regions investigated, and 26 occurred in at least one third of these regions. Basedow *et al.* (1976) monitored carabid assemblages in wheat crops in four

The Agroecology of Carabid Beetles

European countries (Belgium, Germany, The Netherlands, Sweden). They came to the same conclusion and stressed the importance of that fauna for integrated control programmes. From 1977 to 1985, Hance *et al.* (1989) collected 82 carabid species in agroecosystems in Belgium, among which 16 were only sampled in adjacent habitats such as hedges, rows of trees and uncropped areas. Lövei and Sarospataki (1990) have completed the list for Eastern Europe, and pointed out that in Eastern Europe agricultural fields have more common species and also more large species and phytophagous/mixed species than in Western Europe fields. They conclude that it is perhaps related to the influence of the steppe environment. *Carabus* spp. or *Cychrus* spp. are seldom recorded in Western Europe agricultural fields, probably because of their two-years cycle which is incompatible with the annual timing of husbandry practices and their sensitivity to perturbation (Kleinert, 1987). Indeed, in Romania, characterized by a low-input agriculture, Mircea *et al.* (1989) and Ciochia and Donescu (1997) have found some large species in potato and in sugar beet fields such as *Cicindela germanica*, *Carabus cancellatus*, *Carabus scabriusculus* and *Carabus violaceus*. The observation that larger species prefer less disturbed habitat was well described by Blake *et al.* (1994). Luff (1987) reported a list of species currently caught in North America and discusses carabid assemblages further in Chapter 2.

Husbandry practices are part of habitat management and may be used to promote conservation biological control (Landis *et al.*, 2000). In this chapter, the aim is to investigate the influence of crop husbandry practices on carabid abundance and assemblage composition, in order to answer two questions:

1. What are the indicators of good assemblage health (species composition, abundance, densities, indicator species, range of sizes, level of interaction)?
2. Can carabid populations be enhanced in order to maximize pest control and promote biodiversity?

Crop field organization in space and time: landscape level

The study of husbandry practices on carabid populations suffers from major methodological difficulties. Because most agricultural practices are usually combined, very few studies analyse only one factor and, in most of them, treatments represent more than one husbandry practice. Pitfall trapping is currently used for carabid activity monitoring. However, this method has certain serious well-known drawbacks and cannot distinguish between dispersing beetles from neighbouring plots and resident beetles, especially when the size of plot is limited. Moreover, larvae are never taken into account because of sampling difficulties, the limited knowledge of their taxonomy (Barney and Pass, 1986a; Traugott, 1998) and the difficulty of identification. Moreover, it is nearly impossible to distinguish between species that became more abundant after a certain cultivation method because cultivation took place during the larval life stage or because they could rapidly re-colonize from nearby habitats (Clark *et al.*, 1997).

INFLUENCE OF CROP TYPES

Large differences in assemblage composition and dominance ratios are observed according to crop species (Thiele, 1977; Hance and Grégoire-Wibo, 1987; Hance *et al.*, 1990; Booij, 1994; Carcamo and Spence, 1994; Dammer and Heyer, 1997; Szél *et al.*, 1997; Kromp, 1999, Honek and Jarosik, 2000). Booij (1994) stated that species richness (overall abundance and species composition) was primarily determined by crop type. However, species richness is the result of many factors, such as the timing of soil cultivation, insecticide and herbicide treatments and differences in microclimatic conditions at the soil level. Microclimatic conditions are governed by crop cover development, which will influence the speed of soil warming and the level of relative humidity. In winter wheat, the vegetation already covers the soil in April, whereas it is still bare in May for most root crops. Booij (1994) indicated that crops with an early and persistent ground cover were richer in species than were late-sown and more open crops. Hance *et al.* (1990) have indeed shown that the activity of some abundant species is earlier in cereal crops than in sugar beet crops. For instance, in their study, the peak of activity of *Agonum dorsale* took place in May in winter wheat and not until July in sugar beet. These results confirmed those of Pietrasko and DeClercq (1980). In carrot crops, the densities and variety of carabids increased as the crop matured (Boivin and Hance, unpublished results) because of new species emergence, but also as a consequence of a change in crop cover.

According to Baker and Dunning (1975), nocturnally active carabids should be positively influenced by ground cover because the shaded conditions provided by the foliage allow a larger period of activity. This, they proposed, explained the large catch on bare soil of *Bembidion* species because it has a diurnal habit. Kegel (1990) analysed the diurnal activity of carabid beetles using automatic time-sorting pitfall traps in a winter rye field. He concluded that species living in cereals show a greater proportion of daytime activity. He reported that diurnal activity pattern was correlated to overwintering behaviour, with adult overwintering species exhibiting diurnal activity and highly reflective colours. Other authors have observed that spring breeders are less abundant in root crops than in winter wheat fields (Lövei, 1984; Hance and Grégoire-Wibo, 1987; Hance, 1990).

A change in dominant species according to crop species was found by Ellsbury *et al.* (1998), with *Cyclotrachellus alternans* being dominant in corn and alfalfa, and *Pterostichus lucublandus* in wheat. The rank of abundance may change between crops, but also within fields for the same crop species during the season. Carabid data are often pooled for an entire year of catching. This may mask seasonal differences, and particularly the influence of local temperature on early activity. Hance and Grégoire-Wibo (1987) observed that, in winter wheat in spring, *A. dorsale* represented 30% of all captures, followed by *Asaphidion flavipes* (22%) and *Pterostichus melanarius* (less than 15%). In summer, due to the development cycles of the different species and vegetation growth, even if *A. dorsale* and *A. flavipes* remained abundant, *P. melanarius* and *Loricera pilicornis* increased in percentage. Such changes must be taken into account, as they will influence the impact of carabid assemblages on agricultural pests at different periods of the season. Abundance may also change from year to year, as pointed out by Dammer and Heyer (1997) for *P. melanarius*. In that case, comparisons between plots or treatments between years is critical. Dammer and Heyer (1997) proposed statistical methods to separate the influences of such factors on the occurrence of selected species.

CROP ROTATION

In many farming systems, crop rotation is widely applied as a useful way to avoid soil exhaustion and to reduce diseases and pest problems. Each year after the harvest, the carabids present in the fields will experience a change in husbandry practices and their timing. By comparing maize fields under different farming systems, Lövei (1984) concluded that non-rotational cropping was more favourable to carabid communities than those with crop rotations, at least for the year maize was grown in the field. In this rotation system, the previous crop was wheat that was harvested in midsummer. The field consequently remained without any plant cover for eight months until the next crop, maize, had reached 20 cm high. On the other hand, in the non-rotational systems, a well-established carabid community was present, and autumn breeders were not disturbed by harvesting and soil cultivation. Using a split plot design, Ellsbury *et al.* (1998) demonstrated that the total number of carabid beetles also varied in relation to rotational systems for corn and soybean. In their case, the level of input did not influence carabid abundance, except in one year for soybean. They obtained the highest Shannon-Weaver Index (H') for the corn–soybean rotation followed by continuous corn. Hance *et al.* (1990) sampled two fields during three consecutive years in order to observe changes in carabid assemblages through a rotation of sugar beet – winter wheat – winter barley. They recorded dramatic changes in rank of abundance, but a gradual increase in total population number from sugar beet (581 individuals in field 1 and 861 in field 2) through winter wheat (868 in field 1 and 949 in field 2) and to winter barley (2202 in field 1 and 2425 in field 2). This increase may be due to the similarity of wheat and barley crops, which ensures some continuity for the carabid fauna, whereas different husbandry practices in sugar beet crop negatively affect populations. *Nebria brevicollis* was only found in winter barley, whereas *Nebria salina* was mainly present in sugar beet and in wheat. Cereals supported species such as *Bembidion ustulatum* and *Agonum mulleri*.

Thiele (1977) and Hance *et al.* (1990) also pointed out the influence of the preceding crop. Sekulic *et al.* (1987) related carabid densities with the previous crop. They found the highest densities (1.2 individual/m^2) in wheat monoculture followed by 0.7 individuals/m^2 when sunflower was the preceding crop, 0.4 individuals/m^2 for maize and only 0.3 individuals/m^2 for sugar beet. They attributed the low values in sugar beet crops to the high inputs of pesticides.

A cover crop may reduce the negative impact of bare soil on carabid assemblages between two cultures. Laub and Luna (1992) studied the consequences of different winter cover crop suppression practices on *Pseudaletia unipunctata* (Lepidoptera: Noctuidae) in a no-till corn system. They compared two techniques of rye cover crop suppression: herbicide spraying or mowing. They concluded that the abundance of *Scarites subterraneus* and *Pterostichus* spp. was 1.5–2.5 higher in the mowed than in the sprayed treatment. It is, however, difficult to separate the effect of cover crop and the green manure it brings to the next culture. This point will be discussed in the soil level section.

CROP DENSITY

Information is seldom available on the consequences of an increase in crop plant density on carabids. Baker and Dunning (1975) did not record significant differences between low and high sugar beet plant densities. In maize fields with two different crop densities, Alderweireldt and Desender (1990) reported significantly higher catches of *A. mulleri* and *Bembidion quadrimaculatum* with lower plant densities. This is again, probably, linked to their preference for low humidity and warmth. Wallin (1985) assigned differences in pitfall catches in the same field to a difference in sun exposition; spring breeders were caught more often in warm sun-exposed parts of the fields than autumn breeders. Honek (1988) also found significantly more carabids in cereal fields with missing rows than in fields with denser stands, except *Trechus quadristriatus*, which preferred dense stands. He attributed the higher level of catches in sparse stands to a greater running activity or an increase in population density. In the first case, it may be the consequence of microclimate change in the sparse stands, as daily catches increased with temperature (Honek, 1988). He also suggested that ground surface warmed by sunshine may attract species like *L. pilicornis*, especially on cool days after rain. In the same context, Honek and Jarosik (2000) showed that the density of *Pseudoophonus rufipes* and *Harpalus distinguendus* varied with rape stand density and maturation. They attributed that trend to large variation in microclimate. In their study, small plots of bare soil were established within the investigated stands. In pea and rape, they also found that carabids preferred to shelter within plant stands at low crop densities, but became aggregated on bare ground plots when surrounding crops became dense. However, all studies concerning crop density were done using pitfall trapping and currently neglected the consequence of change in stand density on ground beetle trapability.

INTERCROPPING

Intercropping is rarely used in Europe as it is more labour intensive and hampers intensive agricultural production (Theunissen, 1997). In two successive years investigating intercropping of cabbage and clover, Booij *et al.* (1997) recorded a higher activity-density and diversity of carabids, especially early in the growing season, in the intercropped cabbage plots when compared to the cabbage monocrop. This difference, they suggested, was a consequence of the small plot sizes used, which led to an aggregation of ground beetles in the intercropped plots. They also proposed that such studies may be difficult to interpret because carabid movement, and thus trapability, may be influenced by the structure and abundance of the vegetation. In a similar design under-sowing cabbages with clover, Amstrong and McKinlay (1997) also reported that a positive effect on carabid abundance was more marked at the beginning of the growing season. In their case, species such as *T. quadristriatus* were more abundant in the hoed treatment. They concluded that the response must be analysed on a species basis, taking into account each species' ecological requirements. Potts and Vickerman investigated the relationships between faunal diversity and under-sowing in cereal crops in 1974. They reported that 38% more taxa emerged from under-sown fields than from cultivated fields. This value may be even higher for the predatory insects, which reached 42.8% for the under-sown sites versus 27.6% for the other fields. In a more recent study of under-sowing, the activity-density of carabids was compared in ten under-sown and ten conventional spring barley crops in the United Kingdom. Carabid numbers were similar in such systems during May and June, but when the under-sown ley had become established, there were more carabids in the under-sown crops (Holland, unpublished).

WEED COVER INFLUENCE

Weed cover and composition may influence the presence and abundance of carabid species because of microclimatic change, but also diversification of the niche and food availability. Honek (1997) compared the activity of ground beetles on bare and weedy soil. He showed that preferences for dense stands increased with temperature; six carabid species preferred shaded ground, although some other species preferred areas of bare soil. Many carabid species are at least partially phytophagous (Johnson and Cameron, 1969). While some species are sometimes considered as agricultural pests (Thiele, 1977), others are also weed seed consumers (see Tooley and Brust, this volume). Booij and Noorlander (1991) explained the increase of seed-feeding species in integrated-farming systems by the greater availability of weeds. Ellsbury *et al.* (1998) explained the high value of H' calculated from pitfall trap data in low input plots by the presence of a more diverse and abundant weed cover than in high input plots. Purvis and Curry (1984) compared sugar beet treated with pre- and post-emergence herbicide versus no herbicide application. They caught more herbivores in the weedy plots, mainly Chrysomelidae and non-pest aphid species. However, the total number of Carabidae was slightly lower in these weedy plots. Oppositely, Hawthorne and Hassall (1995) observed a decrease in abundance of most carabid species by removing vegetation in uncropped strips bordering on a cereal field. In an attempt to quantify the influence of vegetation fragmentation and composition, Banks (1999) could not find any relation between *P. melanarius* catches and vegetation cover. He suggested that the high movement capacity of the beetles concealed any relationship between cover and catches. Using analysis of spatial distribution, Holland *et al.* (1999) only found associations with weed cover for the phytophagous beetles (*Amara* spp.).

CONSERVATION HEADLANDS

The lack of permanent undisturbed areas is also a characteristic of agroecosystems. Several investigations have studied the carabid fauna of field edges and adjacent habitats and their possible role as refuge habitats, notably during winter (Riedel, 1995). Using soil samplings in December, Riedel found a strong edge preference for crop-inhabiting species, possibly indicating overwintering in these sites. However, using a directional pitfall trapping design, Kromp and Nitzlader (1995) could only detect spring dispersal from a hedge border to the field for *B. lampros* and not for the other species. The kind of edge and its management could change carabid composition and abundance. Hassall *et al.* (1992) compared the carabid fauna in uncropped headlands (6 m wide), conservation headlands, sprayed headlands and arable fields using pitfall trapping during two weeks in June–July, 1988. Uncropped headlands were not ploughed but rotovated three out every four years in the autumn, they were not sown and did not receive any pesticide or fertilizer treatment. Conservation headlands were ploughed and received only selected herbicides. Hassall *et al.* (1992) caught more individuals in uncropped headlands than in conservation headlands and sprayed headlands. Moreover, they suspected that this result is probably an underestimate because stem density and soil cover might have reduced carabid trapability in the uncropped headlands. They also observed that carabids were twice as abundant in crops adjacent to the uncropped and conservation headlands as in crops adjacent to the fully sprayed headlands, probably indicating that untreated headlands play an important role as a refuge for carabids. This observation is encouraging from a biological control perspective. Species richness was also positively influenced by uncropped and conservation headlands. However, these conclusions were only supported by a two-week pitfall trap sampling effort, limiting the interpretation of the role of uncropped headlands as overwintering sites and as refuges for dispersing species in spring. Hawthorne and Hassall (1995) continued with the same experiment in 1990, 1991 and 1992 with a longer period of sampling. They came to the same conclusion and demonstrated that species richness was most strongly correlated with total vegetation cover, Collembola, aphids and general invertebrate abundance. Indeed, an experimental reduction of the vegetation in the uncropped strip provoked a decrease in abundance of most carabid species except *B. lampros*. Lys (1994) obtained similar results when examining carabid activity-density in a sown weed strip established in a cereal field compared to a control field, followed over three consecutive years. He found activity-density to be 5 to 20 times higher in the sown weed strip area than in the field. In the strip-managed areas, 54 species were found compared to only 44 species in the control area, some only close to the edge of the field. Using mark–release–recapture, he showed that a significantly larger number of *Poecilus cupreus* and *Pterostichus melanarius* moved from the control area to the strip-managed area than *vice versa*. Moreover, the density of overwintering ground beetles was more than three times higher in the grassy strips than in the cereal fields. So, it clearly indicates that uncultivated and unsprayed edges of fields could provide refuges and overwintering sites for carabid beetles. Similar results were obtained by Hance *et al.* (1989). They showed that assemblages were more diverse with less over-dominant species nearby the unsprayed and uncropped edges of fields than in their centre. Thomas *et al.* (1991) proposed the creation of 'island' habitats (grassy raised earth banks 1.5 m wide, now known as 'beetle banks') in farmland to promote beneficial arthropods. They found very high densities of predators in these ridges when compared to the adjacent fields and to bare soil. They suggested that the ridges provided a nucleus for predator populations at the field centre. Their most striking increase was obtained for *Demetrias atricapillus*, which accounted for the major part of the total carabid catches. Further discussion on the importance of non-crop habitats is provided in Lee and Landis (this volume).

The soil level

CULTIVATION METHODS

Cultivation methods have several objectives, such as preparing the soil for seed germination, incorporation of organic matter, facilitation of water percolation, weed and some pest control. However, deep tillage encourages erosion and, to combat this, the use of soil conservation techniques is becoming more widespread. In literature reviews (Thiele, 1977; House and All, 1981; Luff, 1987; Stinner and House, 1990; Kromp, 1999; Holland and Luff, 2000), several authors have pointed out that deep tillage influences carabid beetle populations by reducing both abundance and diversity of assemblages, even though it may encourage some species.

Using both a quadrat method and pitfall trapping, Brust *et al.* (1985) showed that deep tillage reduced the absolute density of carabid predators. At the same time, they observed a negative correlation between carabid densities and the damage caused by the black cutworm, *Agrotis ipsilon*. Sometimes, no-tillage treatment increases pest incidence, for instance in the case of the Stalk Borer on maize (Levine, 1993). However, in this study no data was available on the predatory fauna. In the same way, Gallo and Pekar (1999) observed that most pest species occurred more abundantly on plots with shallow ploughing than on plots following deep ploughing. They mentioned that natural enemies were also more abundant with shallow ploughing, but did not analyse the carabid fauna. Symondson *et al.* (1996) stressed that deep ploughing is usually performed in order to reduce slugs, but it also adversely affects *P. melanarius*, which is an important slug predator. House *et al.* (1989) observed an increase in carabid populations in no-tillage plots. However, in their study, the amount of damage caused by insect pests was also higher in no-tillage plots than in the conventional tillage system. Ferguson and McPherson (1985) argued that carabids are more diverse and abundant in a soybean system when no tillage is carried out because of the greater amount of organic matter on the soil surface, which also increases niche availability. On a reduced tillage farm, Carcamo (1995) found a higher diversity (Shannon–Wiener indices) than on a conventional tillage farm. Barney and Pass (1986b) obtained contradictory results, but they only monitored selected species. Indeed, Carcamo (1995) and Carcamo *et al.* (1995) have shown that the species responses he observed were inconsistent in the two systems, with some species being more abundant in the conventional tillage regime.

The timing of cultivation depends largely on whether or not the crop is to be spring or autumn sown. Species may be present in the soil either as larvae or already as adult surface dwellers (Hance *et al.*, 1990). Tillage may influence carabid populations in different ways. Larvae present in the soil may be killed mechanically by ploughing or hoeing, be affected by the soil structure, porosity water flow and temperature modifications, or be exposed to predators. Moreover, organic residues are incorporated into the soil, whereas in no-tillage systems these remain on the surface and are thus available for the ground-dwelling fauna (Brust *et al.*, 1985; Cromar *et al.*, 1999). In a reduced tillage experiment without any herbicide treatment, Andersen (1999) observed that weeds were unusually numerous and probably influenced more ground beetle occurrence than the tillage itself, especially in the case of *Amara* spp.

Contradictory results are reported from the literature concerning the influence of spring cultivation on spring and autumn breeders. In a comparative study of three cultivation methods (30 cm depth, 15 cm depth and no-tillage) in spring (maize and sugar beet) and winter crops (winter wheat and winter barley), Baguette and Hance (1997) reported that total numbers of carabids were higher in deep and semi-deep ploughed plots than in no-tillage plots. This observation was mainly because of the dominant species, *P. melanarius*, which was considerably more numerous in the ploughed plots. The authors explained that this species is an autumn breeder and is thus at the larval stage in the spring, when ploughing takes place in maize and sugar beet. In winter wheat and winter barley, even if many adults are still present in September when ploughing occurs, eggs and first larval stages are already in the soil and thus less disturbed by the soil cultivation. Skuhravy (1958) had exactly the opposite opinion. He stated that autumn ploughing decimates the larvae and pupae of autumn breeders. His hypothesis was based on the fact that *P. melanarius* was found mainly in winter crop fields, whereas *Poecilus cupreus*, a spring breeder, was much more frequent in root crops than in winter cereals. However, Thiele (1977) pointed out that those two species also diverge in their reaction to soil cover. Results from Fadl *et al.* (1996) correspond well to those of Skuhravy (1958). In their case, catches made during the main emergence phases of the new generation of *P. melanarius* were lower in spring cultivated fields compared with uncultivated or autumn cultivated fields. They observed a gradual increase in abundance from late spring cultivation, early spring cultivation, uncultivated field, and finally autumn-cultivated field. In their study, however, previous crop cultivation timing, the fields and the rotation also influenced numbers of catches, making results difficult to interpret. Fadl *et al.* (1996) suggested that large late-instar larvae and pupae are likely to be prone to physical injury during cultivation. Their results were supported by the experiment of Purvis and Fadl (1996). Using emergence arenas, they obtained an emergence density of 13.8 *P. melanarius* per m^2 on winter-sown plots compared with only 2.5 individuals per m^2 on spring-cultivated plots.

By the release and recapture of marked beetles, Powell and Ashby (1995) demonstrated that the difference in *P. melanarius* pitfall catches between ploughed and tine cultivated crops may be caused, at least partly, by differences in beetle activity. Purvis and Fadl (1996) explained that the failure to detect any influence of field type on total seasonal catches was because of redistribution among field types. These contradictory results highlight the need for detailed studies involving only one factor and taking into account direct measures of carabid density at both the larval and adult stages directly preceding and immediately following use of different cultivation methods.

Using multivariate analysis, Baguette and Hance (1997) identified four groups of species in relation to cultivation. In deep ploughing areas, the most abundant species were autumn breeders (*P. melanarius*, *T. quadristriatus* and *Clivina fossor*). Small species like *Bembidion lampros*, *B. obtusum* and *B. quadrimaculatum* were most abundant in areas with a low ploughing intensity, mainly in sugar beet and maize. Another group of species were found living in winter barley and wheat without ploughing (*A. mulleri*, *Harpalus affinis*, *Notiophilus biguttatus* and *Pterostichus niger*). Spring breeders (*A. flavipes*, *A. dorsale* and *L. pilicornis*) depended on crop type (winter barley and wheat), no matter what soil treatment was performed. Clark *et al.* (1997) did not agree with these conclusions because in their analysis spring breeders dominated in tilled plots regardless of the period when tillage occurred. However, they did not exclude the possibility of preferential re-colonization of the field after cultivation by spring breeders which were present in the adult stage at that time. In their case, *Pterostichus lucublandus* and *Agonum placidum* were the most common in conventional systems, whereas *P. melanarius* and *Cyclotrachelus sodalis* were abundant in perennial, no tillage-crops. Tonhasca (1993) also found that *C. sodalis* and *Harpalus pensylvanicus* were more abundant in no-tillage crops, whereas *Pterostichus chalcites* and *Scarites substriatus* were more common in conventional tillage crops in Iowa.

The impact of cultivation on adults or larvae is an old debate, as already pointed out by Thiele (1977). Moreover, Fadl and Purvis (1998) plead in favour of a re-discussion of the classification of the breeding periods since they observed that, for many species, breeding may be spread over several months or can be bimodal, indicating two generations per year. This is probably partly the source of the contradictions in the literature. Future studies should be focused on the immediate impact of mechanical intervention in the few days just following cultivation on both larvae and adult populations. Moreover, conservation tillage is often associated with mulching, as in the study of House and All (1981). In this case, it is also difficult to distinguish between the consequence of the tillage itself and of the mulching.

FERTILIZATION AND ORGANIC MATTER

Pietrasko and DeClercq (1982) observed that the addition of 40 t of organic matter to sandy soil increased the number of carabid beetles caught to 127% of the control plot. However, some species did not increase, such as *B. lampros*, or were caught more frequently on the control plot, such as *P. melanarius*. Hance and Grégoire-Wibo (1987) showed that the level of organic manure added to the crop is the main factor determining carabid species composition. Using a cage experiment, Grégoire-Wibo (1980, 1983) distinguished three categories of ground beetles in relation to organic matter content: 1) spring breeders positively influenced by organic matter (*B. lampros*, *B. quadrimaculatum*, *Bembidion femoratum*, *L. pilicornis*, *Harpalus rufipes*, *C. fossor*); 2) spring breeders unaffected by organic matter (*A. dorsale*, *A. mulleri*, *A. flavipes*); 3) autumn breeders unaffected by organic matter (*Harpalus rufipes*, *T. quadristriatus*). In the case of group 1, organic matter reduced the impact of insecticide treatments. Humphreys and Mowat (1994) confirmed Grégoire-Wibo's observations. They found more individuals in manured plots than in control plots, *B. lampros* comprising a large proportion of their counts. This could be due to an increase of alternative prey species available in manured plots. The number of captures in straw-mulched plots was lower, probably because straw impeded movement, and thus trapability, of carabids. Purvis and Curry (1984) also concluded that farmyard manure increased the activity of species common early in the season, such as *Pterostichus strenuus* and *B. lampros*. According to Helenius and Tolonen (1994), the production of new generation adults increased by 50% using ryegrass as manure in spring cereals. Winder *et al.* (1999) reported that the use of sewage sludge with moderate Zn content for agricultural soil amendment produces a concentration effect on the third trophic level. They indeed showed that Zn concentration in the wheat plants, aphids and *B. lampros* reached respectively 31.7, 116.0 and 242 $\mu g\ g^{-1}$ dry weight in a short, nine-day feeding trial. In conclusion, they stressed the importance of gathering information on the tri-trophic transfer of trace metals through the soil–plant–herbivore–predator pathway. They also underlined the lack of knowledge we have on the consequence of trace metals on the survival and fecundity of predatory insects.

The influence of chemical fertilization on carabids is poorly documented (Kromp, 1999). Utrobina (1976) compared the effects of treatment with ammonia-water (200–250 l/ha) in a potato crop. He could not detect any influence on carabid populations before and after fertilizer application. Pekar (1997) tried to quantify the short-term effect of liquid fertilizer (urea and ammonium nitrate 100 l/ha) on carabid and spider fauna. He observed that, a week after the application, the overall abundance was significantly lower on fertilized plots than on control plots, with *Harpalus rufipes* being the most affected carabid species. However, after seven weeks, all effects disappeared, probably because of re-colonization. Kleinert (1987) suggested that synthetic fertilizers, and other elements of agriculture intensification, caused the decline of *Carabus cancellatus* in Slovakia. Tietze (1987) also linked agriculture intensification and the uniformity in grassland composition by fertilization. This simplification of the grassland community led to a change in dominance structure of the carabid fauna. Idinger and Kromp (1997) did not detect an influence of fertilization using organic compost or inorganic fertilizer on total carabid number in comparison to an unfertilized plot, because their results varied from year to year.

Pesticide application

The pesticide market is continuously growing by approximately 1.4% per annum. According to Silvy (1999), total sales reached 31 billion US $ in 1997, among which herbicides represent 49%, insecticides 27% and fungicides 20%. The negative influence of pesticides on carabid populations has been thoroughly studied since the 1970s and is well understood (Edwards and Thompson, 1975; Chambon, 1982; Basedow, 1983; Shelton and Edwards, 1983; Hance and Grégoire-Wibo, 1987; Vickerman *et al.,* 1987) and was recently reviewed by Holland and Luff (2000). However, the consequences of pesticide treatments are currently linked with other factors of agroecosystems, as illustrated in a diagram by Holland and Luff (2000), and it is valuable to analyse them by categories.

INSECTICIDES

Chambon (1982) pointed out that daily pitfall trap catches dropped by up to 80% just after an insecticide treatment. He explained that the re-colonization of the field by a diverse entomofauna is possible only if the area treated is restricted. Sekulic *et al.* (1987) observed that, after a lindane application (1 kg/ha), the carabid population needed two months to regain its initial density. Grégoire-Wibo (1983) underlined the fact that the catches of *B. lampros*, *B. quadrimaculatum*, *B. femoratum* and *C. fossor* were reduced less in insecticide-treated plots when organic matter is also applied. Hance and Grégoire-Wibo (1987) confirmed this point. Using a correspondence analysis, they showed that 54% of the species assemblages composition can be explained by the organic matter application versus 28.2% for the insecticide treatment. In cage experiments, Gyldenkærne *et al.* (2000) have observed a severe effect of a dimethoate application in June on *B. lampros* and *A. dorsale*, compared to an application in autumn. They hypothesized that, because both species are univoltine and spring breeding, they will have finished oviposition at the end of June and, being at the end of their cycle, are probably more sensitive to insecticide at this moment. The higher toxicity of dimethoate applied in June compared to the autumn occurred, despite the lower levels of pesticide reaching the soil surface, because crop cover was higher in June. They also showed that cypermethrin had little effect on carabids, but affected *Tachyporus hypnorum* (Coleoptera: Staphylinidae) populations at the maximum application rate. Sekulic *et al.* (1987), analysing the dynamics of overwintering stages of carabids over a period of 24 years, observed a drop in the global mean densities per m^2 during the period 1971–1980. They linked that observation to an increased use of pesticide during that period. Quinn *et al.* (1991) related the immediate change of composition of ground beetles to specific habitat characteristics.

They showed that the habitats with clay-rich soils and a high coverage of *Bouteloua gracilis* and a low coverage of *Buchloe dactyloides* were more likely to undergo carabid population changes after an insecticide application, regardless of the type of treatment. However, the most diverse communities were associated with soils containing a high percentage of sand and a low percentage of clay. Many factors must therefore be considered when analysing the consequences of insecticide applications. Whether carabids are suitable as bioindicators, and their response to insecticides and farming systems, is considered in more detail in Chapter 9.

WEED CONTROL AND HERBICIDES

It is difficult to distinguish between the direct influence of weed control and the consequences of weed removal. For instance, in 1964, Fox concluded that a decrease in the weed cover because of herbicide treatments primarily affected the abundance of soil invertebrates, rather than acute toxicity or even changes in floristic composition. In a very original experiment, Dierauer and Pfiffner (1993) tried to determine the influence of flame weeding on carabid beetles. They observed that, in maize plots when flame weeding was applied to an existing dense weed cover, carabid activity-density was lower two days after flaming than in the control plots, probably because under weed cover the carabid beetles were more active at the time of flaming. However, in nine other crops, such as vegetables, flame weeding was mainly used in pre-emergence treatment and, in that case, no negative effect was recorded. They recommend that flaming should be conducted at noon, when the activity of carabids is lower and when they are protected in soil crevices. Petersen *et al.* (1996) did not find any adverse effect of mechanical weed control on carabid abundance in a two-year study in spring barley.

The use of herbicides has increased substantially. For example, in the USA herbicide use has increased since 1971 by 180% to about 173 million kg in 1987, whereas insecticide use declined by about 35% (Freemark and Boutin, 1995). Chiverton and Sotherton (1991) described the same trends for Great Britain. Herbicide applications also have numerous indirect effects on agricultural landscape. As stated by Freemark and Boutin (1995), herbicides have reduced the need for crop rotation for weed control, and thus have favoured monocultures and continuous cropping. Moreover, field margins and other non-crop habitats are currently sprayed by broad-spectrum herbicides. In a review, Holland and Luff (2000) stated that herbicides and fungicides are currently not directly toxic to carabids, but may influence survival through habitat modification or food removal. Numerous studies have shown that carabid activity-density is higher in weedy plots than in herbicide-treated plots. In the future, the use of transgenic plants tolerant to herbicide should accentuate the oversimplification of agroecosystems by allowing farmers to increase treatments with broad-spectrum herbicides, with dramatic consequences on invertebrate fauna. This point must certainly be investigated in risk assessment and when determining the impact of transgenic crops on environmental quality.

Conclusions

Analysing the consequences of husbandry practices is hampered by the difficulty in isolating the effect of different factors in complex agroecosystems. In addition, while at first glance carabid species distribution seems to be quite homogenous, it is actually composed of species exhibiting very different ecological requirements. Most authors discuss assemblages because we do not yet know what interactions exist between species and whether they are functioning as a community as defined by Liss *et al.* (1986). Multifactorial analyses give interesting results and allow interpretation of differences in carabid activity and densities in the assemblages in relation to complex environmental factors (Sanderson, 1994). However, knowledge of species life cycles and ecological characteristics is also essential if we want to understand the spatio-temporal changes so frequent in arable land.

In the introduction, two questions were asked. Can these now be answered? Some general conclusions may be drawn. Reducing the frequency and the intensity of disturbances will lead to a more diverse carabid fauna, with fewer overly-dominant species. Diversity of natural enemies cannot be increased selectively. Landis *et al.* (2000) stressed the point that key elements must be identified. Organic matter, both in and on the soil, is one of those key factors because it provides shelter, increases alternative food sources, buffers local climatic variations, and positively influences soil texture. Organic matter may be provided by animal or green manure, or as a result of intercropping and weed cover. Weed cover increases the heterogeneity of microclimatic conditions and niche availability. It may change species assemblage composition in favour of some species because of their soil temperature and humidity requirements. Weed cover increases the diversity of food, both in preys and weed seeds. Plant cover positively affects species richness.

Deep and frequent soil cultivation must be avoided. Ploughing could, for instance, take place at the beginning of a four- or five-years' rotation and then be replaced by non-inversion tillage. Bare soil in winter is not favourable to ground beetles and increases the risk of erosion. Cover crops and conservation headlands provide improved overwintering sites. In the same way, perennial habitats are necessary in an agricultural landscape to facilitate re-colonization of crop fields after detrimental husbandry practices. Finally, insecticide treatments may affect carabid assemblages in the long term and, when needed, specific and spatially localized applications should be chosen.

If carabids are to be used as indicator species, further knowledge is still required. Relationships between size and ecological behaviour are unclear, and generalizations cannot be made at this time. This is particularly true for the biological control potential of selected species. For instance, Finch (1996) showed that most of the carabid species he caught, both in cultivated fields or in other habitats, accepted and ate cabbage root fly eggs. He observed a clear relationship between the size of the beetles and the number of eggs eaten. Finch (1996) drew attention to the fact that the pest control potential is probably linked to other ecological requirements. Indeed, *T. quadristiatus* is probably more efficient as an egg predator than *B. lampros* as it prefers the shady area around the base of the spaced *Brassica* plants, rather than the intra-plant areas favoured by *B. lampros*. While the breeding periods and requirements of adults of abundant species are rather well known (Fadl and Purvis, 1998), the egg and larval stages are still poorly studied. Nevertheless, they constitute an important, and probably vulnerable, part of the life cycle and no information is available about their sensitivity to husbandry practices and soil conditions. Information is also lacking about the population dynamics of most species on natural mortality factors and on the growth potential of carabid populations. The relative importance of diurnal versus nocturnal activity also needs to be investigated. Finally, while a number of studies have investigated the feeding behaviour of a few abundant species, such as *P. melanarius* (Pollet *et al.*, 1986), studies of multi-trophic interactions and intra-guild predation have been neglected, but these are needed if we are to understand the community status of well known assemblages.

Agriculture is continually changing towards sustainable production, integrated, and even organic, farming. This may lead to new developments in carabid science, which at the same time must help farmers to make the best choices, given their potential to contribute to pest control (Chapters 6 and 7).

Acknowledgements

The author is grateful to Renate Wesselingh, Georges Van Impe, John Holland and anonymous reviewers for their comments and the revision of the manuscript.

References

ALDERWEIRELDT, M. AND DESENDER, K. (1990). Microhabitat preference of spiders (Araneae) and carabid beetles (Coleoptera, Carabidae) in maize fields. *Mededelingen, Faculteit Landbouwwetenschap Rijksuniversiteit, Gent* **55**, 501–510.

AMSTRONG, G. AND MCKINLAY, R.G. (1997). The effect of undersowing cabbages with clover on the activity of carabid beetles. *Biological Agriculture and Horticulture* **15**, 269–277.

ANDERSEN, A. (1999). Plant protection in spring cereal production with reduced tillage. II. Pest and beneficial insects. *Crop Protection* **18**, 651–657.

BAGUETTE, M. AND HANCE, TH. (1997). Carabid beetles and agricultural practices: influence of soil ploughing. *Biological Agriculture and Horticulture* **15**, 185–190.

BAKER, A.N. AND DUNNING, R.A. (1975). Some effects of soil type and crop density on the activity and abundance of the epigeic fauna, particularly Carabidae, in sugar beet fields. *Journal of Applied Ecology* **12**, 809–818.

BANKS, J.E. (1999). Differential response of two agrosystem predators, *Pterostichus melanarius* (Coleoptera: Carabidae) and *Coccinella septempunctata* (Coleoptera: Coccinellidae), to habitat-composition and fragmentation-scale manipulations. *The Canadian Entomologist* **131**, 645–657.

BARNEY, R.J. AND PASS, B.C. (1986a). Pitfall trap collections of ground beetles larvae (Coleoptera: Carabidae) in Kentucky alfafa fields. *The Great Lakes Entomologist* **9**, 147–151.

BARNEY, R.J. AND PASS, B.C. (1986b). Ground beetles (Coleoptera: Carabidae) population in Kentucky alfalfa and influence of tillage. *Journal of Economic Entomology* **79**, 511–517.

BASEDOW, T. (1983). *Agonum dorsale* Pont. (Col. Carabidae) an important predator of cereal pest: the effects of insecticides on its population in a cereal field in Northern Germany. *Proceedings of the 8th International Congress of Soil Zoology. New trends in soil biology*. Eds. Ph. Lebrun, H.M. André, A. De Medts, C. Grégoire-Wibo and G. Wauthy, pp 583–587. Ottignies Louvain-la-Neuve, Belgium: Dieu-Brichart.

BASEDOW, T.H., BORG, A., DE CLERCQ, R., NIJVELDT, W. AND SCHERNEY, F. (1976). Untersuchungen über das Vorkommen der Laufkäfer (Col.: Carabidae) auf europäischen Getreidefeldern. *Entomophaga* **21**, 59–72.

BLAKE, S., FOSTER, G.N., EYRE, M.D. AND LUFF, M.L. (1994). Effect of habitat type and grassland management practice on the bodysize distribution of carabid beetles. *Pedobiologia* **38**, 502–512.

BOOIJ, C.J.H. AND NOORLANDER, J. (1991). The impact of integrated farming on Carabid beetles. *Proceedings of Experimental and Applied Entomology* **2**, 16–21.

BOOIJ, C.J.H., NOORLANDER, J. AND THEUNISSEN, J. (1997). Intercropping cabbage with clover: effects on ground beetles. *Biological Agriculture and Horticulture* **15,** 261–268.

BOOIJ, K. (1994). Diversity pattern assemblages in relation to crops and farming systems. In: *Carabid beetles, ecology and evolution*. Eds. K. Desender, M. Dufrêne, M. Moreau, M.L. Luff and J.-P. Maelfait, pp 425–431. Dordrecht: Kluwer Academic Publishers.

BRUST, G.E., STINNER, B.R. AND MCCARTNEY, D.A. (1985). Tillage and soil insecticide effects on predator-black cutworm (Lepidoptera: Noctuidae) interactions in corn agrosystems. *Journal of Economic Entomology* **78**, 1389–1392.

CARCAMO, H.A. (1995). Effect of tillage on ground beetles (Coleoptera: Carabidae): a farm-scale study in Central Alberta. *The Canadian Entomologist* **127**, 631–639.

CARCAMO, H.A. AND SPENCE, J. (1994). Crop type effects on the activity and distribution of ground beetles. *Environmental Entomology* **23**, 684–692.

CARCAMO, H.A., NIEMALA, J.K. AND SPENCE, J. (1995). Farming and ground beetles: effect of agronomic practice on populations and community structure. *The Canadian Entomologist* **127**, 123–140.

CHAMBON, J.P. (1982). Recherche sur les biocénoses céréalières. II Incidence des interventions insecticides sur les composantes de l'entomofaune. *Agronomie* **2**, 405–416.

CHIVERTON, P.A. AND SOTHERTON, N.W. (1991). The effects of beneficial arthropods of the exclusion of herbicide from cereal crop edges. *Journal of Applied Ecology* **28**, 1027–1039.

CIOCHIA, V. AND DONESCU, D. (1997). Quelques aspects de l'entomofaune épigée des cultures de betterave sucrière, de pomme de terre et de chicorée en Roumanie. *Bulletin Academy of Agricultural and Sylvicultural Sciences 'Gheorghe Inonescu – Sissesti'* **24**, 90–95.

CLARK, M.S., GAGE, S.H. AND SPENCE, J.R. (1997). Habitat and management associated with common ground beetles (Coleoptera: Carabidae) in a Michigan agricultural landscape. *Environmental Entomology* **26**, 519–527.

CROMAR, H.E., MURPHY, S.D. AND SWANTON, C.J. (1999). Influence of tillage and crop residue on postdispersal predation of weed seed. *Weed Science* **47**, 184–194.

DAMMER, K.H. AND HEYER, W. (1997). Quantifying the influence of the cultivated plant species on the occurrence of carabid beetles within certain species using contingency table analysis. *Environmental and Ecological Statistics* **4**, 321–336.

DIERAUER, VON H.U. AND PFIFFNER, L. (1993). Auswirkungen des Abflammens auf Laufkäfer. *Gesunde Pflanzen* **45**, 226–229.

EDWARD, C.A. AND THOMPSON, A.R. (1975). Effects of insecticides on predatory beetles. *Annals of Applied Biology* **79**, 132–134.

ELLSBURY, M.M., POWEL, J.E., FORCELLA, F., WOODSON, W.D., CLAY, S.A. AND RIEDELL, W.E. (1998). Diversity and dominant species of ground beetles assemblages (Coleoptera: Carabidae) in crop rotation and in chemical input systems for the Northern Great Plains. *Annals of the Entomological Society of America* **91**, 619–625.

FADL, A. AND PURVIS, G. (1998). Field observations on lifecycles and seasonal activity patterns of temperate carabid beetles (Coleoptera: Carabidae) inhabiting arable land. *Pedobiologia* **42**, 171–183.

FADL, A., PURVIS, G. AND TOWEY, K. (1996). The effect of time of soil cultivation on the incidence of *Pterostichus melanarius* (Illig.) (Coleoptera: Carabidae) in arable land in Ireland. *Annales Zoologici Fennici* **33**, 207–214.

FERGUSON, H.J. AND MCPHERSON, R.M. (1985). Abundance and diversity of adult Carabidae in four soybean cropping systems in Virginia. *Journal of Entomological Sciences* **20**, 163–171.

FINCH, S. (1996). Effect of beetle size on predation of cabbage root fly eggs by ground beetles. *Entomologia Experimentalis et Applicata* **81**, 199–206.

FOX, C.J.S. (1964). The effects of five herbicides on the numbers of certain invertebrate animals in grassland soil. *Canadian Journal of Soil Sciences* **44**, 405–409.

FREEMARK, K. AND BOUTIN, C. (1995). Impacts of agricultural herbicide use on terrestrial wildlife in temperate landscapes: A review with special reference to North America. *Agriculture, Ecosystems and Environment* **52**, 67–91.

GALLO, J. AND PEKAR, S. (1999). Winter wheat pests and their natural enemies under organic farming system in Slovakia: effect of ploughing and previous crop. *Journal of Pest Science* **72**, 31–36.

GRÉGOIRE-WIBO, C. (1980). Etude de l'effet de certains pesticides betteraviers sur certains ravageurs (atomaires) et sur la faune endogée et épigée participant à la fertilité du sol et au contrôle des populations nuisibles (acariens, collemboles, carabides). *Institut Belge pour l'amélioration de la Betterave* **III**, 133–165.

GRÉGOIRE-WIBO, C. (1983). Incidences écologique de traitement phytosanitaires en culture de betteraves sucrières. II. Acariens, Polydesme, Staphylins, Cryptophagides et Carabides. *Pedobiologia* **25**, 93–108.

GYLDENKÆRNE, S., RAVN, H.P. AND HALLING-SØRENSEN, B. (2000). The effect of dimethoate and cypermethrin on soil-dwelling beetles under semi-field conditions. *Chemosphere* **41**, 1045–1057.

HANCE, T. (1990). Relationships between crop types, carabid phenology and aphid predation in agroecosystems. In: *The role of ground beetles in ecological and environmental studies.* Ed. N.E. Stork, pp 55–63. Andover: Intercept.

HANCE, T. AND GRÉGOIRE-WIBO, C. (1987). Effect of agricultural practices on carabid populations. *Acta Phytopathologica et Entomologica Hungarica* **22**, 147–160.

HANCE, T., GRÉGOIRE-WIBO, C., STASSART, P. AND GOFFART, F. (1989). Etude de la taxocénose des Carabidae (Coleoptera) dans les écosystèmes agricoles du Brabant Wallon. In: *Invertébrés de Belgique*, pp 315–324. Bruxelles: Institut Royal des Sciences Naturelles de Belgique.

HANCE, T., GRÉGOIRE-WIBO, C. AND LEBRUN, P.H. (1990). Agriculture and ground-beetles populations: The consequence of crop types and surrounding habitats on activity and species composition. *Pedobiologia* **34**, 337–346.

HASSALL, M., HAWTHORNE, A., MAUDSLEY, M., WHITE, P. AND CARDWELL, C. (1992). Effects of headlands management on invertebrate communities in cereal fields. *Agriculture, Ecosystems and Environment* **40**, 155–178.

HAWTHORNE, A. AND HASSALL, M. (1995). The effect of cereal headland treatments on carabid communities. In: *Arthropod natural enemies in arable land. I. Density, spatial heterogeneity and dispersal.* Eds. S. Toft and W. Riedel, pp185–198. Aarhus: Aarhus University Press.

HELENIUS, J. AND TOLONEN, T. (1994). Enhancement of generalist aphid predators in cereals: effect of green manuring on recruitment of ground beetles (Col., Carabidae). *IOBC WPRS Bulletin* **17**, 201–210.

HOLLAND, J.M. AND LUFF, M.L. (2000). The effects of agricultural practices on Carabidae in temperate agroecosystems. *Integrated Pest Management Reviews* **5**, 109–129.

HOLLAND, J.M., PERRY, J.N. AND WINDER, L. (1999). The within-field spatial and temporal distribution of arthropods in winter wheat. *Bulletin of Entomological Research* **89**, 499–513.

HONEK, A. (1988). The effect of crop density and microclimate on pitfall trap catches of Carabidae, Staphylinidae (Coleoptera) and Lycosidae (Aranae) in cereal fields. *Pedobiologia* **32**, 233–242.

HONEK. A. (1997). The effect of plant cover and weather on the activity density of ground surface arthropods in a fallow field. *Biological Agriculture and Horticulture* **15**, 203–210.

HONEK, A. AND JAROSIK, V. (2000). The role of crop density; seed and aphid presence in diversification of field communities of Carabidae (Coleoptera). *European Journal of Entomology* **97**, 517–525.

HOUSE, G.J AND ALL, J.N. (1981). Carabid beetles in soybean agrosystems. *Environmental Entomology* **10**, 194–196.

HOUSE G.J. AND DEL ROSARIO ALZUGARAY, M. (1989). Influence of cover cropping and no-tillage practices on community composition of soil arthropods in a North Carolina agroecosystem. *Environmental Entomology* **18**, 302–307.

HUMPHREYS, I.C. AND MOWAT, D.J. (1994). Effect of some organic treatments on predators (Coleoptera: Carabidae) of cabbage root fly, *Delia radicum* (L.) (Diptera: Anthomyiidae), and on alternative prey species. *Pedobiologia* **38**, 513–518.

IDINGER, J. AND KROMP, B. (1997). Ground photoeclector evaluation of different arthropod groups in unfertilized, inorganic and compost-fertilized cereal fields in eastern Austria. *Biological Agriculture and Horticulture* **15**, 171–176.

JOHNSON, N.E AND CAMERON, R.S. (1969). Phytophagous ground beetles. *Annals of the Entomological Society of America* **62**, 909–914.

KEGEL, B. (1990). Diurnal activity of Carabid beetles living on arable land. In: *The role of ground beetles in ecological and environmental studies.* Ed. N.E. Stork, Chapter 7, pp 65–76. Andover: Intercept.

KLEINERT, J. (1987). Changes in the distribution of *Carabus cancellatus* (Coleoptera: Carabidae) in Slovakia. *Acta Phytopathologica et Entomologica Hungarica* **22**, 161–163.

KROMP, B. (1999). Carabid beetles in sustainable agriculture: a review on pest control efficacy, cultivation impacts and enhancement. *Agriculture, Ecosystems and Environments* **74**, 187–228.

KROMP, B. AND NITZLADER, M. (1995). Dispersal of ground beetles in a rye field in Vienna, Eastern Austria. In: *Arthropod natural enemies in arable land. I. Density, spatial heterogeneity and dispersal.* Eds. S. Toft and W. Riedel, pp 269–277. Aarhus: Aarhus University Press.

LANDIS, D.A., WRATTEN, S.D. AND GURR, G.M. (2000). Habitat management to conserve natural enemies of arthropod pests in agriculture. *Annual Review of Entomology* **45**, 175–201.

LAUB, C.A. AND LUNA, J.M. (1992). Winter cover crop suppression practices and natural enemies of armyworm (Lepidoptera: Noctuidae) in no-till corn. *Environmental Entomology* **21**, 41–49.

LEVINE, E. (1993). Effect of tillage practices and weed management on survival of stalk borer (Lepidoptera: Noctuidae) eggs and larvae. *Journal of Economical Entomology* **86** (3), 924–928.

LISS, W.J., GUT, L.J., WESTIGARD, P.H. AND WARREN, C.E. (1986). Perspectives on arthropod community structure, organisation and development in agricultural crops. *Annual Review of Entomology* **31**, 455–478.

LÖVEI, G.L. (1984). Ground beetles (Coleoptera: Carabidae) in two types of maize fields in Hungary. *Pedobiologia* **26**, 57–64.

LÖVEI, G.L. AND SAROSPATAKI, M. (1990). Carabid beetles in agricultural fields in Eastern Europe. In: *The role of ground beetles in ecological and environmental studies.* Ed. N.E. Stork, Chapter 9, pp 87–93. Andover: Intercept.

LUFF, M.L. (1987). Biology of polyphagous ground beetles in agriculture. *Agricultural Zoology Reviews* **2**, 237–278.

LYS, J.A. (1994). The positive influence of strip-management on ground beetles in a cereal field: increase, migration and overwintering. In: *Carabid beetles, ecology and evolution.* Eds. K. Desender, M. Dufrêne, M. Loreau, M.L. Luff and J.-P. Maelfait, pp 451–455. Dordrecht: Kluwer Academic Publishers.

MIRCEA, V., DONESCU, D. AND DASCALU, A. (1989). Coenological observations on the populations of Carabidae (Insecta, Coleoptera) from some potato crops in Moldavia (Romania). *Anaieie stiinfice aie Universitatii All Cuza, Iasi s; Biologie animala*, Tom. **44–45**, 105–116.

ODUM, E.P. (1971). *Fundamentals of ecology*. Philadelphia: W.B.Saunders Company.

PEKAR, S. (1997). Short-term effect of liquid fertilizer (UAN) on beneficial arthropods (Araneae, Opilionida, Carabidae, Staphylinidae) in winter wheat. *Ochrana Rostlin* **33**, 17–24.

PETERSEN, M.K., RASMUSSEN, J. AND RASMUSSEN, K. (1996). Does mechanical weed control affect the carabid fauna in arable field (in Danish). *Sp Rapport – Statens Plantevaerns Forsog* **4**, 91–98.

PETIT, S. AND USHER, M.B. (1998). Biodiversity in agricultural landscape: the ground beetle communities of woody uncultivated habitats. *Biodiversity and Conservation* **7**, 1549–1561.

PIETRASKO, R. AND DECLERCQ, R. (1980). Etude de la population d'arthropodes épigés dans les cultures agricoles au cours de la période 1974–1978. *Revue de l'agriculture* **4**, 719–731.

PIETRASKO, R. AND DECLERCQ, R. (1982). Influence of organic matter on epigeic arthropods. *Mededelingen, Faculteit Landbouwwetenschap Rijksuniversiteit, Gent* **47**, 721–728.

POLLET, M., DESENDER, K. AND MALFAIT, J.-P. (1986). Aspects of the feeding ecology of *Pterostichus melanarius* in a heavy grazed pasture. *Annales de la Société Royale Zoologique de Belgique* **116**, 110–111.

POTTS, G.R. AND VICKERMAN, G.P. (1974). Studies on the cereal ecosystems. *Advances in Ecological Research* **8**, 107–197.

POWELL, W. AND ASHBY, J. (1995). Using mark–release–recapture techniques to study the influence of spatial heterogeneity on carabids. In: *Arthropod natural enemies in arable land. I. Density, spatial heterogeneity and dispersal*. Eds. S. Toft and W. Riedel, pp 165–173. Aarhus: Aarhus University Press.

PURVIS, G. AND CURRY, J.P. (1984). The influence of weeds and farmyard manure on the activity of Carabidae and other ground-dwelling arthropods in a sugar beet crop. *Journal of Applied Ecology* **21**, 271–283.

PURVIS, G. AND FADL, A. (1996). Emergence of Carabidae (Coleoptera) from pupation: a technique for studying the 'productivity' of carabid habitats. *Annales Zoologici Fennici* **33**, 215–223.

QUINN, M.A., KEPNER, R.L., WALGENBACH, D.D., FOSTER, R.N., BOHLS, R.A., POOLER, P.D., REUTER, K.C. AND SWAIN, J.L. (1991). Effect of habitat characteristics and perturbation from insecticides on the community dynamics of ground beetles (Coleoptera: Carabidae) on mixed-grass rangeland. *Environmental Entomology*, **20** (5), 1285–1294.

RIEDEL, W. (1995). Spatial distribution of hibernating polyphagous predators within field boundaries. In: *Density, spatial heterogeneity and dispersal*. Eds. S. Toft and W. Riedel. *Acta Jutlandica* **70** (2), 221–226.

SANDERSON, R.A. (1994). Carabidae and cereals: a multivariate approach. In: *Carabid beetles, ecology and evolution*. Eds. K. Desender, M. Dufrêne, M. Loreau, M.L. Luff and J.-P. Maelfait, pp 457–463. Dordrecht: Kluwer Academic Publishers.

SEKULIC, R., CAMPRAG, D., KERESI, T. AND TALOSI, B. (1987). Fluctuation of carabid population density in winter wheat fields in the region of Backa, Northeastern Yugoslavia (1961–1985). *Acta Phytopathologica et Entomologica Hungarica* **22**, 265–271.

SHELTON, M.D. AND EDWARDS, C.R. (1983). Effects of weeds on the diversity and abundance of insects in soybeans. *Environmental Entomology***12**, 296–298.

SILVY, C. (1999). Quantifions le phytosanitaire III. *Les dossiers de l'environnement de l'INRA* **19**, 203–212.

SKUHRAVY, V. (1958). Einfluss landwirtschaftlicher Massnahmen aud die Phänologie der Feldcarabiden. *Folia Zoologicae* **7**, 325–338.

STINNER, B.R. AND HOUSE, G.J. (1990). Arthropods and other invertebrates in conservation-tillage agriculture. *Annual Review of Entomology* **35**, 299–318.

SYMONDSON, W.O.C., GLEN, D.N., WILTSHIRE, C.W., LANGDON, C.J. AND LIDDELL, J.E. (1996). Effects of cultivation techniques and methods of straw disposal on predation by *Pterostichus melanarius* (Coleoptera: Carabidae) upon slugs (Gastropoda: Pulmonata) in an arable field. *Journal of Applied Ecology* **33**, 741–753.

SZÉL, G.Y., KADAR, F. AND FARAGO, S. (1997). Abundance and habitat preference of some adult overwintering ground beetle species in crops in Western Hungary (Coleoptera: Carabidae*). Acta Phytopathologica et Entomologica Hungarica* **32**, 369–376.

THEUNISSEN, J. (1997). Application of intercropping in organic agriculture. *Biological Agriculture and Horticulture* **15**, 251–259.

THIELE, H.-U. (1977). *Carabid beetles in their environments*. Berlin: Springer-Verlag.

THOMAS, M.B., WRATTEN, S.D. AND SOTHERTON, N.W. (1991). Creation of 'island' habitats in farmland to manipulate populations of beneficial arthropods: predator densities and emigration. *Journal of Applied Ecology* **29**, 906–917.

TIETZE, F. (1987). Changes in the structure of carabid taxocenoses in grasslands affected by intensified management and industrial air pollution. *Acta Phytopathologica et Entomologica Hungarica* **22**, 305–319.

TONHASCA, A. (1993). Carabid beetle assemblage under diversified agroecosystems. *Entomologia Experimentalis et Applicata* **68**, 279–285.

TRAUGOTT, M. (1998). Larval and adult species composition, phenology and life cycles of carabid beetles (Coleoptera: Carabidae) in an organic potato field. *European Journal of Soil Biology* **34**, 189–197.

UTROBINA, N.M. (1976). Einfluss von Ammoniakwasser auf den Bodentierbesatz unter Kartoffeln. *Pedobiologia* **16**, 206–218.

VICKERMAN, G.P., COOMBES, D.S., TURNER, G., MEAD-BRIGSS, M. AND EDWARDS, J. (1987). The effect of pirimicarb, dimethoate, and deltamethrin on Carabidae and Staphylinidae in winter wheat. *Mededelingen, Faculteit Landbouwwetenschap Rijksuniversiteit, Gent* **52**, 213–223.

WALLIN, H. (1985). Spatial and temporal distribution of some abundant carabid beetles (Coleoptera: Carabidae) in cereal fields and adjacent habitats. *Pedobiologia* **28**, 19–34.

WINDER, L., MERRINGTON, G. AND GREEN, I. (1999). The tri-trophic transfer of Zn from the agricultural use of sewage sludge. *Science of the Total Environment* **229**, 73–81.

9

Carabids as Indicators within Temperate Arable Farming Systems: Implications from SCARAB and LINK Integrated Farming Systems Projects

JOHN M. HOLLAND[1], GEOFF K. FRAMPTON[2] AND PAUL J. VAN DEN BRINK[3]

[1]*The Game Conservancy Trust, Fordingbridge, Hampshire SP6 1EF, UK,* [2]*Biodiversity and Ecology Division, School of Biological Sciences, University of Southampton, Bassett Crescent East, Southampton, Hampshire SO16 7PX, UK and* [3]*Alterra Green World Research, Department of Water and the Environment, PO Box 47, 6700 AA Wageningen, The Netherlands*

Introduction

In experimental studies of farming systems, the environmental impact is often measured by monitoring the abundance and diversity of arthropods. However, the suitability of arthropods, and in particular carabids, as indicators has not been adequately addressed by evaluating the results of such studies. Criteria have been proposed for the selection of bioindicator species (e.g. Çilgi, 1994; Pearson, 1994) or groups of species (e.g. Kremen, 1992), and many studies have used these to justify, *a priori*, the suitability of particular taxa as indicators, although the aims and meaning of bioindication are not always clear (McGeoch, 1998). Carabids appear to fulfil many of the selection criteria. For instance, they are widely distributed and relatively abundant (Luff, this volume), and many species are easy to sample and identify. Most species found in agricultural areas are permanent residents, spending most of their life cycle within or adjacent to cropped fields, meaning that they are intimately associated with the agricultural ecosystem. Indeed, a quarter of all temperate carabid species reside predominantly in arable or managed grassland areas (Thiele, 1977). Carabids are also relatively immobile, dispersing mainly by walking (Thomas *et al.*, this volume), which allows them to be studied at relatively small spatial scales. However, carabids are predominantly predaceous and, consequently, are not representative of

---

Abbreviations: ANOVA, Analysis of Variance; CFP, conventional farming practice or current farm practice; IFS, Integrated Farming Systems; RIA, reduced input approach; RDA, redundancy analysis; PRC, Principal Response Curves; SCARAB, Seeking Confirmation About Results At Boxworth

The Agroecology of Carabid Beetles

many functional groups. It is therefore questionable whether carabids can be used as indicators to predict the influence of farming systems on a wider range of arthropod taxa, or on the overall community of epigeic arthropods. This is difficult to evaluate as the ecology of other macro-arthropods has not been studied in as much detail as that of carabids.

Many of the studies in which carabids were used as indicators of the environmental impact of farming practices or systems have failed to produce conclusive results (Hance, this volume). This does not necessarily imply that carabids are unsuitable as indicators because other factors need to be taken into account. For instance, in some studies the spatial or temporal scales of experimental treatments might have been inappropriate for the aims of the work (as treatment effects are influenced by the plot size; Duffield and Aebischer, 1994; Duffield *et al.*, 1996). Sampling methods also have to be considered carefully as pitfall traps, with their associated limitations (Adis, 1979), are frequently the only method used to estimate carabid numbers. The choice of taxonomic units is another important consideration because analyses restricted to higher taxa could mask the effects of treatments upon individual species (Büchs *et al.*, 1997). Alternatively, subtle responses to the treatments by a range of species might only be detectable using multivariate analysis of the species composition. The possibility that individual species could mask the effects of treatments on the species composition also needs to be taken into account as the typical carabid assemblage of arable crops is dominated by relatively few species (Luff, this volume), and dominant species such as *Pterostichus melanarius* could bias the results of community analyses.

In this chapter we explore the suitability of pitfall-captured carabids as bioindicators of the effects of farming systems, using data from two major long-term, multi-site studies that involved intensive programmes of arthropod sampling: (1) the LINK Integrated Farming Systems study that compared the effects on arthropods of integrated and conventional management systems; and (2) the 'SCARAB' Project in which the impact on arthropods of conventional and reduced-input pesticide regimes was compared. The responses of individual species and communities of Carabidae to these management regimes are compared with those of other farmland arthropods, and the effect of dominant species on the interpretation of the community-level response is investigated.

Carabid communities as indicators of farming system effects: evidence from the LINK Integrated Farming Systems Project

In this project, which ran from 1992–1997, integrated and conventional farming were compared using six farms located in the main arable production areas of the UK (Ogilvy *et al.*, 1994). At each site there was a minimum of seven pairs of plots consisting of split or quartered fields in which an integrated farming system (IFS) and conventional farming practice (CFP) were compared. Each plot was a minimum of 2.5 ha and had a minimum width of 72 m. A five-course rotation of cereals and break crops, with rotational set-aside at some sites, was chosen with management that followed local practice. Each course of the rotation was present at each site each year, with some replication. A range of husbandry measures was implemented in the integrated plots to reduce the need for agrochemical treatment. Details of the crop husbandry approaches are given in Ogilvy (2000). The integrated methods adopted included: lower and more selective use of pesticides; closer crop monitoring and strict adherence to spray thresholds; soil monitoring and judicious use of fertilizers; use of resistant varieties; weed control maximizing use of cultural control and including choice of rotation and sowing dates; choice of soil cultivations designed to meet crop needs, but also considering cost and environmental impact; and some field margin manipulation. In each field, ground active arthropods were monitored using two transects of five pitfall traps, spaced at 10 m intervals, extending into the field starting at 30 m from a common boundary. The pitfall traps were partly filled with water and detergent and were operated for 5-day periods, at monthly intervals, throughout each crop's growing period. Sampling during the pre-treatment year (1992) was used to determine the species most often found at the six sites, and 19 carabid, 12 staphylinid and 8 linyphiid taxa were then identified in all further sampling.

Data was analysed using three approaches: (1) univariate analysis of the impact of the two regimes on individual arthropod taxa; (2) contingency table analysis to investigate any overall effects of the cropping systems; and (3) a multivariate analysis using Principal Response Curves (PRC) to investigate community-level effects of the farming regimes.

(1) UNIVARIATE APPROACH

Data was pooled from the months of April to July. Carabid taxonomic richness, counts of total Carabidae, and counts of some individual species within the family within each site, except for Pathhead, were analysed using a nested ANOVA design. All data were $\log_{10}$ transformed. Only means are presented from Pathhead because replicate fields were not sampled and the data could not be analysed using the above model. Further details of the analysis are given in Chapman *et al.* (2000).

(2) OVERALL EFFECTS OF CROPPING SYSTEM

The ANOVA approach is only suitable for comparing a few taxa because of the risk of generating a multiple testing error. Moreover, interaction effects made the results difficult to interpret. To evaluate the value of carabids as indicators in relation to the other arthropods captured and to concentrate on the effects of the farming system at each site whilst also comparing individual species, the total number of each species captured at each site was compared using a Chi squared test, and the species were then ranked according to the result. The percentage difference between the integrated and conventional systems was calculated, and the number of species favoured by each system counted. The percentage difference in total numbers captured at each site and crop was also calculated, and the number of species favoured by each system counted within the different crop types.

(3) MULTIVARIATE APPROACH

Principal Response Curves (PRC) analysis, a relatively new method based on Redundancy analysis (RDA) (Davis and Tso, 1982; Van den Brink and Ter Braak, 1998, 1999), was used to investigate community-level effects of the two farming systems in the LINK IFS study, and also the effects of the two pesticide regimes in the

SCARAB Project (see below). The application of PRC analysis to terrestrial arthropod communities, and examples of how PRC diagrams can be interpreted quantitatively, are given by Frampton *et al.* (2000, 2001a,b).

In PRC analysis, one treatment level is nominated as a 'control' or reference against which deviations in the overall community response through time under other treatment levels can be displayed graphically (PRC diagram). The abundance data are modelled as a mean of the control or reference (here IFS) and a deviation for each treatment (here CFP). For a set of species k, treatments d, and sampling dates t, PRC analysis models the response pattern of each species, T_{dtk}, as a multiple of one basic response pattern (c_{dt}), i.e. $T_{dtk} = b_k \times c_{dt}$. The arthropod counts are thus modelled as a count in the reference treatment level plus a deviation, which is calculated for each treatment level at each sampling date (c_{dt}). By definition, $c_{0t} = 0$ for every t, so the response curve for the reference treatment is a straight horizontal line against which the fitted values of c_{dt} for the other treatment(s) can be plotted; this shows the temporal change in species composition relative to that of the reference treatment level, and is referred to as the 'principal response' of the community. The 'species weight' b_k indicates how closely the response of each species follows that of the overall community response. PRC diagrams have an advantage over traditional ordination methods in that they can more clearly display temporal changes in treatment effects on biological communities, and are easy to interpret quantitatively (values of b_k and c_{dt} can be used to obtain the fitted relative abundance of individual species under different treatment contrasts (Frampton *et al.*, 2000a,b)).

PRC diagrams were tested statistically in two ways. First, Monte Carlo permutations of whole time series in the partial RDA from which the PRC was obtained (Van den Brink and Ter Braak, 1999) gave an *F*-type statistic to test the null hypothesis that the PRC diagram does not display the treatment variance (i.e. $T_{dtk} = 0$ for all d, t and k). Where a PRC diagram displayed a significant proportion of the treatment variance, a second series of permutation tests was performed separately upon treatments and replicates within each sampling date to test the null hypothesis that the species composition did not differ between CFP and IFS within each sampling date. All analyses were performed on $\ln(x + 1)$-transformed counts, x, of Carabidae from pitfall catches using the software program CANOCO 4 (Ter Braak and Šmilauer, 1998).

Results from the LINK Integrated Farming Systems Project

Of the 241 666 invertebrates which were identified, 50% were Carabidae, 18% Staphylinidae and 32% Araneae, although the relative proportions reflect more the efficiency of the pitfall traps for capturing these different groups rather than their relative abundance. The numbers and species richness of all invertebrates varied considerably between the sites, crops, fields, systems and years, and no single species was captured in sufficient numbers to allow a comparison across all sites. For this reason, families were analysed, although it was recognized that, by grouping species taxonomically, underlying trends may be masked. Wherever possible, individual species within sites were examined. Full results for all groups are given in Holland (2000).

UNIVARIATE APPROACH

There was considerable variation between the sites in total numbers of carabids. Comparing the five-year average, numbers were highest in all crops at Boxworth, intermediate at Manydown and Lower Hope, and lowest at Sacrewell and High Mowthorpe (*Figure 9.1*). The fauna was dominated by a few taxa: *Pterostichus melanarius*, *Bembidion lampros*, *Agonum dorsale*, *Nebria brevicollis* and *Amara* species made up 73% of the total capture. These are all typically dominant in the arable areas of northern Europe, as found in other studies (reviewed in Kromp, 1999; Holland and Luff, 2000; Luff, this volume).

The type of farming system, crop or year had no effect on the numbers of carabids at three of the sites. Diversity differed significantly at the remaining three sites. At Boxworth, there were significantly lower numbers and diversity of carabids in the integrated linseed crop (*Figures 9.1* and *9.2*). At High Mowthorpe, a greater number and diversity of carabids were captured in the integrated wheat crop, but only in two years. At Lower Hope, carabid diversity was lower in integrated spring beans compared to conventional winter oilseed rape, but only in two years. Carabid numbers and diversity were lowest in potatoes, and often in the following crop. Analysis of Staphylinidae and Aranaeae in the LINK IFS Project also showed few differences between the farming systems (Holland *et al.*, 1998; Holland, 2000).

Individual species appeared to be favoured by either the integrated or conventional system, with relatively few species not favoured by either system (*Table 9.1*). *P. melanarius* was the most ubiquitous and frequently caught ground beetle species and occurred in sufficient numbers during July for analysis at each site. Overall, *P. melanarius* was only slightly favoured by the integrated approach (*Table 9.1*). This species is of particular interest because of its abundance and potential as a predator of aphids and slugs (Symondson and Sunderland, this volume). At Sacrewell, Boxworth, and Lower Hope, there was a three-way interaction of *P. melanarius* catches between the phase of the rotation, year and farming system. Significantly more *P. melanarius* were caught on average in the integrated system at High Mowthorpe when compared across all crops. No differences occurred at Manydown. *P. madidus*, which has a very similar life history, was favoured by the integrated approach, especially in set-aside. *B. lampros* was found in sufficient numbers for analysis at three sites during April, May and June each year. Significantly higher numbers occurred in integrated spring beans compared to conventional winter oilseed rape, and especially low numbers were found in potatoes at High Mowthorpe and Lower Hope (Holland, 2000). The

Table 9.1. Mean over 5 years for *Pterostichus melanarius* per pitfall trap per day during July at High Mowthorpe and Lower Hope in the conventional and integrated plots. (CFP = Conventional Farm Practice; IFS = Integrated Farming System; WOSR = Winter Oilseed Rape).

Rotation	High Mowthorpe		Lower Hope	
	CFP	IFS	CFP	IFS
Winter wheat	0.05	0.10	0.62	1.02
Set-aside	0.05	0.11	0.77	0.26
WOSR/Spring beans	0.10	0.52	0.66	1.06
Winter wheat	0.36	0.87	0.71	0.6
Potatoes	0.71	0.78	0.17	0.23

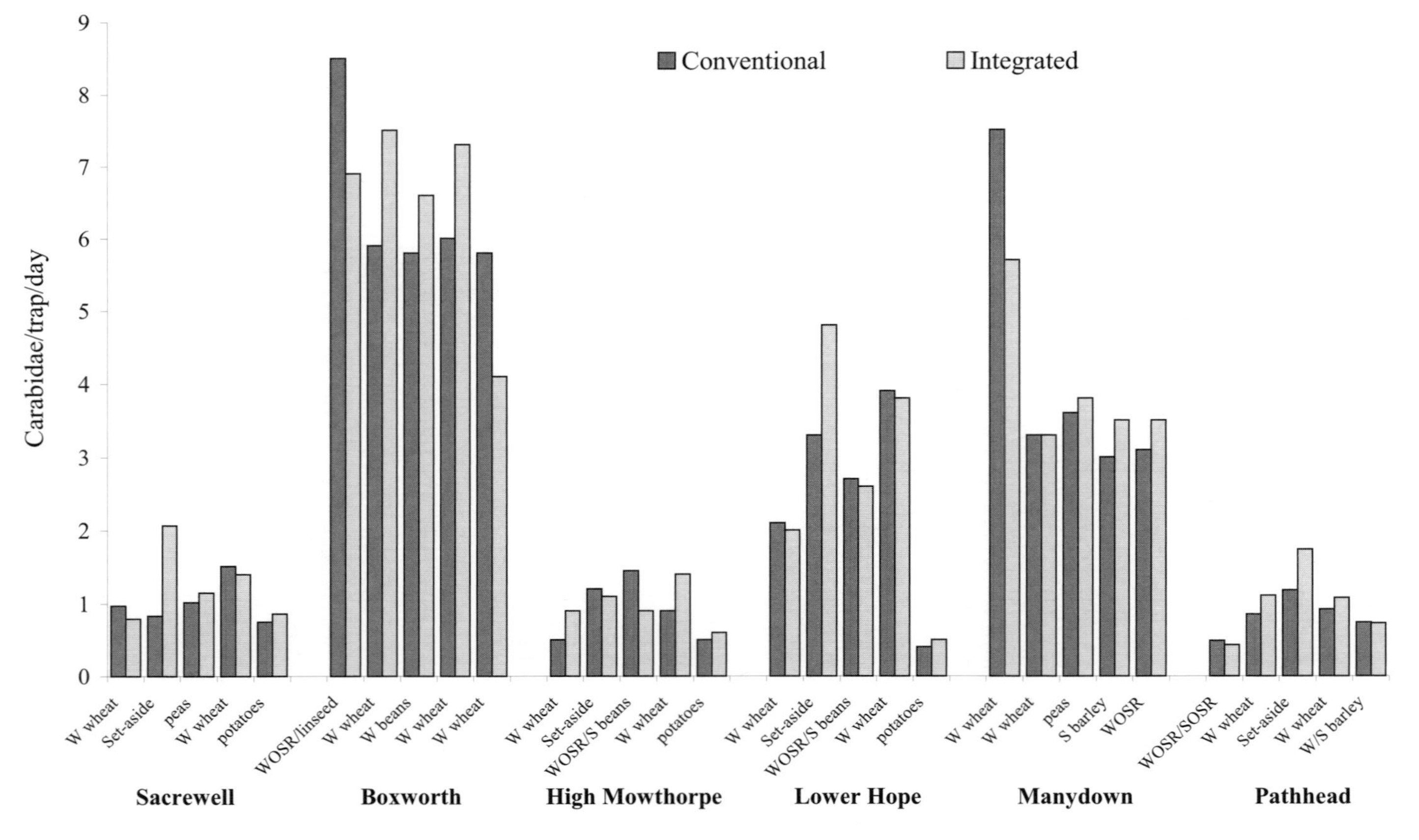

Figure 9.1. Mean number of pitfall-captured Carabidae for April to July over 5 years in the LINK IFS Project within each course of the rotation and farming system (S: spring sown; W: winter sown; OSR: oilseed rape).

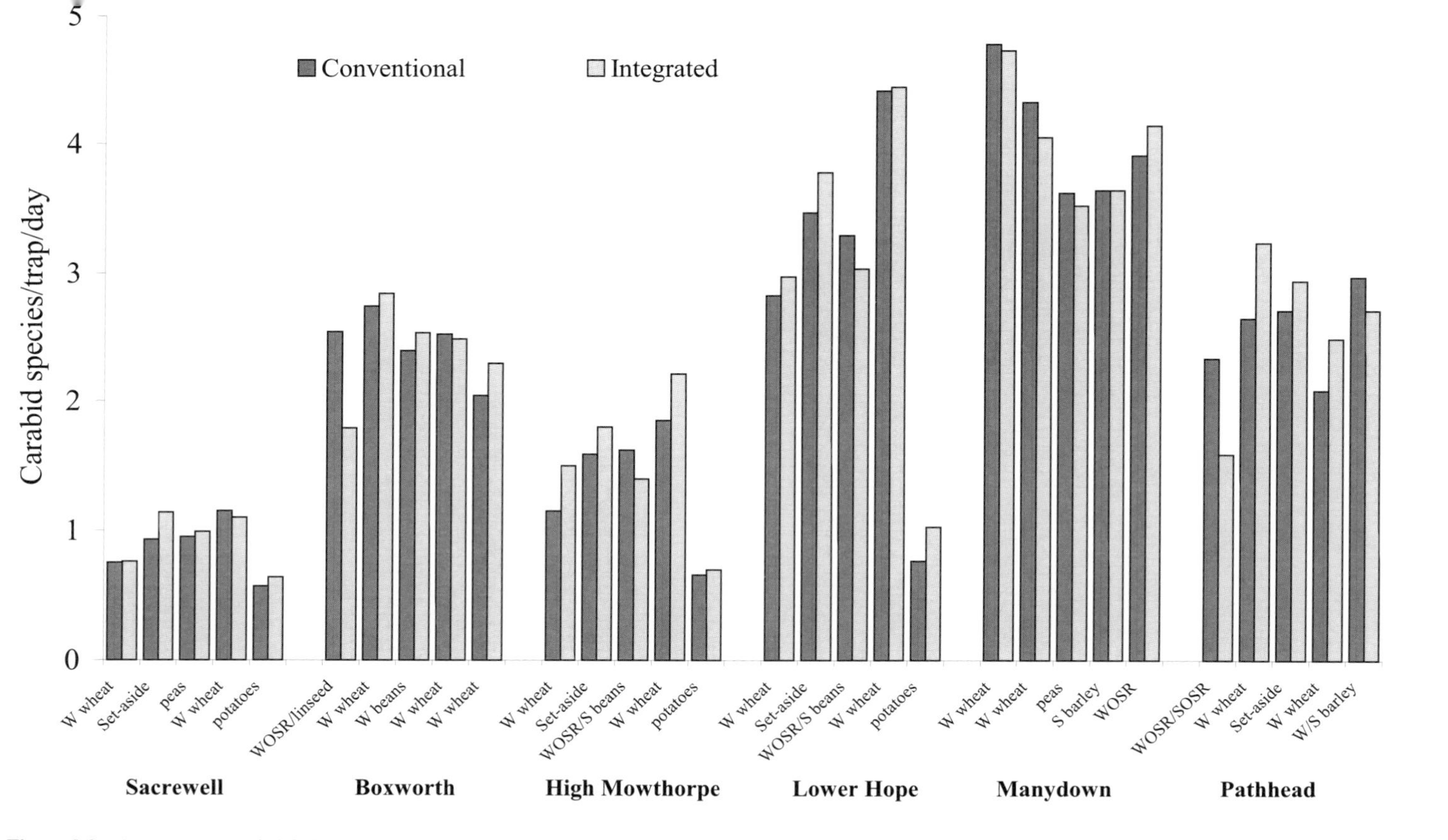

Figure 9.2. Mean number of pitfall-captured carabid species for April to July over 5 years in the LINK IFS Project within each course of the rotation and farming system (S: spring sown; W: winter sown; OSR: oilseed rape).

predominantly seed-feeding *Amara* species were also less frequently captured in the integrated break crops of linseed and spring beans at Boxworth and High Mowthorpe. No other consistent trends were found across the sites for individual species.

Table 9.2. Summary of effects on total numbers trapped, their relative proportions, and their ranking according to difference between the two farming systems. (CFP = Conventional Farm Practice; IFS = Integrated Farming System).

	CFP		IFS		Chi sq	Rank	% difference
	Number	% of total	Number	% of total			IFS/CFP
Carabidae							
Agonum dorsale	6705	5.6	6925	5.7	3.6	35	103
Amara spp.	6021	5.0	4810	3.9	135.4	8	80
Asaphidion flavipes	1450	1.2	821	0.7	174.2	5	57
Bembidion lampros	6765	5.7	6922	5.7	1.8	38	102
B. obtusum	1674	1.4	2378	1.9	122.3	9	142
Calathus fuscipes	275	0.2	205	0.2	10.2	25	75
C. melanocephalus	4	0.0	11	0.0	3.3	36	275
Carabus spp.	99	0.1	67	0.1	6.2	30	68
Harpalus affinis	194	0.2	158	0.1	3.7	34	81
H. rufipes	1177	1.0	1671	1.4	85.7	11	142
Loricera pilicornis	2562	2.1	1988	1.6	72.4	13	78
Nebria brevicollis	6161	5.2	6996	5.7	53.0	15	114
Notiophilus biguttatus	872	0.7	976	0.8	5.8	31	112
Pterostichus melanarius	12603	10.6	14360	11.7	114.5	10	114
P. madidus	3290	2.8	4507	3.7	189.0	2	137
P. niger	489	0.4	319	0.3	35.8	18	65
P. nigrita	70	0.1	61	0.0	0.6	39	87
Trechus quadristriatus	4276	3.6	3246	2.7	141.0	7	76
Other spp.	4326	3.6	5693	4.7	186.5	4	132
Total Carabidae	59013	49.5	62114	50.8			
Staphylinidae							
Aleocharinae	6320	5.3	6152	5.0	2.3	37	97
Anotylus spp.	4190	3.5	2419	2.0	474.6	1	58
Philonthus cognatus	2547	2.1	2406	2.0	4.0	33	94
Philonthus spp.	692	0.6	580	0.5	9.9	26	84
Staphylinus caeseus	52	0.0	24	0.0	10.3	24	46
S. olens	47	0.0	80	0.1	8.6	28	170
Tachinus signata	599	0.5	208	0.2	189.4	3	35
Tachyporus chrysomelinus	259	0.2	269	0.2	0.2	41	104
T. hypnorum	4124	3.5	3349	2.7	80.4	12	81
T. nitidulus	199	0.2	187	0.2	0.4	40	94
T. obtusus	195	0.2	125	0.1	15.3	23	64
Xantholinus glabratus	43	0.0	76	0.1	9.2	27	177
Other spp.	3923	3.3	3351	2.7	45.0	17	85
Total Staphylinidae	23190	19.4	19226	15.7			
Araneae							
Bathyphantes spp.	2316	1.9	2736	2.2	34.9	19	118
Erigone spp.	15225	12.8	17401	14.2	145.1	6	114
Lepthyphantes spp.	2566	2.2	2373	1.9	7.5	29	92
Meionetus spp.	919	0.8	1188	1.0	34.3	20	129
Milleriara spp.	897	0.8	1291	1.1	70.9	14	144
Oedothorax spp.	6057	5.1	6622	5.4	25.2	22	109
Savignya frontata	576	0.5	367	0.3	46.3	16	64
Immature Linyphiidae	572	0.5	567	0.5	0.0	42	99
Other spp.	870	0.7	1123	0.9	32.1	21	129
Total Linyphiidae	29998	25.1	33668	27.5			
Lycosidae	7087	5.9	7370	6.0	5.5	32	104

OVERALL EFFECTS OF FARMING SYSTEM

When the sensitivity to the two farming systems was compared for all arthropod taxa, 15 were favoured by the IFS and 17 by the CFP approach. However, for some taxa these results were based on the capture of only a few individuals giving low Chi squared values and, therefore, these must be treated with some caution. For those taxa captured in sufficient numbers to give a reliable response, six carabid taxa were favoured by the IFS and five by the CFP approach (*Table 9.2*). The Staphylinidae showed a different pattern, with the five most numerous taxa occurring most in the CFP plots. In contrast, the Araneae showed the opposite effect, with the most numerous taxa favoured by the IFS plots (5) compared to only one in CFP. Total numbers of Carabidae, Staphylinidae and Araneae were similar in the two systems, and seven relatively numerous taxa (> 400 individuals captured) were not favoured by either system.

These epigeal invertebrates also showed a strong preference for the different crop types. When the numbers of species favoured by one system or the other were compared across the crop rotation (*Table 9.3*), more species were favoured by each course of the integrated rotation, with one exception. When a spring crop was grown in the integrated system as opposed to a winter sown crop in the conventional plot, more taxa were found in the winter sown plots. Approximately a third of the taxa, however, showed no difference in counts between the two farming systems.

MULTIVARIATE APPROACH

The results presented here are for PRC analyses performed on summer (May–July) pitfall counts of 19 carabid taxa obtained during 1993 to 1997 at four of the six farms (Manydown, Rosemaund, Boxworth and High Mowthorpe), where visual inspection of the data suggested the largest differences in carabid abundance between the farming systems would be found. An additional analysis of the High Mowthorpe data that excluded the dominant species, *Pterostichus melanarius*, was carried out, in case the response of this species was masking the impact on less abundant carabids. The number of sampling dates analysed differed between the farms, but samples were included from at least one pitfall-trapping date per month in each year. To minimize heterogeneity of carabid counts resulting from the effects of cropping, whilst maximizing the availability of data in each year, the analyses were restricted to winter wheat, which was the most frequently grown crop in the experiment design.

PRC analyses generally did not yield significant effects of the conventional versus

Table 9.3. Number of invertebrate species favoured by CFP, IFS or no difference pooling data from all sites (% difference >10%, otherwise assumed to be equal; CFP = Conventional Farm Practice; IFS = Integrated Farming System; W = Winter; S = Spring; SAS = Set Aside).

	WWheat1	WWheat2	SAS	Peas	Potatoes	W break	W/S crop
CFP > IFS	9.1	10	8.5	10.5	6.3	10.7	12.3
IFS > CFP	14.9	17	12.3	12	12.7	14.7	8.5
CFP = IFS	10.9	11	13.3	14.5	17.3	12.0	11.5

Table 9.4. Variance allocation in PRC analyses and results of the accompanying permutation tests for carabid communities at four LINK IFS Project farms. Bracketed results for High Mowthorpe are from a PRC analysis that excluded the dominant species, *Pterostichus melanarius*.

	All farms	Manydown	Rosemaund	Boxworth	High Mowthorpe
Sampling date	47%	79%	52%	67%	48% (48%)
Between replicates	52%	13%	40%	28%	42% (41%)
Farming system	1%	8%	8%	5%	10% (11%)
Significance of farming system	$P > 0.10$	$P > 0.10$	$P > 0.10$	$P > 0.10$	$P > 0.05 (P > 0.05)$
Significance of PRC diagram	$P > 0.10$	$P > 0.10$	$P > 0.10$	$P > 0.10$	$P > 0.05 (P = 0.08)$

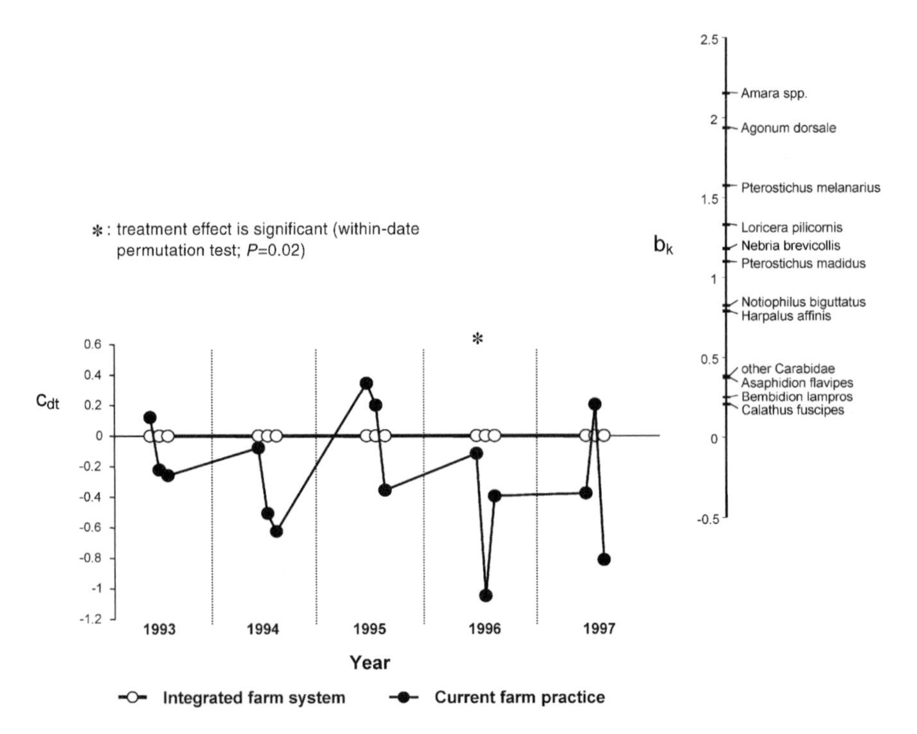

Figure 9.3. PRC diagram showing the response of the pitfall-sampled carabid community (c_{dt}) at High Mowthorpe under a conventional (current farm practice) relative to an integrated (reference) farming system in the LINK IFS Project. Species weights (b_k) indicate how closely individual taxa follow the fitted community response (Note: species weights in the range 0 ± 0.25 are not shown).

integrated farming systems on the carabid community (*Table 9.4*). However, at High Mowthorpe the CFP regime appears to have negatively affected the carabid species composition relative to IFS in the long term, although the effect was significant only in one year (*Figure 9.3*). The exclusion of *P. melanarius* data from the High Mowthorpe carabid community analysis made virtually no difference to the results.

Carabid communities as indicators of pesticide effects: long-term evidence from the SCARAB Project

BACKGROUND

Objectives and design of the SCARAB Project

The MAFF 'SCARAB' Project was set up in 1990 to investigate the responses of farmland arthropods in a variety of arable crops to two regimes of pesticide use, representing conventional and reduced inputs. The primary aim of the project was to investigate whether long-term adverse pesticide effects on non-target arthropods that had been observed in commercial wheat production in eastern England during the 1980s (Greig-Smith *et al.*, 1992) would be likely to occur in a wider range of crops and at different geographical locations. SCARAB was sited at three research farms and comprised a total of eight fields, to which the two pesticide regimes were applied in a split-field design. The crop rotation in each field was typical for the locality of each farm, being a rotation of mixed cereals, winter oilseed rape and spring beans at High Mowthorpe in north-east England (three fields on calcareous loam), mixed cereals, sugar beet and potatoes at Gleadthorpe in central England (three fields on stony sand), and a rotation of grass and winter wheat at Drayton in the south Midlands (two fields on calcareous clay).

Pesticide use

The conventional regime (current farm practice, CFP) aimed to mimic actual pesticide use for a given crop, based on Pesticide Usage Survey Reports (e.g. Thomas *et al.*, 1997), with the proviso that at least one insecticide was applied to each arable crop (except grass) to ensure that arthropods were exposed to some of the insecticides typically applied in commercial practice. The reduced input approach (RIA) regime employed pesticides only when required to avert a yield loss exceeding 10% or a serious reduction in crop quality (Ogilvy, 2001). A major difference between the regimes is that insecticide use was avoided entirely under the RIA regime in all fields and years. Overall, 57.7% fewer label-recommended-rate pesticide applications were applied under RIA compared with CFP (note that the SCARAB Project CFP regime is distinct, and was managed differently, to that of the LINK IFS Project described above).

During the pre-treatment 'baseline monitoring' year, 1990, both halves of each SCARAB Project field received identical pesticide inputs. Subsequently, during 1991–1996, the two halves of each field continuously received the contrasting pesticide regimes. Pesticide inputs differed between crops, and hence between years, but higher numbers of CFP than RIA pesticide applications were maintained in all fields and years (Ogilvy, 2001). Arthropods were monitored at matched locations in both halves of each field, both during the baseline year (1990) and the phase of contrasting treatments (1991–1996), using suction sampling (Frampton, 1997) and pitfall trapping, following a standard sampling protocol (Ogilvy, 2001). The latter sampling method employed white plastic 9 cm diameter pitfall traps located in the CFP and RIA halves of each field, 75 m from a hedgerow that was common to both experimental areas. In each half of the field, four traps were positioned 10 m apart.

The traps were left open for 7-day periods and contained water, plus a drop of detergent to break the surface tension so that captured arthropods sank.

Statistical analysis

The analyses reported here focus on pitfall catches because suction sampling captured relatively few species of Carabidae; these were principally *Trechus quadristriatus* and species of *Bembidion*, which occurred sporadically and at low densities in samples, precluding a meaningful long-term comparison of the responses of carabids and other arthropods to the pesticide regimes. Multivariate analysis of suction catches nevertheless revealed long-term negative effects of the CFP regime on the overall arthropod community, which were most severe upon Collembola (Frampton, 2000, 2001a,b). These effects occurred in one of the eight SCARAB Project fields, 'Field 5', under a rotation of grass and winter wheat at the Drayton site, where the CFP pesticide regime included repeated use of organophosphorus insecticides in consecutive years. The CFP regime in 'Field 5' probably represents a worst-case scenario of pesticide use in Britain (Frampton, 2001a). Accordingly, the Carabidae community of 'Field 5' is used here to investigate the relative reactions of pitfall-trapped carabids and other arthropods to long-term regimes of pesticide use.

The method of Principal Response Curves (PRC; see above) was used to investigate the response of carabid and other pitfall-trapped arthropod communities under the CFP and RIA regimes of pesticide use in 'Field 5' (Van den Brink and Ter Braak, 1999). The PRC analyses were each carried out using the software program CANOCO 4 (Ter Braak and Šmilauer, 1998), following a similar procedure to analyses reported by Frampton *et al.* (2000a,b, 2001) and as described above for the LINK IFS Project, except that RIA was nominated as the reference treatment in the analysis, against which the effects of CFP were compared. For each taxon, the arthropod count, x, from each trap was $\ln(x + 1)$-transformed prior to analysis.

Four separate analyses of the pitfall catch data were carried out: (1) the catch of the most abundant arthropods in pitfall samples, excluding Acari and Collembola; (2) the carabid community; (3) the carabid community excluding *Pterostichus melanarius*; and (4) the Collembola community. The first analysis aimed to represent the response of the overall arthropod community, using the taxa that are most frequently identified in pitfall catch studies (i.e. with the notable exclusion of Acari and Collembola), against which the response of carabids alone could be compared. The third analysis was carried out to investigate whether domination of the carabid catch by *P. melanarius* (which on average comprised 45% of the carabids captured) influenced the detection of pesticide effects on other species. The fourth analysis was to allow a comparison of the carabid community response with that of Collembola, which in suction samples was the group most vulnerable to effects of the CFP regime in 'Field 5' (Frampton, 1997, 2000, 2001a,b). Analyses 1–3 utilized the same pitfall catches, which were obtained predominantly from the summer months in each year, whereas Collembola catches were obtained on different sampling dates (*Table 9.5*). The analyses performed on Carabidae used data from 19 species of adults, plus the larvae of three of these species, while analyses of Collembola used data from 22 taxa (all age classes pooled), most of which were identified to species or genus. Analyses performed on the total arthropod catch comprised the above 22 carabid taxa, together

Table 9.5. SCARAB Project pitfall-trap sampling dates (7-day catch ending) used in data analyses. For Collembola only, an additional sample in 1997 (04 March) was included. Asterisks denote pre-treatment samples.

	1990	1991	1992	1993	1994	1995	1996
Collembola	18 Sep*	15 Jan*	18 Feb	27 Apr	10 May	30 May	05 May
	23 Oct*	21 Jan	12 May	20 Jul	26 Jul	08 Aug	14 May
	13 Nov*	05 Mar	07 Jul	17 Aug	27 Sep	17 Oct	09 Jul
	31 Dec*	25 Jun					
		10 Dec					
Other Arthropods	25 Sep*	07 May	26 May	13 Apr	24 May	16 May	14 May
	27 Nov*	11 Jun	09 Jun	06 Jul	21 Jun	13 Jun	18 Jun
		09 Jul	21 Jul		05 Jul	11 Jul	23 Jul

with adult Staphylinidae (19 taxa), larval Staphylinidae (3), adult Chrysomelidae (1), Coccinellidae (2), Cryptophagidae (1), Curculionidae (2), Hydrophilidae (3), Lathridiidae (4), Nitidulidae (1), Leiodidae, (1), Elateridae (1), larvae of Coleoptera (4) and Diptera (2), individual species of adult Linyphiidae (10) (sexes recorded separately), total Lycosidae (1) and total Opiliones (1) (altogether 88 taxa, including both linyphiid sexes).

Permutation tests were used to determine the significance of the PRC diagrams and of the overall effect of pesticide regime, as described above for the LINK IFS Project. However, the lack of replication within 'Field 5' precluded within-date statistical testing of treatment effects. This aspect of the experiment design is considered in the Discussion (see below).

Results from the SCARAB Project

Cropping in 'Field 5'and the pesticide applications made under the RIA and CFP regimes are listed in *Table 9.6*. All four PRC analyses indicate significant overall effects of the pesticide regimes on the species composition of pitfall-captured arthropods (*Table 9.7*).

The PRC diagram for the arthropod community response based on the most commonly occurring arthropods in pitfall traps, except Acari and Collembola, shows that the effects of the SCARAB Project 'Field 5' CFP regime were mostly negative. The largest differences in species composition followed the use of chlorpyrifos in three years, especially in 1994 and 1995 (*Figure 9.4*). The PRC diagram obtained by restricting the analysis to carabids alone also shows that the impacts of the CFP regime were mostly negative, with a particularly large decline in overall CFP abundance following the application of chlorpyrifos in 1994 (*Figure 9.5*). Relatively high species weights for *Pterostichus melanarius* and *Poecilus cupreus* mean that these species most closely followed the overall community response pattern for carabids (*Figure 9.5*). The PRC diagram obtained by excluding the dominant species, *P. melanarius*, from the analysis of the carabid community response shows that a substantial difference in CFP and RIA species composition, which was only weakly evident from the two previous analyses, followed the use of chlorpryifos in 1991 (*Figure 9.6*). The three analyses of macro-arthropod communities (*Figures 9.4–9.6*) broadly agree that the influences of the CFP regime relative to RIA were generally negative, but substantial differences in species composition were mostly transient,

Table 9.6. SCARAB Project insecticide (I), fungicide (F) and herbicide (H) applications made to the CFP and RIA areas of 'Field 5' during 1990–1997. No pesticides were applied prior to the study during 1984–1990.

Crop	Date	Active ingredient	Application rate g.a.i. ha^{-1} CFP	RIA
Grass 1990–1991	28 Jan 91	chlorpyrifos (I)	720	0
Winter wheat1991–1992	26 Nov 91	diflufenican (H)	100	50
		isoproturon (H)	1000	500
	29 Apr 92	propiconazole (F)	125	62.5
		fluoroxypyr (H)	200	100
		metsulfuron-methyl (H)	6	3
	09 Jun 92	propiconazole (F)	125	62.5
	23 Jun 92	dimethoate (I)	340	0
Winter wheat1992–1993	04 Nov 92	paraquat (H)	800	400
	05 Feb 93	diflufenican (H)	100	50
		isoproturon (H)	1000	500
	25 Mar 93	fenoxaprop-P-ethyl (H)	69	34
	22 Apr 93	cyproconazole (F)	60	30
		prochloraz (F)	400	200
		metsulfuron-methyl (H)	6	3
		mecoprop-P	1380	690
	02 Jun 93	fenpropimorph (F)	375	187
		propiconazole (F)	125	62.5
	23 Jun 93	dimethoate (I)	340	0
		propiconazole (F)	125	62.5
Grass1993–1994	14 Jun 94	chlorpyrifos (I)	720	0
Grass1995	22 Mar 95	chlorpyrifos (I)	720	0
	03 Apr 95	propiconazole (F)	125	0
Grass1996–1997	09 Apr 96	propiconazole (F)	125	0

Table 9.7. Variance allocation in PRC analyses and results of the accompanying permutation tests for arthropod communities under Current Farm Practice and Reduced Input Approach pesticide regimes in SCARAB Project 'Field 5'.

	Total arthropods excluding Acari and Collembola (Fig. 9.4)	Carabidae (Fig. 9.5)	Carabidae excluding *Pterostichus melanarius* (Fig. 9.6)	Collembola (Fig. 9.7)
Sampling date	56%	61%	56%	61%
Pesticide regime	13%	12%	13%	27%
Pesticide regime variance displayed in PRC diagram	27%	36%	32%	60%
Significance of pesticide regime	$P < 0.05$	$P < 0.05$	$P < 0.05$	$P < 0.05$
Significance of PRC diagram	$P < 0.05$	$P < 0.05$	$P < 0.05$	$P < 0.05$

not persisting for more than one year. In contrast, the pitfall-sampled Collembola community exhibited a very clear long-term negative effect of the CFP regime, without recovery during the Project (*Figure 9.7*). This is consistent with the collembolan community response observed previously using suction sampling (Frampton, 2000, 2001a,b).

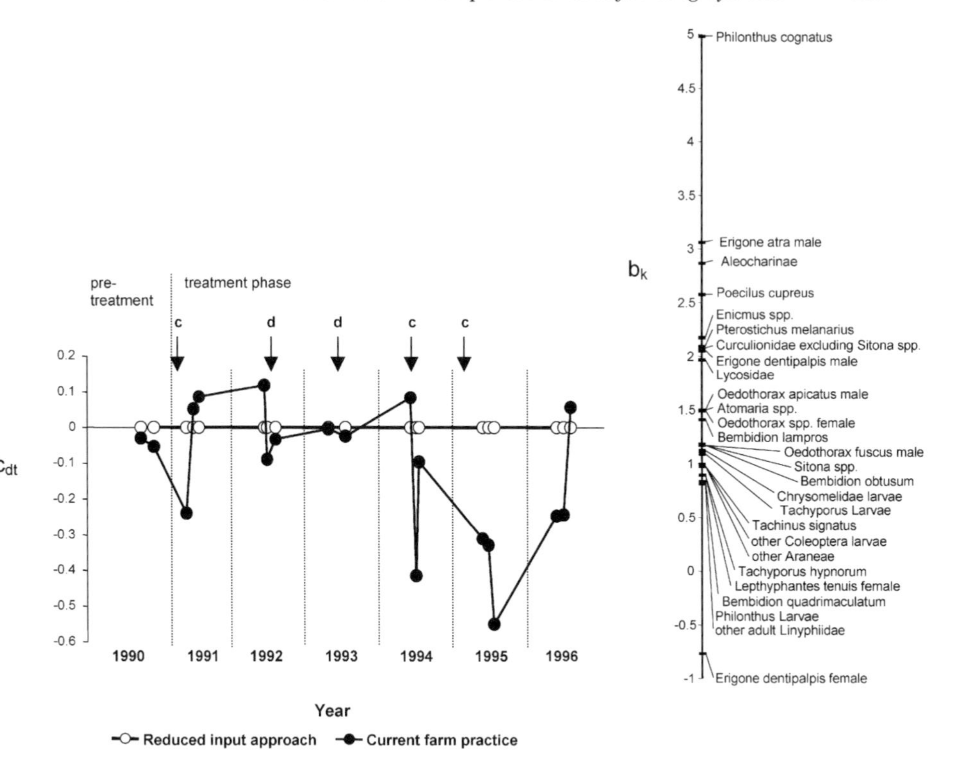

Figure 9.4. PRC diagram showing the response (c_{dt}) of the pitfall-sampled arthropod community, excluding Acari and Collembola, in SCARAB Project 'Field 5' under current farm practice (CFP) and reduced-input (reference) pesticide regimes. Arrows indicate the timing of chlorpyrifos (c) and dimethoate (d) applications that were made only under the CFP regime. Species weights (b_k) indicate how closely individual taxa follow the fitted community response (Note: species weights in the range 0 ± 0.25 are not shown).

Discussion

EFFECTS OF LINK FARMING SYSTEMS AND SCARAB PESTICIDE REGIMES

Responses of carabids to conventional and integrated farming systems

The LINK IFS Project was one of the most extensive farming system studies ever conducted, with 96 plots of over 5 ha each, distributed across six geographically diverse sites. The diversity of rotation and husbandry practices incorporated the majority of farming approaches underway in the UK at that time. As a consequence, the study provided an extensive comparison of conventional and integrated farming. However, the complexity of the study ultimately meant that the results were difficult to interpret, particularly those from the multi-factorial analyses of variance. A number of approaches were tried and, overall, the results were relatively consistent: the largest differences in the abundance and diversity of carabids were found between sites and years, then crops and, to the least extent, between the farming systems. This pattern was also found in a number of other similar long-term farming system studies in western Europe (Booij and Noorlander, 1992; Winstone *et al.*, 1996; Zhang *et al.*,

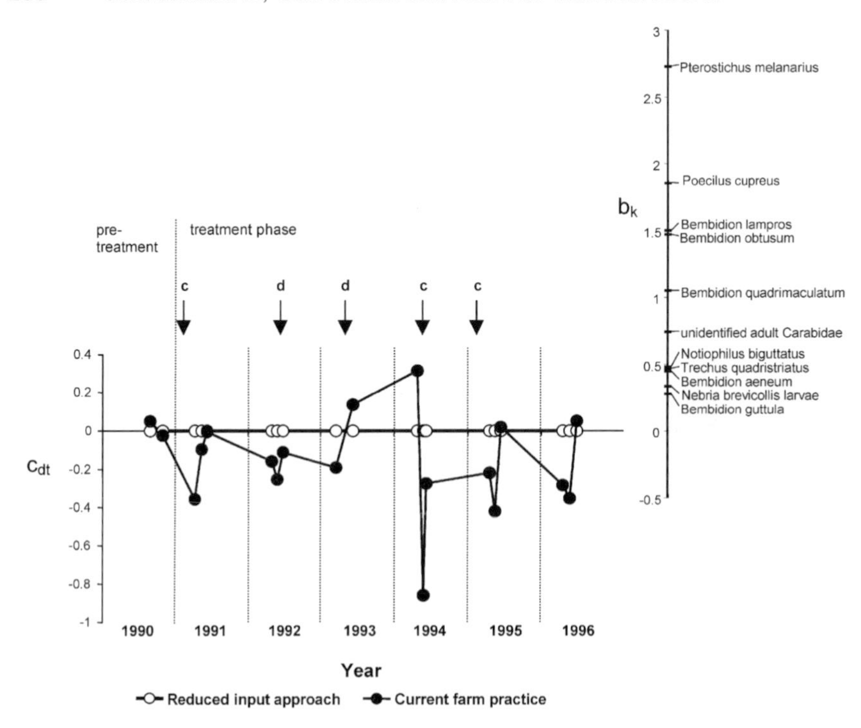

Figure 9.5. PRC diagram showing the response (c_{dt}) of the pitfall-sampled Carabidae community in SCARAB Project 'Field 5' under current farm practice (CFP) and reduced-input (reference) pesticide regimes. Arrows indicate the timing of chlorpyrifos (c) and dimethoate (d) applications that were made only under the CFP regime. Species weights (b_k) indicate how closely individual taxa follow the fitted community response (Note: species weights in the range 0 ± 0.25 are not shown).

1997; Gardner *et al.*, 1999). In contrast, some studies reported increases in arthropod numbers and diversity where an integrated or low-input approach was used (Basedow *et al.*, 1991; El Titi, 1991; Cárcamo *et al.*, 1995; Büchs *et al.*, 1997; Basedow, 1998; Ellsbury *et al.*, 1998; Huusela-Veistola, 1996). However, when results from projects examining the impact of different farming systems were summarized for individual species or genera of Carabidae, the number of species encouraged by the integrated approach was the same as those favoured by conventional farming (Holland and Luff, 2000).

Long-term responses of carabids to pesticide regimes

The absence of any long-term insecticide effects in both projects, with the exception of 'Field 5' in SCARAB, suggests that current insecticide usage (in which synthetic pyrethroids are the most frequently used group; e.g. Thomas *et al.*, 1997) is not damaging to non-target arthropods in field crops in the long term. Alternatively, it may be that even the relatively large spatial and temporal scales of the LINK IFS and SCARAB Projects were inadequate for the detection of any effects of such pesticide use (see below). In the LINK IFS Project, insecticide usage was relatively infrequent within both systems (except in potatoes), and use of broad-spectrum organophosphate (OP) products was avoided. The dicotyledonous crops received more synthetic

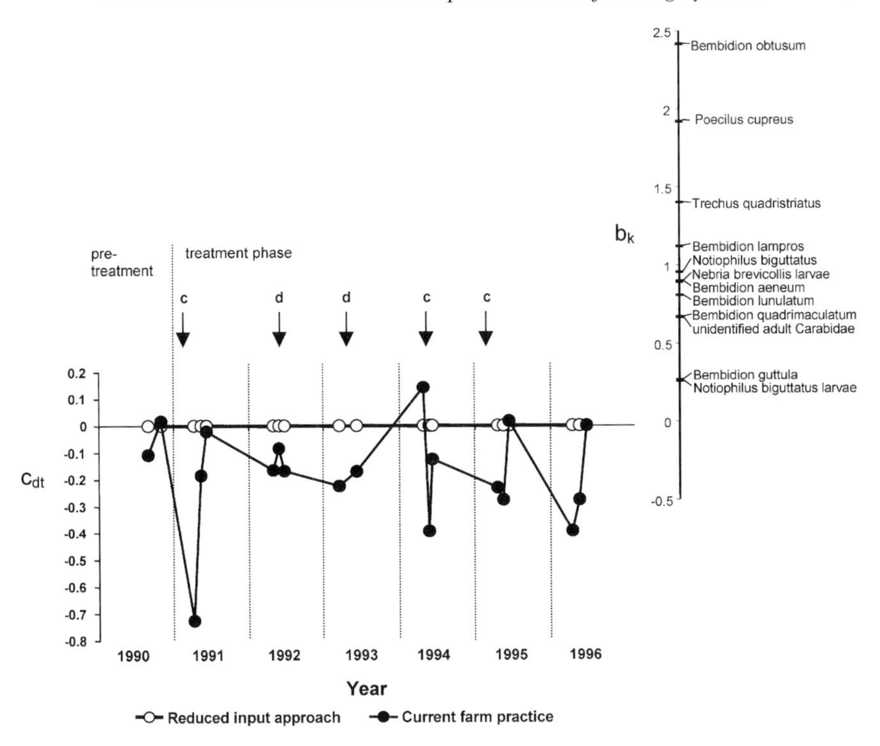

Figure 9.6. PRC diagram showing the response (c_{dt}) of the pitfall-sampled Carabidae community, excluding *Pterostichus melanarius*, in SCARAB Project 'Field 5' under current farm practice (CFP) and reduced-input (reference) pesticide regimes. Arrows indicate the timing of chlorpyrifos (c) and dimethoate (d) applications that were made only under the CFP regime. Species weights (b_k) indicate how closely individual taxa follow the fitted community response (Note: species weights in the range 0 ± 0.25 are not shown).

pyrethroid applications than the cereal crops but, when the impact of insecticides applied during the first four years was examined, there was little compelling evidence that they were reducing numbers of Carabidae (Holland, 1998). Similarly, in the SCARAB Project, synthetic pyrethroids were the main insecticide group used in most fields, except 'Field 5'. The results from the worst-case scenario of insecticide use in SCARAB Project 'Field 5' suggest that, compared with Collembola, field catches of Carabidae may be relatively insensitive in the long term, even to broad-spectrum and persistent OP insecticides. However, the choice of taxonomic unit for monitoring and analysis can have a key bearing on the detection of any pesticide effects (see below).

Possible explanations for the apparent lack of effects of pesticides on pitfall-captured carabids are discussed by Holland and Luff (2000). An obvious factor to consider is inherent susceptibility to pesticides. Although carabids are well known to be susceptible to certain insecticides in laboratory and field studies (e.g. Brown *et al.*, 1983; Vickerman *et al.*, 1987; Floate *et al.*, 1989; Çilgi *et al.*, 1996), when the response to different pesticides were compared for a wide range of natural enemies, Carabidae were found overall to be the least susceptible arthropod family (Theiling and Croft, 1988). Even for susceptible arthropods, effects of a pesticide in the field may vary considerably depending upon the extent to which the arthropods are

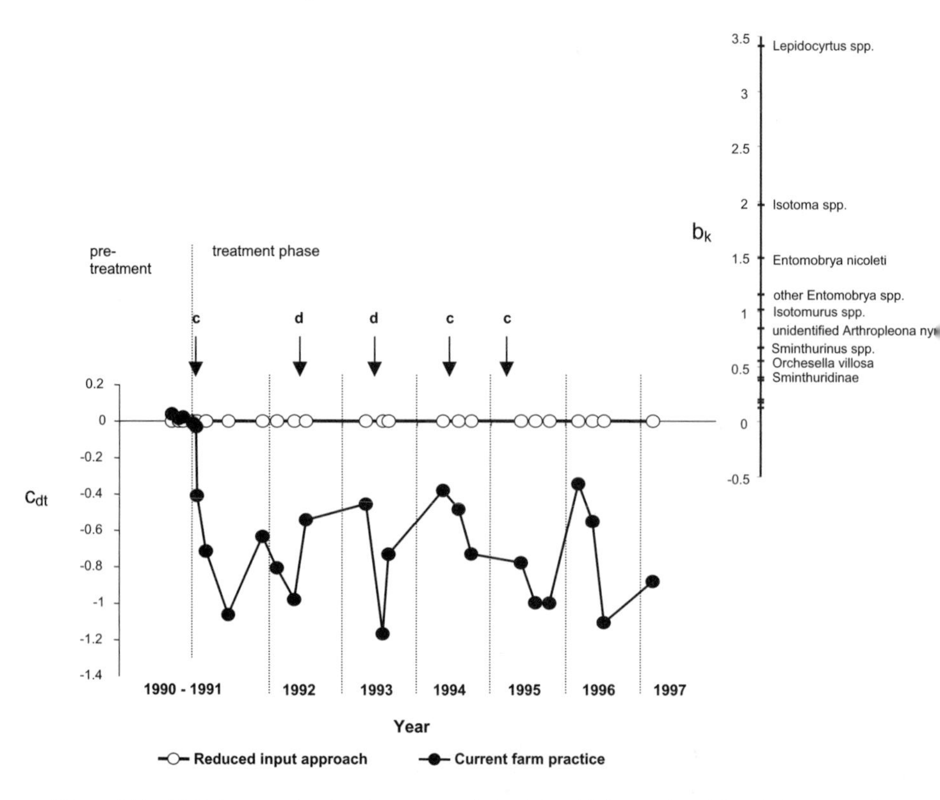

Figure 9.7. PRC diagram showing the response (c_{dt}) of the pitfall-sampled Collembola community in SCARAB Project 'Field 5' under current farm practice (CFP) and reduced-input (reference) pesticide regimes. Arrows indicate the timing of chlorpyrifos (c) and dimethoate (d) applications that were made only under the CFP regime. Species weights (b_k) indicate how closely individual taxa follow the fitted community response (Note: species weights in the range 0 ± 0.25 are not shown). See page 265.

exposed. Exposure and the behaviour of a pesticide are in turn determined by numerous interacting variables relating to the organism's biology, the application dynamics and the environmental conditions, as schematized in Holland and Luff (2000). Recovery is also an important factor to consider (Kelly and Harwell, 1990), as rapid recovery could make the impact of a pesticide difficult to detect. Thus, it may be necessary to examine many fields to ensure that an appropriate range of treatment and crop conditions, pesticide application timings, exposure scenarios and recovery opportunities is encompassed. In practice, this means using experiment designs that mimic agricultural practice as realistically as possible in terms of spatial scale. It is less clear, however, what the optimal temporal scale of a farming systems or pesticide effects study should be, given the resource constraints on long-term studies (e.g. McAninich and Strayer, 1988). The LINK IFS and SCARAB Projects could both be classed as long term (Woiwod, 1991), but there is evidence that even these studies may have been temporally and/or spatially inadequate for the detection of subtle effects of certain types of pesticide use. For instance, the Game Conservancy Trust's Sussex Study, in which 100 cereal fields have been suction sampled annually for the last 30 years, has revealed effects on arthropods of synthetic pyrethroid applications in the year following application (Ewald and Aebischer, 2000). The fact that changes in abundance and diversity may only be detectable after several years, or with extensive replication, could explain the rather contradictory results found between farming system studies and between years (e.g. Fan *et al.*, 1993).

Sampling considerations

Pitfall trapping has the advantage of readily capturing a wide range of carabid species, but is not ideal for detecting the impact of pesticide applications because the activity and abundance components of catches cannot easily be separated. Thus, short-term sub-lethal or indirect effects of a pesticide on activity may be difficult to distinguish from direct effects on mortality (e.g. Dixon and McKinlay, 1992; Sunderland *et al.*, 1995). Pitfall capture efficiency of carabids differs among species, meaning that certain species, such as *Demetrias atricapillus*, are under-represented in pitfall catches (Halsall and Wratten, 1988). Emergence traps, fenced pitfall traps or suction sampling are more suitable for estimating abundance of carabids (Sunderland *et al.*, 1995) but the efficiency of trapping individual species with these methods is largely unknown. Where emergence traps and pitfall traps were used together, the responses of Staphylinidae to insecticide usage were more consistent using the emergence traps (Krooss and Schaefer, 1998). The use of pitfall traps in the LINK and SCARAB Projects is justified on the grounds that it can provide an indication of changes in carabid abundance in long-term monitoring programmes (Baars, 1979; Den Boer, 1985). It is also justified because it was not the sole sampling method employed, and so comparison with other methods can aid the interpretation of pitfall catches; in the SCARAB Project, for instance, both suction and pitfall catches of Collembola indicated the same long-term response pattern to the 'Field 5' CFP regime, giving confidence that pitfall sampling can be used for detecting at least effects of intensive pesticide use in the long term. Although pitfall trapping is not an ideal sampling method for carabids, it is used very widely and, at present, would be very difficult to improve upon without invoking a considerable increase in sampling effort (Sunderland *et al.*, 1995). Therefore, one of our aims is to consider whether carabid sampling using this imperfect but convenient method can provide useful information for the detection of long-term effects of overall farming systems, or of specific types of pesticide use (see below).

Realism versus replication in the SCARAB Project

The split-field design of the SCARAB Project was chosen to allow arthropods to be monitored under conditions that were close to agricultural realism in terms of spatial scale, but it precluded replication of the pesticide regimes (Cooper, 1990). With such a 'quasi-experimental' approach (Parker, 1988), evidence other than statistical testing must be called upon to confirm that the changes in arthropod catches were caused by the treatment regimes. Two key lines of evidence are: (1) substantial changes in arthropod counts occurred when the pre-treatment pesticide inputs were changed to the contrasting pesticide regimes in 1991; (2) subsequently, after the SCARAB Project (during 1997–1999), the pesticide regimes were reversed spatially, and this caused a concurrent spatial reversal of the long-term patterns of arthropod abundance in 'Field 5'. These findings prove unequivocally, in the absence of formal replication, that the pesticide regimes in 'Field 5' were responsible for the patterns of arthropod abundance shown by the PRC analyses of suction and pitfall catches (Frampton, 2000, 2001b).

CARABIDS AS INDICATORS

Bioindicators can be used for a number of purposes, and so it is important to be clear what the indicative role of a particular arthropod species or group is, and to be aware of its limitations, as relatively few bioindicators have been formally tested (McGeoch, 1998). Rather than assume that pitfall catches can detect long-term effects of experimental treatments and then analyse catches to conclude whether such effects have occurred or not, our emphasis here is different: we accept that pitfall trapping is not an ideal method but, being convenient, it is used widely, and is likely to continue in widespread use despite its limitations. Our focus is therefore on whether this popular (albeit imperfect) method for sampling carabids can be used to detect long-term effects of overall farming systems, or of specific systems of pesticide use. For this purpose, the LINK IFS Project provides a realistic, well-replicated comparison of two contemporary farming systems, while the SCARAB CFP regime of 'Field 5' is a worst-case scenario of pesticide use that can be used to check the capability of pitfall sampling at detecting long-term treatment effects. The intensity of sampling, in terms of the number of traps used in the LINK IFS and SCARAB Projects outlined, is assumed here to be representative of the sampling intensity of a typical experimental field study. However, when Vickerman (1985) calculated the number of sample units required for a 20% level of sampling precision, he concluded that, for Carabidae, a much higher number of pitfall traps would be needed than are commonly used. For example, if 10 pitfall traps were used, as in the LINK IFS Project, such sampling precision would only be achieved for species with at least five individuals per trap; a larger number of samples would be needed for less abundant species.

One of the essential requirements of an indicator group is that it is widely distributed and sufficiently abundant to be readily sampled. Carabids are considered to be one of the most abundant arthropod groups in arable crops, with a well characterized assemblage (Luff, this volume). Despite this, only *P. melanarius* was abundant at each of the LINK IFS and SCARAB sites. At some sites, the carabid fauna was impoverished and not suitable as an indicator, given the sampling effort invested (*sensu* Vickerman, 1985). Such inherent variation between sites, but also between fields, always creates difficulties in field-based investigations. Typically, this is overcome by using replication to increase the chance that enough taxa are present in sufficient plots for analysis (Perry and Anon., 1999). However, guidelines for appropriate intensity of sampling and replication for studying carabids and other polyphagous predators in field trials are currently inadequate (Kennedy *et al.*, 2001).

The comparison of the effects of the LINK farming systems and SCARAB pesticide regimes suggests that, at least for pitfall catches, carabids are not the most sensitive indicator available for detecting the environmental effects of pesticide use. It may be that no single arthropod group alone is an ideal bioindicator of the adverse effects of pesticide use, as (for instance) Collembola are sensitive to OP insecticides but apparently not to synthetic pyrethroids (Frampton, 1999), whereas Linyphiidae are sensitive to synthetic pyrethroids in the short term but, in our analyses, did not exhibit long-term effects of either OP or synthetic pyrethroid insecticides. The fact that major community-level effects occurred only where there was high organophosphate usage, in one SCARAB field, suggests that any effects of synthetic pyrethroid use in the long term are probably relatively difficult to detect (cf. Ewald and Aebischer, 2000). On balance, it seems that carabids are not appropriate as 'early warning' indicators of pesticide effects, but the question remains as to whether they could be used in monitoring studies as indicators of the overall macro-arthropod response to long-term effects of farming systems. PRC analyses show that only under the SCARAB Project worst-case pesticide regime was a carabid community response pattern obtained that would be consistent with an effect of the experimental treatments. However, even this response was not strictly a long-term effect, as negative effects of the CFP regime were interspersed with periods of recovery. It is notable that the community response of carabids, though less clearly affected than Collembola by the CFP regime, nevertheless appears representative of the overall macro-arthropod community response. Macro-arthropods alone, and hence carabids, could be a useful bioindicator if the aim is to detect any effects of agroecosystem management practices on the availability of arthropod prey to insectivorous wildlife; even short-term depletion of prey could be ecologically important, e.g. if it coincides with chick feeding by insectivorous birds (Potts, 1986).

Using the carabid community as an indicator of the overall macro-arthropod community response would considerably reduce the effort required in sorting pitfall catches. Results from the SCARAB Project suggest that a potentially productive combination to monitor the influence of pesticide use on pitfall catches could be Carabidae and Collembola, to represent macro-arthropod and micro-arthropod communities respectively. Collembola have hitherto been neglected in studies of the arable fauna, but resources to monitor this group might be justified by the effort saved in monitoring Carabidae instead of all macro-arthropods.

We have focused here on the use of carabids for detecting the long-term effects of farming systems or pesticide regimes. If, however, the aim is to investigate the impact of management practices on agroecosystem function, carabids would be unsuitable on account of their trophic bias. In arable ecosystems, the diversity of carabids can be relatively low and is dominated numerically by relatively few species, the majority of which are polyphagous (Luff, this volume).

Taxonomic resolution of analysis

To some extent, the results reported in the literature reflect whether the analysis was conducted for the family Carabidae or for individual carabid species. If only total numbers are compared between farming systems, the effects on individual species may be missed and can give misleading conclusions. For example, in one study, *B. quadrimaculatum* abundance was suppressed by higher pesticide and fertilizer inputs, whereas that of *Clivina fossor* and *Calathus melanocephalus* increased. These opposite reactions effectively cancelled one another out and were not detectable at the family level (Büchs *et al.*, 1997). This also means that neither species' individual response pattern would predict that of the overall community. Use of individual species or functional groups as indicators is more likely to be of relevance in studies where relatively few explanatory variables are of interest, e.g. in evaluations of the effects of a specific type of pesticide use. For instance, some carabid species are predominantly spermophagous and may therefore be suitable indicators of floral abundance/diversity, and thereby indicators of herbicide impact.

Information on the choice of indicator species or groups can be gained using the species weights from PRC analysis, which indicate those species that most closely follow the fitted response pattern for the overall community. In the analysis of SCARAB 'Field 5' data, *P. cupreus* and *Bembidion obtusum* have relatively high species weights, whether or not the dominant *P. melanarius* is included in the analysis, suggesting that the group of *P. melanarius*, *P. cupreus* and *B. obtusum* could have indicator value for predicting effects of the overall CFP regime. In LINK IFS Project also, *B. obtusum* and *P. melanarius* responded strongly to the different farming systems. This is consistent with the finding from the long-term Boxworth Project that *B. obtusum* is a sensitive indicator of the effects of organophosphate insecticide-based regimes of pesticide use (Vickerman, 1992). However, the inclusion of *Pterostichus melanarius* in the analysis of the SCARAB Project carabid community appears to have masked an adverse effect of chlorpyrifos during 1991. Chlorpyrifos is among the most broad-spectrum and persistent of the OP insecticides approved for use in European agriculture (e.g. Van Straalen and Van Rijn, 1998). The ability of *P. melanarius* to mask the effects of such a potent chemical on other species clearly should be taken into consideration when deciding upon the species to include in a community analysis. A problem, however, is that the effects of excluding the dominant species from the analysis were not the same following chlorpyrifos sprays in SCARAB in 1991 and 1994; in the latter year, the effect of chlorpyrifos on the overall carabid community appears to have been larger when *P. melanarius* was included in the analysis. It is also the case that the inclusion or exclusion of this species made virtually no difference to the overall carabid community response to IFS and CFP farming systems in the High Mowthorpe data set of the LINK IFS study. In the Nagele Project in The Netherlands, *P. melanarius* was also the most abundant species but, along with the low density species, was encouraged by the integrated system (Booij and Noorlander, 1992), indicating that it may sometimes be responsive to farming inputs. These findings lead to the recommendation that it is prudent to conduct carabid community analyses both with and without the dominant species included in the data set, rather than performing only one analysis. In practice, this would involve a relatively small increase in the analytical effort.

Community-level analysis such as PRC has the advantage of using the data from all species and reflects the actual temporal change in community sensitivity that occurs as the proportions of sensitive species change through time. Even relatively uncommon species could together contribute to the overall community response if they all respond in the same way, but such species are usually not analysed separately on account of their rarity or temporal transience. Community-level analysis therefore has the potential to detect certain subtle effects of farming practice that might otherwise go unnoticed. However, the benefits of community analysis must be weighed carefully against the possible problems outlined above, particularly if species differ widely in their response patterns.

Diversity indices are frequently used as indicators of environmental impact and have been shown to detect differences between farming systems. There are some examples for carabids (Fan *et al.*, 1993; Cárcamo *et al.*, 1995; Ellsbury *et al.*, 1998) and spiders (Stippich, 1994; Huusela-Veistola, 1998) where numbers and species richness increased when agrochemical inputs were reduced. In the LINK IFS Project, the species richness of Carabidae did not differ between the integrated and conventional systems. However, compared with multivariate community analyses, diversity indices are relatively uneconomical in that they require individual species' data to calculate but do not themselves provide information on individual species responses nor their relative importance. Turnover or loss of species is particularly important to bear in mind when interpreting results from long-term monitoring studies as many invertebrate groups, including Carabidae, have undergone substantial declines over the last 30 years (Aebischer, 1991; Holland, this volume), with implications for the availability and reliability of particular species or groups of carabids as bioindicators.

Conclusions

Pitfall catches of carabids are unlikely to detect long-term effects of farm management practices, except in cases where high usage of OP insecticides occurs. In such cases, micro-arthropods are a more sensitive indicator of pesticide effects than macro-arthropods.

The response of the pitfall-captured carabid community to farm management practices could be suitable as an indicator of the overall macro-arthropod community response, particularly where high usage of OP insecticides occurs. Even short-term changes in macro-arthropod abundance and activity could have implications for prey availability to insectivorous wildlife.

Monitoring carabids instead of all macro-arthropods in pitfall catches would permit greater effort to be directed to additionally monitoring micro-arthropods such as Collembola, which are more sensitive at detecting the effects of certain types of pesticide use.

Exclusion of the dominant carabid species affects the outcome of carabid community analyses. It is therefore prudent to conduct community-level analyses both with and without the dominant species.

Abundance and diversity of carabids varies considerably between localities, even within individual farms, and this may mask subtle differences in response to changes in farming practices. Long-term changes in species composition limit the suitability of individual species of carabids or diversity indices as indicators of the impact of farm management practices, thereby favouring community-level analyses for the detection of treatment effects.

Acknowledgements

Financial support was provided for the LINK IFS Project by the Ministry of Agriculture, Fisheries and Food (MAFF), Scottish Office Agriculture and Fisheries Department, Zeneca Agrochemicals, Home-Grown Cereals Authority (Cereals and Oilseeds) and the British Agrochemicals Association. We also thank all colleagues involved at each site, and especially Dr Neil Fisher for carrying out the cropping system comparisons. Dr Peter Chapman of Syngenta (previously Zeneca Agrochemicals) provided statistical advice for the LINK IFS Project. The Prestonhall

Farming Co., Lower Hope Estate and Manydown Co. allowed access and provided considerable co-operation with the farming inputs. MAFF funding of the SCARAB Project is also gratefully acknowledged, together with the invaluable contribution made by colleagues at each of the SCARAB farms and at meetings of the SCARAB and TALISMAN Steering Group. Comments on this chapter were provided by Dr Guy Poppy and Dr Peter Chapman.

References

ADIS, J. (1979). Problems of interpreting arthropod sampling with pitfall traps. *Zoologischer Anzeiger Jena* **202**, 177–184.

AEBISCHER, N.J. (1991). Twenty years of monitoring invertebrates and weeds in cereal fields in Sussex. In: *The ecology of temperate cereal fields*. Eds. L.G. Firbank, N. Carter, J.F. Darbyshire and G.R. Potts, pp 305–331. Oxford: Blackwell.

BAARS, M.A. (1979). Catches in pitfall traps in relation to mean densities of carabid beetles. *Oecologia* **41**, 25–46.

BASEDOW, TH. (1998). The species composition and frequency of spiders (Araneae) in fields of winter wheat grown under different conditions in Germany. *Journal of Applied Entomology* **122**, 585–590.

BASEDOW, TH., BRAUN, C., LÜHR, A., NAUMAN, J., NORGALL, TH. AND YANES YANES, G. (1991). Abundanz, Biomasse und Artenzahl epigäischer Raubarthropoden auf unterschiedlich intensive Weizen- und Rübenfeldern: Unterschiede und ihre Ursachen. Ergebnisse eines dreistufigen Vergeichs in Hessen, 1985 bis 1988. *Zoologischer Jahrbücher Systematik* **118**, 87–116.

BOOIJ, C.J.H. AND NOORLANDER, J. (1992). The impact of integrated farming on carabid beetles. *Proccedings of Experimental and Applied Entomology* **2**, 16–21.

BROWN, K.C., LAWTON, J.H. AND SHIRES, S.W. (1983). Effects of insecticides on invertebrate predators and their cereal aphid (Hemiptera: Aphididae) prey: laboratory experiments. *Environmental Entomol*ogy **12**,1747–1750.

BÜCHS, W., HARENBERG, A. AND ZIMMERMANN, J. (1997). The invertebrate ecology of farmland as a mirror of the intensity of the impact of man? – an approach to interpreting results of field experiments carried out in different crop management intensities of a sugar beet and an oil seed rape rotation including set-aside. *Biological Agriculture and Horticulture* **15**, 83–108.

CÁRCAMO, H.A., NIEMALA, J.K. AND SPENCE, J.R. (1995). Farming and ground beetles: effects of agronomic practice on populations and community structure. *Canadian Entomologist* **127**, 123–40.

CHAPMAN, P., MCINDOE, E. AND LENNON, M. (2000). Experimental design and statistical analysis. In: *LINK integrated farming systems (a field scale comparison of arable rotations). Volume 1: Experimental work*. Project Report No. 173. Ed. S.E. Ogilvy, pp 22–26. London: HGCA.

ÇILGI, T. (1994). Selecting arthropod 'indicator species' for environmental impact assessment of pesticides in field studies. *Aspects of Applied Biology* **37**, 131–140.

ÇILGI, T., WRATTEN, S.D., ROBERTSON, J.L., TURNER, D.E., HOLLAND, J.M. AND FRAMPTON, G.K. (1996). Residual toxicities of three insecticides to four species (Coleoptera: Carabidae) of arthropod predators. *Canadian Entomologist* **128**, 1115–1124.

COOPER, D.A. (1990). Development of an experimental programme to pursue the results of the Boxworth Project. *Proceedings, 1990 British Crop Protection Conference – Pests and Diseases*, pp 153–162.

DAVIS, P.T. AND TSO, M.K.-S. (1982). Procedures for reduced-rank regression. *Applied Statistics* **31**, 244–255.

DEN BOER, P.J. (1985). Fluctuations of density and survival of carabid populations. *Oecologia* **67**, 322–330.

DIXON, A.F.G. AND MCKINLAY, R.G. (1992). Pitfall trap catches of and aphid predation by

Pterostichus melanarius and *Pterostichus madidus* in insecticide and untreated potatoes. *Entomologia Experimentalis et Applicata* **64**, 63–72.

DUFFIELD, S.J. AND AEBISCHER, N.J. (1994). The effect of spatial scale of treatment with dimethoate on invertebrate population recovery in winter wheat. *Journal of Applied Ecology* **31**, 263–281.

DUFFIELD, S.J., JEPSON, P.C., WRATTEN, S.D. AND SOTHERTON, N.W. (1996). The spatial changes in invertebrate predation rate in winter wheat following treatment with dimethoate. *Entomologia Experimentalis et Applicata* **78**, 9–17.

ELLSBURY, M.M., POWELL, J.E., FORCELLA, F., WOODSON, W.D., CLAY, S.A. AND RIEDELL, W.E. (1998). Diversity and dominant species of ground beetle assemblages (Coleoptera: Carabidae) in crop rotation and chemical input systems for the Northern Great Plains. *Annals of the Entomological Society of America* **91**, 619–625.

EL TITI, A. (1991). The Lautenbach project 1978–89: integrated wheat production on a commercial arable farm, south-west Germany. In: *The ecology of temperate cereal fields*. Eds. L.G. Firbank, N. Carter and G.R. Potts, pp 399–411. Oxford: Blackwell.

EWALD, J.A. AND AEBISCHER, N.J. (2000). Trends in pesticide use and efficacy during 26 years of changing agriculture in southern England. *Environmental Monitoring and Assessment* **64**, 493–529.

FAN, Y., LIEBMAN, M., GRODEN, E. AND ALFORD, A.R. (1993). Abundance of carabid beetles and other ground-dwelling arthropods in conventional versus low-input bean cropping systems. *Agriculture, Ecosystems and Environment* **43**, 127–139.

FLOATE, K.D., ELLIOTT, R.H., DOANE, J.F. AND GILLOTT, C. (1989). Field bioassay to evaluate contact and residual toxicities of insecticides to carabid beetles (Coleoptera: Carabidae). *Journal of Economic Entomology* **82**, 1543–1547.

FRAMPTON, G.K. (1997). The potential of Collembola as indicators of pesticide usage: evidence and methods from the UK arable ecosystem. *Pedobiologia* **41**, 179–184.

FRAMPTON, G.K. (1999). Spatial variation in non-target effects of the insecticides chlorpyrifos, cypermethrin and pirimiarb on Colembola in winter wheat. *Pesticide Science* **55**, 875–886.

FRAMPTON, G.K. (2000). Recovery responses of epigeic Collembola after spatial and temporal changes in pesticide use. *Pedobiologia* **44**, 489–501.

FRAMPTON, G.K. (2001a). SCARAB: Effects of pesticide regimes on arthropods. In: *Reducing agrochemical use on the arable farm*. Eds. J.E.B. Young, M.J. Griffin, D.V. Alford and S.E. Ogilvy, pp 219–254. London: DEFRA.

FRAMPTON, G.K. (2001b). Large-scale monitoring of non-target pesticide effects on farmland arthropods in England. In: *Pesticides and wildlife,* ACS Symposium Series 771. Ed. J.J. Johnston, pp 54–67. Washington DC: American Chemical Society.

FRAMPTON, G.K., VAN DEN BRINK, P.J. AND GOULD, P.J.L. (2000a). Effects of spring precipitation on a temperate arable collembolan community analysed using Principal Response Curves. *Applied Soil Ecology* **14**, 231–248.

FRAMPTON, G.K., VAN DEN BRINK, P.J. AND GOULD, P.J.L. (2000b). Effects of spring drought and irrigation on farmland arthropods in southern Britain. *Journal of Applied Ecology* **37**, 865–883.

FRAMPTON, G.K., VAN DEN BRINK, P.J. AND WRATTEN, S.D. (2001). Diel activity patterns in an arable collembolan community. *Applied Soil Ecology* **17**, 63–80.

GARDNER, S.M., LUFF, M.L., RIDING, A. AND HOLLAND, J.M. (1999). *Evaluation of carabid beetle populations as indicators of normal field ecosystems*. MAFF Project Report. London: MAFF.

GREIG-SMITH, P.W., FRAMPTON, G.K. AND HARDY, A.R. (Eds.) (1992). *Pesticides, cereal farming and the environment.* London: HMSO.

HALSALL, N.B. AND WRATTEN, S.D. (1988). The efficiency of pitfall trapping for polyphagous predatory Carabidae. *Ecological Entomology* **13**, 293–299.

HOLLAND, J.M. (1998). Comparative impact of insecticide treatments on beneficial invertebrates in conventional and integrated farming systems. In: *Ecotoxicology; pesticides and beneficial organisms*. Eds. P.T. Haskell and P. McEwen, pp 267–278. Dordrecht: Kluwer Academic Publishers.

HOLLAND, J.M. (2000). Invertebrates (beetles and spiders). In: *LINK integrated farming systems*

(a field scale comparison of arable rotations). Volume 1: Experimental work. Project Report No. 173. Ed. S.E. Ogilvy, pp 46–55. London: HGCA.
HOLLAND, J.M. AND LUFF, M.L. (2000). The effects of agricultural practices on Carabidae in temperate agroecosystems. *Integrated Pest Management Reviews* **5**, 105–129.
HOLLAND, J.M., COOK, S.K., DRYSDALE, A., HEWITT, M.V., SPINK, J. AND TURLEY, D. (1998). The impact on non-target arthropods of integrated compared to conventional farming: results from the LINK Integrated Farming Systems Project. *1998 Brighton Crop Protection Conference – Pests and Disease* **2**, 625–630.
HUUSELA-VEISTOLA, E. (1996). Effects of pesticide use and cultivation techniques on ground beetles (Col., Carabidae) in cereal fields. *Annales Zoologici Fennici* **33**, 197–205.
HUUSELA-VEISTOLA, E. (1998). Effects of perennial grass strips on spiders (Araneae) in cereal fields and impact on pesticide side-effects. *Journal of Applied Entomology* **122**, 575–583.
KELLY, J.R. AND HARWELL, M.A. (1990). Indicators of ecosystem recovery. *Environmental Management* **14**, 527–545.
KENNEDY, P.J., CONRAD, K.F., PERRY, J.N., POWELL, D., AEGERTER, B., TODD, A.D., WALTERS, K.F.A. AND POWELL, W. (2001). Comparison of two field-scale approaches for the study of effects of insecticides on polyphagous predators in cereals. *Applied Soil Ecology* **17**, 253–266.
KREMEN, C. (1992). Assessing the indicator properties of species assemblages for natural areas monitoring. *Ecological Applications* **2**, 203–217.
KROMP, B. (1999). Carabid beetles in sustainable agriculture: a review on pest control efficacy, cultivation impacts and enhancement. *Agriculture, Ecosystems and Environment* **74**, 187–228.
KROOSS, S. AND SCHAEFER, M. (1998). The effect of different farming systems on epigeic arthropods: a five-year study on the rove beetle fauna (Coleoptera: Staphylinidae) of winter wheat. *Agriculture, Ecosystems and Environment* **69**, 121–133.
MCANINICH, J.B. AND STRAYER, D.L. (1988). What are the tradeoffs between the immediacy of management needs and the longer process of scientific discovery? In: *Long term studies in ecology. Approaches and alternatives.* Ed. G.E. Likens, pp 203–205. New York: Springer-Verlag.
MCGEOCH, M.A. (1998). The selection, testing and application of terrestrial insects as bioindicators. *Biological Reviews* **73**, 181–201.
OGILVY, S.E. (2000). *LINK Integrated Farming Systems (a field scale comparison of arable rotations), Volume 1: Experimental work.* Project Report No. 173. London: HGCA.
OGILVY, S.E. (2001). Experiment design, treatments, monitoring and pesticide use. In: *Reducing agrochemical use on the arable farm.* Eds. J.E.B. Young, M.J. Griffin, D.V. Alford and S.E. Ogilvy, pp 199–218. London: DEFRA.
OGILVY, S.E., TURLEY, D.B., COOK, S.K., FISHER, N.M., HOLLAND, J.M., PREW, R.D. AND SPINK, J. (1994). Integrated farming – putting systems together for farm use. *Aspects of Applied Biology* **40**, 53–60.
PARKER, G.G. (1988). Are currently available statistical methods adequate for long-term studies? In: *Long term studies in ecology. Approaches and alternatives.* Ed. G.E. Likens, pp 199–200. New York: Springer-Verlag.
PEARSON, D.L. (1994). Selecting indicator taxa for the quantitative assessment of biodiversity. *Philosophical Transactions of the Royal Society of London B* **345**, 75–79.
PERRY, J.N. AND ANON. (1999). EPPO guideline for the efficacy evaluation of plant protection products: design and analysis of efficacy evaluation trials. In: *EPPO standards, guidelines for the efficacy evaluation of plant protection products, Volume 1, Introduction, general & miscellaneous guidelines, new & revised guidelines*, pp 37–51. Paris: EPPO/OEPP.
POTTS, G.R. (1986). *The partridge: pesticides, predation and conservation.* London: Collins.
STIPPICH, G. (1994). Extensivierung im Ackerbau: I. Auswirkungen auf Spinnen und Laufkäfer. *Mitteilungen Deutsche Geschichtsforschung Allgemeine Angewandte Entomologie* **9**, 125–129.
SUNDERLAND, K.D., DE SNOO, G.R., DINTER, A., HANCE, T., HELENIUS, J., JEPSON, P., KROMP, B., SAMU, F., SOTHERTON, N.W., ULBER, B. AND VANGSGAARD, C. (1995). Density estimation for beneficial predators in agroecosystems. *Acta Jutlandica* **70**, 133–164.

TER BRAAK, C.J.F. AND ŠMILAUER, P. (1998). CANOCO 4. Microcomputer Power, Ithaca, New York.

THEILING, K.M. AND CROFT, B.A. (1988). Pesticide side-effects on arthropod natural enemies: a database summary. *Agriculture, Ecosystems and Environment* **21**, 191–218.

THIELE, H.-U. (1977). *Carabid beetles in their environments*. Berlin: Springer-Verlag.

THOMAS, M.R., GARTHWAITE, D.G. AND BANHAM, A.R. (1997). *Arable farm crops in Great Britain 1996. Pesticide Usage Survey Report 141*. London: MAFF.

VAN DEN BRINK, P.J. AND TER BRAAK, C.J.F. (1998). Multivariate analysis of stress in aquatic ecosystems by principal response curves and similarity analysis. *Aquatic Ecology* **32**, 163–178.

VAN DEN BRINK, P.J. AND TER BRAAK, C.J.F. (1999). Principal Response Curves: Analysis of time-dependent multivariate responses of biological community to stress. *Environmental Toxicology and Chemistry* **18**, 138–148.

VAN STRAALEN, N.M. AND VAN RIJN, J.P. (1998). Ecotoxicological risk assessment of soil fauna recovery from pesticide application. *Reviews of Environmental Contamination and Toxicology* **154**, 83–141.

VICKERMAN, G.P. (1985). Sampling plans for beneficial arthropods in cereals. *Aspects of Applied Biology* **10**, 191–198.

VICKERMAN, G.P. (1992). The effects of different pesticide regimes on the invertebrate fauna of winter wheat. In: *Pesticides, cereal farming and the environment*. Eds. P.W. Greig-Smith, G.K. Frampton and A.R. Hardy, pp 82–108. London: HMSO.

VICKERMAN, G.P., COOMBES, D.S., TURNER, G., MEAD-BRIGGS, M. AND EDWARDS, J. (1987). The effects of pirimicarb, dimethoate and deltamethrin on Carabidae and Staphylinidae in winter wheat. *Mededelingen Facultat Landbouwwhogeschool Rijksuniversitat Gent* **52,** 213–223.

WINSTONE, L., ILES, D.R. AND KENDALL, D.J. (1996). Effects of rotation and cultivation on polyphagous predators in conventional and integrated farming systems. *Aspects of Applied Biology* **47**, 111–117.

WOIWOD, I.P. (1991). The ecological importance of long-term synoptic monitoring. In: *The ecology of temperate cereal fields*. Eds. L.G. Firbank, N. Carter, J.F. Darbyshire and G.R. Potts, pp 275–304. Oxford: Blackwell.

ZHANG, J.X., DRUMMOND, F.A. AND LEIBMAN, M. (1997). Effect of crop habitat and potato management practices on the population abundance of adult *Harpalus rufipes* (Coleoptera: Carabidae) in Maine. *Journal of Agricultural Entomology* **15**, 63–74.

10
Non-Crop Habitat Management for Carabid Beetles

JANA C. LEE[1] AND DOUGLAS A. LANDIS[2]

[1]*Department of Entomology, University of Minnesota, 219 Hodson Hall, 1980 Folwell Ave., St. Paul, MN 55108, USA and* [2]*Department of Entomology, Center for Integrated Plant Systems, Michigan State University, East Lansing, MI 48824, USA*

Introduction

Current agricultural management involves intensive mechanical operations such as tillage, often with high inputs of pesticides and large fields to maximize efficiency. These practices can reduce habitat quality and remove the necessary habitat structure that is important to many natural enemies, including predatory and weed seed-eating carabid beetles (Lesiewicz *et al.*, 1982; Boac and Popisil, 1984; Burn, 1989; House and Del Rosario Alzugaray, 1989). Carabids usually require a year to complete their life cycle, and their recovery from such agricultural disturbances is often slow (Jepson and Thacker, 1990). Many carabid species exhibit a 'cyclical colonization' pattern, where they overwinter in more stable habitats and colonize ephemeral habitats, including annual crops when conditions are favourable (Wissinger, 1997). Their cyclical behaviour makes natural areas (woodland, pastures, meadows) an important source of carabids that will colonize agricultural fields (Gravesen and Toft, 1987; Duelli *et al.*, 1990; Bedford and Usher, 1994; Kajak and Lukasiewicz, 1994). However, agricultural landscapes are being increasingly simplified, with natural areas being fragmented and replaced altogether by large monocultural fields. As a consequence, carabids and other natural enemies that depend on such areas for overwintering need to disperse further to reach summer feeding areas. This fragmentation and loss of critical habitats has caused assemblages of natural enemies to decline in abundance and diversity, and has even resulted in extinctions (Gut *et al.*, 1982; Fahrig, 1997). Given these trends, managing agricultural landscapes to restore habitats that favour natural enemies is critical to their conservation in agroecosystems (Landis *et al.*, 2000b).

Habitat management entails manipulating the environment to enhance the survival of natural enemies and to improve their efficiency as pest control agents. This is usually done by providing them with refuges and/or resources (Gurr *et al.*, 1998;

Landis *et al.*, 2000a). This chapter focuses on the use of non-crop habitats to augment predatory carabid beetle populations. These habitats may consist of:

1. Existing field margins such as grassy margins, hedgerows, shelterbelts, fencerows, and riparian buffers.
2. Margins established to prevent soil erosion and run-off pollution.
3. Those created to benefit gamebirds (Pollard and Relton, 1970; Best, 1983; National Research Council, 1993; Boatman, 1998).

Non-crop habitats can be established within fields. One such example is that of 'beetle banks', which consist of raised, grassy strips of *ca.* 2 m width intersecting crop fields (Thomas *et al.*, 1991). Non-crop habitats can also include less managed areas such as cereal field edges where weeds are allowed to establish, as seen in uncropped wildlife strips and conservation headlands (Rands, 1985). In this chapter, we review how non-crop habitats benefit predatory carabids and the impact on their health, abundance and species community structure. In addition, we discuss how non-crop areas affect invertebrate pests and weeds in nearby fields and review the current recommendations for managing these habitats. Understanding the long-term effects of non-crop habitats on carabid community structure and pest abundance is important if habitat manipulation is to be a viable strategy for controlling pests.

Functions of non-crop habitat

FOR OVERWINTERING CARABIDS

Non-crop vegetation has been widely documented as an important overwintering site for carabids (e.g. Sotherton, 1984, 1985; Thomas *et al.*, 1991; Dennis and Fry, 1992; Collins *et al.*, 1996; Thomas *et al.*, 2000a). 'Beetle banks' sown with the tussock-forming grass *Dactylus glomerata* have supported as many as 1100 overwintering adult carabids per m^2 (Thomas *et al.*, 1991). Sotherton (1984) and Thomas and Marshall (1999) found significantly more adult carabids overwintering in hedgerows and grassy strips than in bare ground and crop fields. Moreover, some spring breeding carabids, e.g. *Agonum dorsale* and *Demetrias atricapillus,* which disperse into the crop early in the season, overwintered almost entirely in field boundaries (Sotherton, 1984). Non-crop vegetation not only harbours higher densities of overwintering carabids, but can also be highly speciose. More species might be expected to hibernate in non-crop habitats such as hedgerows and woodland because these areas are utilized by forest inhabiting, as well as crop inhabiting, species (Bedford and Usher, 1994). Yet species richness has also been found to be high in other herbaceous, non-crop habitats. Lys and Nentwig (1994) collected adults of 14 species overwintering in weedy strips, whereas only two species overwintered in the nearby cereal field. Pfiffner and Luka (2000) found five times more species overwintering in meadows and sown wildflower strips, compared to the cereal field and strips of bare ground.

The consistently high densities of overwintering adult carabids found within non-crop refuges may be the result of active selection for them by mobile adults, as well as differential overwintering mortality within different habitats (Dennis *et al.*, 1994). For instance, late season harvest or tillage practices may kill many carabids diapausing in the field, while those diapausing within non-crop habitats are left undisturbed.

Moreover, non-crop vegetation may protect carabids from temperature extremes; Dennis *et al.* (1994) reported that winter survival of adult *D. atricapillus* burrowed in tussock grass was 36–44% higher than those burrowed in bare earth. Although the mean temperature within *D. glomerata* was not different from the mean temperature on bare ground, the temperature within *D. glomerata* fluctuated between narrower limits (Thomas *et al.*, 1991).

The abundance of diapausing larvae in refuges has also been studied, but to a lesser extent. About twice as many carabid larvae were found per unit area in weed strips than in the cereal field (Lys and Nentwig, 1994). Likewise, overwintering larvae have been found in higher numbers in grassy margins than within a maize field (Desender and Alderweireldt, 1988). Notably, larvae are often less mobile than adults and thus an understanding of the ovipositional behaviour of carabids will help determine any impact refuges may have on larval overwintering carabids.

FOR ACTIVE CARABIDS

Non-crop habitats have several characteristics of benefit to active adult carabids. Firstly, the vegetation structure in non-crop habitats may create a favourable microclimate and thus serve as shelter during adverse weather. Some carabid species often remain burrowed in the soil during the day to avoid desiccation and emerge to forage at the night when temperature and humidity are optimal (Rivard, 1966; Jones, 1979). Luff (1965) demonstrated that the microclimate within *D. glomerata* vegetation remained humid and fluctuated little in temperature during hot weather. Also, humidity within weedy headlands was higher than in herbicide-sprayed headlands and, correspondingly, more carabids were active in them (Chiverton and Sotherton, 1991). Speight and Lawton (1976) demonstrated that beetles foraged more and exerted higher predation pressure in weedy, and presumably more humid, fields. Therefore, non-crop vegetation may provide an attractive foraging and resting site, which is particularly valuable prior to crop canopy closure, although the specific environmental conditions created will influence which species are attracted.

Secondly, non-crop habitats may also provide steady and suitable food sources for carabids. This may be especially important for sustaining beetles early in the season before pests have colonized the crop. Hawthorne and Hassall (1995) demonstrated that carabid beetle density in variously treated headlands (cereal crop edges with or without weed regeneration) was positively correlated with prey abundance. Headlands with the highest densities of aphids and Collembola likewise contained the highest carabid density and diversity. Zangger *et al.* (1994) observed that female *Poecilus cupreus* found in the strip-managed plot possessed fuller digestive tracts and weighed more than those in the monoculture cereal plot, thus implying that favourable feeding conditions existed in the non-crop areas.

Thirdly, non-crop vegetation may be a favoured oviposition (Desender and Alderweirelt, 1988) or aggregation site (Lyngby and Nielsen, 1981); however, these aspects are not well understood.

Finally, it has been suggested that non-crop habitats act as a refuge, protecting subpopulations of carabids when the field is subjected to disruptive farming practices (Kareiva, 1990; Frampton *et al.*, 1995). Carabid populations commonly move between fields and semi-natural areas via flight or walking (Duelli *et al.*, 1990) and are thereby capable of recolonizing insecticide disturbed fields. Duffield and Aebischer (1994) demonstrated that, following an insecticide application, carabid populations recovered gradually, starting at the edge and moving towards the centre of a treated plot. However, the exact source of carabids was not investigated, but carabids were suggested to be recolonizing from adjacent untreated crop areas. This recovery pattern was also demonstrated in 4 and 16 ha arable fields surrounded by hedges and herbaceous banks (Holland *et al.*, 2000). Here, reinvasion of the field occurred after an insecticide application, starting from the edges with evidence of a rapid and more extensive reinvasion from the margins protected by an unsprayed buffer zone. In a more recent study, carabids from perennial strips were observed to recolonize crop areas treated with insecticide after the toxicity declined (Lee *et al.*, 2001). Carabids have been observed to leave areas recently treated with insecticide and move rapidly into untreated areas, suggesting that non-crop areas could be a refuge for those actively escaping field disturbances (Chen and Wilson, 1997). However, carabids moving through treated areas were exposed to insecticide residues and suffered very high mortality rates. Thus, the ability of carabids to successfully flee insecticide-treated areas requires further evaluation.

Relevance of non-crop habitats for carabid beetles

INTERACTION WITH CARABID LIFE HISTORIES

Because of differences in phenology and habitat requirements, the benefits of non-crop habitats for carabids may be species-specific. Ground beetles have been differentially classified as being ‘open field’ or ‘boundary’ carabids (Sotherton, 1984, 1985; Thomas, 1990). ‘Open field’ carabids are normally present in the field crop during winter, whereas ‘boundary’ species primarily overwinter in undisturbed field margins. Thus, non-crop habitats are considered to benefit greatly ‘boundary’ carabids during the winter. Only some species have been classified in this way, and other species will require greater understanding of their life history traits before such classification is possible.

A beetle’s breeding strategy and life stage (larva or adult) may affect its ability to select an overwintering habitat. Since non-crop habitats are often important overwintering habitats, some species are more likely to benefit from them than others. Spring breeders, i.e. beetles that overwinter as adults and breed upon emerging in the spring (Den Boer and Den Boer-Daanje, 1990), are often mobile before entering diapause, and may actively select a burrowing site. High densities of overwintering spring breeders have been consistently found in non-crop vegetation (Desender, 1982; Sotherton, 1984, 1985; Lys and Nentwig, 1994). These results suggest that adults are selecting these sites for diapause; some carabids are known to select habitats using olfactory or mechanical cues from vegetation (Evans, 1982; Andersen, 1985). In contrast, autumn breeders, beetles that primarily overwinter as larvae and breed in the following autumn, often have restricted mobility and probably overwinter close to where they were oviposited. The offspring are therefore dependent on the adults selecting the most appropriate site for oviposition. There is evidence that some species prefer to oviposit in the field boundary, as opposed to the cultivated area (Desender and Alderweireldt, 1988) and some autumn breeding species have been found to overwinter in non-crop areas in greater numbers than in the crop (Lys and Nentwig, 1994). Therefore, to understand how a particular species might benefit from non-crop habitat during winter, the breeding strategy, as well as the habitat selection behaviour, of adults, and especially ovipositing females, are important factors to consider.

Although it is unclear whether spring and autumn breeders differentially benefit from non-crop habitats during the winter, during the active season both spring and autumn breeders appear to commonly use non-crop habitats. Studies using pitfall sampling have found both spring and autumn breeders to be highly prevalent in such areas (Lys *et al.*, 1994; Asteraki *et al.*, 1995; Hawthorne and Hassall, 1995). How much individual species use non-crop habitats during the season may depend on the species' food and microclimatic preferences, with species with higher humidity requirements possibly gaining the most benefit.

EFFECTS ON THE MORPHOLOGY AND PHYSIOLOGY OF CARABIDS

Given that non-crop habitats can provide favourable food resources and microclimate, carabid beetles that have convenient access to them should have a better nutritional status than carabids living in simplified agroecosystems with little variety of resources. Some studies have looked at the effect of habitat complexity and non-crop vegetation on the body size, feeding condition, and fecundity of adult carabids. Chiverton and Sotherton (1991) evaluated *P. melanarius* and *A. dorsale* in two differently managed headlands, Bommarco (1998) evaluated *Poecilus cupreus* in conventional and organic farms, Zangger *et al.* (1994) evaluated *P. cupreus* in a cereal monoculture and cereal field with weedy strips, and Lee (2000) evaluated *P. melanarius* in enclosed plots with and without perennial refuge strips and the adjacent cornfield with and without an application of a soil insecticide (*Table 10.1*). In three of these studies, habitat complexity increased the size of adult carabids. *P. cupreus* adults were larger in landscapes where field sizes were small (Bommarco, 1998) and females were larger in the strip-managed cereal field (Zangger *et al.*, 1994). *P. melanarius* females from plots with refuge areas and untreated cornfields (refuge-untreated crop) were slightly larger than females from plots without refuge areas and cornfield treated with insecticide (control-treated crop) (*Figure 10.1*) (Lee, 2000), but there was no effect for males. Since larval feeding and development determines adult size, monitoring adult body size allows an evaluation of the past conditions experienced by larvae (Nelemans, 1988; Van Dijk, 1994). Based on these studies, it appears that complex habitats and certain non-crop habitats may provide developing larvae with more food resources.

Since non-crop habitats are expected to increase food availability to adults, these studies also monitored how habitat diversification affected the current feeding condition of adults (in terms of mass/size indices or gut content examination). Food limitation experienced by adults has important implications for fat reserve build-up, longevity, and fecundity (Wallin *et al.*, 1992). As expected, a greater number of *P. melanarius* and *A. dorsale* had taken more meals in weedy headlands than herbicide sprayed headlands (Chiverton and Sotherton, 1991). *P. cupreus* experienced better feeding conditions in landscapes dominated by small, organic farms and strip-managed cereal fields, as opposed to large monocultural fields (Bommarco, 1998;

Table 10.1. Studies on the effects of habitat complexity on carabid beetle morphology and physiology.

Reference	Species and system	Body size	Feeding condition	Fecundity
Chiverton and Sotherton, 1991	*P. melanarius* and *A. dorsale* were monitored in headlands (cereal field edges) with and without herbicide treatment.	Not studied.	Using gut dissections, significantly more beetles in weedy headlands (no herbicide) had solid arthropod food remains.	Both female *P. melanarius* and *A. dorsale* contained more eggs in weedy (no herbicide) headlands.
Bommarco, 1998	*P. cupreus* was monitored in 3 sites dominated by small organic farms and in 2 sites with larger conventional farms.	Elytra length was shorter among *P. cupreus* residing in landscapes with larger field sizes than those found in landscapes with small field sizes.	Using the Energy Reserve Index (ERI), beetles found in monotonous landscapes with conventionally-managed anual crops had very low energy reserves.	Egg production of beetles was low in conventionally-managed areas. Fecundity was negatively correlated with the presence of annual crops and was positively correlated with smaller field sizes.
Zangger *et al.*, 1994	*P. cupreus* was monitored in a cereal field with weedy strips and in a cereal monoculture.	Females had longer elytra lengths in the strip-managed cereal as opposed to monoculture. There were no significant differences among males.	Using the Condition Factor (CF) (see Juliano, 1986), females were heavier for their size in the strip-managed cereal field. According to gut content examination, more males and females were satiated in strip-managed areas.	Egg production started earlier and lasted for a longer duration for females in the strip-managed cereal field than in cereal monoculture.
Lee, 2000	*P. melanarius* was monitored in enclosed plots: with and without perennial refuge strips, and adjacent corn field with and without soil insecticide (terbufos).	Females had longer elytra lengths in refuge-untreated crop plots than control-treated crop plots. The effect of refuge on female elytra length was significant at $P = 0.07$. Males were not significantly different.	The condition factors did not significantly differ among females and males found in plots with or without refuge strips and insecticide application.	Not studied.

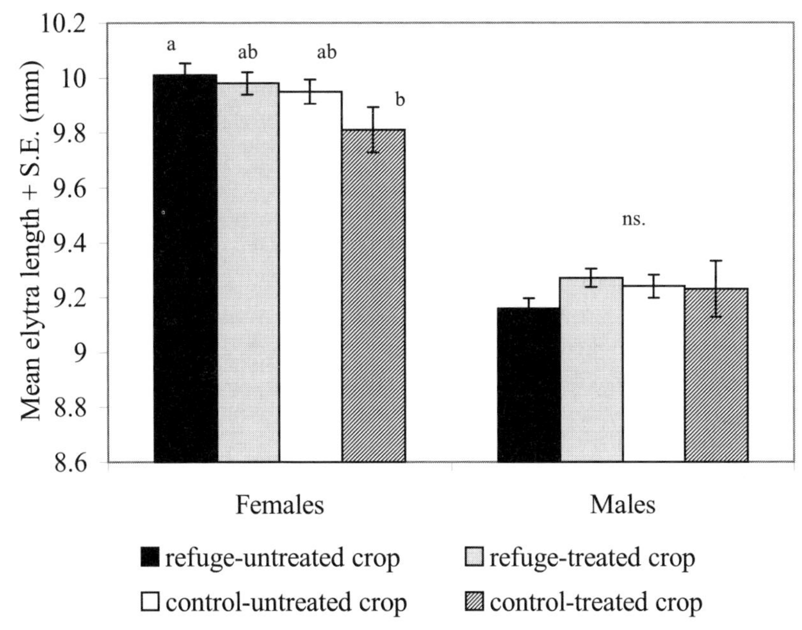

Figure 10.1. Mean elytra length of female and male *P. melanarius* in plots with/without perennial refuge strips and insecticide application in the adjacent crop, Michigan State Entomology Farm, US, 1999. The presence of refuge, insecticide, and refuge × insecticide interactions as sources of significant variation were tested with split-plot analysis of variance. For females (n = 327) and males (n = 457) all effects were non-significant ($P < 0.05$). Multiple comparisons of the four treatments are shown in the figure, different letters denote significant differences at $P < 0.05$ using LSD tests.

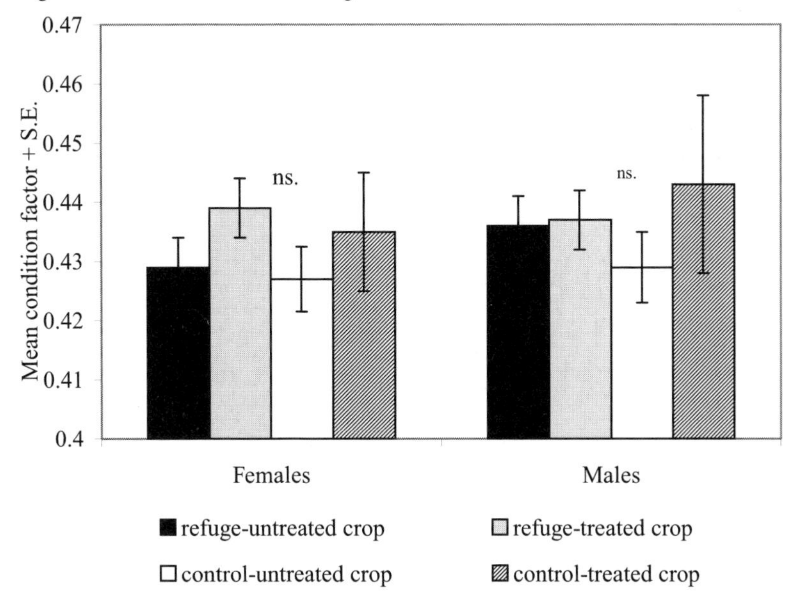

Figure 10.2. Mean condition factor of female and male *P. melanarius* in plots with/without perennial refuge strips and insecticide application in the adjacent crop, Michigan State Entomology Farm, US, 1999. The presence of refuge, insecticide, and refuge × insecticide interactions as sources of significant variation were tested with split-plot analysis of variance. For females (n = 327) and males (n = 457) all effects were non-significant ($P < 0.05$).

Zangger *et al.*, 1994). In contrast, the feeding condition of *P. melanarius* was not clearly influenced by the presence of perennial strips, nor the application of an insecticide (*Figure 10.2*) (Lee, 2000). However, in this study, beetle foraging was restricted to a 15 by 15 m area within which they emerged, and this may have had an unknown impact on food acquisition.

Given that suitable habitat diversification may enhance larval and adult food intake, the final impact on fecundity is critical. *P. melanarius* and *A. dorsale* in weedy headlands had higher egg counts than those from less vegetated headlands (Chiverton and Sotherton, 1991). *P. cupreus* found in conventional fields had low egg production (Bommarco, 1998). Likewise, *P. cupreus* from a strip-managed plot consistently produced more eggs, and did so earlier in the season (Zangger *et al.*, 1994). In April, 90% of females collected in the strip-managed cereal field possessed ripe eggs, whereas only 30% in the cereal monoculture possessed ripe eggs. Since females in strip-managed areas were also more satiated than in monoculture, these results imply that the food resources provided by weedy strips augmented and prolonged the reproduction of this numerically important species.

EFFECTS ON CARABID ACTIVITY-DENSITY: SOURCE OR SINK HABITAT?

Many studies have documented very high activity-densities of adult carabids within hedgerows, field margins or within-field strips (Hassall *et al.*, 1992; Lagerlof and Wallin, 1993; Lys *et al.*, 1994; Thomas and Marshall, 1999). Capture rates were usually highest during the early season, which may result from overwintering adults emerging from these sites (Coombes and Sotherton, 1986; Thomas *et al.*, 1991; Carmona and Landis, 1999). Mark–recapture experiments indicated that weedy strips have an attractive or arrestive effect on roaming beetles (Lys and Nentwig, 1992), with *P. melanarius* and *P. cupreus* moving more frequently from a monocultural cereal field to a strip-managed cereal field than the reverse. However, if pest control is to be improved, these non-crop areas need not only to provide carabids with food and shelter, but also to be a 'source' habitat augmenting activity-density in adjacent fields. If resource availability is very high, the non-crop habitat might potentially reduce feeding activity in the crop. For instance, carabid movement was greater between a grass ley and the crop than between permanent grass and the crop (Kajak and Lukasiewicz, 1994). It was suggested that food availability was higher in the permanent grass, causing fewer beetles to move from it. There is a clear trade-off between resource provisioning and predatory activity in the crop. Corbett and Plant (1993) addressed this concern by observing natural enemy distributions after non-crop vegetation strips were added within a field, using simulation models. Generally, if natural enemies utilize non-crop vegetation before the crop germinates, the vegetation serves as a 'source' of natural enemies to the adjacent field. Following germination, natural enemies are available to exert pressure on pests. However, if the non-crop vegetation and crop germinate simultaneously, the vegetation may serve as a 'sink' of natural enemies by reducing their activity within the crop. The attractiveness of the non-crop vegetation compared to the crop and the mobility of the natural enemy are important factors. Carabid adults are quite mobile (Wallin and Ekbom, 1988; Lys and Nentwig, 1991) and therefore expected to disperse from the strip through time. Although a non-crop habitat may attract carabids, it only becomes a 'sink' if the mortality rate surpasses the reproductive rate (Pulliam, 1988). However, non-crop habitats can enhance the nutritional status and fecundity of carabid beetles (Bommarco, 1998). The greater reproductive rate of carabids with access to non-crop habitats, and the potential of their progeny to disperse, may counteract the retention effects of non-crop habitats. Given that non-crop vegetation frequently harbours high densities of overwintering carabids, enhances reproductive output, and that adults are mobile, this vegetation should serve as a 'source' of predatory carabids to the crop area when food is available. However, this enhancement effect has only been demonstrated near to the non-crop habitat (< 20 m) and demonstration that this can be achieved at the field scale is needed before the technique can be recommended for farmers.

Most evidence is currently indirect or circumstantial on whether non-crop habitats act as a source. Varchola and Dunn (1999) found that, early in the season, carabids were more abundant in roadside vegetation than in a cornfield but, following crop canopy closure, they were more frequently caught within the field. It was postulated that the roadside vegetation served as an early season food source but resources later became more favourable in the crop but this may also have occurred in response to changing environmental conditions, the field becoming more attractive. Coombes and Sotherton (1986) positively correlated densities of *D. atricapillus* during the summer, with those measured concurrently in the adjacent boundary vegetation ($r^2 = 0.76$) and with previous overwintering densities ($r^2 = 0.92$), but only at 5 m into the field. Higher rates of carabid capture were also found in fields with adjacent headland vegetation strips (*Table 10.2*) (Hassall *et al.*, 1992), but again only at 8 m and 14 m into the crop. Repetition of the study by Hawthorne and Hassall (1995) also discovered that the beetle numbers 8 m inside the crop reflected the amount collected in the crop edge. Lys *et al.* (1994) showed that activity-density of carabids was higher in cereal areas between weed strips than in a cereal field. In a study by Lee *et al.* (2001), when carabid populations within fields were depleted with soil insecticide, perennial strips provided more carabids to colonize fields than similar sized crop strips unperturbed by insecticide. While non-crop habitats may enhance abundance within adjacent fields, how far these benefits extend from the boundary has not been adequately addressed. Spatial distribution studies showed that carabids were mainly abundant within 60 m of the boundary (Holland *et al.*, 1999), suggesting that penetration may be limited.

If a negative gradient in beetle activity-density can be detected with increased distance from the refuge, this provides evidence that beetles are dispersing from refuges and may indicate to what extent they penetrate crop habitats. Dennis and Fry (1992) collected more predatory arthropods, including carabids in field margins, with a graduated decline in capture from 1 to 50 m from the margin. Using directional traps, they documented a net movement of carabids into the field during May and June. By late June, densities were homogeneous throughout the field. Likewise, Vitanza *et al.* (1996) found a decreasing gradient of carabids at sample points 1 m, 6 m and 12 m from strips. Coombes and Sotherton (1986) described three carabid species, *A. dorsale*, *D. atricapillus*, and *Bembidion lampros*, as having a 'slow wave' of dispersal into the crop. At the start of the season, high densities existed only in the refuge, then a gradient in densities became apparent, with an eventual uniform distribution throughout the field. Thomas *et al.* (1991) also found activity-density of *D. atricapillus* to be highest in grass banks in April, but during May, proportionately similar numbers

Table 10.2. Studies comparing carabid abundance in fields without adjacent non-crop habitats and with variously vegetated non-crop habitats.

Reference	Years	System	Sampling	Outcome
Hassall *et al.*, 1992	1988	Compared headlands: 3 uncropped wildlife strips (with natural regeneration), 3 conservation headlands (minimal chemical inputs), and 2 sprayed headlands (vegetation similar to crop field) at two farm sites. Winter barley field sizes varied from 3.3–22 ha.	Carabids were sampled with pitfall traps within headlands and 8–14 m in the field away from headlands.	Activity-density of carabids 8–14 m into fields adjacent to uncropped wildlife strips was twice the activity-density in fields adjacent to sprayed headlands. Also, activity-density was higher in fields adjacent to conservation headlands than sprayed headlands.
Hawthorne and Hassall, 1995	1990–1992	Compared headlands (as described above). Each headland treatment was replicated twice in RCBD of an 18.6 ha winter wheat field. Strips were 120 × 6 m.	Carabids were sampled with pitfall traps within headlands and 8 m into the wheat field.	Carabid captures were higher 8 m into crops adjacent to uncropped wildlife strips than conservation or sprayed headlands.
Lys, 1994; Lys *et al.*, 1994	1989–1991	Compared a winter cereal field with weed strips to a cereal monoculture. The 8 ha cereal field had weed strips spaced 12 m, 24 m and 36 m apart, the control area (monoculture) comprised the other half, being 78 m long.	Carabids were sampled with pitfall traps in the control area in 1989, in the strip-managed area in 1990, and both areas in 1991.	Activity-density was higher in cereal areas between weed strips than in the control area in 1991. The maximum distance from weed strips in the strip-managed field was 6–18 m.
Lee, Menalled and Landis, 2001	1998–1999	Compared cornfields with/without soil insecticide and with/without adjacent 3 m wide perennial refuge strips. Four treatments were replicated in four blocks, plots were enclosed in 15 × 15 m plastic barriers.	Carabids were sampled with pitfall traps within strips (refuge or control) and 4 m and 8 m into the cornfield.	Activity-density was higher 4–8 m into insecticide disturbed cornfields when adjacent to perennial strips as opposed to control strips (these strips were planted with corn and did not receive insecticide).

of ground beetles were captured 60 m away from the grassy bank. More recently, the spatial distribution of boundary- and field-overwintering carabids within a field adjacent to a beetle bank was compared with that of a hedgerow (Thomas *et al.*, 2000b). A wave of reinvasion by boundary-overwintering carabids was detected, extending up to 150 m into the field from both types of cover. The boundaries appeared to have no influence on the field-overwintering carabids.

It is important that studies conducted on carabid distribution should carefully consider the spatial pattern of sampling. Holland *et al.* (1999) sampled carabids using grid patterns at three spatial scales: 1.5 m, 7.5 m and 30 m spacings. They found that for most taxa, a 30 m sampling grid was suitable for detecting clusters within a field. The appropriate sampling grid size would depend on the dispersal ability of the species or group of insects under consideration. To determine dispersal ability, Thomas *et al.* (1998) used mark–recapture methods and recorded *P. melanarius* moving 24–28 m^2 per day. Whilst there was considerable movement within the field, most were found within 55 m of the release site over a period of 30 days. This would suggest that non-crop habitats were an insignificant source of *P. melanarius* at this distance into the field. Detailed study of dispersal and careful selection of sampling patterns will help assess the extent that carabids disperse from non-crop habitats into adjacent crops.

EFFECTS ON CARABID COMMUNITY STRUCTURE

In addition to influencing the abundance and activity of carabids, non-crop habitats may have an effect on carabid community composition. Species richness and diversity were greater in non-crop habitats than in crops during the overwintering period (Lys and Nentwig, 1994) and while carabids were active (Kromp and Steinberger, 1992; Lys *et al.*, 1994; Frank, 1997; Fournier and Loreau, 1999). Carmona and Landis (1999) and Lee (2000) found that, over a period of four years, species richness increased in an experimental field with newly sown refuge strips. During the first year after establishment, 14 species were captured within the entire site but in the fourth year, 37 species were captured (*Figure 10.3*). Diversity also increased, although the Simpson diversity index declined in 1999 because *P. melanarius* dominated the catch. This increase in species richness suggests that the non-crop vegetation made the agroecosystem gradually more suitable for a wider variety of species. Moreover, non-crop habitats may specifically enhance species richness in the adjacent crop. Species richness of carabids has been observed to decline with distance from field margins (Fournier and Loreau, 1999). Frank (1997) found that fields adjacent to newly sown weed strips increased in carabid species richness from the first to second year of refuge establishment. Since fields with refuges were not compared to a control field without a refuge, and dispersal was not monitored, only indirect evidence that refuges enhance species richness was obtained.

Ultimately, non-crop habitats will favour some species and thereby alter the abundance of others. For instance, some studies have found autumn breeding *P. melanarius* to be highly abundant in densely vegetated headlands (Hassall *et al.*, 1992) and in perennial strips (Carmona and Landis, 1999), although studies of its spatial distribution showed that it favours the mid-field (Thomas *et al.*, 1998; Holland *et al.*, 1999). *P. melanarius* is a very voracious predator with the ability to outcompete

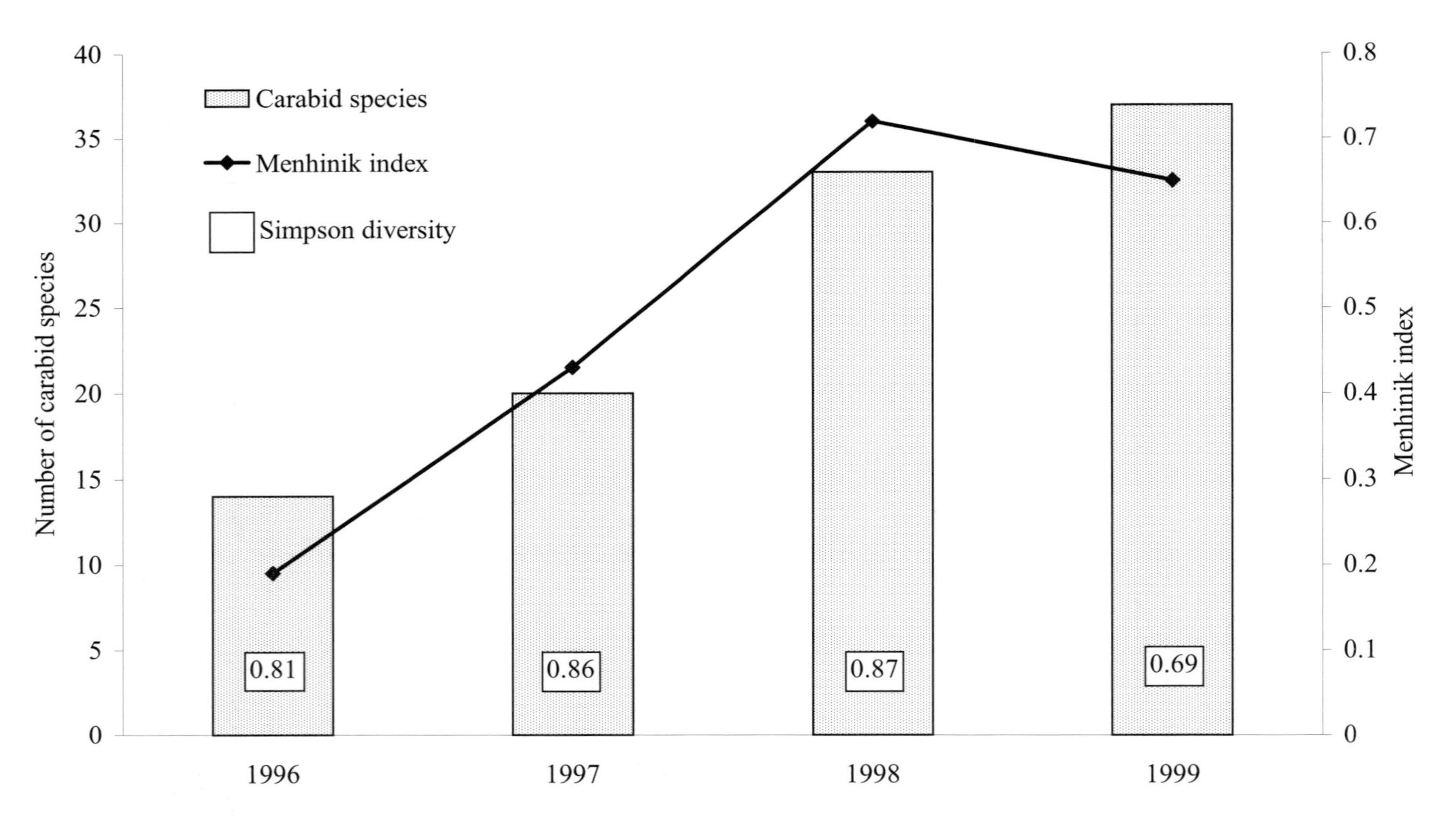

Figure 10.3. Total number of species and diversity indices for carabids found in a 1.4 ha annual crop field from 1996 to 1999, Michigan State Entomology Farm, US. Perennial refuge strips were established in 1995. Total number of individuals collected were 5517, 2114, 2126 and 3234, respectively. Species richness was adjusted for sample size with the Menhinick index (number of species/(sample size)$^{1/2}$).

other carabid species (Currie and Digweed, 1996; Currie *et al.*, 1996). Thus, non-crop areas may not only enhance the abundance of *P. melanarius* but also reduce the abundance of other carabid species, although this has not been evaluated to date.

How non-crop vegetation influences community composition has only rarely been addressed. Asteraki *et al.* (1992) revealed that carabid communities within a hedge were different from communities 10 m into the crop field, and that various non-crop habitats harboured distinct communities (Asteraki *et al.*, 1995). Not surprisingly, habitats with varying densities and types of vegetation supported different carabid communities, but a number of questions remain unanswered. For example, the extent that non-crop habitat influences community composition in the adjacent field is not well known; for example, it may promote the dominance of certain species within fields over the long term, i.e. those with certain breeding strategies, habitat selection preferences and feeding behaviour best suited to the conditions created. If so, such community changes may influence pest control.

The influence of non-crop habitats on pest abundance

ROLE IN REDUCING PEST ABUNDANCE

Given that refuges influence predator populations in the field, the subsequent effect on prey populations is of economic concern. Several researchers have described correlations between non-crop habitats and decreased pest populations or densities in the crop. Hausammann (1996) found significantly fewer aphids and cereal leaf beetles in field crops with weedy strips. In contrast, a field without a weedy strip reached the economic threshold requiring insecticide treatment. Hawthorne and Hassall (1995) observed that crops adjacent to uncropped wildlife strips contained higher carabid densities 8 m into the field, and likewise significantly lower aphid densities than where the headlands were sprayed. Gradients in pest densities at varying distances from vegetation strips have also been recorded. Vitanza *et al.* (1996) revealed significantly lower insect pest densities and damage to cotton 1 m away from grassy strips than at 6 m and 12 m away, with a reverse gradient in carabid densities. Hausammann (1996) observed significantly lower insect pest densities at 3 m versus 50 m and 75 m away from strips early in the season. Similarly, densities of *Sitobion avenae* were lower at 8 and 33 m from a beetle bank, compared to 83 m away (Collins *et al.*, 1997).

Knowing the distance at which predators exert significant pest control would allow guidelines to be produced describing optimal spatial arrangements of non-crop habitats to maximize pest control. However, not all studies have found significant pest gradients in the field; for example, Thomas' *Drosophila* pupae removal study (1990) and Hausammann's first year study (1996) with observations 3 m, 6 m and 10 m away from strips. Moreover, the influence of non-crop habitats on predators has only been demonstrated over relatively short distances. A lack of predation rate gradients may result not only from little influence of the nearby non-cropped area, but also from other factors. Carabid beetles and other natural enemies may disperse very actively, creating only an emphemeral gradient in pest predation rates, which is difficult to detect. A gradient may also not be found if the pest is not a preferred food item. Aphids are frequently monitored in such studies, but Toft and Bilde (this volume) revealed that these prey are avoided. The role of carabids in pest control is reviewed in Sunderland (this volume) and he concludes that carabids alone may not be sufficient to control pest outbreaks. Such research may help to explain the inconclusive results described above.

Whilst carabids play a role in weed seed predation in agroecosystems (Brust and House, 1988), the influence of non-crop vegetation on weed seed removal by carabids is largely unexplored, although is discussed more fully in Tooley and Brust (this volume). Herbaceous filter strips have been found to harbour about two to three times more seed-eating carabids than the crop and, correspondingly, seed removal within strips was sometimes four to eight times greater than in the crop (Menalled *et al.*, in press). In addition to seed-eating carabids, other animals associated with non-crop habitats such as birds, rodents or other invertebrates may play an important role in weed seed removal; for example, the substantial removal of weed seeds in field margins was attributed to small mammals (Povey *et al.*, 1993). Carmona *et al.* (1999) found that herbaceous filter strips contained large numbers of common field cricket *Gryllus pennsylvanicus*, which consumed many weed seeds in the laboratory. However, Marino *et al.* (1997) examined whether the presence of hedgerows enhanced seed removal in maize fields but only found patchy distributions of removal rather than a gradient.

ROLE IN CONTRIBUTING TO PEST PROBLEMS

Non-crop habitats may cause negative effects on crop production if they harbour weeds and accumulate a seed bank that can invade fields. Rew *et al.* (1996) addressed whether weed seeds were spread from non-crop habitats and into adjacent fields by farm machinery under normal management. Ninety-nine percent of marked *Bromus sterilis* were found within 1 m of the source. When a combine harvester was driven parallel and flush with the field boundary, the potential for the combine to disperse seeds into the centre of a field was minimal. Where annual weeds proliferate in margins, the naturally generated swards may be replaced with a sown wildflower mixture (Smith *et al.*, 1994). In addition, the new herbaceous vegetation may encourage a greater diversity of invertebrate species. Alternatively, a grass mixture, which is highly competitive with annual weeds, can be used (Smith and MacDonald, 1989). Lee (2000) found newly sown *D. glomerata* (commonly known as orchardgrass or cocksfoot) and orchardgrass/clover strips had very few invading weed species compared to red clover strips (*Trifolium pratense*). These studies indicated that, by careful selection of non-crop vegetation, potential weed problems may be prevented or alleviated.

Non-crop habitats may also harbour invertebrate pests that can damage crops. For instance, oilseed rape fields adjacent to wildflower strips have been shown to suffer greater crop damage by slugs (Frank and Friedli, 1997; Frank, 1998). Nevertheless, this may be overcome through the selection of appropriate species that deter slugs from feeding on oilseed rape (Cook *et al.*, 1997; Frank and Barone, 1999), and represent a viable way to control slugs encouraged by wildflower strips.

A modelling approach was used by Bhar and Fahrig (1998) to demonstrate the value of non-crop habitats for pest control. They postulated that woody borders around fields would enhance activity of natural enemies locally and reduce pest abundance at a landscape level. Holland and Fahrig (2000) tested this further by surveying landscapes with varying amounts of woody borders. They found that landscapes with woody borders contained a higher diversity, but similar density, of herbivorous insects compared to landscapes devoid of borders. This study supported the conclusion that woody borders do not exacerbate pest problems. Likewise, studies conducted on sown weed strips found little evidence that such non-crop habitats created major pest problems. Lethmayer *et al.* (1997) observed a few coleopteran pests inside weed strips, but all except one species existed in low numbers. These weed strips also contained a number of aphid species, but few of them were crop pests (Lethmayer, 1995). Nentwig *et al.* (1998) stated that aphid pests were unlikely to be augmented by these strips since 50% of aphid species are monophagous, feeding only on the crop.

The possible effects on pests should always be addressed when developing management techniques for non-crop habitats. Pearson (1990) and Gurr *et al.* (1998) provide selection criteria for choosing the appropriate vegetation to establish. The benefits that non-crop vegetation may bring are weighed against the risks of it interfering with the primary crop and being a source of crop pathogens. While non-crop habitats can harbour arthropod pests or weed seeds, the same pests and weed seeds can have a role in enhancing natural enemy populations such as carabid beetles, by providing alternative feeding resources.

Managing non-crop habitats

The evidence that non-crop habitats could enhance predators, and especially carabid populations, with the assumption that biological control of pests would be improved, led to the development of a number of non-crop management practices. These techniques varied in composition, size and arrangement; however, before recommendations can be made to growers, we need to establish whether, and to what extent, each type of non-crop habitat influences pest control and whether it is economically viable. Unfortunately, the impact on pest control has rarely been investigated. although it is the answer most farmers want.

VEGETATION COMPOSITION

Many of the previously mentioned non-crop habitats were comprised of perennial grasses. The tussock-forming *D. glomerata* has been cited to harbour more beetles than other grass species (Thomas, 1990; Dennis and Fry, 1992), and winter survival of carabids within this species was higher than that of other grasses (Dennis *et al.*, 1994). *D. glomerata* strips have sometimes yielded 2000 overwintering predators per m^2 (Wratten and Thomas, 1990). In addition, establishing this grassy vegetation on raised banks improved drainage and increased the survival and abundance of overwintering carabids (Dennis and Fry, 1992; Dennis *et al.*, 1994). Now called ‘beetle banks’, these strips are eligible for grant aid in the United Kingdom. Lagerlof and Wallin (1993) discovered that flowering strips were helpful for natural enemies in general. Yet clover mixture strips failed to harbour any overwintering carabid adults and larvae. In some cases, hedgerows were better habitats than grassy strip as they contained higher species diversity (Asteraki *et al.*, 1995), including more overwintering beetles (Sotherton, 1985). In contrast, Thomas *et al.* (2000a) reported that the density and diversity of carabids was similar in beetle banks compared to the bases of hedgerows.

The type and amount of vegetation may influence prey availability and soil quality, thereby influencing predator abundance. Hawthorne and Hassall (1995) found that percentage cover of dicotyledonous plants and abundance of invertebrates was positively correlated with overall carabid abundance. Similarly, predator species diversity was also positively correlated with total vegetation cover, with weedier headlands containing higher densities of non-pest species and predators (Chiverton and Sotherton, 1991). The plants growing in non-crop habitats may help create a deep sod layer, known to harbour higher densities of overwintering adult (Desender, 1982) and larval carabids (Lagerlof and Wallin, 1993).

SIZE AND SPACING

Most research on carabid beetle conservation has been conducted on vegetation type and the number of beetles it supports. However, the characteristics of refuges which affect beetle dispersal are also of interest and relevant to management. The width of a refuge and its location (bordering the crop or intersecting the field) may influence carabid movement. For example, Frampton *et al.* (1995) and Thomas *et al.* (1998) demonstrated that the presence of grassy banks impeded carabid movement between fields, with the inhibitory effect increasing with greater bank width. Hedgerows have been shown to have similar effects (Mauremooto *et al.*, 1995); although not consistently, as *Nebria brevicollis* moved readily through hedgerows (Joyce *et al.*, 1999). In designing refuge habitats, the size should be large enough to support beetles without greatly inhibiting movement into the field.

One of the most commonly asked questions regarding non-crop habitat management for carabids is how the habitats, strips in most cases, should be spaced in the field. The answer requires an understanding of the dispersal capabilities of carabids and other natural enemies. Corbett and Plant (1993) modelled how the spatial distribution of natural enemies would change if a central vegetational strip was used. Their model depended on predator mobility, and carabids were calculated documented to have a diffusion rate of 10 m^2 per day. They predicted that predators with this diffusion rate would be notably enhanced in a 50 m or 100 m wide field with a central 10 m vegetational strip compared to a monocultural field. This suggests that a vegetational strip could augment carabids by 20 or 40 m into the field. In addition to theoretical models, the author compared carabid abundance in fields with various headland treatments or mid-field habitats to that of control areas in an attempt to verify the extent that vegetation strips augment field populations (*Table 10.2*). In general, these non-crop habitats were shown to increase levels of carabids only at relatively short distances into the field (8–18 m). Hausammann (1996) focused on predatory arthropods other than carabid beetles and found similarly that augmentative effects of weed strips extended only 10–25 m away. Whether non-crop habitats augment carabids and enhance pest control at further distances requires further study. Carabids have been monitored over longer distances (50–200 m) from the non-crop habitat and there is evidence that they penetrate this far into the field (Coombes and Sotherton, 1986; Dennis and Fry, 1992; Kromp and Steinberger, 1992; Thomas *et al.*,

1992; Thomas *et al.*, 2000b). Large-scale studies where fields with non-crop habitats are compared to control fields are underway (Holland, pers. com.) and will provide additional information.

MECHANICAL AND CHEMICAL INPUTS

Non-crop habitats are subjected to fewer disturbances than the field and this may make them attractive to some carabid species. Ideally, mechanical disruption and chemical inputs should be minimal. However, non-crop areas may require maintenance, especially for weed control. While herbicides may eliminate unwanted weeds from hedge bottoms, they also have reduced carabid beetle abundance (Chiverton and Sotherton, 1991). Herbicides combined with the mis-application of fertilizer to the strip or field margin can encourage weed establishment (Rew *et al.,* 1995; Theaker *et al.*, 1995). This occurs because the herbicide creates an open area where annual weeds can easily establish without competition. Moreover, annual weeds respond more to high nitrogen levels compared to perennials, giving them an advantage following fertilizer mis-application (Muller and Garnier, 1990). To keep vegetation manageable, hedge bottoms and grassy/weedy strips are sometimes mowed. Removing the seed heads of annuals has been suggested as an alternative to herbicides for weed control (Fielder and Roebuck, 1987), but mowing is ineffective at reducing the abundance of annual weeds (Smith and MacDonald, 1989). Also, mowing can reduce natural enemy populations (Kajak and Lukasiewicz, 1994) and needs to be carefully timed. Specific recommendations on the types and frequency of mechanical and chemical inputs that may be necessary for the appropriate management of non-crop habitats are in development. Beetle banks, however, appear to be able to maintain their cover of perennial grasses with little management for over a decade (Thomas *et al.*, 2000a).

TIME TO ESTABLISH

The length of time required until a new refuge habitat becomes as equally attractive as existing non-crop habitat is also important. In a review of a long-term set-aside, Corbet (1995) found that mostly multivoltine predators and some parasitoids inhabit newly created refuges. Also, *r*-selected herbivorous insects may colonize rapidly and can damage adjacent crops. Crop pollinators and univoltine polyphagous predators (carabids) usually require more time. Thomas *et al.* (1992) found mostly 'open field carabids' dominating newly sown grassy banks. After a year, the banks conserved carabids that otherwise would not be found overwintering in the field and, during the 2nd and 3rd years, such 'boundary carabids' dominated the banks during winter. Likewise, in Collins *et al.* (1996), a newly sown grass bank developed similar densities of overwintering carabids to established hedgerows within three years of planting. Species richness across the entire field had increased from 14 to 37, four years after the establishment of a perennial strip (*Figure 10.3*) (Carmona and Landis, 1999; Lee, 2000). Similarly, Frank (1996) found that, where weedy strips had been established, the dispersal of carabids into the field was greater during the 2nd year. Also, Lys *et al.* (1994) found higher beetle activity-density within strips one year after strip establishment. Yet, during the following year, activity-densities within strips and the cereal areas in between were similar, suggesting carabids were dispersing into the field. Establishment of natural enemies in non-crop habitats is a gradual process, but most studies have shown a noticeable impact on abundance and species composition within two to three years of implementation. Interestingly, while many studies have addressed the speed at which newly sown vegetation will establish, few have focused on the long-term viability of sown vegetation. A weak positive correlation was detected between beetle bank age and the density of overwintering carabids, but there was no effect of age on carabid diversity (Thomas *et al.*, 2000a). More long-term research is needed to discover whether sown vegetation needs to be replaced and at what point.

ECONOMICS

Wratten (1988) and Thomas *et al.* (1991) discussed the costs of implementing 'beetle banks' on a farm. The initial cost of establishment, as well as the subsequent loss of yield, need to be considered. They considered that the potential of these grassy banks to keep aphid populations low would reduce pesticide inputs and the associated costs, but this has not been adequately demonstrated. Semple *et al.* (1994) reviewed the economic implications of changing hedgerow management on a large arable farm, smaller mixed farm, and dairy farm. The costs depended on amount of land loss, labour and machine usage, the benefits of providing shelter for crops and livestock, and the loss of crop yield as a result of shade. They noted that the more costly practices had the most conservation value. Non-crop habitats may benefit a wide diversity of farmland wildlife by providing additional nesting sites and foraging (Pollard and Relton, 1970; Best, 1983; Holland and Thomas, 1996; Thomas *et al.*, 2000a). In addition, conservation buffers, filter strips and windbreaks established around farms to prevent soil erosion, run-off pollution and sedimentation of waterways (National Research Council, 1993; Boatman, 1998) may create similar valuable habitat. Government programmes in the United States and United Kingdom compensate farmers for the loss of crop income when establishing conservation areas (Baldock, 1993; Dwyer, 1994; USDA, 1997). While specific cost–benefit assessments of non-crop habitat management are still being developed, the current economic incentives make implementing these habitats possible and attractive.

Future directions

As agricultural landscapes become increasingly simplified, implementing non-crop habitats, such as field margins and within-field vegetation strips, for natural enemies becomes increasingly critical. The strikingly high densities of active and overwintering carabids found in non-crop sites emphasizes the importance of these sites in providing undisturbed shelter and resources in otherwise inhospitable agricultural fields. These non-crop habitats have been shown to increase fecundity, populations and species diversity of carabids. Whether such areas may improve pest control in adjacent fields has been rarely demonstrated clearly. Lack of this information is currently limiting their adoption by farmers unless compensation payments are available. Despite the potential benefits, non-crop habitats will not become more common until we understand how to best manage them to maximize pest control.

One of the most common questions is whether these habitats contribute to weed, pest and pathogen problems. A few studies have addressed this issue and found little or no evidence of non-crop habitats contributing to major pest problems (Rew *et al.*, 1996; Frank, 1998; Nentwig *et al.*, 1998). Understanding how enhancing structural diversity influences pest–natural enemy dynamics and having the management techniques available to deter potential pest problems will provide the necessary assurance for farmers. Another aspect of concern is the cost of establishing and maintaining vegetation strips and the resultant potential loss of income. Cost–benefit ratios are not widely available for most crop systems, but increasingly available economic incentives to implement headlands and vegetative buffers will make these practices attractive. We have little knowledge on how these habitats may change species community dynamics in the field in the long term. While this is not a concern for most growers, community change does need to be addressed, as it has consequences for the preservation of biodiversity and future pest suppression. Finally, practices to maximize natural enemy effectiveness and pest control are still being developed. Research needs to determine which vegetation is most suitable for certain fields, which factors govern carabid dispersal into the field and how to optimally space these habitats. The question of spacing has been difficult to address, mainly because most comparative experiments have been conducted at relatively small scales. As more non-crop habitats are established through the incentive of government programmes, larger, farm-scale studies may be required to clarify these pressing questions.

Population models indicate that increasing the proportion of non-crop habitats within a landscape can increase overall diversity of beneficial species, while reducing pest abundance (Bhar and Fahrig, 1998). Studies at the landscape level support the concept that field margins (Thies and Tscharntke, 1999), hedgerows and woodland (Marino and Landis, 1996) increase natural enemy abundance and pest control. Non-crop habitats not only carry tremendous potential to proliferate natural enemy populations, but also are an important component of an integrated biological control strategy (Landis *et al.*, 2000b). For instance, the implementation of conservation practices can enhance the success rates of released classical or augmentative agents. Also, other farming practices that are benign to natural enemies are needed to complement non-crop habitat management such as those discussed by Hance (this volume). While non-crop habitats can mitigate the negative consequences of agricultural disturbances, a reduction of disruptive practices should take place to enable long-term changes. Given the immediate benefits of non-crop habitats to wildlife, natural enemy and soil conservation, it would appear important that growers establish and manage more of these habitats. Non-crop habitat management represents an attractive step for growers to take in developing a more sustainable and ecologically based pest control strategy.

Acknowledgements

We thank Fabian Menalled, Dora Carmona and Foster Purrington for their assistance in many of the studies described here. Comments from Drs John Holland, George Thomas and Sue Thomas have helped to improve this manuscript. Funding provided by USDA SARE grant LNC 95-85 and USDA NRI grant 99-35316-7911 and by the Michigan Agricultural Research Station made much of the work possible.

References

ANDERSEN, J. (1985). Low thigmo-kinesis, a key mechanism in habitat selection by riparian *Bembidion* (Carabidae) species. *Oikos* **44**, 499–505.

ASTERAKI, E.J., HANKS, C.B. AND CLEMENTS, R.O. (1992). The impact of two insecticides on predatory ground beetles (Carabidae) in newly-sown grass. *Annals of Applied Biology* **120**, 25–39.

ASTERAKI, E.J., HANKS, C.B. AND CLEMENTS, R.O. (1995). The influence of different types of grassland field margin on carabid beetle (Coleoptera, Carabidae) communities. *Agriculture, Ecosystems and Environment* **54**, 195–202.

BALDOCK, D. (1993). Incentives for environmentally sound farming in the European community. *Interim report of the Countryside Commission*. Cheltenham, England.

BEDFORD, S.E. AND USHER, M.B. (1994). Distribution of arthropod species across the margins of farm woodlands. *Agriculture, Ecosystems and Environment* **48**, 295–305.

BEST, L.B. (1983). Bird use of fencerows: implications of contemporary fencerow management practices. *Wildlife Society Bulletin* **11**, 343–347.

BHAR, R. AND FAHRIG, L. (1998). Local vs. landscape effects of woody field borders as barriers to crop pest movement. *Conservation Ecology* **2**, 3.

BOAC, J. AND POPISIL, J. (1984). Ground (Coleoptera: Carabidae) and rove (Coleoptera: Staphylinidae) beetles of wheat and cornfields and their interaction with surrounding biotopes. *Ekologiya* **3**, 22–33.

BOATMAN, N.D. (1998). The value of buffer zones for the conservation of biodiversity. *1998 Brighton Crop Protection Conference – Pests & Diseases* **63**, pp 939–950.

BOMMARCO, R. (1998). Reproduction and energy reserves of a predatory carabid beetle relative to agroecosystem complexity. *Ecological Applications* **8**, 846–853.

BRUST, G.E. AND HOUSE, G.J. (1988). Weed seed destruction by arthropods and rodents in low-input soybean agroecosystems. *American Journal of Alternative Agriculture* **3**, 19–25.

BURN, A.J. (1989). Long-term effects on natural enemies of cereal crop pests. In: *Pesticides and non-target invertebrates*. Ed. P.C. Jepson, pp 177–193. Andover: Intercept.

CARMONA, D. AND LANDIS, D.A. (1999). Influence of refuge habitats and cover crops on seasonal activity-density of ground beetles (Coleoptera: Carabidae) in field crops. *Environmental Entomology* **28**, 1145–1153.

CARMONA, D.A., MENALLED, F.D. AND LANDIS, D.A. (1999). *Gryllus pensylvanicus* (Orthoptera: Gryllidae): laboratory weed seed predation and within field activity-density. *Journal of Economic Entomology* **92**, 825–829.

CHEN, Z.Z. AND WILSON, H.R. (1997). The impact of insecticides on ground beetles. *Midwest Biological Control News* **5**, http://www.entomology.wisc.edu/mbcn/field405.html#grd

CHIVERTON, P.A. AND SOTHERTON, N.W. (1991). The effects on beneficial arthropods of the exclusion of herbicides from cereal crop edges. *Journal of Applied Ecology* **28**, 1027–1039.

COLLINS, K.L., WILCOX, A., CHANEY, K. AND BOATMAN, N.D. (1996). Relationships between polyphagous predator density and overwintering habitat within arable field margins and beetle banks. *Brighton Crop Protection Conference – Pests & Diseases*, pp 635–640.

COLLINS, K.L., WILCOX, A., CHANEY, K., BOATMAN, N.D. AND HOLLAND, J.M. (1997). The influence of beetle banks on aphid population predation in winter wheat. *Aspects of Applied Biology* **50**, 341–346.

COOK, R.T., BAILEY, S.E.R. AND MCCROHAN, C.R. (1997). The potential for common weeds to reduce slug damage to winter wheat: laboratory and field studies. *Journal of Applied Ecology* **34**, 79–87.

COOMBES, D.S. AND SOTHERTON, N.W. (1986). The dispersal and distribution of polyphagous predatory Coleoptera in cereals. *Annals of Applied Biology* **108**, 461–474.

CORBET, S.A. (1995). Insects, plants and succession: advantages of long-term set-aside. *Agriculture, Ecosystems and Environment* **53**, 201–217.

CORBETT, A. AND PLANT, R.E. (1993). Role of movement in response of natural enemies to agroecosystem diversification: a theoretical evaluation. *Environmental Entomology* **22**, 519–531.

CURRIE, C.R. AND DIGWEED, S.C. (1996). Effect of substrate depth on predation of larval *Pterostichus adstrictus* Eschscholtz by adults of *P. melanarius* (Illiger) (Coleoptera: Carabidae). *Coleopterists Bulletin* **50**, 291–296.

CURRIE, C.R., SPENCE, J.R. AND NIEMELA, J. (1996). Competition, cannibalism and intraguild predation among ground beetles (Coleoptera: Carabidae): a laboratory study. *Coleopterists Bulletin* **50**, 135–148.

DEN BOER, P.J. AND DEN BOER-DAANJE, W. (1990). On life history tactics in carabid beetles: are there only spring and autumn breeders? In: *The role of ground beetles in ecological and environmental studies.* Ed. N.E. Stork, pp 247–258. Andover: Intercept.

DENNIS, P. AND FRY, G.L.A. (1992). Field margins: can they enhance natural enemy population densities and general arthropod diversity on farmland? *Agriculture, Ecosystems and Environment* **40**, 85–115.

DENNIS, P., THOMAS, M.B AND SOTHERTON, N.W. (1994). Structural features of field boundaries which influence the overwintering densities of beneficial arthropod predators. *Journal of Applied Ecology* **31**, 361–370.

DESENDER, K. (1982). Ecological and faunal studies on Coleoptera in agricultural land II. Hibernation of Carabidae in agro-ecosystems. *Pedobiologia* **23**, 295–303.

DESENDER, K. AND ALDERWEIRELDT, M. (1988). Population dynamics of adult and larval carabid beetles in a maize field and its boundary. *Journal of Applied Entomology* **106**, 13–19.

DUELLI, P., STUDER, M., MARCHAND, I. AND JAKOB, S. (1990). Population movements of arthropods between natural and cultivated areas. *Biological Conservation* **54**, 193–207.

DUFFIELD, S.J. AND AEBISCHER, N.J. (1994). The effect of spatial scale of treatment with dimethoate on invertebrate population recovery in winter wheat. *Journal of Applied Ecology* **31**, 263–281.

DWYER, J. (1994). An overview of UK policy and grant schemes. In: *Field margins: integrating agriculture and conservation, Monograph No. 58.* Ed. N. Boatman, pp 359–365. Farnham: British Crop Protection Council.

EVANS, W.G. (1982). *Oscillatoria* sp. Cyanophyta mat metabolites implicated in habitat selection in *Bembidion obtusidens* (Coleoptera: Carabidae). *Journal of Chemical Ecology* **8**, 671–678.

FAHRIG, L. (1997). Relative effects of habitat loss and fragmentation on population extinction. *Journal of Wildlife Management* **61**, 603–610.

FIELDER, A.G. AND ROEBUCK, J.F. (1987). Weed control at field margins: Experimental techniques and problems. *British Crop Protection Conference – Weeds*, pp 299–305.

FOURNIER, E. AND LOREAU, M. (1999). Effects of newly planted hedges on ground-beetle diversity (Coleoptera, Carabidae) in agricultural landscape. *Ecography* **22**, 87–97.

FRAMPTON, G.K., CILGI, T., FRY, G.L.A. AND WRATTEN, S.D. (1995). Effects of grassy banks on the dispersal of some carabid beetles (Coleoptera: Carabidae) on farmland. *Biological Conservation* **71**, 347–355.

FRANK, T. (1996). Species diversity and activity densities of epigaeic and flower visiting arthropods in sown weed strips and adjacent fields. *IOBC/WPRS Bulletin* **19**, 101–105.

FRANK, T. (1997). Species diversity of ground beetles (Carabidae) in sown weed strips and adjacent fields. *Biological Agriculture and Horticulture* **15**, 297–307.

FRANK, T. (1998). Slug damage and numbers of the slug pests, *Arion lusitanicus* and *Deroceras reticulatum*, in oilseed rape grown beside sown wildflower strips. *Agriculture, Ecosystems and Environment* **67**, 67–78.

FRANK, T. AND BARONE, M. (1999). Short-term field study on weeds reducing slug feeding on oilseed rape. *Journal of Plant Diseases and Protection* **106**, 534–538.

FRANK. T. AND FRIEDLI, J. (1997). Application of metaldehyde against slug damage in oilseed rape along sown wildflower strips. *Mededelingen van de Faculteit Landbouwwetenschappen, Rijksuniversiteit Gent* **62**, 2b.

GRAVESEN, E. AND TOFT, S. (1987). Grass fields as reservoirs for polyphagous predators (Arthropoda) of aphids (Homopt., Aphididae). *Journal of Applied Entomology* **104**, 461–473.

GURR, G.M., VAN EMDEN, H.F. AND WRATTEN, S.D. (1998). Habitat manipulation and natural enemy efficiency: implications for the control of pests. In: *Conservation biological control.* Ed. P. Barbosa, pp 155–183. New York: Academic Press.

GUT, L.J., JOCHUMS, C.E., WESTIGARD, P.H. AND LISS, W.J. (1982). Variations in pear psylla (*Psylla pyricola* Foerster) densities in southern Oregon orchards and its implications. *Acta Horticulterea* **124**, 101–111.

HASSALL, M., HAWTHORNE, A., MAUDSLEY, M., WHITE, P. AND CARDWELL, C. (1992). Effects of headland management on invertebrate communities in cereal fields. *Agriculture, Ecosystems and Environment* **40**, 155–178.

HAUSAMMANN, A. (1996). The effects of sown weed strips on pests and beneficial arthropods in winter wheat fields. *Journal of Plant Diseases and Protection* **103,** 70–81.

HAWTHORNE, A. AND HASSALL, M. (1995). The effect of cereal headland treatments on carabid communities. In: *Arthropod natural enemies in arable land I.* Eds. S. Toft and W. Riedel, pp 185–198. Denmark: Aarhus University Press.

HOLLAND, J. AND FAHRIG, L. (2000). Effect of woody borders on insect density and diversity in crop fields: A landscape-scale analysis. *Agriculture, Ecosystems and Environment* **78**, 115–122.

HOLLAND, J.M. AND THOMAS, S.R. (1996). *Phacelia tanacetifolia* flower strips: their effect on beneficial invertebrates and game bird chick food in an integrated farming system. In: *Arthopod natural enemies in arable land II.* Eds. C.J.H. Booij and L.J.M.F. den Nijs, pp 171–182. Denmark: Aarhus University Press.

HOLLAND, J.M., PERRY, J.N. AND WINDER, L. (1999). The within-field spatial and temporal distribution of arthropods in winter wheat. *Bulletin of Entomological Research* **89**, 499–513.

HOLLAND, J.M., WINDER, L. AND PERRY, J.N. (2000). The impact of dimethoate on the spatial distribution of beneficial arthropods in winter wheat. *Annals of Applied Biology* **136**, 93–105.

HOUSE, G.J. AND DEL ROSARIO ALZUGARA, M. (1989). Influence of cover cropping and no-tillage practices on community composition of soil arthropods in a North Carolina agroecosystem. *Environmental Entomology* **18**, 302–307.

JEPSON, P.C. AND THACKER, J.R.M. (1990). Analysis of the spatial component of pesticide side-effects on non-target invertebrate populations and its relevance to hazard analysis. *Functional Ecology* **4**, 349–355.

JONES, M.G. (1979). The abundance and reproductive activity of common Carabidae in a winter wheat crop. *Ecological Entomology* **4**, 31–43.

JOYCE, K.A., HOLLAND, J.M. AND DONCASTER, C.P. (1999). Influences of hedgerow inter-sections and gaps on the movement of carabid beetles. *Bulletin of Entomological Research* **89**, 523–531.

JULIANO, S.A. (1986). Food limitation of reproduction and survival for populations of *Brachinus* (Coleoptera: Carabidae). *Ecology* **67**, 1036–1045.

KAJAK, A. AND LUKASIEWICZ, J. (1994). Do semi-natural patches enrich crop fields with predatory epigean arthropods. *Agriculture, Ecosystems and Environment* **49**, 149–161.

KAREIVA, P. (1990). Population dynamics in spatially complex environments: theory and data. *Philosophical Transaction of Royal Society of London* **330**, 175–190.

KROMP, B. AND STEINBERGER, K.H. (1992). Grassy field margins and arthropod diversity: a case study on ground beetles and spiders in eastern Austria (Coleoptera: Carabidae; Arachnida: Aranei, Opiliones). *Agriculture, Ecosystems and Environment* **40**, 71–93.

LAGERLOF, J. AND WALLIN, H. (1993). The abundance of arthropods along two field margins with different types of vegetation composition: an experimental study. *Agriculture, Ecosystems and Environment* **43**, 141–154.

LANDIS, D.A., MENALLED, F.D., LEE, J.C., CARMONA, D.M. AND PEREZ-VALDEZ, A. (2000a). Habitat management to enhance biological control in IPM. In: *Emerging technologies for integrated pest management.* Eds. G.G. Kennedy and T.B. Sutton, pp 226–239. Minnesota: APS Press.

LANDIS, D.A., WRATTEN, S.D. AND GURR, G.M. (2000b). Habitat management to conserve natural enemies of arthropod pests in agriculture. *Annual Review of Entomology* **45**, 175–201.

LEE, J.C. (2000). The conservation of ground beetles (Coleoptera: Carabidae) in annual crop systems using refuge habitats. M.Sc. Thesis, Michigan State University, United States.

LEE, J.C., MENALLED, F.D. AND LANDIS, D.A. (2001). Refuge habitats modify impact of insecticide disturbance on carabid beetle communities. *Journal of Applied Ecology* **38**, 472–483.

LESIEWICZ, D.S., VAN DUYN, J.W. AND BRADLEY, J.R., JR. (1982). Foliar carabids of field corn in northeastern North Carolina. *Coleopterists Bulletin* **36**, 162–165.

LETHMAYER, C. (1995). Effects of sown weed strips on pest insects. PhD. Thesis. University of Berne, Switzerland.

LETHMAYER, C., NENTWIG, W. AND FRANK, T. (1997). Effects of weed strips on the occurrence of noxious coleopteran species (Nitidulidae, Chrysomelidae, Curculionidae). *Journal of Plant Diseases and Protection* **104**, 75–92.

LUFF, M.L. (1965). The morphology and microclimate of *Dactylis gomerata* tussocks. *Journal of Ecology* **53**, 771–787.

LYNGBY, J.E. AND NIELSEN, H.B. (1981). The spatial distribution of carabids (Coleoptera: Carabidae) in relation to a shelterbelt: ecologic aspects, plant pests, Denmark. *Entomolgiske Meddelelser* **44**, 133–140.

LYS, J.A. (1994). The positive influence of strip-management on ground beetles in a cereal field: increase, migration and overwintering. In: *Carabid beetles: ecology and evolution.* Ed. K. Desender, pp 451–455. Netherlands: Kluwer Academic Publishers.

LYS, J.A. AND NENTWIG, W. (1991). Surface activity of carabid beetles inhabiting cereal fields: seasonal phenology and the influence of farming operations on five abundant species. *Pedobiologia* **35**, 129–138.

LYS, J.A. AND NENTWIG, W. (1992). Augmentation of beneficial arthropods by strip-management 4: Surface activity, movements and activity density of abundant carabid beetles in a cereal field. *Oecologia* **92**, 373–382.

LYS, J.A. AND NENTWIG, W. (1994). Improvement of the overwintering sites for Carabidae, Staphylinidae and Araneae by strip-management in a cereal field. *Pedobiologia* **38**, 238–242.

LYS, J.A., ZIMMERMANN, M. AND NENTWIG, W. (1994). Increase in activity density and species number of carabid beetles in cereals as a result of strip-management. *Entomologia Experimentalis et Applicata* **73**, 1–9.

MARINO, P.C. AND LANDIS, D.A. (1996). Effect of landscape structure on parasitoid diversity and parasitism in agroecosystems. *Ecological Applications* **6**, 276–284.

MARINO, P.C., GROSS, K.L. AND LANDIS, D.A. (1997). Weed seed loss due to predation in Michigan maize fields. *Agriculture, Ecosystems and Environment* **66**, 189–196.

MAUREMOOTO, J.R., WRATTEN, S.D., WORNER, S.P. AND FRY, G.L.A. (1995). Permeability of hedgerows to predatory carabid beetles. *Agriculture, Ecosystems and Environment* **52**, 141–148.

MENALLED, F.D., LEE, J.C. AND LANDIS, D.A. Herbaceous filter strips in agroecosystems: implications for carabid beetle conservation and invertebrate weed seed predation. *Great Lakes Entomologist* (in press).

MULLER, B. AND GARNIER, E. (1990). Components of relative growth rate and sensitivity to nitrogen availability in annual and perennial species of *Bromus*. *Oecologia* **84**, 513–518.

NATIONAL RESEARCH COUNCIL. (1993). *Soil and water quality: an agenda for agriculture.* Washington, D.C.: National Academy Press.

NELEMANS, M.N.E. (1988). Surface activity and growth of larvae of *Nebria brevicollis* (F.) (Coleoptera, Carabidae). *Netherlands Journal of Zoology* **38**, 74–95.

NENTWIG, W., FRANK, T. AND LETHMAYER, C. (1998). Sown weed strips: artificial ecological compensation areas as an important tool in conservation biological control. In: *Conservation biological control*. Ed. P. Barbosa, pp 133–153. New York: Academic Press.

PEARSON, A.W. (1990). The management of research and development. In: *Technology and management*. Ed. R. Wild, pp 28–41. New York: Nichols Publications.

PFIFFNER, L. AND LUKA, H. (2000). Overwintering of arthropods in soils of arable fields and adjacent semi-natural habitats. *Agriculture, Ecosystems and Environment* **78**, 215–222.

POLLARD, E. AND RELTON, J. (1970). Hedges: V. A study of small mammals in hedges and uncultivated fields. *Journal of Applied Ecology* **7**, 549–557.

POVEY, F.D., SMITH, H. AND WATT, T.A. (1993). Predation of annual grass weed seeds in arable field margins. *Annals of Applied Biology* **122**, 323–328.

PULLIAM, R.H. (1988). Sources, sinks, and population regulation. *American Naturalist* **132**, 652–661.

RANDS, M.W.R. (1985). Pesticide use on cereals and the survival of grey partridge chicks: a field experiment. *Journal of Applied Ecology* **22**, 49–54.

REW, L.J., FROUD-WILLIAMS, R.J. AND BOATMAN, N.D. (1995). The effect of nitrogen, plant density and competition between *Bromus sterilis* and three perennial grasses: the implications for boundary strip management. *Weed Research* **35**, 363–368.

REW, L.J., FROUD-WILLIAMS, R.J. AND BOATMAN, N.D. (1996). Dispersal of *Bromus sterilis* and *Anthriscus sylvestris* seed within arable field margins. *Agriculture, Ecosystems and Environment* **59**, 107–114.

RIVARD, I. (1966). Ground beetles (Coleoptera: Carabidae) in relation to agricultural crops. *Canadian Entomologist* **98**, 189–195.

SEMPLE, D.A., BISHOP, E.C. AND MORRIS, J. (1994). An economic analysis of farm hedgerow management. In: *Field margins: integrating agriculture and conservation, BCPC Monograph No. 58*. Ed. N. Boatman, pp 161–166. Farnham: British Crop Protection Council.

SMITH, H. AND MACDONALD, D.W. (1989). Secondary succession on extended arable field margins: its manipulation for wildlife benefit and weed control. *Brighton Crop Protection Conference – Weeds*, pp 1063–1069.

SMITH, H., FEBER, R.E. AND MACDONALD, D.W. (1994). The role of wild flower seed mixtures in field margin restoration. In: *Field margins: integrating agriculture and conservation, BCPC Monograph No. 58*. Ed. N. Boatman, pp 289–294. Farnham: British Crop Protection Council.

SOTHERTON, N.W. (1984). The distribution and abundance of predatory arthropods overwintering on farmland. *Annals of Applied Biology* **105**, 423–429.

SOTHERTON, N.W. (1985). The distribution and abundance of predatory Coleoptera overwintering in field boundaries. *Annals of Applied Biology* **106**, 17–21.

SPEIGHT, M.R. AND LAWTON, J.H. (1976). The influence of weed-cover on the mortality imposed on artificial prey by predatory ground beetles in cereal fields. *Oecologia* **23**, 211–233.

THEAKER, A.J., BOATMAN, N.D. AND FROUD-WILLIAMS, R.J. (1995). The effect of nitrogen fertilizer on the growth of *Bromus sterilis* in field boundary vegetation. *Agriculture, Ecosystems and Environment* **53**, 185–192.

THIES, C. AND TSCHARNTKE, T. (1999). Landscape structure and biological control in agroecosystems. *Science* **285,** 893–895.

THOMAS, C.F.G. AND MARSHALL, E.J.P. (1999). Arthropod abundance and diversity in differently vegetated margins of arable fields. *Agriculture, Ecosystems and Environment* **72**, 131–144.

THOMAS, C.F.G., PARKINSON, L. AND MARSHALL, E.J.P. (1998). Isolating the components of activity-density for the carabids beetle *Pterostichus melanarius* in farmland. *Oecologia* **116**, 103–112.

THOMAS, M.B. (1990). The role of man-made grassy habitats in enhancing carabid populations in arable land. In: *The role of ground beetles in ecological and environmental studies*. Ed. N.E. Stork, pp 77–85. Andover: Intercept.

THOMAS, M.B., WRATTEN, S.D. AND SOTHERTON, N.W. (1991). Creation of island habitats in farmland to manipulate populations of beneficial arthropods: predator densities and emigration. *Journal of Applied Ecology* **28**, 906–917.

THOMAS, M.B., WRATTEN, S.D. AND SOTHERTON, N.W. (1992). Creation of 'island' habitats in farmland to manipulate populations of beneficial arthropods: predator densities and species composition. *Journal of Applied Ecology* **29**, 524–531.

THOMAS, S.R., GOULSON, D. AND HOLLAND, J.M. (2000a). The contribution of beetle banks to farmland biodiversity. *Aspects of Applied Biology* **62**, 31–38.

THOMAS, S.R., GOULSON, D. AND HOLLAND, J.M. (2000b). Spatial and temporal distributions of predatory Carabidae in a winter wheat field. *Aspects of Applied Biology* **62**, 55–60.

USDA (1997). Buffers: common-sense conservation. Washington DC: US Department of Agriculture.

VAN DIJK, TH.S. (1994). On the relationship between food, reproduction and survival of two carabid beetles: *Calathus melanocephalus* and *Pterostichus versicolor*. *Ecological Entomology* **19**, 263–270.

VARCHOLA, J.M. AND DUNN, J.P. (1999). Changes in ground beetle (Coleoptera: Carabidae) assemblages in farming systems bordered by complex or simple roadside vegetation. *Agriculture, Ecosystems and Environment* **73**, 41–49.

VITANZA, S., SORENSON, C.E. AND BAILEY, W.C. (1996). Impact of warm season grass strips on arthropod populations in Missouri cotton fields. *Proceedings of Beltwide Cotton Conference, Memphis TN* **1**, pp 174–176.

WALLIN, H. AND EKBOM, B.S. (1988). Movements of carabid beetles (Coleoptera: Carabidae) inhabiting cereal fields: a field tracing study. *Oecologia* **77**, 39–43.

WALLIN, H., CHIVERTON, P.A., EKBOM, B.S. AND BORG, A. (1992). Diet, fecundity and egg size in some polyphagous predatory carabid beetles. *Entomologia Experimentalis et Applicata* **65**, 129–140.

WISSINGER, S.A. (1997). Cyclic colonization in predictably ephemeral habitats: a template for biological control in annual crop systems. *Biological Control* **10**, 4–15.

WRATTEN, S.D. (1988). The role of field boundaries as reservoirs of beneficial insects. In: *Environmental management in agriculture: European perspectives*. Ed. J.R. Park, pp 144–150. London: Belhaven Press.

WRATTEN, S.D. AND THOMAS, C.F.G. (1990). Farm-scale spatial dynamics of predators and parasitoids in agricultural landscapes. In: *Species dispersal in agricultural habitats*. Eds. R.G.H. Bunce and D.C. Howard, pp 219–237. London: Bellhaven Press.

ZANGGER, A., LYS, J.A. AND Nentwig, W. (1994). Increasing the availability of food and the reproduction of *Poecilus cupreus* in a cereal field by strip-management. *Entomologia Experimentalis et Applicata* **71**, 11–120.

11
The Spatial Distribution of Carabid Beetles in Agricultural Landscapes

C.F. GEORGE THOMAS[1], JOHN M. HOLLAND[2] AND NICOLA J. BROWN[3]

[1]Seale-Hayne Faculty of Land, Food and Leisure, University of Plymouth, Newton Abbot, Devon TQ12 6NQ, UK, [2]The Game Conservancy Trust, Fordingbridge, Hampshire SP6 1EF, UK and [3]IACR Long Ashton Research Station, Department of Agricultural Sciences, Long Ashton, Bristol BS41 9AF, UK

Introduction

This chapter reviews the spatial distribution of carabids: that is, how population density varies in space and time, among and within habitats. Studying patterns of population aggregations can provide important and useful insights into the ecology of a species. By identifying the location of patches of high and low population density in relation to the distribution of biotic and abiotic factors, it may be possible to identify the underlying causes of population distribution patterns. This may facilitate the management of natural, semi-natural and agricultural habitats for biodiversity, conservation or pest control on the farm. Similarly, observations of the spatial and temporal dynamics of distribution patterns can help us to understand the movements of individuals and populations between habitats in response to seasonal factors and other natural or man-made disturbances. Observations of carabid distribution patterns can also help us to understand how different species share resources in the same space or react to environmental factors. The practical application of this knowledge is of central importance to the development of sustainable agriculture through the design of integrated pest management regimes. In particular, the rapidly developing field of precision agriculture is dependent on knowing the spatial distribution of target areas. An ability to describe the two-dimensional distribution of pests and their predators and parasitoids, even in general terms, will be essential if insecticides are to be applied with precision techniques that minimize their detrimental effects on beneficial insect populations.

In many regions of the world, mankind has cleared and cultivated extensive areas of native forests and grasslands over the last few thousand years. In some parts of the New World, these changes to the landscape have occurred rapidly over the last few

---

Abbreviations: SADIE, Spatial Analysis by Distance IndiciEs

The Agroecology of Carabid Beetles

centuries (Rackham, 1995). Regrettably, the process of creating fragmented agricultural landscapes continues in parts of the world exemplified by the Amazon rain forest. Viewed from an aircraft, the result is a landscape mosaic with patches and networks evident at several spatial scales. In some areas, for example much of lowland Britain, a sea of irregular fields surrounds small islands of woodland and other habitats. In other regions, for example Finland, the inverse is apparent, with isolated patches of farmland existing as islands in an extensive sea of forest. In yet other parts of the world, tracts of regular fields appear to stretch from horizon to horizon. Superimposed on this multiple-scaled and shifting mosaic are further networks of rivers, roads and railways; and the whole overlays patterns of solid and drift geology, with its related topography and hydrology of watersheds, drainage, aspect, altitude and soil types. Within the cultivated elements, each field has its own history of tillage, planting and agrochemical treatments. Much of the farmed areas comprise fields of annual crops. In these fields, the soil environment, at and below the surface, changes throughout the year in terms of cover and disturbance in ways that affect microclimate, food availability, and the risks and resources to which different carabid life stages are exposed. In areas where livestock is a major concern, there may be substantial areas under permanent grassland, disturbed only by grazing. However, from the perspective of carabids, even at the within-field scale, beetles' preferences for specific environmental requirements and dietary needs will render different areas more or less favourable. This heterogeneity at all possible spatial scales in the landscape ultimately leads to variation in carabid reproductive, survival, and mortality rates from a large number of possible factors, thus creating spatial variation in population density.

It is in this context of spatial fragmentation and temporal disturbance that carabid beetles are distributed and ecological studies are conducted. The population density at any location in such a heterogeneous landscape is a function of local birth and death rates; the intrinsic ability of species to disperse and move within and between areas; and the accessibility of sites to dispersing individuals. Consequently, in order to understand the spatial distribution of carabids, it is important to identify the spatial and temporal distribution of risks and resources, and understand how individual species respond to them.

There are also important problems for scientific methodology and experimental design to be considered. What we observe as a spatial pattern in the field is a discontinuity in the distribution of insects, usually as a snapshot in time, overlaid on a discontinuous environment. It is a truism worth emphasizing that our ability to resolve discontinuities in space depends on the spatial scale of our observations. Similarly, temporal discontinuities, or the dynamics of the distributions, are only revealed by a series of snapshot observations, the frequency of which determines the level of resolution available to us. The spatial scale and intensity at which sampling can be physically conducted, therefore, limit the spatial resolution at which data can be analysed. Furthermore, since no two places are the same, by definition, true replication of sites is impossible to achieve. The factors that might affect any of the quantities – birth, death and dispersal – and the interactions between them, seem innumerable and, for the most part, relationships between variables are likely to be non-linear, with many stochastic influences. Furthermore, many carabid beetles of interest have more or less annual life cycles. Thus, very few successive generations can be monitored within the time-scale of funding for an experimental study or series of observations, making it difficult to differentiate between natural influences on population dynamic processes and the influence of agricultural disturbance. Our ability to make accurate, quantitative observations of carabid populations in the field is thus compromised; not only by the necessity to limit the scale and intensity of sampling, but also by the use of indirect sampling methods; namely the unreliable and inconsistent (but nevertheless indispensable) pitfall trap. Finally, analysis and interpretation of data (often pooled into a taxon of 'total carabids') are inevitably problematical; cause and effect are not always readily apparent when many relevant variables are either unknown or unmeasured.

Many of the above difficulties could be resolved if the movement of individual beetles could be monitored more frequently. Marking and recapture can provide some information on dispersal, but only between a limited number of recapture points and only over a relatively short distance (> 100–200 m), owing to the logistics of operating a trapping grid. Harmonic radar was initially used with some success (Mascanzoni and Wallin, 1986; Wallin and Ekbom, 1988); however, the detection range is low (1–10 m) and individuals cannot be recognized. More recently, advances in radio telemetry have produced equipment small enough to fit on the back of a large beetle (Riecken and Raths, 1996). This was used to successfully track the movement of *Carabus coriaceus* individuals through a variety of natural habitats. The study demonstrated that information provided by pitfall trapping can give misleading results about real habitat preferences.

Given these difficulties, it may seem impossible to be able to make any precise predictions about the distribution and abundance of particular carabid species. Certain predictions can be made at some spatial scales using indices of habitat disturbance and productivity (Eyre, 1994), and multivariate approaches to characterize community structure in particular habitats overcome many replication problems (Sanderson, 1994). These and other approaches are valuable methods that can go some way towards identifying species' basic ecological requirements and determining the probability of a species presence in, or absence from, a particular site. They are more useful at larger geographical scales and, as such, they are dealt with more fully elsewhere (Luff, this volume). However, we are still largely ignorant of much of the ecology and biology of many of the species that we deal with (Luff, 1987; Lövei and Sunderland, 1996). Much needs to be learnt, especially about those factors that act at smaller spatial and temporal scales governing spatial distributions of population density within an occupied habitat and within a single generation.

This review will be illustrative rather than exhaustive. We hope to extract some generalizations from the literature rather than to detail all of the evidence. Structuring the review around the theme of scale and landscape, we begin by considering spatial distributions at the larger scale in some of the more stable habitats found in agricultural landscapes, such as forest and woodland. We then consider the interstitial habitats of hedgerows and field boundaries that are derived from the larger elements or are related to them in form and structure. We then consider the way in which spatial distributions of some species shift between the interstitial habitats and cropped fields. At the smallest scale, we consider spatial distributions within fields because most field studies are conducted at this scale; it is within this section that we consider the treatment of data and some of the potential causes of aggregation that might operate at this scale and above. Spatial distributions at the farm-scale and movement between fields leads to the role of field boundaries as barriers to movement and the metapopulation concept of carabids in farmland. Finally, we briefly discuss future directions for research, including the potential role of population genetics in answering some of the seemingly intractable questions we pose.

Stable natural habitats: heathland, forests and woodland

The classic, long-term studies by Den Boer (1990a) on the population dynamics of carabids, mainly in heathland habitats, and the analysis of long-term dynamics from historical data sets (Desender *et al.*, 1994), both indicate long-term cycling in population size. Peaks and troughs of abundance seem to occur with a frequency of a few years in many populations. Compared with agricultural habitats, especially arable fields, those of heathland, forest and woodland represent relatively undisturbed habitats. Daily and annual fluctuations of the microclimate on forest floors are generally small compared with open field environments (Geiger, 1957). Studying carabids in such stable environments can provide useful information on the underlying factors that influence their distributions, their spatial stability, and the scale of year-to-year fluctuations of population size in the absence of any anthropogenic influence.

CARABID DISTRIBUTION PATTERNS IN WOODLAND

Given the heterogeneous nature of the landscape, it is not surprising that most carabid distributions are aggregated, even when studied at relatively small spatial scales. For example, Niemelä *et al.* (1992) analysed two-dimensional data of carabid distributions based on a 12 × 25 grid of traps over 1.3 ha on forest floor, which included five habitat types. They found that the small-scale distributions of the most abundant species among the habitat types were best explained by active micro-habitat selection by carabids. All species except one had highly aggregated distributions. Some species were associated (although not exclusively) with certain micro-habitats; two species were negatively associated with the presence of ants; only one species (*Notiophilus biguttatus*) was evenly distributed over the grid. However, at larger scales, comparisons with other data from similar habitats at different locations within the same region seemed to indicate that local population sizes did not fluctuate in synchrony. Regional-scale influences appeared to affect local dynamics differently.

In another forest floor study, Brunsting (1981) investigated several populations of *Pterostichus oblongopunctatus* from two areas over a number of years. A number of soil and vegetation factors were also measured. The distribution patterns within populations, and at the landscape level, changed little between years. In common with the findings of Niemelä *et al.* (1992), Brunsting (1981) found fluctuations in population size within an area ran in parallel, while between different areas in the landscape, population density fluctuations were unsynchronized.

In a study conducted in three forest stands a few kilometres apart, Koivula *et al.* (1999) examined carabid assemblages and distributions in plots with and without leaf litter. They found variation due to litter effects to be significant, but less than that due, to the different forest stands and years. There was some temporal stability with stands remaining ‘species rich’ or ‘species poor’ over the course of four years. Heterogeneous distributions and fluctuating population size are not universal, however. Loreau and Nolf (1993), studying *Abax parallelepipedus* in a beech wood, found that population size remained fairly constant over a 13-year period. Using a grid of traps, the spatial distribution of *A. parallelepipedus* was found to be only weakly aggregated, and then only at the smallest scales. The low level of aggregation could not be correlated with small-scale environmental heterogeneity in terms of temperature or food availability, and when aggregation was analysed over larger sampling units, the population tended to be more randomly distributed.

At larger spatial scales, clues as to the driving factors behind heterogeneous distributions can become more apparent. Baguette (1987, 1993) found that the distributions of carabids sampled from over a large area of southern Belgium were mostly related to the acidity, organic content and water-holding capacity of the soils. Baguette (1987) suggests that pH possibly acts through its indirect effects on the degradation and availability of soil organic content, which in turn affects prey availability to carabids. Similarly, Vansteenwegen (1987), in a factorial analysis of carabid species from a number of sites, found that clusters separated mainly according to damp and dry sites. It is possible, however, to describe the requirements of a single species in more detail. Comparisons of a number of populations in the study by Brunsting (1981) mentioned above, revealed that significantly more *P. oblongopunctatus* were caught than expected if the following environmental conditions were fulfilled: cover of tree layer > 50%; cover of shrub layer > 5%; cover of herb <= 50%; mosses absent; litter layer < 4 cm; soil loam (10% < lutum < 25%); drainage bad to moderate.

THE INFLUENCE OF WOODLAND IN ARABLE AREAS

Some species are common both to forest or woodland and to field habitat types, although the composition of species lists can be dependent on the scale at which observations are made, or where and how populations are sampled. Farm woodlands are, by definition, intimately associated with agricultural habitats. As such, they have been found to contain a mixture of woodland species and field species (Usher *et al.*, 1993). Thus, many carabids of the field seem to find their way into farm woodland by choice or accident. These habitats may therefore have an important function as passive refuges. At any time of the year, those members of the population that happen to wander into neighbouring woodland will be protected from any direct, adverse effects of agricultural operations. Sampling in transects through farm woodland and into arable fields has shown that only four or five species were not found in all habitat types (Bedford and Usher, 1994).

In a slightly larger scale study, Mader and Muhlenberg (1981) monitored the distribution of carabids in a small forest fragment, surrounding arable fields, and the edge and centre of a large spruce forest 500 m away. Pitfall traps collected most individuals in the arable field. Approximately half as many were caught at the forest edge, in the forest fragment, while about a fifth as many were trapped in the forest centre. Of 41 species in total, 19, 23, 16 and 20 were found in the above habitats respectively, and in contrast to the findings of Bedford and Usher (1994) cited above, only four of these species were found in all four habitats. Thus, forest centres appear to support low population density but high diversity, while arable fields support higher population density and low diversity. Forest or woodland edges and small isolated fragments of forest in arable land appear to support intermediate population densities. However, although Bedford and Usher (1994) found relatively few carabid species specific to either woodland or field habitats, they did find, in common with Mader and Muhlenberg (1981), that species richness was generally higher at the ecotone than in the centre of either woodland or field. High diversity in the ecotone between habitats is simply a result of the co-occurrence of members of the field and forest groups raising the total species number. However, the extent to which field species penetrate woodland to affect species richness falls off rapidly, with little effect beyond 5 m into the wood. The inverse effect of woodland species on diversity in the field could not be evaluated, due to limited sampling in the field. Consequently, woodland shape, in terms of the ratio of edge to interior, can also be an important factor that can affect the spatial distribution of some carabid species.

Burel (1989), in her study of forest carabids in the bocage landscapes of north-western France, defined three groups of forest species. These were forest core species, which were limited to the forest and the first 50 m adjacent to it; forest peninsular species, whose abundance declines to zero at 500 m to 700 m from the forest edge; and forest corridor species. The latter are abundant near to forests, decline rapidly with distance, but remain constant within the network of corridor habitats comprising hedgerows and double-banked lanes, which typify the bocage landscape. Niemelä and Halme (1992), investigating carabids on the Åland islands, were able to classify the most frequently caught species as generalists, field and pasture species, open habitat species or forest species. As with many other studies, most species occurred to some extent in all habitats. In general, abundance and diversity were highest in fields and pastures, and lowest in forests. A similar study at the landscape scale (Holland and Fahrig, 2000) has also shown that the presence of woody borders to crop fields can enhance the diversity of invertebrate groups other than carabids. However, the effects of woodland on the enhancement of population density in adjacent crop fields could not be demonstrated in their study.

Forests and woodlands can therefore be important components of agricultural landscapes, and may help define the spatial distribution of certain species. At the large scale, soil factors and other regional-scale effects on population dynamics may determine the presence or absence of ground beetles. At smaller scales, spatial distributions are likely to be determined by micro-habitat selection where carabids aggregate in areas where factors that affect micro-climate and food availability, etc. meet species' requirements (see below and Holland, this volume). At 'rich' and 'poor' sites, there is evidence of relatively stable carabid assemblages over time. Similarly, although long-term fluctuations in population size can be observed, some species in stable habitats have relatively stable population size and even distributions. Edge effects and the shape of farm woodland can influence carabid diversity, and many carabid species of woodland are also found in agricultural land and hedgerow habitats.

Semi-stable, semi-natural habitats: hedges, field boundaries and other interstitial habitats

The similarity between woodland edge habitats and hedgerows has long been recognized (Pollard *et al.*, 1974). Although hedges are characteristic of specific regional landscapes, most fields are usually bordered by some form of semi-natural habitat that is more stable than the cropped areas of fields, for example grassy margins. Some form of management is normally applied to these areas (Lee *et al.* this volume; Maudsley *et al.*, 2000) but, in general, they are much less disturbed than the adjacent cropped areas. Consequently, in recent decades a lot of research has focused on the distribution of carabids in these interstitial (*sensu* Elton, 1966) or network habitats in agricultural landscapes. Moreover, semi-natural habitats have been recognized as potentially important refuges for species that may have a role in the suppression of crop pests. Attempts to enhance the population size of these species, and carabid diversity in general, have therefore centred on increasing the area and distribution of refuge habitats in farmland. Approaches include the reintroduction of hedges (Fournier and Loreau, 1999), the construction of beetle banks (Thomas *et al.*, 1992b), the introduction or widening of field margins, often sown with grasses and wild flowers (Marshall, 1988; Frank, 1997; Telfer *et al.*, 2000) and similar habitats as strips adjacent to or within fields (Nentwig, 1989; Lys and Nentwig, 1992; Rodenhouse *et al.*, 1992; Lys, 1994; Wade French and Elliott, 1999a). While most of these are linear habitats, other adjacent refuge habitats, including crops, have also attracted interest (Kajak and Lukasiewicz, 1994; Carmona and Landis, 1999; Wade French and Elliott, 1999b).

Interstitial habitats may influence the spatial distributions of different carabid species for a number of reasons. Hedgerows and other margin habitats may contain forest and field species and are frequently the places in farmland where diversity is found to be the highest (Kromp and Steinberger, 1992). Although some forest species may wander into cropped areas to a limited extent (Burel, 1989), some species are found exclusively in hedgerows (Fournier and Loreau, 1999) or show a marked preference for one field boundary habitat over another. In a study comparing hedged with fenced field margins adjacent to grassland, Asteraki *et al.* (1995) found nearly 30% of species, including *Leistus fulvibarbis* and *Trechus obtusus*, exclusively in the hedged margins. Compelling evidence that *Carabus coriaceus* prefers forest edges and ecotones was found by Riecken and Raths (1996) using radio telemetry. Beetle behaviour and habitat use was measured by attaching radio transmitters to their elytra and monitoring their movements in a large forest area and in a small river valley with more diverse habitat. Some individuals used strips of alder forest as a guideline for dispersal, while the meadow was shown to be unsuitable; beetles released in a meadow moved into the nearby forest.

The spatial distribution of some species within a given landscape is therefore likely to be defined to some extent by the pattern of interstitial habitats. However, no studies have been conducted on a large enough scale and at a resolution fine enough to map and determine the causes of continuities and discontinuities in the distributions of specific carabid species. Although there are many plausible arguments for the function of interstitial habitats as corridors for insect movement, there is very little empirical evidence, with the exception of that for *C. coriaceus* obtained using radio telemetry (Riecken and Raths, 1996). However, where large-scale studies have been conducted (Burel, 1989; Petit and Burel, 1993; Petit, 1994), some species do appear to be able to use hedgerows as corridors, e.g. *Abax parallelepipedus* and *Carabus granulatus*, and can be classified as forest corridor species. Higher movement rates and lower densities of *A. parallelepipedus* within hedgerows compared with woods or nodes at hedgerow intersections indicate that this species at least uses hedgerows as corridors for movement between woodland fragments (Petit, 1994).

HABITATS FOR OVERWINTERING

The importance of hedgerows to forest species has been mentioned above. However, interstitial habitats may also be important overwintering and aestivation sites for some field species. Evidence that hedges were of less importance to those field species that might have a predatory impact on field pests led Desender (1982) to investigate the role of grassy margins as hibernation sites for field species of spring breeding (adult overwintering) carabids. Sampling these areas indicated that they did, indeed, contain high densities of carabids (up to 900 m^{-2}) during the winter. Grassy margins around cropped fields, and fields of more permanent pasture, appeared to have characteristic faunas related to adjacent field populations. The most abundant species in the edges of cropped fields were *Bembidion lampros, Agonum dorsale, Pterostichus vernalis* and *Agonum muelleri*; while around pastures *Pterostichus vernalis, Bembidion properans, Clivina fossor, Pterostichus strenuus, Bembidion lampros* and *Amara aenea* were most abundant.

Sotherton (1984) compared a wider range of habitats, including field boundaries, woodland, and a variety of cropped fields – cereal and turnip stubbles, winter cereals, permanent pasture and temporary grass leys. He found the spatial distribution of overwintering predators (including Staphylinidae and Dermaptera) among the different habitats to vary only by a factor of about two when the mean density was adjusted to take account of the proportion of each habitat on the farm. Field boundaries, winter sown cereals, and established leys held approximately 1.4 predators per square metre; cereal stubbles and first year grass leys held approximately 0.6 predators per square metre; woodland and stubble turnips were intermediate at about 0.8 predators per square metre. However, among the carabids, different species were differently distributed. *Agonum dorsale* and *Demetrias atricapillus* overwintered almost entirely in field boundaries; *Amara aenea* were found predominantly in established leys, *Bembidion lampros* in field boundaries and established leys, *Bembidion obtusum* in field boundaries, established leys and winter cereals, and *Pterostichus melanarius* were predominantly in winter cereals. Among the carabids, between about 70% and 90% of the total catch of *D. atricapillus, A. dorsale, B. lampros, Amara familiaris, A. plebeja* and *Asaphidon flavipes* came from the field boundary habitats. Interestingly, Sotherton's (1984) study found *Notiophilus biguttatus* at similar densities among all overwintering habitats (although slightly higher in winter cereals and first year leys). This species was also found to be evenly distributed among micro-habitats on forest floor (Niemelä *et al.*, 1992) and among woodland, edge and field habitats (Bedford and Usher, 1994). This species, therefore, appears to have non-specific habitat requirements.

In a similar study conducted in Norway (Andersen, 1997), in which grass and ploughed fields were sampled during winter, together with their associated field boundaries, the mean density of overwintering adult carabids in ploughed fields and their margins were about 8 and 120 beetles m^{-2} respectively. The spatial distribution of overwintering carabids between grass fields and their margins was less dramatic, but still significantly different with about 23 beetles m^{-2} in the field and about 63 m^{-2} in the margins.

Carabids are known to exhibit preferences for a limited range of temperature, humidity, light levels and other abiotic factors when exposed to gradients. Natural variation of these factors in the field is therefore likely to explain carabid spatial distributions to some extent. However, the large number of interacting factors means that it is extremely difficult to devise experiments that will allow a conclusive elucidation of species preferences in the field (Thiele, 1979). Nevertheless, the structural, biotic and abiotic features of field margin habitats vary along their length, and this may influence the spatial distribution of overwintering carabids in them. Desender (1982) found that carabid distributions in grassy margins were significantly aggregated, and clusters of high density were generally associated with areas that had a well-developed sod layer. It is normally thought that overwintering adult carabids enter a period of quiescence in which feeding does not occur. However, Thomas *et al.* (1992a), studying the carabid *D. atricapillus* overwintering in beetle banks, found evidence, from an analysis of their distributions and gut contents, to suggest that food availability was important. Correlations were also found between the abundance of *D. atricapillus* and the presence of tussock-forming grasses (e.g. *Dactylis glomerata* and *Holcus lanatus*). Lower densities of the carabid were found in those parts of the beetle bank with mat-forming grasses. The buffering effect of tussock-forming grasses on temperature fluctuations was considered a likely cue for site selection by the beetles. Structural features are also important since a correlation was found between vegetation biomass and carabid numbers. The survival of overwintering *D. atricapillus* has also been found to be significantly higher in grass tussocks compared with bare earth (Dennis *et al.*, 1994).

Although the studies mentioned above imply that mat-forming grasses are not of particular benefit to overwintering survival of carabids, other studies have indicated that they can be favoured sites at other times of the year, especially for carabid larvae. Lagerlöf and Wallin (1993) investigated the autumn abundance of carabids in field margins with different vegetation composition. In four types of margin – naturally florally diverse, couch grass dominated, leguminous dominated, and ploughed – nearly twice as many adult carabids were trapped in pitfall traps in the florally diverse areas, compared with all other treatments. However, of the carabid larvae, more than twice as many were caught in the couch grass dominated margin as in the other plots. By using suction sampling, which collects more of the smaller carabid species than pitfall traps, mean densities were found to increase from lowest in the ploughed plots, through clover dominated and florally diverse, to highest in the couch grass dominated plots.

In addition to their role as overwintering sites for some species, field boundaries can also be used as aestivation sites, thus influencing the spatial distributions of carabid species at different times of the year. In contrast to most other farmland carabids, *N. brevicollis* has a bimodal period of activity. Adults appear in May and June. After a brief spell of activity following emergence, the adults enter a period of aestivation, until they emerge again in September. Reproduction occurs during the autumn (Nelemans, 1987, 1988; Nelemans *et al.*, 1989). Spatial distributions of *N. brevicollis* populations in arable fields have been shown to aggregate in a hedgerow habitat between fields in June, with activity declining to zero by early July. Aestivation appears to occur exclusively in these habitats, with subsequent migration into the cropped area completed by late September (Fernández García *et al.*, 2000).

Seasonal movement between field boundaries and field centres

Arable fields self-evidently do not provide a constant habitat throughout the year. From a tilled field of bare soil, the cover provided by annual crops gradually increases to maturity. In many crops, especially cereals, ripening and senescence create a more open cover, which is then entirely removed at harvest. An aftermath of trash may cover the soil for a period before the field is tilled again. At the soil surface, the microclimate and food availability clearly undergo great changes throughout the season. To exploit this temporary habitat and its resources, those species in field boundaries and adjacent habitats need to shift their spatial distributions by dispersing into and out of fields at the appropriate times of year.

Many carabids are able to walk at considerable speeds, for example *Pterostichus niger* has been observed travelling at up to 20 m h^{-1} in a cereal field (Wallin and Ekbom, 1988). Although some species found in arable fields are incapable of flight, for example *P. madidus*, more than 80% of carabid species are demonstrably or probably capable of flight, or have macropterous forms in the population (Lindroth, 1992). However, although their powers of flight are generally weak, such mobility (either by flying or walking) means that the spatial distributions of carabids can shift rapidly at the scale of fields and their boundaries. Flight activity by carabids in the genera *Harpalus*, *Pterostichus*, *Agonum*, *Bembidion* and *Bradycellus* has been shown for a number of species colonizing potato fields in Canada (Boiteau, 1983).

Wratten and Thomas (1990) list five main spatially dynamic processes of relevance to developing integrated control programmes. These are: seasonal movements between crops and non-crop habitats; movement between phenologically asynchronous crops; colonization of new habitats; recolonization of areas previously depopulated by insecticides; and aggregative movement to areas of high prey density. The detection of spatially dynamic processes is a difficult feat to accomplish in the field. Spatial distributions need to be sampled frequently in order to achieve high temporal and spatial resolution. Various methods and techniques have been used to quantify these dynamics, including radar tracing, mark–release–recapture, time series analysis of data from transects of pitfall traps, and directional pitfall trap studies (Coombes and Sotherton, 1986; Wallin, 1986; Wallin and Ekbom, 1988; Dennis and Fry, 1992; Kromp and Nitzlader, 1995). Investigations that have focused on the spring dispersal from field boundaries into cropped areas, in species that overwinter as adults, have been the most successful. Species such as *A. dorsale*, *D. atricapillus* and *B. lampros* have provided evidence of waves of dispersal from overwintering sites in field boundaries into the cropped areas of fields (Coombes and Sotherton, 1986). Using lines of pitfall traps at various locations between field boundaries and field centres, species were shown to differ in the timing of their dispersal. *B. lampros* began to migrate into fields in late March and were fully dispersed by early May. *D. atricapillus* and *A. dorsale* began dispersing early to mid-April with dispersal completed by late May or early June. A similar study (Jensen *et al.*, 1989) of dispersal of marked and unmarked *A. dorsale* also showed rapid spring invasion of cropped areas from field boundaries (grass strips). Dispersal up to several hundred metres into the cropped areas was completed within a few weeks.

In other studies, the spring dispersal of some species from overwintering quarters appears to have been so rapid that the process was not detectable. Kromp and

Nitzlader (1995), using directional pitfall traps, found evidence of directional movement only for *B. lampros* in early April. Other species, for example *A. dorsale* [*Platynus dorsalis*] and *Poecilus cupreus*, were found to be evenly distributed over fields almost as soon as their activity started in early spring. Dispersal into fields from boundaries does not always result in populations penetrating all the way to field centres. For example, *Amara* species, which overwinter in boundaries, were found mainly within 60 m of the field edge after the spring dispersal phase (Holland *et al.*, 1999a). This limited penetration of the field might be expected for spermophagous *Amara* species since weed cover was greatest at the field edges. The centres of arable fields are unlikely to provide sufficient seed resources for these species, except in the weedier areas. *Bembidion lampros* overwinter in the field (Riedel, 1995) and the margins (Sotherton, 1984), and have been shown to disperse into the field during the summer (Coombes and Sotherton, 1986), returning to the field edges late in the season prior to hibernation (Wallin, 1985, 1987). However, although *Bembidion lampros* has been found to be relatively evenly distributed by early summer in some studies (Coombes and Sotherton, 1986; Kromp and Nitzlader, 1995), Holland *et al.* (1999a) found this predatory species mainly within 60 m of the field edge, as well as the spermophagous *Amara* species. In this case, two different mechanisms may have resulted in similar distributions of the different species: *Amara* responding to food in the form of weed seeds, *Bembidion* responding to a preferred microclimate under weed cover.

Another type of overwintering refuge is the 'beetle bank' (Thomas *et al.*, 1992b). Thomas *et al.* (2000) used a grid of pitfall traps spanning a field from a hedgerow on one side to a beetle bank within the field on the other side of the grid. They provide clear evidence of the spatial patterns of two classes of carabids emerging in the spring and early summer. Boundary overwintering carabids exhibited a clear wave of dispersal predominantly emanating from the beetle bank and the hedgerow. The wave was first apparent close to the beetle bank during mid- to late April. By early to mid-May, the emerging wave fronts of carabids from the beetle bank and the hedgerow had made considerable progress across the field. By mid-June, carabids were more evenly distributed across the field. In the same study, Thomas *et al.* (2000) showed field-inhabiting species, mainly *Pterostichus melanarius*, to emerge *en masse* within the field during the first week of June. The highest activity-density was in the field centre. This pattern of emergence, however, was preceded in mid-May by a small amount of activity aggregated around the overwintering habitats, especially the beetle bank. This may represent a small cohort of late adults from the previous season that successfully overwintered as adults.

The spatial dynamics of the autumn emigration of carabids from fields into field boundaries has received less attention than spring dispersal in the opposite direction. This is probably due in part to the fact that many field studies, especially those employing pitfall traps, are necessarily terminated at harvest time. There is also much variation in field conditions in the autumn, with some fields ploughed, harrowed and drilled with winter crops within a few days. Other fields may be left ploughed and unharrowed for some considerable time, and in this state it is not easy to set pitfall traps in the ground. Studies of the spatial distribution of overwintering carabids at this time clearly indicate that densities vary between fields and field boundaries (Sotherton, 1985) and that newly-formed field margins are rapidly utilized (Thomas and Marshall,

1998). However, without spatial dynamics studies, it is uncertain to what extent the distributions of carabids in the autumn and winter are due to migration to preferred habitats or to differential survival in tilled field centres compared with field edges.

The spatial distribution of carabids can thus alternate between fields and field boundaries seasonally. Adult overwintering species, such as *P. cupreus*, *A. dorsale*, *B. lampros* and *D. atricapillus*, move from overwintering sites at field boundaries into fields in the spring (Wallin, 1985, 1986; Coombes and Sotherton, 1986). Larval overwintering species are less dependent on field boundary habitats, although *P. melanarius* and *Harpalus rufipes* have been reported to accumulate at field boundaries during the summer (Wallin, 1985, 1986). The latter distributions may reflect movement out of crops towards field boundaries when the crop no longer provides a suitable environment in terms of microclimate and food resources.

Spatial distributions within fields

Despite the apparently homogeneous nature of drilled crops in arable fields, carabid spatial distributions are rarely, if ever, uniform in these habitats. Once a species has dispersed into a field from its boundaries, or emerged from pupation directly into a field, spatial distributions may stabilize in particular aggregations or continue to shift in a dynamic manner. Even in permanent pasture, where spatial dynamics between field centres and marginal habitats are less likely to be necessary for survival, spatial distributions of a number of carabid species can be significantly aggregated, but to a degree that depends on season and species (Desender, 1988). In his study, Desender assessed aggregation from randomly placed pitfall traps. Other studies have used traps located at field edges and at a few other positions at various distances from the field edge, for example Dennis and Fry (1992), Coombes and Sotherton (1986), Wade French and Elliott (1999a,b) and Kromp and Steinberger (1992). While these types of study can provide valuable data on broad-scale distributions, because data are gathered at a coarse level of resolution, results sometimes contradict other findings reported in the literature (e.g. Boiteau, 1983).

It is at the field scale that most experimental and observational data are gathered. In this section, therefore, we first discuss some of the methods for displaying and analysing spatial data before reviewing evidence for population aggregation at this scale.

DISPLAYING SPATIAL DATA

The most informative studies have employed grids of pitfall traps. In addition to statistical measures of aggregation, the use of grids of traps can allow correlations to be tested between population density and a range of biotic and abiotic variables suspected of being driving factors. Such spatially structured sampling designs using grids of pitfall traps can also provide a simple way of obtaining two-dimensional information on carabid distributions. Various methods have been used to display these types of data to give an immediate visual impression of field distributions and how they change over time. Greenslade (1964) and Ericson (1978) illustrated spatial distributions in their papers by drawing a circle at each grid position of a size proportional to the total number of beetles captured. Hengeveld (1979) plotted activity-density as a trend-surface drawn as a series of contours, and established correlations between distribution patterns of carabids and abiotic factors. For most species, he found the spatial distribution patterns to be stable between years. With the advent of widespread access to computing facilities, these types of data can now be assessed visually. A range of contour-plotting programs incorporating various smoothing and interpolation methods are readily available to draw maps of activity-density. Best *et al.* (1981) examined the distribution of three carabid species in a grid covering part of a cornfield. All species were shown to be aggregated using conventional statistics to derive an index of dispersion. The contour plots, however, provide a visual representation of species occurrence in space. Visual inspection of these plots indicated that some parts of the grid were shared by different species. Some areas remained sites of aggregation throughout the study, while other clusters shifted to different parts of the grid throughout the season. Determining the reasons for aggregation in some areas was still based on a visual analysis of the plots, combined with knowledge of the field site. Low densities seemed related to poor crop coverage and factors relating to soil quality.

ANALYSIS OF SPATIAL DATA

Statistical analyses of aggregated populations are generally restricted to descriptions of frequency distributions of counts, e.g. the negative binomial, or relationships between means and variances, such as Taylor's power law (Southwood, 1978; Taylor, 1984), giving a measure of the degree of aggregation. A major constraint of these methods is that they generally make little or no use of the spatial information within a data set (Perry and Hewitt, 1991). Some spatial information can be used in autocorrelation methods (Cliff and Ord, 1981) and these techniques can be used to test for independence of samples as a function of distance between sample units (Midgarden *et al.*, 1993). A new class of analytical techniques, the SADIE method (Spatial Analysis by Distance IndiciEs), has been introduced recently (Perry *et al.*, 1996), which is applicable to experimental designs where data, in the form of counts, are available together with the spatial co-ordinates of each sample unit. The SADIE method provides an index of aggregation (Ia) for spatial data and a probability (Pa) that the data are not distributed randomly. Perry (1995) gives a clear and concise summary of the method.

Statistical tests for differences in the degree of aggregation between two populations generally depend on testing for significant differences between parameters of their two distribution functions, without explicitly using any spatial information. Therefore, it is not possible to differentiate between two population distributions with the same level of aggregation but which occupy different space because of different behavioural or other ecological traits. If significance tests indicate differences or similarities in the level of aggregation between two populations, the best options to pursue for interpretation still rely on textual or graphical reference to the original data (Syrjala, 1996). Similarly, principal components and detrended correspondence analyses of multivariate data can provide axis scores that can be mapped for visual comparison as an aid to interpretation (Sanderson *et al.*, 1995).

The SADIE method has been further developed and extended to enable the derivation of indices of spatial association or dissociation between two data sets

(Korie *et al.*, 2000). The two data sets may be counts of the same species taken at different times, or of two species sharing habitat space, or a species and an environmental variable that is quantifiable in counts. Perry (1997, 1998) gives details of the methods. These techniques can be used to test for spatial associations between the distributions of predators and prey, or the spatial distribution of a species and environmental variables. When the spatial distribution of a population is quantified at different times throughout a season or between years, these techniques can also be used to test for the spatio-temporal stability of population patches. The most recently developed version of the SADIE algorithm (Perry *et al.*, 1999) provides an index of clustering for each sample unit that quantifies the contribution made by each sample to the overall measure of aggregation. Clusters or gaps can also be identified that stand in high contrast to neighbouring areas, enabling the size of population patches to be estimated.

EVIDENCE OF AGGREGATIONS AT THE FIELD SCALE

One of the earliest examples of using grids of traps to sample carabids (Greenslade, 1964) examined distributions of *Nebria brevicollis*, *P. madidus*, *A. parallelepipedus* and *Calathus piceus*. Grids were set over nearly 5000 m^2 in an area of woodland containing sub-habitats of litter, scrub, bracken and grass, and an area of grassy heath surrounded on some sides by woodland. Three different spatial distribution patterns were apparent among the four species. Mark–recapture experiments within the grid also revealed the extent of movement within and between the habitats. Ericson (1978), studying distributions of *P. cupreus*, *P. melanarius* and *H. rufipes* in an arable field, used a similar technique. The grids were smaller than Greenslade's (1964) study at 500 m^2 and 700 m^2 in a wheat field. Species distributions were patchy over the grid and apparently related to areas of seemingly nitrogen deficient wheat that gave poor cover – a finding similar to that of Best *et al.* (1981) mentioned above. Activity-density was generally lowest in these more exposed areas. The first study employing a grid of pitfall traps that extended over an entire arable field was conducted on the Dutch Ijsselmeer polders (Hengeveld, 1979). He showed that, in a very uniform cereal habitat, carabid activity-density could be related to the moisture content of the soil inferred from the height of the field surface above the water table.

Other studies have also employed grids of pitfall traps, often combined with mark–recapture experiments, although the distributions of carabids have not always been specifically analysed (Lys and Nentwig, 1991). Recent studies involving frequent sampling in a grid of pitfall traps covering an area of over 8000 m^2 have examined the spatial distributions of six carabid species in parts of two adjacent cereal fields and a field margin (Thomas *et al.*, 1998; Fernández García *et al.*, 2000; Thomas *et al.*, 2001). Contour plots of density distributions, mark–recapture and SADIE analyses in these studies revealed evidence that suggested species packing in space. *Amara* spp. were confined to a few small aggregations within the field boundary with little evidence of movement into the cropped areas of the field. *H. rufipes* were distributed in field boundary aggregations but were also present at low densities within the cropped areas of the field. *A. dorsale* were found in small aggregations in both the field and the field boundary. By testing successive and time-lagged distributions with SADIE for spatial association, the patch structure of all species through time was surprisingly stable. Spatial stability may be expected among seed-feeding species that are dependent on fixed resources in the hedgerow habitats. However, *P. melanarius* and *P. cupreus* also had highly stable field distributions, even though both these species are highly mobile (Wallin and Ekbom, 1988, 1994; Thomas *et al.*, 1998).

In another study, total carabids, *N. brevicollis* and *B. lampros*, also exhibited relatively stable distributions across a 4 ha and a 16 ha field (Holland *et al.*, 1999a). The data from these studies suggest that there must be a strong attraction to certain areas, or repulsion from others, for the patch structure to have persisted in the way observed. However, scale and field size may be an important factor to take into consideration. When the distribution of *P. melanarius* was measured on a number of occasions using a 16 × 16 grid with 12 m spacing, a considerable change in its distribution was found, with patches apparently moving across the field (Holland *et al.*, 1999a). It is likely that mobile species are able to respond rapidly to changing environmental conditions and food availability, tracking resources as they become depleted and available at different places within a field. Distributions will thus fluctuate accordingly or remain stable, provided suitable conditions persist.

Small differences in the spatial stability of male and female *P. melanarius* distributions (Thomas *et al.*, 1998) suggested behavioural differences that might be related to mate finding and oviposition behaviour. A lack of spatial association between male and female distributions early in the season may also reflect behavioural differences. In the same study, the distribution of *P. cupreus*, although stable for most of the time, shifted markedly at the end of the summer as one generation died off and the following generation emerged. A possible explanation might be differential larval survival in the two fields, but studies need to include the monitoring of putative causal factors to fully understand the driving forces determining spatial distributions.

INTRICACIES OF SCALE AND AREA

Knowing the scale of aggregation in patchy distributions is important for the design of sampling regimes (Kuno, 1991) in order to obtain reasonable estimates of mean population density. Similarly, the sampling scale is also an important consideration when trying to detect aggregation. However, physical limits on the number of pitfall traps that can be attended during a study necessitate a trade-off between the scale of sampling and the total area sampled. Some of the work cited above (Thomas *et al.*, 1998; Fernández García *et al.*, 2000; Thomas *et al.*, 2001) employed a high density trapping grid (~ 5 m × 10 m spacing over 8000 m^2) and frequent sampling (alternate days over 2–3 months), but the trapping grids covered only a portion of the field. Recent years have produced the most extensive and detailed studies in which whole fields have been sampled. This has not been achieved at the expense of fine-scale spatial resolution since small-scale grids (1.5 m × 1.5 m spacing) and medium-scale grids (7.5 m × 7.5 m spacing) were nested within the whole field grid (~ 30 m × 30 m spacing) (Holland *et al.*, 1999a). At a second site, two fields of different size (4 ha and 16 ha) were used. Weed cover was assessed at both sites over two years and data were analysed using the most recently developed version of the SADIE algorithm (Perry *et al.*, 1999). The spatial distributions of several arthropod groups were reported, including carabids. *Amara* spp. and *Bembidion lampros*, and some other carabids, were mainly found in temporally stable regions within 60 m of the field edge. Within the field, *N. brevicollis* occurred mainly in a 1 ha patch and *P. madidus* in a 2 ha patch. Where weed cover was shown to be aggregated, there was also association between the distribution of carabids and the location of areas of high-density weed cover.

A similar study (Holland *et al.*, 1999b) used the same approach to investigate the effects of a buffer zone on the reinvasion of arthropods into a field following an insecticidal spray. Although arthropod populations (predominantly carabids) were considerably aggregated, the detail afforded by grid sampling and SADIE analysis showed that arthropods within the unsprayed buffer zone were not protected at the time of spraying. However, reinvasion of the field from surrounding marginal habitats was more extensive where buffer zones were in place.

To date, the most extensive observations of carabid spatial distributions within fields have been conducted on 24 m spaced grids over the whole area of two pairs of adjacent fields (Brown, 2000; N.J. Brown and C.F.G. Thomas, unpublished data). The accumulated monthly distributions of the most abundant carabid species in one pair of fields are given for 1997 (four species) and 1998 (six species) in *Figures 11.1* and *11.2* respectively. The total area of these two fields was approximately 7.5 ha. In 1997, beans and wheat were grown in the left and right fields respectively, and this cropping pattern was reversed in 1998. The fields were separated only by a strip of bare ground one metre wide between the two cropped areas. This is marked on the figures by the vertical dashed line; the shaded area in the top left of the figures is part of an adjacent field that was not sampled. These data, and other more extensive and detailed data, are dealt with fully by Brown (2000) and Brown and Thomas (in prep.).

Figures 11.1 and *11.2* highlight several features of spatial distributions that are only discernible when sampling is performed over extensive grids. It is immediately clear that all distributions are aggregated. The different species appear to be aggregated into patches of different size. The overall position of the patches remains fairly static, but the precise locations appear to shift, amoeba-like, with time. A clear edge to many of the distributions coincides with the location of the bare strip between the beans and the wheat. The highest density areas of most distributions occur in the field planted with beans. There is a clearly aggregated distribution of *P. melanarius* in the bean field in August 1997. Throughout 1998, the population seems to drift from left to right to aggregate predominantly in the field on the right by August 1998. Many other phenomena are also apparent in these data; for instance, close associations between the distributions of *Trechus quadristriatus* and Collembola in 1998, but these data require further analysis. It seems, therefore, that this approach is the way ahead. Data with this level of spatial and temporal resolution can spawn new insights into the mechanisms and processes that underpin the spatial distributions of carabid abundance, and may generate new ideas about ways to manage populations of beneficial carabids in the field.

Possible causes of spatial variation in population density within fields

There are a large number of potential factors that might cause carabid spatial distributions to become aggregated in particular patterns within a field. Many of these are likely to act together, or be positively or negatively correlated in non-linear relationships. As such, designing field experiments to identify the interactions between causal factors and to quantify the extent to which different factors affect spatial distribution patterns is highly problematical. This is more of a logistic problem than an analytical one. The large number of variables that need to be measured, the large number of samples that need to be taken in space, and the frequency at which the samples need to be taken in time are generally beyond the resources of any one research group. Analytically, the problems are not impossible to resolve, given the computing power that is now widely available. However, when most sampling of carabids still relies on pitfall traps, interpretation of catches is still a challenge (Luff, 1975; Adis, 1979), particularly when factors that affect spatial distributions of carabids, such as vegetation density, also affect the efficiency of pitfall traps (see Holland, this volume).

It is now generally accepted that carabid populations often exist as a series of local interaction groups or sub-populations (Den Boer, 1990a) that together constitute a metapopulation (Levins, 1970; Gilpin and Hanski, 1991). Baars and Van Dijk (1984) estimated that individuals of *Pterostichus versicolor* (size 9–12 mm) were moving around within an area of about 12.5 ha. The smaller *Calathus melanocephalus* (6–9 mm) was estimated to move around an area of 2 ha. Because non-adjacent, individual fields were sampled in this study, it was not possible to verify the presence of any metapopulation structure; although within fields, clusters were smaller than those found above the field scale. An extensive study by Thomas *et al.* (1998) using mark–recapture demonstrated limited dispersal; the majority of *P. melanarius* were captured within 55 m of their release site after 30 days. Areas of high activity-density were also detected, often only 10 m^2 in extent, even though they are capable of moving much greater distances (Wallin and Ekbom, 1994).

ARTEFACTS OF SAMPLING

The number and placement of traps and the scale of sampling in relation to the scale of patchiness in carabid distributions is an important factor that has been mentioned above. Trapping at scales below the size of a patch may suggest uniform distributions when these do not, in reality, exist at higher spatial scales. Trapping in transects may miss important features in the pattern of patchiness that may only be revealed in regular grids of traps. To test the extent to which grid size influenced mean capture and ability to detect spatial pattern, Holland *et al.* (1999a) used nested grids at different spatial scales. The placement of the grids proved to be critical because the taxa total carabids and *N. brevicollis* occurred in clusters of at least 1 ha in size. Thus, grids of less than 1 ha had lower mean captures unless they were, by chance, nested within patches that were over 30 m wide. This size of cluster is similar to that detected in other studies. The smallest grid (1.5 m spacing) was inappropriate for most taxa because the mean was often smaller than for the other scales. They concluded that a 30 m grid spacing was sufficient for sampling the spatial distributions of most species.

A beetle's preference for the field edges may also make it difficult to identify the size of clusters. In 1996, Holland *et al.* (1999a) found some isolated clusters of up to 90 m width for *N. brevicollis*. However, in 1997 within a large field, clusters of carabid species extended over greater distances – either linearly along field edges or across several hectares. Brust (1990) suggested that carabids establish burrows and foraging areas and remain within these unless the food supply diminishes. Larger

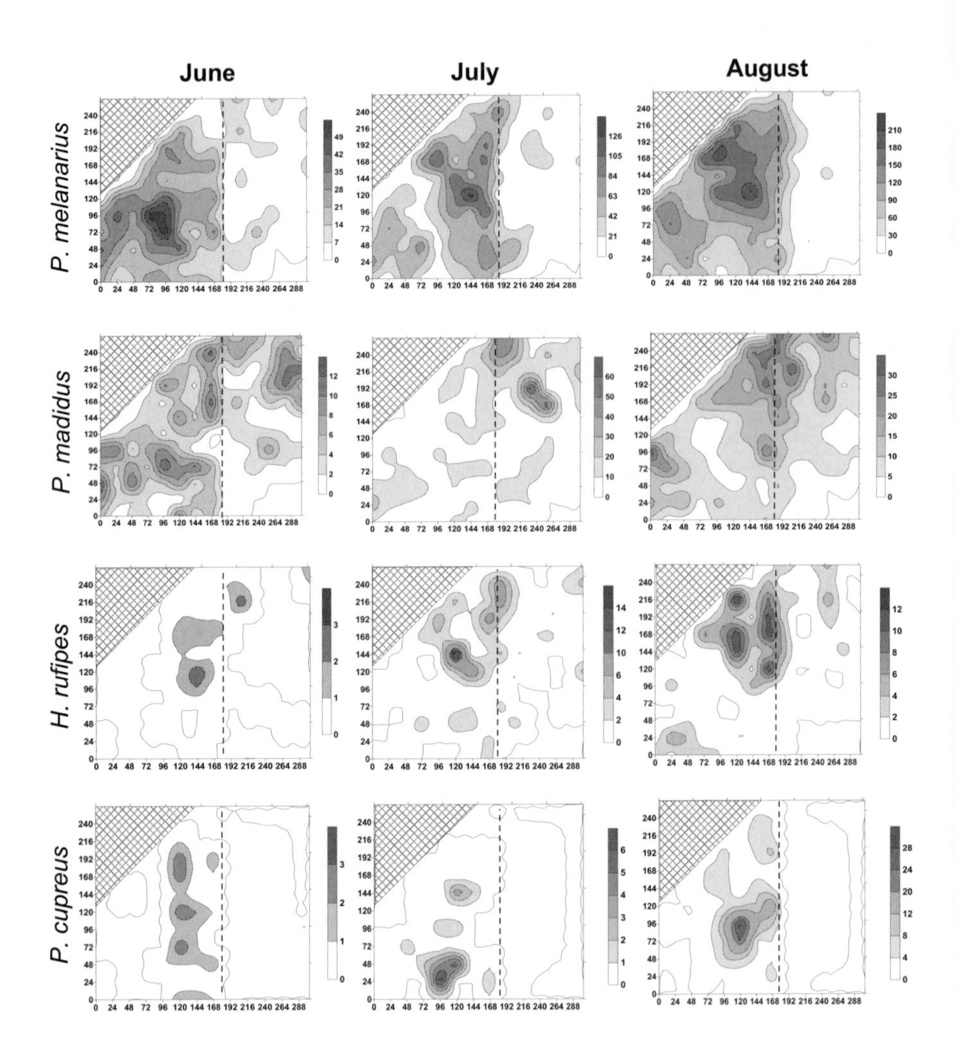

Figures 11.1 and 11.2. Spatial distributions of the monthly activity-densities of four carabid species in 1997 (*Figure 11.1*) and six carabid species in 1998 (*Figure 11.2*) in two adjacent fields during June, July and August. The vertical dashed line marks a 1 m bare strip between the crops. In 1997 (*Figure 11.1*), beans and wheat were grown in the left and right fields respectively. In 1998 (*Figure 11.2*), the crops were reversed with wheat in the left field and beans in the right field. The triangular hatched area indicates an area of an adjacent unsampled field. The axes show the distance (m) from the trapping station in the bottom left corner of the trapping area (0–312 m along the x-axis; 0–264 m along the y-axis). Pitfall traps were installed in a regular grid pattern with 24 m spacing between trapping stations. The monthly activity-density was determined by plotting the accumulated number of carabids caught per trapping station during a month. Increasing colour densities indicate higher activity-densities.

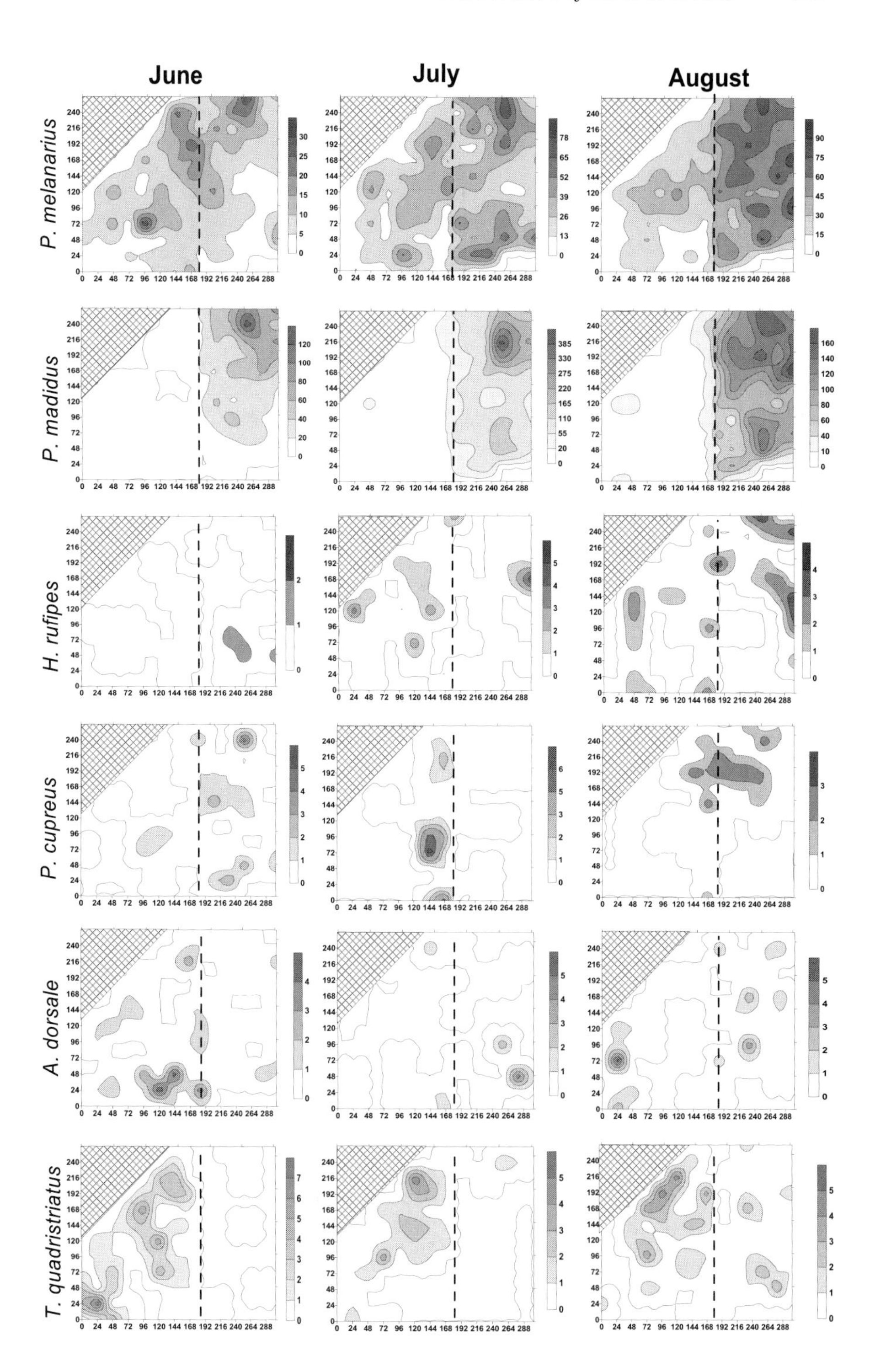
June
July
August
P. melanarius
P. madidus
H. rufipes
P. cupreus
A. dorsale
T. quadristriatus

species are generally more mobile and would be expected to be able to forage over and occupy a greater area than smaller species.

The extent to which the number of carabids caught in pitfall traps accurately reflects the population size in the immediate vicinity is a question still open to debate. The problem of interpreting pitfall trap catches in terms of absolute density against activity-density continues. This problem can be overcome to some extent by additional mark–recapture studies (e.g. Thomas *et al.*, 1998), using barriered pitfall traps or photoeclectors (Desender and Maelfait, 1986; Idinger *et al.*, 1996; Holland and Smith, 1999), or by using other, more direct means of sampling, such as quadrat sampling (Desender, 1982). These methods, however, often entail a more labour-intensive approach, which necessarily compromises the number and frequency of samples that can be taken in an area. Other interpretative problems relate to the fact that carabids can respond to defensive secretions and sexual pheromones given off by conspecifics. Thus, once an individual has fallen in a trap, it may attract others to the same trap and so give a false measure of aggregation that does not necessarily reflect the true level of aggregation, if any, in the local population (Luff, 1986).

PREY AVAILABILITY

Aggregation patterns may arise from attraction or repulsion between conspecifics (Turchin, 1989) or by attraction to direct or indirect olfactory cues from prey items (Evans, 1983; Kielty *et al.*, 1996; Longley and Jepson, 1996; Monsrud and Toft, 1999). Aggregation of carabids in areas of high prey density is of particular interest and importance in agricultural fields when the prey species are also crop pests. Aphid distributions are often shown to be aggregated, especially around field edges (Winder *et al.*, 1999). Non-pest groups such as Collembola, which are thought to be important alternative prey for some carabids (Bilde *et al.*, 2000), are also often aggregated within fields (Alvarez *et al.*, 1997; Holland *et al.*, 1999a). Carabids have been shown to respond to these patches of high prey density. When aphid density was experimentally manipulated in the field, some carabid species (*A. dorsale, Amara plebeja, B. lampros* and *B. obtusum*) were shown to aggregate in areas with high aphid counts (Bryan and Wratten, 1984). *N. brevicollis* has been shown to respond to odours emitted by Collembola in olfactometers (Kielty *et al.*, 1996), and populations of carabids in the field have been correlated with the spatial distribution of Collembola (Niemelä *et al.*, 1986).

Carabids may respond to aggregations of cereal aphids either because they are attracted to the aphids as prey or because they are attracted to alternative prey which aggregate around the honeydew produced by the aphids (Monsrud and Toft, 1999). To test this, these authors compared the aggregative response of carabids and staphylinids to high densities of aphids and to areas sprayed with honey to simulate honeydew. Adults in the taxa *Agonum dorsale* and 'all carabids' were attracted to the honey-sprayed plots, while their larvae responded directly to the aphids. The authors speculated that this difference of response was related to the ability of larvae and adults to tolerate and utilize aphids in their diet: adults showing poor ability compared to their larvae (Toft and Bilde, this volume). The response found by Bryan and Wratten (1984) may have been to the honeydew as much as the aphids, especially as the aphid colonies were particularly high in the crop canopy and out of reach of ground predators.

Movement and searching behaviour are known to change in response to hunger level and the presence of prey items (Chiverton, 1988; Wallin and Ekbom, 1994). Spatial relationships between prey and predator distributions are therefore likely to be highly complex and result in spatially dynamic aggregations. As prey densities increase in a patch, carabids may aggregate, deplete the population, and subsequently disperse before aggregating again in another patch of high prey density in a different part of a field. To detect such a dynamic process would require extensive and frequent sampling of all the relevant variables. At larger regional scales, movements between areas of high prey density are not possible. Nevertheless, relationships between the abundance of carabids and their prey can also be apparent at these scales. Guillemain *et al.* (1997), sampling over 19 widely scattered forest sites around Paris, found a positive relationship between *Abax parallelepipedus* and the abundance in the soil of its main prey of lumbricid worms. However, this relationship is more likely to be due to variation in local productivity levels rather than any form of behavioural aggregation from movement during prey searching behaviour.

Bohan *et al.* (2000) found the distribution patterns of *P. melanarius* with slug remains in their guts to be spatially associated with the distribution of slugs themselves, one of its known items of food. Although both organisms may have been independently associated with a third variable such as weed density that affects the microclimate at the soil surface, the only environmental variables measured were soil temperature (accumulated day degrees one week before slugs were sampled), soil moisture (at an unknown time since rain last fell) and crop biomass (weight of wheat ears). A lack of association between either predator or prey and any of these environmental variables led Bohan *et al.* (2000) to infer that predation of slugs by *P. melanarius* represented a specific trophic association and was not simply opportunistic. Caution is required, however, when trying to determine the nature of causal relationships and infer dependence between species. An appropriate range of putative causative variables needs to be measured and tested before conclusions are drawn. While claims of causality from a single coincidence may be doubtful, such results may be of value in directing and focusing future work.

MICROCLIMATE

Carabid species are well known to have preferred or optimal microclimatic ranges (Thiele, 1977). The most important factors are temperature, humidity and light, all of which are affected by factors associated with soil and vegetation. These can vary at the spatial scale of an individual field. The shade provided by vegetation cover can lower the range of temperature fluctuations beneath the canopy and maintain high humidity at the soil surface. Speight and Lawton (1976) demonstrated that higher levels of predation by carabids occurred in the areas of a field where weed cover was highest. Several examples already cited above have also shown that carabids aggregate in weedy parts of cereal fields (e.g. Holland *et al.*, 1999a). However, other research does not always corroborate these findings. Purvis and Curry (1984) found that, although weed cover enhanced the diversity and abundance of some arthropod groups, carabids were not obviously affected. It should be noted, however, that this study was conducted in a sugar beet crop which, compared with cereals, provides good cover from its broad leaves. The presence of weeds in sugar beet may not have improved microclimatic conditions for carabids above that naturally occurring beneath the crop canopy. In a more open crop of corn in the USA, Pavuk *et al.* (1997) found the responses of carabids to weed cover to differ in each of the two years of their study. In a year of heavy drought, neither broad-leaved nor grassy weeds influenced carabid activity-density. However, in a year of normal rainfall, carabid activity-density was significantly higher in areas of corn with broad-leaved weeds.

In addition to weed cover, the cover provided by the crop itself can affect the distribution of carabids. Work cited above has indicated that fewer ground beetles are trapped in areas where the stand of corn or small grain cereals is thin (e.g. Ericson, 1978; Best *et al.*, 1981). Again, findings from different studies are sometimes inconsistent. Honek (1988) found increased captures of some carabid species in those parts of a cereal field with lower crop cover and higher amounts of bare ground. In this case, higher catches in the more exposed areas were attributed to increased activity due to higher temperatures in these areas. Temperature is known to affect the activity of many carabid species (Honek, 1997); responses to the thermal environment have been suggested as the best explanation for carabid distribution and abundance in a range of mown and unmown grass plots with different ground temperatures (Crist and Ahern, 1999). In addition, temperature has also been shown to have quite subtle effects on carabid activity patterns on different days depending on the weather. On fine days when midday temperatures were limiting, the carabid, *Angoleus nitidus*, had bimodal activity, while on cloudy days activity was unimodal (Atienza *et al.*, 1996). Thus, for mobile species spatial distributions within fields may change dynamically with the weather in quite complex ways. On some days, or at some times of the year, certain carabids may aggregate under the cover of dense vegetation. At other times, they may actively seek out the warmth of more exposed areas or be uninfluenced by crop or weed cover. Such confounding interactions are likely to create complex dynamics in the spatial distributions of carabids within fields which may be difficult to interpret.

SOIL FACTORS

Less dynamic, more stable spatial distributions of carabids are likely to arise from the influence of more stable features of the local environment. Soil factors are known to be related to carabid spatial distributions at regional scales and field scales (see also Holland, this volume). Several papers cited above have indicated correlations between carabid distributions and various edaphic factors, including soil moisture, organic content and pH (Hengeveld, 1979; Best *et al.*, 1981; Brunsting, 1981; Baguette, 1987, 1993). Carabid community structure was found to be more related to soil factors than space or vegetation in a 117 ha upland moor (Sanderson *et al.*, 1995). Soil factors can also affect the quality of a site in terms of its suitability for overwintering (Leather *et al.*, 1993), which may in turn affect the winter distribution of carabids in fields and boundary habitats.

SOIL MOISTURE

Soil moisture is frequently cited as one of the most important factors influencing carabid abundance, diversity and community structure (e.g. Hengeveld, 1979; Lindroth, 1992; Holopainen *et al.*, 1995; Van Dijk, 1996). It seems reasonable to hypothesize that variation in the holding capacity of soils, together with topography and patterns of drainage, can create areas (at the within-field scale) that remain moist for longer than other parts of a field following rain. Such areas may contract and expand with patterns of rainfall and drought, but all the while maintaining stable foci of patches in which adult carabids may aggregate. Adequate levels of moisture in the soil may create localized conditions of humidity that mobile adult carabids can seek out. However, such patches are likely to be even more important to less mobile, more soil-bound larval life stages, and so affect development and survival in different parts of the same field. Systematic studies are required to investigate these factors. Soil moisture may also be an important factor for oviposition. Some beetles other than carabids, which are also heavily dependent on the soil, e.g. the wireworm, *Ctenicera destructor*, lay few eggs outside the range of 11–16% soil moisture in oviposition choice chambers (Doane, 1967).

Hengeveld (1979) attributed the patterns of carabid distributions that he detected within fields to soil moisture. To confirm whether this held true for other species, the distribution of *P. melanarius* and *P. madidus* was intensively monitored using a 16 × 16 grid with a 12 m spacing, and five barriered pitfall traps at each sampling location within a field of winter wheat. Sampling was conducted five times throughout the season from the time when the beetles started to emerge. Soil moisture was measured in July. *P. melanarius* adults demonstrated strong clustering following emergence and during later redistribution through the field (Winder *et al.*, 2000). However, despite a distinct gradient in soil moisture, their aggregations were not spatially correlated with soil moisture at any time (Holland and Winder, unpublished).

CHEMICAL FACTORS

Chemical properties of the soil may affect the distribution of carabids. Some species show preferences, when given the choice, between soils that differ in, for example, the concentration of sodium and calcium salts (Thiele, 1977; Lindroth, 1992). Industrial pollutants can also influence the structure of carabid communities (Holopainen *et al.*, 1995). Evidence for the importance of soil pH is inconsistent and it is uncertain whether soil pH is a decisive factor or an indicator of other characteristics. In canonical correspondence analysis of carabids in 16 fields, Holopainen *et al.*, (1995) found pH to be the least important factor separating carabid communities after fluoride pollutants, clay content, soil type, water content and organic content. As cited above, Baguette (1987) suggests that pH possibly acts through its indirect effects on the degradation and availability of organic material in the soil, which in turn might affect prey availability to carabids. Nevertheless, in laboratory tests Paje and Mossakowski (1984) found five out of seven species to exhibit a preference for soils with a particular pH and that these matched the pH of the soil in the field where the beetles were originally captured.

MECHANICAL PROPERTIES

For ground beetles that live in intimate contact with the soil as larvae or burrowing adults, mechanical properties of the soil can be of paramount importance, affecting their abundance and distribution. Apart from purely mechanical properties, soil structure and particle size are also directly related to the water-holding capacity of soils. In general, carabid abundance and diversity are higher on clay rather than sandy soils (Thiele, 1977). Studies on beetles other than carabids (Tenebrionidae) have shown that the mechanical properties of soil have a strong influence on the seasonal distribution of some species (Krasnov and Shenbrot, 1997). In a laboratory study, oviposition by the wireworm, *Ctenicera destructor*, was found to be greatest in fine sand–silt–clay soils, with none laid in coarse sand. Oviposition was also enhanced by the presence of deep cracks in the soil, permitting access to lower, moister layers (Doane, 1967).

Where soil texture and other relevant properties vary within a field, there are likely to be preferred areas for oviposition, and larval development and survival. Such areas might be observed as stable patches within fields that consistently yield high numbers of emerging adults and ovipositing females, year after year. Few data exist, as yet, to demonstrate this hypothetical phenomenon systematically. However, Helenius (1995) using a grid of emergence traps in a field of spring cereals over two successive years, showed that the highest numbers of *Bembidion guttula* emerged from the same area of the field in both years. Many other possible mechanisms may create similar patterns of spatial distribution, but much further work is required to determine them. Stable and mobile distributions of other species have also been observed between successive years, as cited above (Brown, 2000; Thomas *et al.*, 2001) but longer time series and simultaneous measurements of soil properties and other variables are required for definitive results and to identify the relevant mechanisms.

TOPOGRAPHY AND ASPECT

The gross effects of topography and aspect appear to have been little studied for their effect on carabid distributions. Slope can affect soil depth and drainage, and aspect can influence the microclimate through variation in the amount of sunlight, drainage and soil moisture evaporation rate. These may prove fruitful areas for further research. Spatial distribution patterns of carabids sampled from a grid of traps were found to be very uniform on the flat fields of reclaimed land on the Dutch polders (Booij *et al.*, 1995). Where aspect has been studied as a factor in the distribution of invertebrates overwintering in field boundaries, higher numbers of some species have been found in those boundaries oriented east–west (Dennis *et al.*, 1994).

Spatial distributions at the farm-scale – variation between fields

Soil properties and other environmental factors that might vary spatially and give rise to aggregated distributions within fields, also vary at spatial scales greater than that of an individual field. Similarly, the confounding factors of crop type and crop management that might cause temporal differences in reproduction and mortality rates within a field during the course of a rotation also lead to spatial distribution patterns at the farm-scale and above.

Apart from some early studies (Skuhravy, 1958; Scherney, 1960; Rivard, 1966), surprisingly few studies have systematically examined carabid distributions at the farm-scale among a number of fields growing different crops over several years. This is probably due to the predominance of cereals and grassland in many agricultural landscapes, the perception of the importance of carabids as predators of cereal aphids, and the limited time and effort that can be devoted to field experiments. There are also problems in the numerical interpretation of pitfall trap catches when these are set in habitats differing in their vegetation structure and density (Melbourne, 1999) due to the effects of environmental resistance on activity (Heydemann, 1957).

Where rotations have been followed over a number of years (e.g. Rivard, 1966), spatial distributions, in terms of variation in activity-density among fields, appeared to have areas of consistently high or low density within particular fields, and were not strongly influenced by crop type. Another long-term study (Holland *et al.*, 1998) collected carabids from five fields over seven years. Each field was split, with the two halves being farmed under either conventional or integrated management systems. Regardless of treatments, one field consistently returned higher captures of *P. melanarius* than all the other fields (*Figure 11.3a*). Such a pattern was, however, not apparent in the data for *B. lampros* (*Figure 11.3b*). In Rivard's (1966) study, most carabid beetles were generally captured in cereals and progressively fewer in cultivated, legume, and pasture fields. More commonly, studies report data from samples taken in a number of fields or plots over only one or two seasons or years. In small plot experiments, carabids show preferences for some crop types over others (Carcámo and Spence, 1994). Whole field studies have also shown some carabids to favour bean fields over wheat fields, as cited above (*Figures 11.1* and *11.2*), when able to move freely between crops in adjacent fields with no barrier between (Brown, 2000).

Gross differences in population density between fields, and rapid changes in the abundance of non-flying carabids, are not likely to be due to dispersal when there are barriers between fields (see below) or when fields are widely separated. Variation in reproduction, survival and mortality are mechanisms that are more probable. These factors may vary from field to field because of differences in the timing, extent and frequency of disturbance, with respect to species' phenologies. The carabid species of arable fields are known to breed, emerge as adults, and become active at different times of the year, and hibernate in different life stages (Greenslade, 1965; Thiele, 1977; Fadl and Purvis, 1998). Because of this variation, populations of the different species are exposed to the risks associated with harmful effects of field operations, such as ploughing, to differing degrees (Fadl *et al.*, 1996). Measuring the productivity of arable plots with emergence traps, Purvis and Fadl (1996) demonstrated an 80% lower productivity of the larval overwintering carabid, *P. melanarius*, in plots that were ploughed in the spring compared with autumn cultivated plots. Similarly, patterns and timing of insecticide applications will affect populations with sensitive life stages in the crops being treated, thus depleting particular areas of particular species and changing the pattern of spatial distribution at the farm and landscape scale.

The effects of farming operations on carabids are covered in detail elsewhere (Holland and Luff, 2000; Hance, this volume). A few examples here will illustrate the variation found between the abundance and diversity of carabid populations in fields growing different crops. Den Nijs *et al.* (1996) sampled *P. cupreus* in nine fields over two successive crops. Complex interactions between field, crop sequence and weather resulted in at least tenfold differences in population size. Populations in potato crops were consistently the smallest. Kinnunen and Tiainen (1999) sampled 27 fields between June and August of one year and found highest carabid abundance and

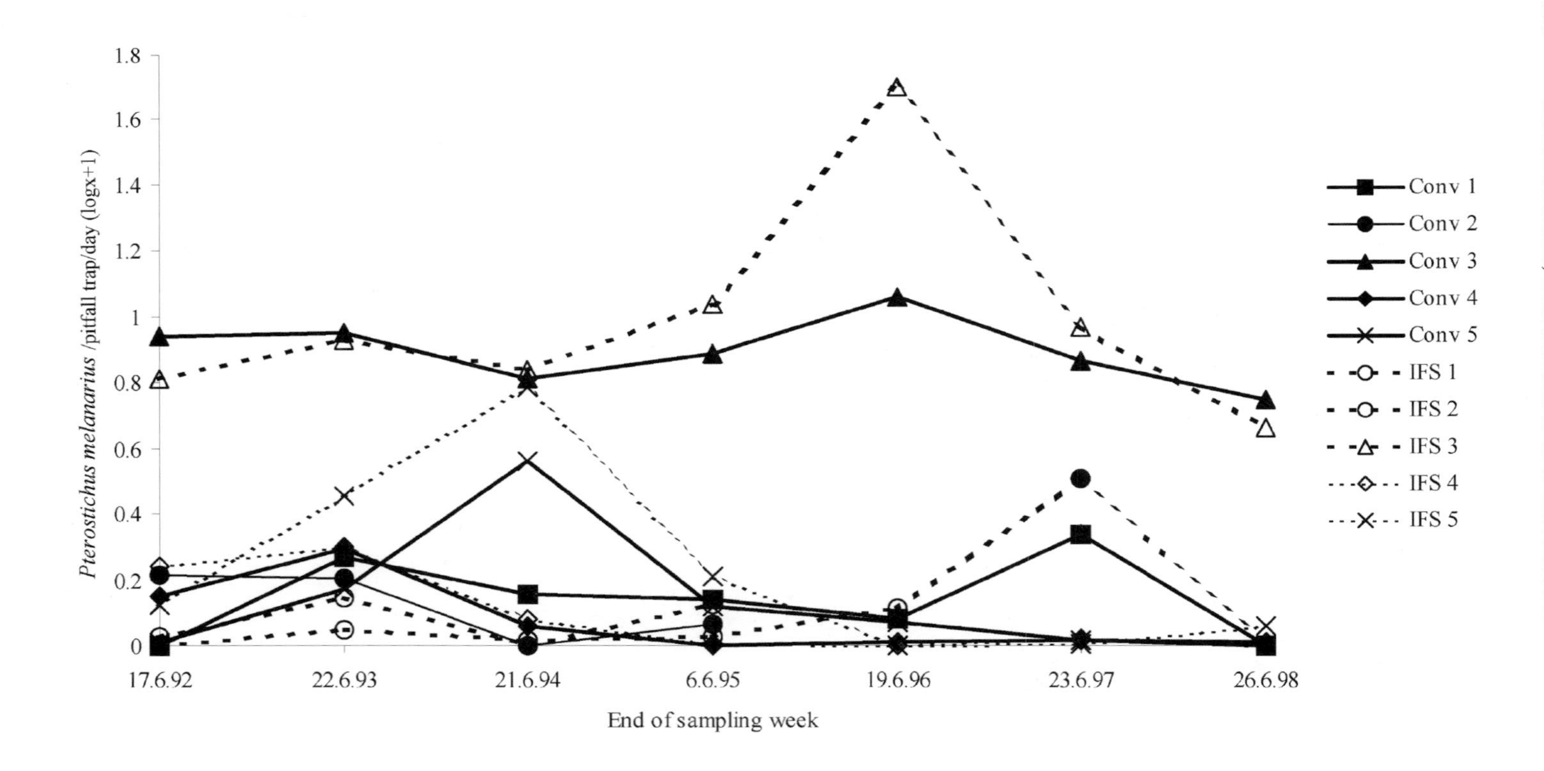

Figure 11.3a. Variation in pitfall capture of *Pterostichus melanarius* between five fields over seven years in the LINK Integrated Farming Systems Project. Fields were farmed either using a conventional (Conv) approach or an integrated farming system (IFS).

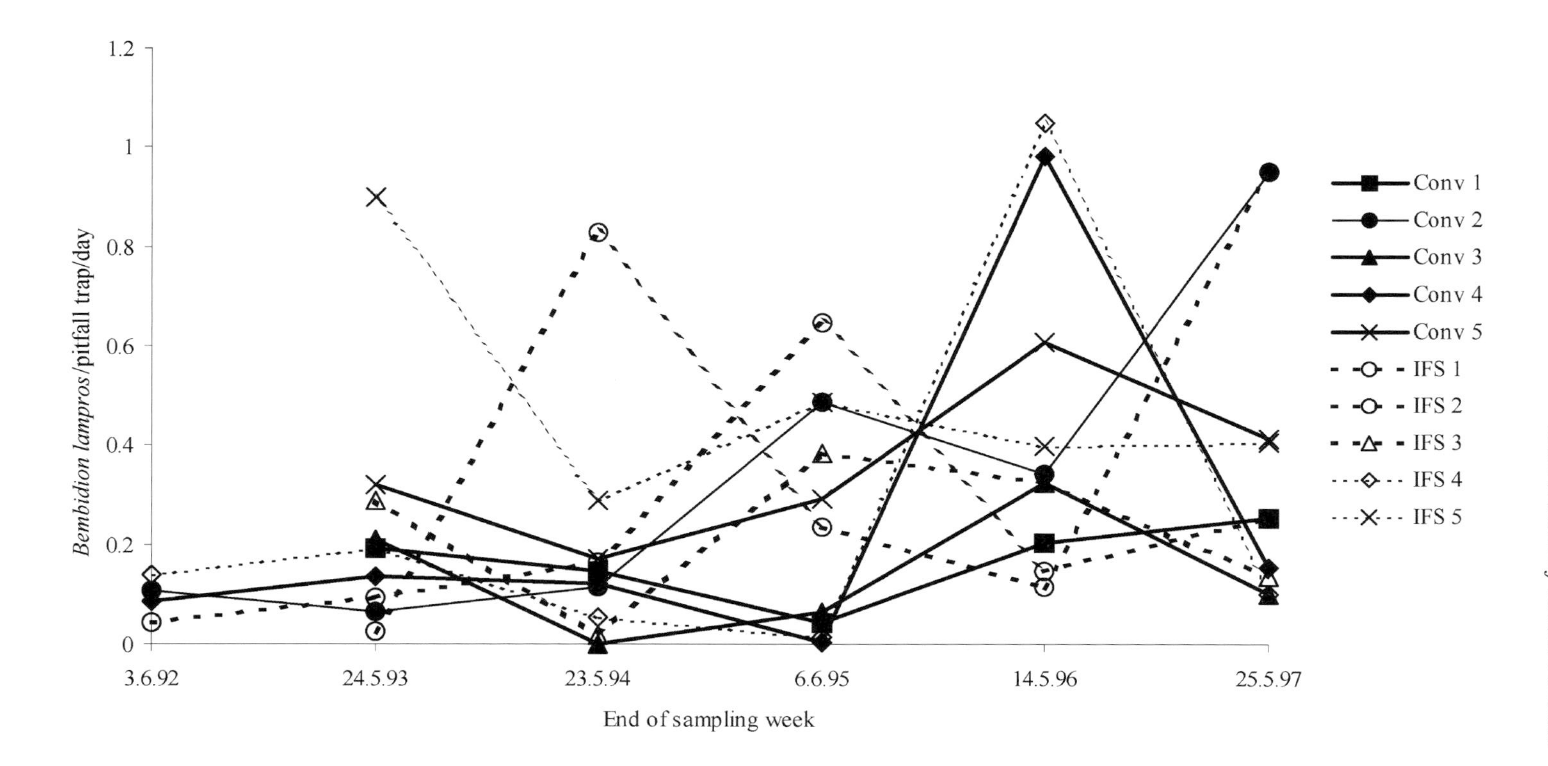

Figure 11.3b. Variation in pitfall capture of *Bembidion lampros* between five fields over seven years in the LINK Integrated Farming Systems Project. Fields were farmed either using a conventional (Conv) approach or an integrated farming system (IFS).

diversity in green set-aside, followed by cereals, sugar beet, potatoes, and oilseed rape, with lowest abundance and diversity in bare set-aside fields. Other studies of the effects of crop type on carabid diversity (Booij, 1994) have found similar results, with significantly more species caught in cereals, peas and sugar beet compared with potato, onion and carrot. Intensive and organic production of wheat can also have different effects on the productivity of different species. Basedow (1994) found no effect of management regime on *A. dorsale* but emergence of, and egg production by *P. melanarius* was delayed in organic wheat by about two weeks, compared with intensively-managed wheat fields.

The spatial distribution of carabids at the landscape scale may also be affected by spatial variation in food availability and its consequent effects on fecundity and survival. Bommarco (1998) sampled *P. cupreus* from a number of sites in organically and intensively farmed landscapes in Sweden. Fecundity of *P. cupreus* was correlated with the degree of landscape heterogeneity (high in organically farmed areas), and feeding and adult body size were also affected. The landscape characteristics that differed between sites that were of most influence seemed to be the number of fields, the amount of field edge, and the proportion of annual and perennial crops within the beetles' dispersal range. It was suggested that the mechanisms involved were related to the ability to disperse between annual and perennial habitats and the distribution of prey available in the two crop types.

Barriers to movement between fields

The extent to which the observed spatial pattern of carabid density among fields at the farm-scale and above is driven by dynamics operating within fields, or by dispersal between fields, is still largely unresolved. Examples given above have shown that very large variation of population size is frequently observed between fields, and large fluctuations often occur within fields, between years. It is apparent that in some cases, where movement between adjacent fields is unimpeded by a physical barrier, virtually a whole population may migrate between fields, although this may be a relatively slow process taking months (*Figures 11.1* and *11.2*) (Brown, 2000). However, a large number of man-made and natural barriers to dispersal exist in the landscape. For carabid species whose movements are limited to walking, hedges, ditches, banks, fences, tracks, roads, railways and rivers may prevent dispersal between patches of habitat, or reduce dispersal rates by varying amounts. Mader *et al.* (1990) showed, using mark–recapture methods, that all of the linear features they tested (various 3 m wide farm tracks and a railway), which were more substantial than a grass track, were a significant barrier for carabid beetles to cross. Although few carabids crossed these features, most of the structures tested stimulated linear movement along them.

Field boundaries may present less of a barrier to movement than metalled tracks and railways. Duelli *et al.* (1990) used directional pitfall traps and long transects through five fields and their boundaries to estimate migration rates between natural and cultivated areas. Although it is uncertain to what extent directional pitfall traps detect directional movement or population density gradients, there appeared to be net movements of many species between habitats. Furthermore, there were sharp transitions in the numbers of trapped beetles (and some other arthropods) where the transect crossed field boundaries. Six categories of boundary were identified, ranging from 'hard edge' in which no population exchange was measurable between adjacent patches, to 'no edge' in which the boundary was hardly perceived by an organism. Intermediate categories were boundaries between habitats across which the populations in the fields had negative, positive or mutual influences on each other depending on the predominant direction of flow of individuals.

Most boundaries between fields at the farm-scale probably permit mutual exchange between populations in adjacent fields. Boundaries such as hedgerows, however, are extremely variable in their structure, and are equally likely to be of variable permeability. Several workers have examined movements of carabids through and along field boundaries using mark–recapture techniques. Mauremooto *et al.* (1995) released a number of marked individuals of several species into plots enclosed by polythene barriers. One plot traversed a hedgerow between fields; another was constructed within one of the adjacent fields. Marked beetles took significantly longer to cross the 4 m wide hedge boundary, compared to equivalent distances over bare ground, barley crop or stubble.

Thomas *et al.* (1998) used a large grid of traps spanning a 5 m wide hedgerow, and intensive sampling of over 2000 individually marked *P. melanarius* to quantify movement through the hedgerow in relation to movement within the crop either side. Although a highly mobile species, during 10 weeks in the summer, only approximately 6% of successive recaptures were made on the opposite side of the hedge to that on which they were released. In the same experiment, from 1077 marked *P. cupreus*, 2332 recaptures were made over the course of 10 weeks. Of these, only one individual had moved through the hedge (C.F.G. Thomas, unpublished data). In the autumn, a similarly low rate of movement through the hedgerow, compared to the amount of mixing of marked populations within the fields, was demonstrated for *N. brevicollis* (Fernández Garcìa *et al.*, 2000). The movement of carabids between fields separated by water-filled ditches does not seem to have been studied; but no matter how impermeable the boundary might be, all fields have gates or bridges for access. If tractors can pass between fields, it is reasonable to assume that beetles are also able to do so. However, such access points do not necessarily form significant breaks in hedgerow networks, interrupting dispersal corridors. When studying the movement of *N. brevicollis* between hedgerow intersections, Joyce *et al.* (1999) found these gaps to be readily crossed, as was the hedgerow.

Metapopulation structure and models

The spatial distribution of a carabid species across a landscape would seem to appear as a mosaic of inhabited and uninhabited patches. At any single point viewed through time, the population will undergo irregular cycles of growth and decline, occasionally disappearing altogether, only to reappear when dispersing individuals from neighbouring areas manage to cross any necessary barriers to re-found a population. Although a metapopulation is an easy image to conjure up, it is more difficult to pin time-scales and spatial-scales on the dynamics. Den Boer's (1977) classic long-term studies show that many carabid local populations undergo density fluctuations between limits, which can be well simulated by random walks (Den Boer, 1991). Graphs of local population sizes of many species appear to cycle with a periodicity of about 5 to 15 years, sometimes out of synchronization with each other, and often oscillating around longer time-scale patterns of growth, decline or stability (Den Boer, 1990a).

Species dispersal power can have important effects on local and metapopulation survival. Most of Den Boer's study from 1959 onwards focused on carabids in forest, heathland and other more or less natural habitats. In the most stable of these, Den Boer (1990b) estimated mean survival time of local populations of species with low powers of dispersal in old deciduous forests to be in excess of 100 years. More generally, regardless of the stability of the habitat, species with high powers of dispersal were estimated to have mean population survival times of 8–10 years; species with low dispersal power had population survival times of about 40 years. Although these latter species appear to have longer population survival times because of low emigration, the same mechanism ensures that in isolated habitats, once extinct, populations are unlikely to be re-founded by immigrants.

It is a pity that similar long-term studies have not yet been conducted in agricultural fields, since the dynamics of habitat disturbance are somewhat different to those of forest and heath. As well as the high level of fragmentation of the overall landscape, agricultural land is further compartmentalized into fields. Although the dispersal power of carabids ensures that few crop habitats are likely to be completely isolated, it is uncertain to what extent individual fields can be considered local populations within a metapopulation or whether, at some scale, fields are simply further subdivisions of local populations that exist at a slightly larger geographical dimension. The size of local populations may differ among species which 'perceive' the landscape and its barriers differently, depending on their powers and modes of dispersal. Whatever the true size of a local population, farmland habitats offer opportunities for species to spread risks (Den Boer, 1968) among fields where different crops create a diversity of accessible habitats with a variety of risks and resources.

Some field evidence cited above suggests that populations of some species periodically appear to become locally extinct in certain fields at some points during crop rotations, for example in potatoes (Den Nijs *et al.*, 1996). Furthermore, a cursory survey of a handful of field studies will show that some species are absent from some lists. It is not always possible, however, to distinguish between zero counts and local population extinction unless there are sufficient data on which to calculate reasonable estimates (Den Boer, 1990b). Nevertheless, Den Boer (1990b) speculates that nearly all local populations of low dispersing species in small, isolated old-habitat fragments in The Netherlands will become extinct during the next two centuries. He also states that the survival times of most interaction groups (i.e. local populations in an area within the dispersal range of the species such that all individuals can potentially interbreed) are too long for an ecologist to directly observe many extinctions. This may be true and highlights the necessity of long-term studies. Evidence for extinctions are therefore rare. However, in the UK many studies of farmland populations have been conducted on the Leckford Estate, Hampshire. Elton conducted the first ecological studies there in 1938. Since then a number of faunal surveys have been completed periodically (Cheesman and Mitchell, 1995). These have persistently shown the widespread distribution of the common carabid, *P. melanarius*, across the farm. However, in recent years, anecdotal evidence suggests that this species seems to have become extinct (apart from one small population) over half the entire area (2000 ha) of the farm situated to the east of a river valley (CFGT pers. obs. 1998–2000).

The only study in which arthropods have been monitored annually across the landscape is the 'Sussex Study' conducted by The Game Conservancy Trust (Potts and Vickerman, 1974). Started in 1970, suction samples have been taken annually in mid-June from 100 cereal fields covering 62 km^2 of arable farmland in West Sussex. Only small, day-active carabids are collected using this technique but it nevertheless provides some insight into what may be occurring at a landscape scale. From 1970 until the early 1990s, the overall density of carabids has declined consistently, but has stabilized more recently (Holland, this volume).

In the absence of much direct evidence for carabid population extinctions in farmland, modelling approaches have been employed to investigate the dynamics from a metapopulation perspective, although even these are scarce. One notable example by Sherratt and Jepson (1993) estimated extinction probabilities for a model organism under a variety of regimes. Population survival was strongly dependent on the number and frequency of toxic insecticide applications and the permeability of field boundaries to carabid dispersal. An optimum dispersal rate mediated by boundary permeability ensured population survival. Too little movement between fields meant that populations were isolated and eventually exposed to toxic insecticides at some point in a rotation sequence; too much movement between fields exposed unsprayed populations to toxic residues in neighbouring fields.

In spite of the recent growth of interest in metapopulation dynamics and modelling, and a deepening understanding of the theoretical aspects, few working models exist. The spatial dynamics of some carabids at the landscape scale have been published (Rushton *et al.*, 1996), but within agricultural land most attention has been focused on crop pests (Sawyer and Haynes, 1985, 1986; Ives and Settle, 1997). The problem is likely to be attributable to a lack of knowledge about much of the relevant ecology and biology of carabid species' population and dispersal dynamics, and a lack of long-term data to parameterize, test, and validate such models.

Summary, conclusions and future directions

Spatial distributions of carabids are nearly always aggregated at some spatial scale, and population densities fluctuate in time, although in some more stable and uniform habitats these features can sometimes be less apparent. Species of forests and woodland are found within cropped habitats to a limited extent, but density falls rapidly with distance from the nearest wood. Similarly, field species are found in farm woodlands but they do not penetrate far enough to affect diversity beyond a few metres. Nevertheless, such habitats may provide important refuges from farming operations for a proportion of the population. Similarly, field boundaries are important refuges for a number of species that migrate from the field to overwintering and aestivation sites in more stable habitats at field edges. Distributions within these habitats are affected by vegetation structure. Within fields, the most detailed information on carabid distributions has come from investigations employing grids of pitfall traps. Carabid distributions within fields are nearly always aggregated, and may be associated mainly with weed cover, crop density, and soil factors – especially moisture, which in turn affect microclimate, prey availability and oviposition. Adjacent habitats also play a role depending on the permeability of field boundaries permitting movement between fields. Novel statistical techniques (SADIE) can quantify the degree and stability of carabid aggregations, and help to identify important associations between species distributions and the biotic and abiotic factors driving them. At the farm and landscape scales, carabids probably exist as metapopulations, although the size and isolation of local populations is uncertain.

Further studies into patterns of population aggregations are likely to provide important insights into the ecology of carabid species. By identifying the location of patches in relation to the distribution of biotic and abiotic factors, it may be possible to identify the underlying causes of population distribution patterns. This may facilitate the management of natural, semi-natural, and agricultural habitats for biodiversity, conservation or pest control. Such investigations should endeavour to obtain data at fine spatial and temporal resolutions over large areas and long periods. Although this is likely to be highly labour intensive, such resolution is needed before accurate observations and descriptions of processes in carabid ecology and spatial dynamics can be made. This is a necessary prerequisite before reliable analysis of causation is undertaken and implemented in effective management to secure a sustainable and productive future for agriculture. It may be a naive hope but, with sufficient resources and willpower, a properly co-ordinated and concerted collaboration could yield invaluable data and provide us with true insights into the factors that affect carabid distributions and abundance.

Work in progress at the time of writing towards these ends includes a four-year study sampling carabids in grids of traps over a contiguous block of six fields totalling ~ 75 ha in the UK (J.M. Holland *et al.*, 2000–2004). A wide range of biotic and abiotic factors are being measured simultaneously, together with large-scale mark–recapture experiments to determine the effects of field boundary structures on isolation of populations within fields. Other work in progress includes an investigation into the population genetics of *P. melanarius* sampled at a range of scales from field, farm, and region to national and European scales (C.F.G. Thomas *et al.*, 1998–2001). Combined with independent data on dispersal rates, this work expects to reveal a relationship between dispersal power and gene flow as a function of landscape structure and fragmentation, and will define the scale of interbreeding interaction groups and local populations.

Acknowledgements

Thanks to Dr Linton Winder for his constructive suggestions on reading an earlier draft. *Figures 11.1* and *11.2* are from data gathered during a BBSRC–CASE studentship with Rhone-Poulenc awarded to NJB. *Figure 11.3* uses data from the LINK Integrated Farming Systems Project which was funded by the Ministry of Agriculture, Fisheries and Food, Scottish Office Agriculture and Fisheries Department, Zeneca Agrochemicals, Home-Grown Cereals Authority (Cereals and Oilseeds) and the British Agrochemicals Association.

References

ADIS, J. (1979). Problems of interpreting arthropod sampling with pitfall traps. *Zoologischer Anzeiger Jena* **202**, 177–184.

ALVAREZ, T., FRAMPTON, G.K. AND GOULSON, D. (1997). Population dynamics of epigeic Collembola in arable fields: the importance of hedgerow proximity and crop type. *Pedobiologia* **41**, 110–114.

ANDERSEN, A. (1997). Densities of overwintering carabids and staphylinids (Col., Carabidae and Staphylinidae) in cereal and grass fields and their boundaries. *Journal of Applied Entomology* **121**, 77–80.

ASTERAKI, E.J., HANKS, C.B. AND CLEMENTS, R.O. (1995). The influence of different types of grassland field margin on carabid beetle (Coleoptera, Carabidae) communities. *Agriculture, Ecosystems and Environment* **54**, 195–202.

ATIENZA, J.C., FARINOS, G.P. AND ZABALLOS, J.P. (1996). Role of temperature in habitat selection and activity patterns in the ground beetle *Angoleus nitidus*. *Pedobiologia* **40**, 240–250.

BAARS, M.A. AND VAN DIJK, T.S. (1984). Population dynamics of two carabid beetles at a Dutch heathland. I. Subpopulation fluctuations in relation to weather and dispersal. *Journal of Animal Ecology* **53**, 375–388.

BAGUETTE, M. (1987). Spring distribution of carabid beetles in different plant communities of Belgian forests. *Acta Phytopathologica Entomologica Hungarica* **22**, 57–69.

BAGUETTE, M. (1993). Habitat selection of carabid beetle in deciduous woodlands of southern Belgium. *Pedobiologia* **37**, 365–378.

BASEDOW, T. (1994). Phenology and egg production in *Agonum dorsale* and *Pterostichus melanarius* (Col., Carabidae) in winter wheat fields of different growing intensity in Northern Germany. In: *Carabid beetles: ecology and evolution*. Eds. K. Desender, M. Dufrêne, M. Loreau, M.L. Luff and J.-P. Maelfait, pp 101–107. Dordrecht: Kluwer Academic Publishers.

BEDFORD, S.E. AND USHER, M.B. (1994). Distribution of arthropod species across the margins of farm woodlands. *Agriculture, Ecosystems and Environment* **48**, 295–305.

BEST, R.L., BEEGLE, C.C., OWENS, J.C. AND ORITZ, M. (1981). Population density, dispersion, and dispersal estimates for *Scarites substriatus*, *Pterostichus chalcites*, and *Harpalus pensylvanicus* (Carabidae) in an Iowa cornfield. *Environmental Entomology* **10**, 847–856.

BILDE, T., AXELSEN, J.A. AND TOFT, S. (2000). The value of Collembola from agricultural soils as food for a generalist predator. *Journal of Applied Ecology* **37**, 672–683.

BOHAN, D.A., BOHAN, A.C., GLEN, D.M., SYMONDSON, W.O.C., WILTSHIRE, C.W. AND HUGHES, L. (2000). Spatial dynamics of predation by carabid beetles on slugs. *Journal of Animal Ecology* **69**, 367–379.

BOITEAU, G. (1983). Activity and distribution of Carabidae, Arachnidae, and Staphylinidae in New Brunswick potato fields. *Canadian Entomologist* **115**, 1023–1030.

BOMMARCO, R. (1998). Reproduction and energy reserves of a predatory carabid beetles relative to agroecosystem complexity. *Ecological Applications* **8**, 846–853.

BOOIJ, C.J.H., DEN NIJS, L.J.M.F. AND NOORLANDER, J. (1995). Spatio-temporal patterns of activity density of some carabid species in large scale arable fields. *Acta Jutlandica* **70**, 175–184.

BOOIJ, K. (1994). Diversity patterns in carabid assemblages in relation to crops and farming systems. In: *Carabid beetles: ecology and evolution*. Eds. K. Desender, M. Dufrêne, M. Loreau, M.L. Luff and J.-P. Maelfait, pp 425–431. Dordrecht: Kluwer Academic Publishers.

BROWN, N.J. (2000). *Carabid ecology in organic and conventional farming systems: Population density, diversity and high resolution spatial dynamics*. PhD Thesis, University of Bristol.

BRUNSTING, A.M.H. (1981). Distribution patterns, life cycle and phenology of *Pterostichus oblongopunctatus* F. (Col., Carabidae) and *Philonthus decorus* Grav. (Col., Staphylinidae). *Netherlands Journal of Zoology* **31**, 418–452.

BRUST, G.E. (1990). Direct and indirect effects of four herbicides on the activity of carabid beetles (Coleoptera: Carabidae). *Pesticide Science* **30**, 309–320.

BRYAN, K.M. AND WRATTEN, S.D. (1984). The responses of polyphagous predators to prey spatial heterogeneity: aggregation by carabid and staphylinid beetles to their cereal aphid prey. *Ecological Entomology* **9**, 251–259.

BUREL, F. (1989). Landscape structure effects on carabid beetles spatial patterns in western France. *Landscape Ecology* **2**, 215–226.

CARCÁMO, H.A. AND SPENCE, J.R. (1994). Crop type effects on the activity and distribution of ground beetles (Coleoptera: Carabidae). *Environmental Entomology* **23**, 684–692.

CARMONA, D.M. AND LANDIS, D.A. (1999). Influence of refuge habitats and cover crops on seasonal activity-density of ground beetles (Coleoptera: Carabidae) in field crops. *Environmental Entomology* **28**, 1145–1153.

CHEESMAN, O. AND MITCHELL, H. (1995). A catalogue of predatory invertebrates collected at Leckford from crop and non-crop habitats by Southampton University personnel, 1984–1995. Leckford Survey Report No. 6. Leckford Estates, Hampshire, UK.

CHIVERTON, P.A. (1988). Searching behaviour and cereal aphid consumption by *Bembidion lampros* and *Pterostichus cupreus*, in relation to temperature and prey density. *Entomologia Experimentalis et Applicata* **47**, 173–182.

CLIFF, A.D. AND ORD, J.K. (1981). *Spatial processes: models and applications*. London: Pion Ltd.

COOMBES, D.S. AND SOTHERTON, N.W. (1986). The dispersal and distribution of polyphagous predatory Coleoptera in cereals. *Annals of Applied Biology* **108**, 461–474.

CRIST, T.O. AND AHERN, R.G. (1999). Effects of habitat patch size and temperature on the distribution and abundance of ground beetles (Coleoptera: Carabidae) in an old field. *Environmental Entomology* **28**, 681–689.

DEN BOER, P.J. (1968). Spreading of risk and stabilization of animal numbers. *Acta Biotheoretica* **18**, 165–194.

DEN BOER, P.J. (1977). Dispersal power and survival. Carabids in a cultivated countryside. *Miscellaneous papers LH Wageningen* **14**, 1–19.

DEN BOER, P.J. (1990a). The survival value of dispersal in terrestrial arthropods. *Biological Conservation* **54**, 175–192.

DEN BOER, P.J. (1990b). Density limits and survival of local populations in 64 carabid species with different powers of dispersal. *Journal of Evolutionary Biology* **3**, 19–48.

DEN BOER, P.J. (1991). Seeing the tree for the wood: random walks or bounded fluctuations of population size. *Oecologia* **86**, 484–491.

DEN NIJS, L.J.M.F., BOOIJ, C.J.H., DAAMEN, R., LOCK, C.A.M. AND NORLANDER, J. (1996). Can pitfall trap catches inform us about survival and mortality factors? An analysis of field data for *Pterostichus cupreus* (Coleoptera: Carabidae) in relation to crop rotation and crop specific husbandry practices. *Acta Jutlandica* **71**, 41–55.

DENNIS, P. AND FRY, G.L.A. (1992). Field margins: can they enhance natural enemy population densities and general arthropod diversity on farmland? *Agriculture, Ecosystems and Environment* **40**, 95–115.

DENNIS, P., THOMAS, M.B. AND SOTHERTON, N.W. (1994). Structural features of field boundaries which influence the overwintering densities of beneficial arthropod predators. *Journal of Applied Ecology* **31**, 361–370.

DESENDER, K. (1982). Ecological and faunal studies on Coleoptera in agricultural land. II. Hibernation of Carabidae in agro-ecosystems. *Pedobiologia* **23**, 295–303.

DESENDER, K. (1988). Spatial distribution of Carabid beetles in a pasture. *Revue d'Ecologie et de Biologie du Sol* **25**, 101–113.

DESENDER, K. AND MAELFAIT, J.-P. (1986). Pitfall trapping within enclosures: a method for estimating the relationship between the abundances of coexisting carabid species (Coleoptera, Carabidae). *Holarctic Ecology* **9**, 245–250.

DESENDER, K., DUFRÊNE, M. AND MAELFAIT, J.-P. (1994). Long-term dynamics of carabid beetles in Belgium: a preliminary analysis on the influence of changing climate and land use by means of a database covering more than a century. In: *Carabid beetles: ecology and evolution*. Eds. K. Desender, M. Dufrêne, M. Loreau, M.L. Luff and J.-P. Maelfait, pp 247–252. Dordrecht: Kluwer Academic Publishers.

DOANE, J.F. (1967). The influence of soil moisture and some soil physical factors on the oviposition behaviour of the Prairie Grain Wireworm, *Ctenicera destructor*. *Entomologia, Experimentalis et Applicata* **10**, 275–286.

DUELLI, P., STUDER, M., MARCHAND, I. AND JAKOB, S. (1990). Population movements of arthropods between natural and cultivated areas. *Biological Conservation* **54**, 193–207.

ELTON, C.S. (1966). *The pattern of animal communities*. London: Methuen.

ERICSON, D. (1978). Distribution, activity and density of some Carabidae (Coleoptera) in winter wheat fields. *Pedobiologia* **18**, 202–217.

EVANS, W.G. (1983). Habitat selection in the Carabidae. *The Coleopterists Bulletin* **37**, 164–167.

EYRE, M.D. (1994). Strategic explanations of carabid distributions in Northern England. In: *Carabid beetles: ecology and evolution*. Eds. K. Desender, M. Dufrêne, M. Loreau, M.L. Luff and J.-P. Maelfait, pp 267–275. Dordrecht: Kluwer Academic Publishers.

FADL, A. AND PURVIS, G. (1998). Field observations on the lifecycles and seasonal activity patterns of temperate carabid beetles (Coleoptera: Carabidae) inhabiting arable land. *Pedobiologia* **42**, 171–183.

FADL, A., PURVIS, G. AND TOWEY, K. (1996). The effect of time of soil cultivation on the incidence of *Pterostichus melanarius* (Illig.) (Coleoptera: Carabidae) in arable land in Ireland. *Annales Zoologici Fennici* **33**, 207–214.

FERNÁNDEZ GARCÍA, A., GRIFFITHS, G.J.K. AND THOMAS, C.F.G. (2000). Density, distribution and dispersal of the carabid beetle *Nebria brevicollis* in two adjacent cereal fields. *Annals of Applied Biology* **137**, 1–9.

FOURNIER, E. AND LOREAU, M. (1999). Effects of newly planted hedges on ground-beetle diversity (Coleoptera, Carabidae) in an agricultural landscape. *Ecography* **22**, 87–97.

FRANK, T. (1997). Species diversity of ground beetles (Carabidae) in sown weed strips and adjacent fields. *Biological Agriculture and Horticulture* **15**, 297–307.

GEIGER, R. (1957). *The climate near the ground*. Cambridge (Mass.): Harvard University Press.

GILPIN, M. AND HANSKI, I. (1991). Metapopulation dynamics: brief history and conceptual domain. *Biological Journal of the Linnean Society* **42**, 3–16.

GREENSLADE, P.J.M. (1964). The distribution, dispersal and size of a population of *Nebria brevicollis* (F.), with comparative studies on three other carabidae. *Journal of Animal Ecology* **33**, 301–309.

GREENSLADE, P.J.M. (1965). On the ecology of some British carabid beetles with reference to life histories. *Transactions of the Society for British Entomology* **16**, 149–179.

GUILLEMAIN, M., LOREAU, M. AND DAUFRESNE, T. (1997). Relationships between the regional distribution of carabid beetles (Coleoptera, Carabidae) and the abundance of their potential prey. *Acta Oecologica* **18**, 465–483.

HELENIUS, J. (1995). Rate and local scale spatial pattern of adult emergence of the generalist predator *Bembidion guttula* in an agricultural field. *Acta Jutlandica* **70**, 101–111.

HENGEVELD, R. (1979). The analysis of spatial patterns of some carabid beetles (Col. Carabidae). In: *Spatial and temporal analysis in ecology*. Eds. R.M. Cormack and J.K. Ord, pp 333–346. Maryland, USA: International Co-operative Publishing House.

HEYDEMANN, B. (1957). Die Biotopstruktur als Raumwiderstand und Raumfülle für die Tierwelt. *Verhandlungen Der Deutschen Zoologische Gesellschaft Fur Hamburg* **1956**, 332–347.

HOLLAND, J. AND FAHRIG, L. (2000). Effect of woody borders on insect density and diversity in crop fields: a landscape-scale analysis. *Agriculture, Ecosystems and Environment* **78**, 115–122.

HOLLAND, J.M. AND LUFF, M.L. (2000). The effects of agricultural practices on Carabidae in temperate agroecosystems. *Integrated Pest Management Reviews* **5**, 109–129.

HOLLAND, J.M. AND SMITH, S. (1999). Sampling epigeal arthropods: an evaluation of fenced pitfall traps using mark–release–recapture and comparisons to unfenced pitfall traps in arable crops. *Entomologia Experimentalis et Applicata* **91**, 347–357.

HOLLAND, J.M., COOK, S.K., DRYSDALE, A., HEWITT, M.V., SPINK, J. AND TURLEY, D. (1998). The impact on non-target arthropods of integrated compared to conventional farming: results from the LINK Integrated Farming Systems Project. *Proceedings of the 1998 Brighton Crop Protection Conference, Pests & Diseases* **2**, 625–630.

HOLLAND, J.M., PERRY, J.N. AND WINDER, L. (1999a). The within-field spatial and temporal distribution of arthropods in winter wheat. *Bulletin of Entomological Research* **89**, 499–513.

HOLLAND, J.M., WINDER, L. AND PERRY, J.N. (1999b). Arthropod prey of farmland birds: their spatial distribution within a sprayed field with and without buffer zones. *Aspects of Applied Biology* **54**, 53–60.

HOLOPAINEN, J.K., BERGMAN, T., HAUTALA, E.-L. AND OKSANEN, J. (1995). The ground beetle fauna (Coleoptera: Carabidae) in relation to soil properties and foliar fluoride content in spring cereals. *Pedobiologia* **39**, 193–206.

HONEK, A. (1988). The effect of crop density and microclimate on pitfall catches of Carabidae, Staphylinidae (Coleoptera) and Lycosidae (Araneae) in cereal fields. *Pedobiologia* **32**, 233–242.

HONEK, A. (1997). The effect of temperature on the activity of Carabidae (Coleoptera) in a fallow field. *European Journal of Entomology* **94**, 97–104.

IDINGER, J., KROMP, B. AND STEINBERGER, K.-H. (1996). Ground photoeclector evaluation of the numbers of carabid beetles and spiders found in and around cereal fields treated with either inorganic or compost fertilizers. *Acta Jutlandica* **71**, 255–267.

IVES, A.R. AND SETTLE, W.H. (1997). Metapopulation dynamics and pest control in agricultural systems. *The American Naturalist* **149**, 220–246.

JENSEN, T.S., DYRING, L., KRISTENSEN, B., NIELSEN, B.O. AND RASMUSSEN, E.R. (1989). Spring dispersal and summer habitat distribution of *Agonum dorsale* (Coleoptera, Carabidae). *Pedobiologia* **33**, 155–165.

JOYCE, K.A., HOLLAND, J.M. AND DONCASTER, C.P. (1999). Influences of hedgerow intersections and gaps on the movement of carabid beetles. *Bulletin of Entomological Research* **89**, 523–531.

KAJAK, A. AND LUKASIEWICZ, J. (1994). Do semi-natural patches enrich crop fields with predatory epigean arthropods? *Agriculture, Ecosystems and Environment* **49**, 149–161.

KIELTY, J.P., ALLEN-WILLIAMS, L.J., UNDERWOOD, N. AND EASTWOOD, E.A. (1996). Behavioural responses of three species of ground beetle (Coleoptera: Carabidae) to olfactory cues associated with prey and habitat. *Journal of Insect Behaviour* **9**, 237–250.

KINNUNEN, H. AND TIAINEN, J. (1999). Carabid distribution in a farmland mosaic: the effects of patch type and location. *Annales Zoologici Fennici* **36**, 149–158.

KOIVULA, M., PUNTTILA, P., HAILA, Y. AND NIEMELÄ, J. (1999). Leaf litter and the small scale distribution of carabid beetles (Coleoptera, Carabidae) in the boreal forest. *Ecography* **22**, 424–435.

KORIE, S., PERRY, J.N., MUGGLESTONE, M.A., CLARK, S.J., THOMAS, C.F.G. AND MOHAMAD ROFF, M.N. (2000). Spatiotemporal associations in beetle and virus count data. *Journal of Agricultural, Biological and Environmental Statistics* **5**, 214–239.

KRASNOV, B. AND SHENBROT, G. (1997). Seasonal variation in spatial organization of a Darkling Beetle (Coleoptera: Tenebrionidae) community. *Environmental Entomology* **26**, 178–190.

KROMP, B. AND NITZLADER, M. (1995). Dispersal of ground beetles in a rye field in Vienna, Eastern Austria. *Acta Jutlandica* **70**, 269–277.

KROMP, B. AND STEINBERGER, K.-H. (1992). Grassy field margins and arthropod diversity: a case study on ground beetles and spiders in eastern Austria (Coleoptera: Carabidae; Arachnida: Aranei, Opiliones). *Agriculture, Ecosystems and Environment* **40**, 71–93.

KUNO, E. (1991). Sampling and analysis of insect populations. *Annual Review of Entomology* **36**, 285–304.

LAGERLÖF, J. AND WALLIN, H. (1993). The abundance of arthropods along two field margins with different types of vegetation composition: an experimental study. *Agriculture, Ecosystems and Environment* **43**, 141–154.

LEATHER, S.R., WALTERS, K.F.A. AND BALE, J.S. (1993). *The ecology of insect overwintering*. Cambridge, England: Cambridge University Press.

LEVINS, R. (1970). Extinction. *Lectures on Mathematics in the Life Sciences* **2**, 77–107.

LINDROTH, C.H. (1992). *Ground beetles (Carabidae) of Fennoscandia. A zoogeographic study*. (English Translation). Andover, UK: Intercept.

LONGLEY, M. AND JEPSON, P.C. (1996). Effects of honeydew and insecticide residues on the distribution of foraging aphid parasitoids under glasshouse and field conditions. *Entomologia Experimentalis et Applicata* **81**, 189–198.

LOREAU, M. AND NOLF, C.-L. (1993). Occupation of space by the carabid beetle *Abax ater*. *Acta Oecologica* **14**, 247–258.

LÖVEI, G.L. AND SUNDERLAND, K.D. (1996). Ecology and behaviour of ground beetles (Coleoptera: Carabidae). *Annual Review of Entomology* **41**, 231–256.

LUFF, M.L. (1975). Some features influencing the efficiency of pitfall traps. *Oecologia* **19**, 345–357.

LUFF, M.L. (1986). Aggregation of some Carabidae in pitfall traps. In: *Carabid beetles: their adaptations and dynamics*. Eds. P.J. den Boer, M.L. Luff, D. Mossakowski and F. Weber, pp 385–397. Stuttgart: Gustav Fischer.

LUFF, M.L. (1987). Biology of polyphagous ground beetles in agriculture. *Agricultural Zoology Reviews* **2**, pp 237–276. Wimborne, Dorset, UK: Intercept.

LYS, J.A. (1994). The positive influence of strip-management on ground beetles in a cereal field: increase, migration and overwintering. In: *Carabid beetles: ecology and evolution*. Eds. K. Desender, M. Dufrêne, M. Loreau, M.L. Luff and J.-P. Maelfait, pp 451–455. Dordrecht: Kluwer Academic Publishers.

LYS, J.A. AND NENTWIG, W. (1991). Surface activity of carabid beetles inhabiting cereal fields. Seasonal phenology and the influence of farming operations on five abundant species. *Pedobiologia* **35**, 129–138.

LYS, J.A. AND NENTWIG, W. (1992). Augmentation of beneficial arthropods by strip management. 4. Surface activity, movements and activity-density of abundant carabid beetles in a cereal field. *Oecologia* **92**, 373–382.

MADER, H.J. AND MUHLENBERG, M. (1981). Species composition and distribution with natural resources of a little habitat island, exemplified by the carabid populations. *Pedobiologia* **21**, 46–59.

MADER, H.J., SCHELL, C. AND KORNACKER, P. (1990). Linear barriers to arthropod movements in the landscape. *Biological Conservation* **54**, 209–222.

MARSHALL, E.J.P. (1988). The ecology and management of field margin floras in England. *Outlook on Agriculture* **17**, 178–182.

MASCANZONI, D. AND WALLIN, H. (1986). The harmonic radar: a new method of tracing insects in the field. *Ecological Entomology* **11**, 387–390.

MAUDSLEY, M.J., WEST, T.M., ROWCLIFFE, H.R. AND MARSHALL, E.J.P. (2000). The impacts of hedge management on wildlife: preliminary results for plants and insects. *Aspects of Applied Biology* **58**, 389–396.

MAUREMOOTO, J.R., WRATTEN, S.D., WORNER, S.P. AND FRY, G.L.A. (1995). Permeability of hedgerows to predatory carabid beetles. *Agriculture, Ecosystems and Environment* **52**, 141–148.

MELBOURNE, B.A. (1999). Bias in the effect of habitat structure on pitfall traps: an experimental evaluation. *Australian Journal of Ecology* **24**, 228–239.

MIDGARDEN, D.G., YOUNGMAN, R.R. AND FLEISCHER, S.J. (1993). Spatial analysis of counts of western corn rootworm (Coleoptera: Chrysomelidae) adults on yellow sticky traps in corn: geostatistics and dispersion indices. *Environmental Entomology* **22**, 1124–1133.

MONSRUD, C. AND TOFT, S. (1999). The aggregative numerical response of polyphagous predators to aphids in cereal fields: attraction to what? *Annals of Applied Biology* **134**, 265–270.

NELEMANS, M.N.E. (1987). On the life-history of the carabid beetle *Nebria brevicollis* (F.). *Netherlands Journal of Zoology* **37**, 26–42.

NELEMANS, M.N.E. (1988). Surface activity and growth of larvae of *Nebria brevicollis* (F.) (Coleoptera, Carabidae). *Netherlands Journal of Zoology* **38**, 74–95.

NELEMANS, M.N.E., DEN BOER, P.J. AND SPEE, A. (1989). Recruitment and summer diapause in the dynamics of a population of *Nebria brevicollis* (Coleoptera: Carabidae). *Oikos* **56**, 157–169.

NENTWIG, W. (1989). Augmentation of beneficial arthropods by strip management. 2. Successional strips in a winter wheat field. *Journal of Plant Diseases and Protection* **96**, 89–99.

NIEMELÄ, J. AND HALME, E. (1992). Habitat associations of carabid beetles in fields and forests on the Åland Islands, SW Finland. *Ecography* **15**, 3–11.

NIEMELÄ, J., HAILA, Y. AND RANTA, E. (1986). Spatial heterogeneity of carabid beetle dispersion in uniform forests the Åland Islands, SW Finland. *Annales Zoologici Fennici* **23**, 289–296.

NIEMELÄ, J., HAILA, Y., HALME, E., PAJUNEN, T. AND PUNTTILA, P. (1992). Small-scale heterogeneity in the spatial distribution of carabid beetles in the southern Finnish taiga. *Journal of Biogeography* **19**, 173–181.

PAJE, F. AND MOSSAKOWSKI, D. (1984). pH-preferences and habitat selection in carabid beetles. *Oecologia* **64**, 41–46.

PAVUK, D.M., PURRINGTON, F.F., WILLIAMS, C.E. AND STINNER, B.R. (1997). Ground beetle (Coleoptera: Carabidae) activity density and community composition in vegetationally diverse corn agroecosystems. *The American Midland Naturalist* **138**, 14–28.

PERRY, J.N. (1995). Spatial analysis by distance indices. *Journal of Animal Ecology* **64**, 303–314.

PERRY, J.N. (1997). Spatial association for counts of two species. *Acta Jutlandica* **72**, 149–169.

PERRY, J.N. (1998). Measures of spatial pattern and spatial association for insect counts. *Proceedings of the 20th International Congress of Entomology – Population and community ecology for insect management and conservation, Florence, 25–31 August 1996*, pp 21–33.

PERRY, J.N. AND HEWITT, M. (1991). A new index of aggregation for animal counts. *Biometrics* **47**, 1505–1518.

PERRY, J.N., BELL, E.D., SMITH, R.H. AND WOIWOD, I.P. (1996). SADIE: software to measure and model spatial patterns. *Aspects of Applied Biology* **46**, 95–102.

PERRY, J.N., WINDER, L., HOLLAND, J.M. AND ALSSTON, R.D. (1999). Red-blue plots for detecting clusters in count data. *Ecology Letters* **2**, 106–113.

PETIT, S. (1994). Diffusion of forest carabid beetles in hedgerow network landscapes. In: *Carabid beetles: ecology and evolution*. Eds. K. Desender, M. Dufrêne, M. Loreau, M.L. Luff and J.-P. Maelfait, pp 337–341. Dordrecht: Kluwer Academic Publishers.

PETIT, S. AND BUREL, F. (1993). Movement of *Abax ater* (Col. Carabidae): Do forest species survive in hedgerow networks? *Vie Milieu* **43**, 119–124.

POLLARD, E., HOOPER, M.D. AND MOORE, N.W. (1974). *Hedges*. London: Collins & Sons.

POTTS, G.R. AND VICKERMAN, G.P. (1974). Studies on the cereal ecosystem. *Advances in Ecological Research* **8**, 107–197.

PURVIS, G. AND CURRY, J.P. (1984). The influence of weeds and farmyard manure on the activity of carabidae and other ground dwelling arthropods in a sugar beet crop. *Journal of Applied Ecology* **21**, 271–283.

PURVIS, G. AND FADL, A. (1996). Emergence of Carabidae (Coleoptera) from pupation: a technique for studying the 'productivity' of carabid habitats. *Annales Zoologici Fennici* **33**, 215–223.

RACKHAM, O. (1995). *The history of the countryside*. London: Weidenfeld & Nicolson.

RIECKEN, U. AND RATHS, U. (1996). Use of radio telemetry for studying dispersal and habitat use of *Carabus coriaceus* L. *Annales Zoologici Fennici* **33**, 109–116.

RIEDEL, W. (1995) Spatial distribution of hibernating polyphagous predators within field boundaries. *Acta Jutlandica*, **70**, 221–226.

RIVARD, I. (1966). Ground beetles (Coleoptera: Carabidae) in relation to agricultural crops. *The Canadian Entomologist* **98**, 189–195.

RODENHOUSE, N.L., BARRETT, G.W., ZIMMERMAN, D.M., AND KEMP, J.C. (1992). Effects of uncultivated corridors on arthropod abundances and crop yields in soybean agroecosystems. *Agriculture, Ecosystems and Environment* **38**, 179–191.

RUSHTON, S., SANDERSON, R., LUFF, M. AND FULLER, R. (1996). Modelling the spatial dynamics of ground beetles (Carabidae) within landscapes. *Annales Zoologici Fennici* **33**, 233–241.

SANDERSON, R.A. (1994). Carabidae and cereals: a multivariate approach. In: *Carabid beetles: ecology and evolution*. Eds. K. Desender, M. Dufrêne, M. Loreau, M.L. Luff and J.-P. Maelfait, pp 457–463. Dordrecht: Kluwer Academic Publishers.

SANDERSON, R.A., RUSHTON, S.P., CHERRILL, A.J. AND BYRNE, J.P. (1995). Soil, vegetation and space: an analysis of their effects on the invertebrate communities of a moorland in north-east England. *Journal of Applied Ecology* **32**, 506–518.

SAWYER, A.J. AND HAYNES, D.L. (1985). Simulating the spatiotemporal dynamics of the cereal leaf beetle in a regional crop system. *Ecological Modelling* **30**, 83–104.

SAWYER, A.J. AND HAYNES, D.L. (1986). Cereal leaf beetle spatial dynamics: simulations with a random diffusion model. *Ecological Modelling* **33**, 89–99.

SCHERNEY, F. (1960). Beitrage zur Biologie und okonomischen Bedeutung rauberisch lebender Kaferarten. Untersuchungen uber da Auftreten von Laufkafern (Carabidae) in Feldkulturen (Teil II). *Zeitschrift fur Angewandte Entomologie* **47**, 231–255.

SHERRATT, T.N. AND JEPSON, P.C. (1993). A metapopulation approach to modelling the long-term impact of pesticides on invertebrates. *Journal of Applied Ecology* **30**, 696–705.

SKUHRAVY, V. (1958). Influence of field cultivation on seasonal occurrence of carabids. *Zoologicke Listy* **7**, 325–328.

SOTHERTON, N.W. (1984). The distribution and abundance of predatory arthropods overwintering on farmland. *Annals of Applied Biology* **105**, 423–429.

SOTHERTON, N.W. (1985). The distribution and abundance of predatory coleoptera overwintering in field boundaries. *Annals of Applied Biology* **106**, 17–21.

SOUTHWOOD, T.R.E. (1978). *Ecological methods*, 2nd edition. London: Chapman & Hall.

SPEIGHT, M.R. AND LAWTON, J.H. (1976). The influence of weed-cover on the mortality imposed on artificial prey by predatory ground beetles in cereal fields. *Oecologia* **23**, 211–223.

SYRJALA, S.E. (1996). A statistical test for a difference between the spatial distributions of two populations. *Ecology* **77**, 75–80.

TAYLOR, L.R. (1984). Assessing and interpreting the spatial distributions of insect populations. *Annual Review of Entomology* **29**, 321–357.

TELFER, M.G., MEEK, W.R., LAMBDON, P., PYWELL, R.F., SPARKS, T.H. AND NOWAKOWSKI, M. (2000). The carabids of conventional and widened field margins. *Aspects of Applied Biology* **58**, 411–416.

THIELE, H.-U. (1977). *Carabid beetles in their environments. A study on habitat selection by adaptations in physiology and behaviour*. Berlin: Springer-Verlag.

THIELE, H.-U. (1979). Relationships between annual and daily rhythms, climatic demands and habitat selection in carabid beetles. In: *Carabid beetles: their evolution, natural history, and classification*. Proceedings of the First International Symposium of Carabidology. Eds. T.L. Erwin, G.E. Ball, D.R. Whitehead and A.L. Halpern, pp 449–470. The Hague: Dr. W. Junk bv Publishers.

THOMAS, C.F.G. AND MARSHALL, E.J.P. (1998). Arthropod abundance and diversity in differently vegetated margins of arable fields. *Agriculture, Ecosystems and Environment* **72**, 131–144.

THOMAS, C.F.G., PARKINSON, L. AND MARSHALL, E.J.P. (1998). Isolating the components of activity-density for the carabid beetle *Pterostichus melanarius* in farmland. *Oecologia* **116**, 103–112.

THOMAS, C.F.G., PARKINSON, L., GRIFFITHS, G.J.K., GARCIA, A.F. AND MARSHALL E.J.P. (2001). Aggregation and temporal stability of carabid beetle distributions in field and hedgerow habitats. *Journal of Applied Ecology* **38**, 100–116.

THOMAS, M.B., MITCHELL, H.J. AND WRATTEN, S.D. (1992a). Abiotic and biotic factors influencing the winter distribution of predatory insects. *Oecologia* **89**, 78–84.

THOMAS, M.B., WRATTEN, S.D. AND SOTHERTON, N.W. (1992b). Creation of 'island' habitats in farmland to manipulate populations of beneficial arthropods: predator densities and species composition. *Journal of Applied Ecology* **29**, 524–531.

THOMAS, S.R., GOULSON, D. AND HOLLAND, J.M. (2000). Spatial and temporal distributions of predatory Carabidae in a winter wheat field. *Aspects of Applied Biology* **62**, 55–60.

TURCHIN, P. (1989). Population consequences of aggregative movement. *Journal of Animal of Ecology* **58**, 75–100.

USHER, M.B., FIELD, J.P. AND BEDFORD, S.E. (1993). Biogeography and diversity of ground-dwelling arthropods in farm woodlands. *Biodiversity Letters* **1**, 54–62.

VAN DIJK, T.S. (1996). The influence of environmental factors and food on life cycle, ageing and survival of some carabid beetles. *Acta Jutlandica* **71**, 11–24.

VANSTEENWEGEN, C. (1987). Distribution of carabid beetles in natural areas within an agricultural region. *Acta Phytopathologica Entomologica Hungarica* **22**, 439–448.

WADE FRENCH, B. AND ELLIOTT, N.C. (1999a). Spatial and temporal distribution of ground beetle (Coleoptera: Carabidae) assemblages in riparian strips and adjacent wheat fields. *Environmental Entomology* **28**, 597–607.

WADE FRENCH, B. AND ELLIOTT, N.C. (1999b). Temporal and spatial distribution of ground beetle (Coleoptera: Carabidae) assemblages in grasslands and adjacent wheat fields. *Pedobiologia* **43**, 73–84.

WALLIN, H. (1985). Spatial and temporal distribution of some abundant carabid beetles (Coleoptera: Carabidae) in cereal fields and adjacent habitats. *Pedobiologia* **28**, 19–34.

WALLIN, H. (1986). Habitat choice of some field-inhabiting carabid beetles (Coleoptera: Carabidae) studied by recapture of marked individuals. *Ecological Entomology* **11**, 457–466.

WALLIN, H. (1987). Dispersal and migration of carabid beetles inhabiting cereal fields. *Acta Phytopathologica Entomologica Hungarica* **22**, 449–453.

WALLIN, H. AND EKBOM, B.S. (1988). Movements of carabid beetles (Coleoptera: Carabidae) inhabiting cereal fields: a field tracing study. *Oecologia* **77**, 39–43.

WALLIN, H. AND EKBOM, B. (1994). Influence of hunger level and prey densities on movement patterns in three species of Pterostichus beetles (Coleoptera: Carabidae). *Environmental Entomology* **23**, 1171–1181.

WINDER, L., PERRY, J.N. AND HOLLAND, J.M. (1999). The spatial and temporal distribution of the grain aphid *Sitobion avenae* in winter wheat. *Entomologia Experimentalis et Applicata* **93**, 277–290.

WINDER, L., WOOLLEY, C., HOLLAND, J.M., PERRY, J.N. AND ALEXANDER, C.J. (2000). The field scale distribution of insects in winter wheat. *2000 Brighton Crop Protection Conference, Pests & Diseases* **2**, 573–578.

WRATTEN, S.D. AND THOMAS, C.F.G. (1990). Farm-scale spatial dynamics of predators and parasitoids in agricultural landscapes. In: *Species dispersal in agricultural landscapes*. Eds. R.G.H.Bunce and D.C.Howard, pp 219–237. London: Belhaven Press.

Index

Abacization, 113
Abax spp.
  generalist carnivores, 83
  predation by birds, 23
Abax parallelepipedus
  chemoreceptors, 122
  detection period for anti-prey monoclonal antibodies, 149
  diet, 95–97, 166
    larval, 85
  digestion rate, 154
  digestive enzymes of, 98
  dispersal ability, 8
  distribution, 309, 311, 318, 325
  pest control by, 195
  prey detection, 86
Abundance
  associated with non-crop headlands, 288
  decline in, 24–28
    see also Population
Abutilon theophrasti, seed predation by carabids, 225
Acquired aversion to sub-optimal prey, 99
Activity-density, effect of habitat on, 286–289
Agonoderus pallipes, see Stenolophus lecontei
Agonum spp., 59–61
  dispersal ability, 8
  ecology of, 63, 64
  locomotion, 115, 314
  predation, 23
Agonum dorsale
  body size, 283
  crop cover preference, 13, 233
  diet, 83, 91, 96, 97, 100, 138, 166, 200, 201
  distribution, 61, 318
  ecology of, 64
  effect of ploughing on, 239
  egg production, 16
  fecundity, 19, 284, 286
  feeding condition, 283, 284
  food limitation, 101
  food preference, 99
  insecticide application and, 241
  life history, 6
  overwintering, 280, 312
  parasitism of, 15
  predator–prey relationships, 193
  response to honeydew, 324
  seasonal abundance, 233
  seasonal movement, 314–316
  soil organic matter and, 240
  soil pH requirements, 13
  in temperate arable farming systems, 255
Agonum muelleri
  crop cover preference, 234, 235
  ecology of, 64
  effect of ploughing on, 239
  overwintering, 312
  soil organic matter and, 240
Agonum placidum
  ecology of, 64
  effect of ploughing on, 239
Agonum punctiform, seed predator, 217
Agricultural habitats
  carabid assemblages in, 231, 232
  ecology of carabids in, 5, 6
  spatial distribution of carabids in, 305–336
  species composition, 57–65
  species dominance, 52–56
  species richness, 44–52
  see also Crop husbandry practices, Farming methods, Habitat management, Spatial distribution of carabids
Agriotes sputator, consumed by carabids, 166
Agrotis ipsilon, see Cutworm
Alopecurus myosuroides, seed predation by carabids, 220
Amara spp., 59–61
  crop cover preference, 14
  diet, 84, 94, 95, 124
  dispersal ability, 8
  distribution, 318, 319
  ecology of, 63
  predation by birds, 23
  proventriculus, 120
  seasonal movement, 315
  in temperate arable farming systems, 255
Amara aenae
  diet, 166, 220, 221
  overwintering, 312
Amara aulica
  adaptations to seed-feeding habit, 218
  ecology of, 63
Amara communis, predation by mammals, 23
Amara cupreolata
  diet, 125, 217
  ecology of, 63
Amara familiaris
  adaptations to seed-feeding habit, 218
  overwintering, 312
  seed choice, 221
Amara impuncticollis, diet, 125
Amara obesa, 63
Amara plebeja
  diet, 125
  locomotion, 115

overwintering, 312
Amara similata, diet, 97, 100
larval, 85
Amara spreta, 62
Amaranthus retroflexus, seed predation by carabids, 221–223
Amarini, seed predators, 217
Ambrosia artemisiifolia, seed predation by carabids, 221, 222
Amphibians, carabid predators, 24
Angoleus nitidus, temperature requirements, 9, 326
Anisodactylus spp., 60, 64
Anisodactylus santaecrucis, 64
Anisodactylus signatus, 64
Aphid alarm pheromone, response of carabids to, 186
Aphids
chemical marking of, 138
consumed by carabids, 166
control by carabids, 189, 192, 193, 198–201
simulation models, 193, 194
DNA techniques to detect predation of, 156
effect of non-crop habitats on populations of, 291
ELISA used to detect predation of, 153
food value of, 89–92, 97, 98
killed by carabids, 174, 186
predator–pest associations, 192
scavenging of, 174
Apple maggots, consumed by carabids, 166
Apristus spp., 60, 65
Archips argyrospila larvae, killed by carabid larvae, 186
Artioposthia triangulata, see New Zealand flatworms
Asaphidion spp., 58
diet, 124
field of vision, 113
Asaphidion curtum, life history, 6
Asaphidion flavipes
dispersal, 8
effect of ploughing on, 239
overwintering, 312
soil organic matter and, 240
Assemblage composition
dominance, 52–56
effects of crop on, 69, 70
effects of environmental factors on, 66–70
effects of management on, 56, 57
sampling, 42–44
species, 57–65
Australian, 61
European, 58–61
Japanese, 58–61
North American, 58–61
species richness, 44–52
Assemblage organization, 41–71
Avena fatua, seed predation by carabids, 220

Barriers to dispersal, 332, 333
Bats, carabid predators, 23
Beetle banks, 237, 280, 293, 294, 311, 315
economics, 296
Beetle eggs, killed by carabid larvae, 185
Bembidion spp., 58, 60
decline in abundance, 25, 28
diet, 83, 95, 217
ecology of, 62, 63
fungal infection of, 14
generalist insectivores, 83
locomotion, 115, 314
nematode infection of, 14
overwintering, 14
predation by birds, 23
seed predators, 217
temperature requirements, 9
Bembidion femoratum
insecticide application and, 241
soil organic matter and, 240
Bembidion guttula, distribution, 328
Bembidion lampros
chemoreceptors, 122
control of aphids by, 92, 194
decline in abundance, 25, 28
diet, 92, 93, 100, 102, 166, 174, 186
distribution, 19, 61, 287, 319, 329, 330
ecology of, 62
effect of ploughing on, 239
egg production, 16
foraging cycle, 86, 87
insecticide applications and, 241
life history, 6
overwintering, 236, 312
in pitfall traps, 3, 4
scavenger, 174
seasonal movement, 314–316
soil organic matter and, 240
in temperate arable farming systems, 255
Bembidion obtusum
bioindicator, 272
ecology of, 62, 63
effect of ploughing on, 239
overwintering, 312
scavenger, 174
Bembidion properans
diet, 166
foraging cycle, 86
overwintering, 312
Bembidion quadrimaculatum
control of onion fly by, 194
crop cover preference, 235
diet, 166, 174. 186
distribution, 61
ecology of, 62
effect of ploughing on, 239
fertilizer input and, 271
insecticide application and, 241
pesticide application and, 271
predator–pest association, 192
soil organic matter and, 240
Bembidion ustulatum, crop cover preference, 234
Bioindicator species, 251–274
see also named species

Birds, carabid predators, 21, 22, 23
Blossom midges, consumed by carabids, 166
Body form, 112–114
  and feeding preferences, 115, 116, 187
Body size
  distribution, 65
  effects of habitat on, 283
  indicator of habitat disturbance, 65
Boundary carabids, 282
Brachinus spp., 60
Brachinus explodens
  diet, 140
  ecology of, 65
Bradycellus spp., flight activity, 314
Brassica pod midge, predator–prey relationships, 193
Broscus spp., 58
Broscus cephalotes
  ecology of, 62
  mandibles, 117
  proventriculus, 121
Bush flies, precipitin test to measure predation of, 152

Cabbage moth, control by pathogens and carabids, 198
Cabbage root fly
  consumed by carabids, 166
  control by carabids, 195, 243
  killed by carabids, 174, 187
  precipitin test to measure predation of, 151
Cabbage stem flea beetle, predator–prey relationships, 193
Calathus spp., 59
  diet, 83, 95
  ecology of, 64
Calathus fuscipes
  distribution, 61
  ecology of, 64
  foraging cycle, 87
  life history, 6
  parasitism of, 15
Calathus melanocephalus
  aphid consumption, 92
  distribution, 321
  ecology of, 64
  egg production, 16
  fertilizer input and, 271
  pesticide application and, 271
  size determination, 19
Calathus mollis, 62
Calathus opaculus, 64
Calathus piceus, distribution, 318
Calleida decora, larval diet, 187
Calleida viridipennis, larval diet, 186
Calosoma spp., 58
  body form, 112, 113
  caterpillar specialists, 84, 94
  ecology of, 61, 62
  larval diet, 112
  maxillae, 118
  prey detection, 86
Calosoma calidum, mouthparts, 119
Calosoma inquisitor, mandibles, 117
Calosoma maximowiczi, pest control by, 188
Calosoma sayi
  larval diet, 186, 187
  pathogen transmission by, 198
Calosoma scrutator, mandibles, 116
Calosoma sycophanta
  control of gypsy moth by, 111, 128, 129, 188, 199
  diet, 94, 95, 102
  foraging, 86
  locomotion, 115
Cannibalism, 7, 17
Capillary ring test, 151
Capsella bursa-pastoris, seed predation by carabids, 220
Carabid research, history of, 1–5
Carabini, diet, 124
Carabus spp., 58
  body form, 112, 113
  diet, 83, 95
    larval, 85
  digestive enzymes of, 98
  distribution, 232
  ecology of, 61, 62
  maxillae, 118
  predation, 23
  prey detection, 86
  proventriculus, 120, 121
Carabus auratus, life history, 6
Carabus auronitens, egg production, 15, 16
Carabus cancellatus
  distribution, 232
  effect of agricultural intensification on, 241
Carabus coriaceus, distribution, 311
Carabus granulatus
  diet, 102
  distribution, 311
Carabus nemoralis, diet, 138
Carabus problematicus, mandibles, 117
Carabus scabriusculus, distribution, 232
Carabus violaceus
  diet, 140
  distribution, 232
Carrot crops, carabid populations in, 233
Carrotfly eggs, killed by carabids, 174
Cassia obtusifolia, seed predation by carabids, 221
Caterpillars, predation by carabids, 84, 94, 151, 166, 174, 185–187
  see also Gypsy moth
Caterpillar specialists, 84, 94
Cereal crops, carabid activity in, 233
Cereal leaf beetles, effect of non-crop habitats on populations of, 291
Chemical labelling of prey, 138, 139, 193
Chemoreceptors, 121–123
Chenopodium album, seed predation by carabids, 222, 223
Chickweed, predation by carabids, 218
Chilo suppressalis, killed by carabid larvae, 185

Chlaenius spp., 60, 120
Chlaenius micans, 64
Chlaenius tricolor, proventriculus, 120
Chlorpyrifos, effect on carabid populations, 272
Chrysomelid beetles, predation by carabids, 166, 174
Cicindela spp.
  diet, 85
  field of vision, 113
  locomotion, 114
  prey detection, 86
Cicindela germanica, distribution, 232
Cicindela hybrida, foraging cycle, 87
Cicindela punctulata, locomotion, 114
Cicindela sexguttata
  body form, 112
  chemoreceptors, 122
  mouthparts, 112
Cicindelini, diet, 124
Cinnabar moth larvae, precipitin test to measure predation of, 151
Circadian rhythm, 10
Clivina spp., 58, 61, 62, 85
Clivina fossor
  diet, 85, 166
  ecology of, 62
  effect of ploughing on, 239
  fertilizer input and, 271
  insecticide application and, 241
  overwintering, 312
  pesticide application and, 271
  soil organic matter and, 240
Clivina impressifrons, 62
Codling moth larvae, predation by carabids, 187
Co-evolution, 219
Collembola, 86, 93
  distribution within fields, 324
  food value of, 91–93
  marking with radioactive isotopes, 139
  pesticide application, effects on populations, 264
Community-level analysis, 272, 273
Conservation headlands, effects on carabid populations, 236, 237, 243, 280
Corn borer
  in carabid diet, 140
  ELISA used to detect predation of, 153
Corn earworms, precipitin test to measure predation of, 152
Corvids, carabid predators, 22
Cratacanthus dubius
  body form, 116
  locomotion, 114
Crop density, effects on carabid populations, 234, 235, 332
Crop husbandry practices, effects on carabid populations, 231–244, 239
  conservation headlands, 236, 237
  crop density, 234, 235, 332
  crop rotation, 234
  crop types, 69, 70, 232, 233, 256, 257, 259, 332
  deep tillage, 237, 238
  fertilizers, 240, 241
  herbicides, 242
  insecticides, 241, 242
  intercropping, 235
  monitoring, 232
  soil cultivation methods, 237–240
  weed control, 242
  weed cover influence, 236
  see also Farming methods, Habitat management
Crop types, effects on carabid populations, 69, 70, 232, 233, 256, 257, 259, 332
Cross-over electrophoresis, 152
Cutworm
  incidence, 238
  larvae, precipitin test to measure predation of, 151, 152
  predation by carabids, 166, 174
Cychrini
  body form, 112
  diet, 83, 124
    larval, 85
Cychrization, 112
Cychrus spp.
  distribution, 232
  maxillae, 118
  pest control by, 188
  prey detection, 86
Cychrus caraboides
  mandibles, 117
  maxillae, 118
  predation by mammals, 23
Cyclical colonization, 279
Cyclotrachellus alternana, crop cover preference, 233
Cyclotrachelus sodalis, effect of ploughing on, 239
Cydia pomonella, larvae, predation by carabids, 187
Cymindis spp., proventriculus, 120
Cymindis limbata, locomotion, 115
Cymindis platicollis, body form, 113
Cypermethrin, effects on carabid populations, 241

Dactylis glomerata, overwintering of carabids in, 280, 281, 293, 313
Dasineura brassicae, predator–prey relationships, 193
Datura stramonium, seed predation by carabids, 221, 222
Deep tillage, effects on carabid populations, 238, 243
Delia antiqua, *see* Onion fly
Delia radicum, *see* Cabbage root fly
Demetrias spp., 61
  decline in abundance, 25, 27
Demetrias atricapillus
  distribution, 287
  ecology of, 65
  overwintering, 14, 18, 280, 281, 312, 313
  populations in beetle banks, 237
  seasonal movement, 314, 316

Diabrotica spp., killed by carabids, 174, 185, 187
Diachromus spp., granivores, 84
Diagnostic techniques for determining diet, 137–158
chemical labelling, 138, 139
immunological techniques, 140–154
molecular detection systems, 155, 156
protein electrophoresis, 139, 140
Diet, 19, 81–103
diagnostic techniques for determining, 137–158
chemical labelling, 138, 139
immunological techniques, 140–154
molecular detection systems, 155, 156
protein electrophoresis, 139, 140
methods of recording, 82
morphology and, 111–131
quality, 87–89
see also Food, named species
Dietary mixing, 100
Dietary specialization, physiological and biochemical correlates of, 98
Digestive system, 119–121
Digitaria sanguinalis, seed predation by carabids, 222
Dimethoate, effects on carabid populations, 241
Diplocheila impressicollis, diet, 166
Diplopods, food value of, 94
Diptera, 192
food value of, 94
see also Fruit fly
Distribution of carabids, *see* Spatial distribution
Ditomus spp., granivores, 84
Diurnal activity, 10, 12
Diversity, 45, 46, 233, 289, 290
DNA techniques, for detection of prey, 155, 156, 158
Dolichus spp., 59
Dolichus halensis
distribution, 61
ecology of, 64
Dromius spp., 61
body form, 113
diet, 85
Dyschirius spp., 58, 62
soil particle size preference, 13

Earthworms
ELISA used to detect predation of, 153
food value of, 90, 96, 150, 151
Elaphrini, diet, 124
Elaphrus spp., 58
field of vision, 113
maxillae, 118
ELISA, 140, 141, 144–147, 152–154, 174
Esterase isoenzyme patterns, 140
Eurygaster integriceps, consumed by carabids, 166
Eurynebria complanata, diet, 96
Euxoa ochrogaster, precipitin test to measure predation of, 151
Evarthrus spp., 60, 64
Exploitative competition, 101

Fall armyworm, control by pathogens and carabids, 198
Farming methods
carabid numbers and, 224, 225, 238, 258, 259, 265, 266
carabid species and, 258–260, 265, 266
see also Crop husbandry practices, Habitat management
Feeding guilds, 82–85, 96–98
Feeding habits, observation of, 123–125
Females, temperature requirements, 9
Fertilizer input, effects on carabid populations, 240, 241
Field margins, 310–313
aestivation sites, 313
overwintering carabids in, 312
see also Crop husbandry practices, Non-crop habitats
Field of vision, 113
Fluid feeders, 116, 117, 121
Food
limitation, 101
preferences, 98, 99, 100
value, 81–103
aphids, 89, 91, 92
collembola, 91–93
diplopods, 94
Diptera, 94
earthworms, 90, 96
fruit fly, 89, 92, 93
gastropods, 90
Lepidoptera, 90, 94
mixed animal diet, 91
mixed insects, 90
pest limitation and, 102
plant material, 95
sciarid midges, 94
seeds, 90, 94, 95
slugs, 95, 96
snails, 95, 96
Foraging, 85, 86
Foraging cycles, 86, 87
Fragment feeders, 82, 116, 117, 121
Fruit fly, food value of, 89, 92, 93

Galerita spp., maxillae, 118
Gastropods, food value of, 90
Generalist carnivores, 83
Generalist insectivores, 83
Genetically modified prey, effect on feeding, 99, 100
Granivores, 84
Grey partridge, carabid predator, 21–24
Ground cover, effect on carabid populations, 233
Gut analysis
limitations, 126
microscopic identification of remains, 126, 127
pest species detected by, 166–174
qualitative aspects, 129–131
quantative aspects, 127–129

rate of food passage, 129, 130
seed remains, 219
Gut dissection, 125–131
Gypsy moth
chemical marking of, 138
control by *Calosoma sycophanta*, 111, 128, 129, 188, 198
detection in diet, 140
ELISA used to detect predation of, 153, 174

Habitat management, 279–297
see also Crop husbandry practices, Farming methods, Non-crop habitats
Harpalini, seed predators, 217
Harpalus spp., 60, 61
crop cover preference, 14
diet, 94, 124
larval, 187
ecology of, 64
flight activity, 314
granivores, 84
hibernation, 7
predation by birds, 23
proventriculus, 120, 121
soil particle size preference, 13
Harpalus aeneus
diet, 128
distribution, 61, 217
ecology of, 64
population size, 8
Harpalus affinis
diet, 97, 166
food preference, 100
foraging cycle, 87
predation by mammals, 23
pathogen transmission by, 198
Harpalus caliginosus, mandibles, 116
Harpalus distinguendus, crop cover preference, 235
Harpalus erraticus, 64
Harpalus fuliginosus, diet, 102
Harpalus pensylvanicus
body form, 116
crop cover preference, 13
diet, 125, 174
ecology of, 64
effect of ploughing on, 239
locomotion, 114, 116
mandibles, 117
pest control by, 192, 198
proventriculus, 120
seed predator, 217
wedge pushing, 113
Harpalus rufipes
adaptations to seed-feeding habit, 218
chemical fertilizers and, 240
crop cover preference, 13
detection of aphids by, 186
diet, 95, 97, 100, 125, 166
larval, 85, 218
dispersal ability, 8
distribution, 61, 217, 318
ecology of, 64
effect of ploughing on, 239
foraging cycle, 87
life history, 6
mortality rate, 7
parasitism of, 15
pathogen transmission by, 198
population size, 8
predation by mammals, 23
predator–prey relationships, 193
prey detection, 86
seasonal movement, 316
seed choice, 221, 222
seed detection, 218
seed predator, 217, 220, 221
soil organic matter and, 240
temperature requirements, 9
Hedgehogs, carabid predators, 23
Hedgerows, carabid distribution in, 310–313
see also Non-crop habitat
Heliothis zea, precipitin test to measure predation of, 152
Herbicides, effects on
carabid populations, 242
larvae, 13, 14
Herbivorous feeding, 84, 124, 125
Honeydew, attraction of carabids to, 92, 189, 324
Hypera postica, 125
antiserum against, 148
Hyphantria cunea larvae, killed by carabid larvae, 186

Immunodot assay, 154
Immunoelectroosmophoresis, 152
Immunological techniques for determining diet, 140–154, 193, 219, 220
assay systems, 150–154
ELISA, 140, 141, 144–147, 150, 152–154, 174
monoclonal antibodies, 148–150
polyclonal antisera, 141, 148
precipitin test, 140–145, 150–152
capillary ring test, 151
raising and characterizing antibodies, 141, 148–150
Indicator species, *see* Bioindicator species
Insecticides, effects on carabid populations, 241, 242, 329
Insects, food value of, 90
Intercropping, effects on carabid populations, 235
Interguild predation, 198, 199
Invertebrate pest control, 165–201
see also Pest control, Pests, named pests

Ladybirds, pest control by, 198
Larval feeding guilds, 84, 85
Leafroller caterpillars, killed by carabid larvae, 186
Leatherjackets, consumed by carabids, 166
Lebia spp.
body form, 113
larval diet, 112
seed predators, 217

Leistus spp.
maxillae, 118
predation by mammals, 23
prey detection, 86
Leistus ferrugineus, microarthropod specialist, 83
Leistus fulvibarbis, distribution, 311
Lepidoptera, 94
food value of, 90
specialists, 188
Leptinotarsa decemlineata, predation by carabids, 166, 174
Leptocarabus kumagaii, diet, 95
Life cycle, factors influencing, 20
Life history in arable land, 5, 6
Light intensity requirements for carabid populations, 10, 12
Lindane, effects on carabid populations, 241
LINK Integrated Farming Systems Project, 252–260, 265–273
Little owl, carabid predator, 23
Locomotion, 114–116, 314
flight, 115, 314
length of legs, 114, 115
Loricera spp., 58
diet, 124
larval, 85
maxillae, 118
Loricera pilicornis
consumption of collembola by, 93
crop cover preference, 235
dispersal ability, 8
ecology of, 62
effect of ploughing on, 239
foraging cycle, 86
hunting technique, 118
life history, 6
mandibles, 116, 117
maxillae, 118
microarthropod specialist, 83
predation by mammals, 23
seasonal abundance, 233
soil organic matter and, 240
trapping prey, 121, 122
L-species, 8
Lycosid spiders, interguid predation and, 199
Lymantria dispar, *see* Gypsy moth
Lymantria obfuscata pupae, killed by carabid larvae, 185

Mammals, carabid predators, 23
Mandibular form and feeding behaviour, 116, 117, 187, 218
Mark–release–recapture experiments, 8, 12, 318, 324, 332, 333
Megacephala spp., 58
Megacephala fulgida, pest control by, 188, 189
Mermis nigrescens, infection of *Pterostichus madidus* by, 14
Metapopulation structure and models, 333–335
Mice, carabid predators, 15, 23
Microarthropod specialists, 83, 84
Microclimate, carabid distribution and, 325–327
Microlestes spp., 60, 65
Microtonus caudatus, parasitism by, 15
Mid-gular apodeme, 118, 119
Mixed feeders, 116, 117, 124
Modified Oakley–Fulthorpe test, 151, 152
Mole crickets, control by *Megacephala fulgida*, 188, 189
Molecular detection of prey, 155, 156, 158
Mollusc specialists, 83, 188
Monoclonal antibodies, used to analyse predator gut contents
advantages of use of, 156, 157
lability of, 149
raising and characterizing, 148–150
specificity of, 149, 150
Morphology and diet, 111–131
Moth pupae, killed by carabids, 174, 185
Mouthparts, 116–119
mandibles, 116, 117
maxillae, 117, 118
setae, 118
Mulberry tiger moth, control by carabids, 198
Musca vetustissima, precipitin test to measure predation of, 152
Myrmecophiles, 84

Nebria spp., 58
maxillae, 118
predation by mammals, 23
Nebria brevicollis
crop cover preference, 234
detection of collembola, 324
diet, 83, 128, 130, 166
dispersal, 8, 333
distribution, 318–321
ecology of, 62
gut analysis, 128
life history, 6
non-crop habitats, 294
parasitism of, 15
population size, 8
prey detection, 86
size determination, 19
soil pH requirements, 13
in temperate arable farming systems, 255
Nebria gyllenhalli, mandibles, 116, 117
Nebria salina
crop cover preference, 234
life history, 6
Neodiprion sertifer, consumed by carabids, 166
Neomyas spp., proventriculus, 120
New Zealand flatworms
chemical marking of, 138
consumed by carabid larvae, 166
ELISA used to detect predation of, 153
Non-crop habitats
effects on carabid
abundance in associated crops, 286–289
activity-density, 286–289
body size, 283
community structure, 289–291
morphology and physiology, 283–286

overwintering, 280–283, 312, 313
species diversity, 289, 290
field boundaries, 310–313
functions of, 280–282
for active carabids, 281, 282
for overwintering carabids, 280–283
hedgerows, 310–313
interaction with carabid life histories, 282, 283
management for carabids, 293–296
economics, 296
mechanical and chemical inputs, 295
size and spacing, 294, 295
time to establish, 295, 296
vegetation composition, 293, 294
models for pest control, 292, 293
role in contributing to pest problems, 292, 293
role in reducing pest abundance, 291, 292
source or sink?, 286–289
woodland in arable areas, 309, 310
No-till farming methods
carabid numbers and, 224, 225, 238
pest damage and, 238
Notiophilus spp., 58
diet, 83, 124
larval, 85
maxillae, 118
prey detection, 86, 113
Notiophilus biguttatus
activity pattern, 10
distribution, 308, 313
ecology of, 62
effect of ploughing on, 239
egg production, 16
food preference, 99
overwintering, 312
size determination, 19

Odontotermes wallonensis pupae, killed by carabids, 166, 174
Omphra pilosa, pest control by, 188
Onion fly
control by carabids, 186, 194
predator–prey relationships, 192
Open-field carabids, 282
Operophthera brumata, precipitin test to measure predation of, 151
Ophionea indica, diet of larvae, 185
Organic matter, effects on carabid populations, 240, 241, 243
Orseolia oryzae pupae, killed by carabid larvae, 185
Ostrinia nubilalis, *see* Corn borer
Otiorhynchus sulcatus, monoclonal antibodies against, 150
Ouchterlony test, 152
Overwintering carabids, importance of non-crop habitats, 280–283, 312, 313

Panicum dichotomiflorum, seed predation by carabids, 222
Parasitism, 14, 15
Parena nigrolineata, diet of larvae, 186
Parena perforata, pest control by, 198
Pathogens
carabid vulnerability to, 14
transmission by carabids, 198
Patrobus spp., 58
Patrobus atrorufus, distribution, 61
Pest abundance, role of non-crop habitats in reducing, 291, 292
Pest control
by assemblages of generalist predators, 199–201
by pathogens and carabids, 198
contribution of carabids to, 19, 21, 187–201
Calosoma maximowiczi, 188
Calosoma sycophanta, 188
carabids acting alone, 187–195
carabids acting with other natural enemies, 195–201
Cychrus sp., 188
Megacephala fulgida, 188
Omphra pilosa, 188
Scaphinotus sp., 188
simulation models, 193, 194
evidence that carabids can kill pests, 174–187
evidence that carabids consume pests, 166–174
hindered through interguild predation, 198, 199
Pesticides
application, current farm practice, 262–265
effect on carabid populations, 266–269
and assemblage composition, 56, 57
Pest limitation, 102, 111
Pests
consumed by carabids, 166
seen to be killed by carabids, 174
Phaenoserphus pallipes, parasitism by, 15
Phaenoserphus viator, infection of carabids by, 15
Pieris rapae larvae
ELISA used to measure predation of, 153
precipitin test to measure predation of, 151
Pitfall traps, 2–4, 23, 42, 43, 61, 192, 252, 253, 261, 262, 269, 270, 273, 307, 316, 318, 321, 329, 332
protocol, 43, 44
Planthoppers, killed by carabids, 174
Plant material, food value of, 95
Plochionus timidis, larval diet, 186
Poa trivialis, seed predation by carabids, 220
Poecilus spp., 59, 60, 63
Poecilus chalcites
ecology of, 63
predator–pest associations, 192
Poecilus corvus, diet, 186
Poecilus cupreus
chemoreceptors, 122
detection of aphids by, 186
diet, 140, 166, 174, 186
distribution, 318, 319, 329, 332
DNA techniques for detection of prey, 156
ecology of, 63
egg production, 16
fecundity, 332
food preference, 99, 186

foraging cycle, 87
habitat and
activity-density, 286
and body size, 283, 284
fecundity, 284, 286
feeding condition, 283, 284
life history, 6
in non-crop habitats, 281
overwintering, 237
population regulation, 18
reproductive capacity, 16
seasonal movement, 315, 316
simulation model for control of aphids by, 194
in temperate arable farming systems, 263
Poecilus lucublandus
diet, 166, 174
ecology of, 63
predator–pest associations, 192
Poecilus sericeus, 63
Poecilus versicolor
egg production, 16
reproductive capacity, 16
Polyclonal antisera, raising and characterizing, 141, 148
Population
density, 309, 310
diversity, 309, 310
extinction, 334, 335
regulation
development, 15, 16
dispersal ability, 8
environmental requirements, 8–14
interspecific competition, 16–19
intraspecific competition, 17, 18
parasitism, 14, 15
pathogens, 14
predation, 15
reproduction, 15, 16
survival of different lifestages, 7, 8
size fluctuations, 308
Power pushers, 114, 115
Precipitin test, 140–145
Predator–pest associations, 192–195
Prey
acquired aversion to sub-optimal, 99
antigens, ELISA tests for, 130
detection, 86
see also Diagnostic techniques for determining diet, Gut analysis, Pest control
size, 130, 131, 140
Procerization, 112
Proctotrupes gladiator, parasitism by, 15
Proventriculus, 119–121
Pserosophus spp., 60
diet, 95
locomotion, 115
seed predators, 217
Pserosophus jessoensis, 65
Pseudoophonus rufipes, crop cover preference, 235
Psila rosae eggs, killed by carabids, 174
Psylloides chrysocephala, predator–prey relationships, 193
Pterostichus spp., 59–61
diet, 94
larval, 85, 187
dispersal ability, 8
ecology of, 63
flight activity, 314
generalist carnivores, 83
hibernation, 7
mandible wear, 125
predation of, 23
prey detection, 86
proventriculus, 120
Pterostichus adstrictus, distribution, 61
Pterostichus aethiops, diet, 166
Pterostichus chalcites
effect of ploughing on, 239
locomotion, 114, 116
mid-gular apodeme, 119
proventriculus, 120
Pterostichus cupreus, seed predation, 220
Pterostichus lucublandus
crop cover preference, 233
effect of ploughing on, 239
Pterostichus madidus
detection periods of anti-prey monoclonal antibodies, 149
diet, 97, 128, 139, 166
larval, 85
distribution, 318, 320, 327
gut analysis, 128
life history, 6
nematode infection of, 14
parasitism of, 15
pest control by, 198
population size, 8
predator–prey relationships, 193
rate of digestion, 154
spatial distribution, 19
in temperate arable farming systems, 255
Pterostichus melanarius
bioindicator, 272
chemoreceptors, 121, 122
crop cover preference, 13
deep ploughing and, 238
detection of aphids by, 186
detection periods for anti-prey monoclonal antibodies, 149, 150
diet, 83, 94–97, 102, 150, 151, 166, 174
digestive enzymes of, 98
dispersal, 289, 333
distribution, 61, 217, 318–321, 327, 329, 330, 334
DNA techniques to detect prey, 156
ecology of, 63
effect of ploughing on, 239, 329
egg production, 16
fecundity, 19, 284, 286
food preference, 99
habitat
and activity-density, 286
and body size, 283–285

and feeding condition, 283, 284, 286
introduction to North America, 29
life history, 6
mandibles, 117
in non-crop headlands, 289–291
overwintering, 237, 312
parasitism of, 15
in pitfall traps, 3, 4
population size, 8
predator–pest associations, 192, 193
prey detection, 86
seasonal abundance, 233
seasonal movement, 315, 316
seed detection, 218
seed predation, 220, 221
slug detection, 325
spatial distribution, 12
in temperate arable farming systems, 255, 263

Pterostichus niger
chemoreceptors, 121, 122
diet, 166
distribution, 61
effect of ploughing on, 239
locomotion, 314

Pterostichus nigrita
diet, 96

Pterostichus oblongopunctatus
diet, 96, 140, 166
dispersal ability, 8
distribution, 308, 309
pupal and larval mortality, 7
temperature requirements, 9

Pterostichus permundus, predator–prey relationships, 192

Pterostichus rhaeticus, diet, 140

Pterostichus strenuus
overwintering, 312
soil organic matter and, 240

Pterostichus vernalis, overwintering, 312

Pterostichus versicolor
distribution, 321
seed predation, 220
size determination, 19

Radioactive markers, 139, 193
Reptiles, carabid predators, 24
Rhagoletis pomonella, 166
Rice gall midge pupae, killed by carabid larvae, 185
Rootflies, *see* Cabbage root fly, Onion fly
Rootworms, killed by carabids, 174, 185, 187

Sampling techniques, 3–5, 42–44
see also Pitfall traps
Sawflies, consumed by carabids, 166

Scaphinotus spp.
mouthparts, 119
pest control by, 188

Scaphinotus marginatus
chemoreceptors, 123
diet, 89
mouthparts, 123

Scaphinotus stratiopunctatus, pest control by, 188
Scapteriscus spp., control by *Megacephala fulgida*, 188, 189
SCARAB Project, 261–273

Scarites spp., 58
ecology of, 62
locomotion, 115
maxillae, 118
prey detection, 86
proventriculus, 120

Scarites subterraneus, 62
body form, 113
crop cover preference, 234
locomotion, 114
mouthparts, 119

Scarites subtriatus, effect of ploughing on, 239
Scavenging, 150, 151, 174
Sciarid midges, food value of, 94
Seasonal movement, 314–316
Secondary predation, 151, 174

Seed predation, 215–226
benefits of, 223, 224
factors affecting seed choice, 221–223
impact of farming systems on, 224–226
methods of detecting, 219, 220
non-crop habitats and, 292
selective predation, 222–224
species composition and, 223, 224

Seed predators, 216, 217
Seed-feeding habit, adaptations to, 217–219

Seeds, 94
food value of, 90, 95

Selenophorus spp., seed predators, 217
Shrews, carabid predators, 15, 23
Shrikes, carabid predators, 22, 23
Sink habitat, 286–289
Sitobion avenae, effect of non-crop habitats on populations of, 291
Sitodiplosis mosellana, *see* Wheat midge
Sitonia lineatus, marking with radioactive isotopes, 139
Sitonia regensteinensis, precipitin test to measure predation of, 151

Slugs
consumed by carabid adults, 166
control by carabids, 194, 195
damage to crops, and non-crop habitats, 292
detection in the diet, 140
detection periods in carabids, 149, 150, 154
effect of deep ploughing on, 238
ELISA use to detect predation of, 153
food value of, 95, 96
killed by carabids, 174
monoclonal antibodies to, 149
precipitin test to measure predation of, 152
predator–pest associations, 192

Snails
consumed by carabid adults, 166
control by carabids, 188
food value of, 95, 96
killed by carabid larvae, 174, 185
monoclonal antibodies to, 149

Sogatella furcifera, killed by carabids, 174
Soil, effects on carabid populations,12, 13, 326, 327
chemical composition, 327
cultivation methods, 237–240
mechanical properties, 327, 328
moisture, 12, 326, 327
pH, 12, 13, 327
temperature, 12, 326
Source habitat, 286–289
Southern corn rootworm larvae, killed by carabids, 187
Spatial distribution of carabids
in agricultural landscapes, 305–336
between fields, 328–333
barriers to carabid movement, 332, 333
in hedgerows and field boundaries, 310–313
overwintering carabids, 312, 313
within fields, 316–320
analysis of data, 317, 318
artefacts of sampling, 321, 324
displaying data, 316, 317
effects of chemicals, 327
factors affecting, 320–328
methodology, 316
microclimate effects, 325, 326
prey availability, 324, 325
soil factors, 326–328
topography and aspect, 328
in woodland, 308–310
Species composition of agricultural habitats, 57–65
Species diversity, 289, 290
Species richness, 44–52, 289, 290
crop type and, 233
curves, 45, 46
Spodoptera frugiperda, control by pathogens and carabids, 198
Springtails, *see* Collembola
Squash bug, 199
Stellaria sp., predation by carabids, 218
Stem borers, killed by carabid larvae, 185
Stenolophus spp., 60
Stenolophus comma
ecology of, 64
proventriculus, 120
Stenolophus lecontei, mandibles, 116, 117
Stenolophus lineola, proventriculus, 120
Stomis spp., maxillae, 118
Stomis pumicatus, life history, 6
Sugar beet crops, carabid activity in, 233
Synuchus spp., 59
granivores, 84
locomotion, 115

Tachyporus hypnorum, foraging cycle, 87
Taraxacum spp., seed predation by carabids, 220
Temperature requirements of carabids, 9, 10, 325–327
Termite specialists, 188
Termites, consumed by carabids, 166, 174
Termitophiles, 84
Thanatarctica imparilis, control by carabids, 198
Tiger beetle, *see Cicindela sexguttata*
Trechus spp., 58, 60
decline in abundance, 25
diet, 95
ecology of, 62
generalist insectivores, 83
Trechus obtusus, distribution, 311
Trechus quadristriatus
chemoreceptors, 122
crop cover preference, 13, 235
diet, 166, 189, 243
distribution, 320
ecology of, 62
effect of hoeing on abundance, 235
effect of ploughing on, 239
foraging cycle, 87
life history, 6, 7
predation by birds, 23
predator–prey relationships, 193
scavenger, 174
soil organic matter and, 240
Trechus secalis, 62
Trichocellus spp., 60
Trichocellus cognatus, distribution, 61
T-species, 8
Tyria jacobaeae, precipitin test to measure predation of, 151

Vegetation cover, carabid requirements, 13,14
see also Weeds, cover
Viola arvensis, seed predation by carabids, 220–222

Wedge pushing, 113, 115
Weeds
control, effects on carabid populations, 242
cover, effects on carabid populations, 236, 243, 325, 326
seed predation by carabids, 215–226
benefits of, 223, 224
factors affecting seed choice, 221–223
impact of farming systems, 224–226
selective predation, 222–224
seed removal, effect of non-crop habitats on, 292
seed spread from non-crop habitats, 292
Weevils, 125
consumed by carabids, 166
control by carabids, 193
monoclonal antibodies against, 150
precipitin test to measure predation of, 148, 151
predation by carabids, 186
radiolabelled, 193
Wheat bugs, consumed by carabids, 166
Wheat midge
consumed by carabids, 166
control by carabids, 193
precipitin test to measure predation of, 152
Wildlife strips, 280
Winter moth pupae, precipitin test to measure predation of, 151

Wireworms, consumed by carabids, 166
Woodland
  carabid distribution patterns in, 308, 309
  influence in arable areas, 309, 310
  see also Non-crop habitats
Woodlice, precipitin test to measure predation of, 151

Xanthium pennsylvaticum, seed predation by carabids, 222

Zabrini, seed predators, 217
Zabrus spp.
  diet, 95
  ecology of, 63
  granivores, 84
Zabrus tenebrioides
  adaptations to seed-feeding habit, 218, 219
  diet, 124
    larval, 218
  ecology of, 63
Zinc levels in carabids, 240
Zoophagous species, 124